**MODERN ZOOLOGY**

# MODERN ZOOLOGY

## Volume 1

## Invertebrates

*By*
**Dr. H.S. Bhamrah**
**Kavita Juneja**

**ANMOL PUBLICATIONS PVT. LTD.**
**NEW DELHI - 110 002 (INDIA)**

**ANMOL PUBLICATIONS PVT. LTD.**
4374/4B, Ansari Road, Daryaganj
New Delhi - 110 002



First Edition, 1999

ISBN 81-261-0401-5 (Set)
ISBN 81-261-0404-X (Vol. 1)



PRINTED IN INDIA

---

Published by J.L. Kumar for Anmol Publications Pvt. Ltd., New Delhi - 110 002 and Printed at Mehra Offset Press, Delhi.

# Contents

# Preface

This book has been designed to approach the morphology, anatomy, physiology and development of selected type in a simple the lucid style. The approach to the discussion of all the types is very simple so as to impact to the students a clear and vivid understanding. According to the scheme of treatment the important animal types of the phylum have been dealt with first, and efforts have have been made to present their elaborate and uptodate account. General characters and classification and brief description of other important types of the phylum have also been dealt with complete, authentic and uptodate account. Further separate chapters on topics of significance and general interest pertaining to the phylum have also been added to make the treatment more elaborate.

It has been the constant endeavour of the authors to furnish maximum substance, keeping in view the limitations of size of the volume. Efforts have been made to condense the matter as far as practicable. The book features both a text and a laboratory guide. It is hoped that this book will not only meet the requirement of Indian students but will also be useful as a guideline to the teachers in their teaching.

The authors express their thanks to their friends and colleagues whose constant inspiration have initiated them to bring out the book.

The authors are fully aware of the shortcomings, errors and misprints that have crept in, and shall be grateful to receive suggestions and constructive criticism for improvement of the book from all the readers.

AUTHORS

# 1

# CHARACTERS & CLASSIFICATION OF PROTOZOA

The protozoa are heterogenous assemblage of some 50,000 acellular or single-cell organisms possessing typical (*eukaryote*) membrane bound cellular organelles. Being the simplest in structure, protozoa, are regarded most primitive or first animals (Gr. *protos* -first; *Zoon* = animal) of nature. The acellular level of organization is the only characteristic by which the protozoa as whole can be described; in all other respects they display extreme display. They exhibit all types of symmetry, a great range of complexity, and adaptations for all types of enviromental conditions. Protozoa were first seen by *Leeuwenhoek* (1677). The other pioneers, in the field of protozoology, are *Joblott* (1718), *Trembley* (1744), *Linnaeus* (1758), *Muller* (1773), *Dujardin* (1845) *Hackel* (1862), *Pasteur* (1870), *Butschli* (1881), *Ronald Ross* (1898), etc. Term Protozoa was coined by *Goldfuss* (1817), *Grassi* (1952), *Hall* (1953), *Kudo* (1954), *Pitelka* (1953-59), and *Allen* (1962) have made their valubale contribution to the protozoology in present century.

## GENERAL CHARACTERS

1. Protozoa occur wherever moisture is present in the sea (marine) in all type of fresh water, and in the soil. These are commensal, mutualistic, and many parasitic species. In fact the sporozoans are entirely parasitic.
2. All though most protozoa occur as solitary individuals, there are numerous colonial forms. Some colonial forms, such as *Volvox*, attain such a degree of cellular interdependence that they approch a true multicellular level of structure. Both solitary and colonial species may be either free moving or sessile.
3. The great majority of protozoa are microscopic. *Anaplasma*, a blood parasite, is so small that it occupies only 1/6 to 1/10th of a red blood corpuscles. On the other hand the fresh water cilliate, *Spirostomum*, may reach a length of 3.0 mm. and be seen with the naked eye. A fossil foraminiferan, *Nummulites*, was 19 cm. across, probably the largest size that has been attained by any protozoa.
4. Body symmetry nonradial, spherical, or bilateral.

5. The body is either naked, or covered by pellicle. Sometimes an exoskeleton is also present.

6. The shape of body is usually constant but in some cases it is unstable and in other it may change with the environment or age.

7. Their locomotory organelles are finger-line pseudopodia, flagella, cilia or absent as in sporozoans.

8. Nutrition may be holozoic (animal like), holophytic (plant like), saprozoic or parasitic. The organelles of ingestion and egestion may be present or absent. Digestion occur intracellularly inside the food vacuoles in most cases.

9. No specific respiratory and excretory organs are present. Both are carried through general body surface by diffusion Excretion sometimes carried with the help of contractile vacuole.

10. In fresh water protozoa excess water is usually eleminated by contractile vacuoles, Thus these are mainly osmoregulatory in function.

11. Asexual reproduction by binary fission, multiple fission or budding occurs at some time in the life history of almost all protozoa. Meiosis, gametes formation, and fertilization have been observed in many species, but the nature of these events and their occurance in the life cycle of the organism is highly variable. Encystment is common.

12. Life-history is often complicated in some cases with alternation of asexual and sexual phases.

13 Their body is not distinctly differentiated into somatoplasm and germplasm, hence natural death does not take place in them.

**Special Characters :**

1 Their acellular body regulates all the life activities thus there is no physiological division of labour. The division of labour is restricted to the organelles which are intracellular or subcellular structures.

2. The animals give their respond to various stimuli. This gives an indication of the beginning of the nervous system of metazoans.

3. The microconjugants and macroconjugants in the conjugation of *Vorticella* indicate the beginning of sexual dimorphism.

## CLASSIFICATION OF PHYLUM-PROTOZOA

There is disagreement among Protozoologits on the ways of classifying Protozoa. Modern workers like *Kazloff* (1972) and *Robert D.Barnes (1980)* tend to believe that this group includes more than one unicellular phyla. *Barnes* (1980) raised the four major groups viz. *Mastigophora, Sarcondina, Sporozoa* and *Ciliophora* to the rank of independent phylum. Here we are following the classification given by "Committee on Taxonomy and Taxonomic Problems of the Protozoologists" (*Honiberge et al*, 1964).

According to *Honiberge et al* Protozoa have been classified into four subphyla.

| | | |
|---|---|---|
| Subphylum | I | Sarcomastigophora |
| Subphylum | II | Sporozoa |
| Subphylum | III | Cnidophora |
| Subphylum | IV | Ciliophora |

## SUBPHYLUM I. SARCOMASTIGOPHORA

1. Organelles of locomotion are pseudopodia or flagella.
2. Nucleus monomorphic (single type).
3. Nutrition holozoic or holophytic.
4. Asexual reproduction by binary or multiple fission.
5. Sexual reproduction by syngamy; no spore formation.

This subphylum is divided into three super classes:

### *SUPER CLASS A-MASTIGOPHORA*

1. Simple and primitive. Body with firm pellicle.
2. One to many flagella for locomotion.
3. Asexual reproduction by binary, more or less sometogenic, fission.
4. Autotrophic or heterotrophic or both.

#### Class-I. Phytomastigophorea

1. Mostly free-living.
2. Plant-like flagellates with or without chromoplasts.
3. Usually one or two flagella, the nucleus vesicular.
4. Reserve food as starch or paramylon.

##### *Order. 1. Chrysomonadina*

1. Small flagellates with yellow or brown chromoplast.
2. One to three flagella are present.
3. Gullet absent; stigma present.
4. Nutrition holophytic, starch absent but leucosin and fat droplets are present.
5. Siliceous cysts. Marine or fresh water.

Examples. *Chromulina, Synura, Ochromonas, Uroglena* etc.

***Order. 2. Coccolithophorida***

1. Tiny marine flagellates covered by calcareous platelets-coccoliths.
2. Two flagella and yellow to brown chromoplast.
3. No endogenous siliceous cysts.

   Examples: *Coccolithus, Rhabdosphaera.*

***Order. 3. Heterochlorida***

1. Two unequal flagella and yellow-green chromoplasts.
2. Silicious cysts.

   Examples: *Heterochloris, Myxochloris.*

***Order. 4. Cryptomonadida***

1. Compressed with a rigid pellicle.
2. Gullet reaches upto the middle of the body.
3. They have two unequal flagella. Green, yellow, brown or colourless chromatophores which form starch.
4. Stigma often present.
5. Marine or fresh water

   Examples: *Chilomonas, Cryptomonas.*

***Order. 5. Dinoflagellida***

1. Small and planktonic. Body naked or with a thick pellicle or cellulose theca.
2. Two flagella, one lying transversely and the other pointing backwards.
3. Chromatophores are green, yellow, or brown.
4. Reserve food is starch or oil.
5. There are complex vacuoles which are not contractile.
6. Some are bioluminescent.
7. Mostly marine, some are parasitic.

   Examples: *Noctiluca, Ceratium; Oodinium, Gonyaulax.*

***Order. 6. Chloromonadida***

1. Small, with a delicate and dorse-ventrally flattened pellicle.
2. Two flagella, one trailing.
3. Chromatophores green and numerous.
4. Reserve food is oil.
5. Stigma absent.

6. Contractile vacuole present.
7. Largely freshwater.

Examples: *Gonyostomum, Vacularia.*

***Order. 7. Euglenida***

1. Large with a thick and firm pellicle.
2. Anterior end with gullet leading into reservoir.
3. One or two flagella arising from an anterior recess.
4. Stigma present in coloured form.
5. Chromatophores green and numerous sometimes colourless.
6. Reserve food is paramylum and oil.
7. Mostly fresh water.

Examples: *Euglena, Phacus, Peranema, Rhabdomonas.*

***Order. 8. Volvocida***

1. Small with rigid cellulose covering theca.
2. Gullet absent.
3. Flagella usually two sometimes more.
4. Body green with, usually single cup-shaped chromoplast.
5. Reserve food is starch and oil.
6. Stigma present.
7. Mostly fresh water forms. Many colonial species.

Examples: *Volvox, Chlamydomonas, Gonium, Pandori*

**Clas-2 Zoomatigophorea (Zoomastigina)**

1. Flagellates with neither chromoplasts nor leucoplasts.
2. One to many flagella, in most cases with basal granule complex.
3. Reserve food glycogen.
4. Sexual stage unknown.
5. Many commensals, symbionts and parasites.

***Order. 1. Choanoflagellida.***

1. Fresh water flagellates, with a single flagellum surrounded by a collar.
2. Sessile, sometimes stalked, sometimes with lorica.
3. Solitary or colonial.

Examples: *Codosiga, Salpingoeca, Proterospongia.*

***Order. 2. Bicosoecida***

1. Largely fresh water flagellate encased within a lorica.
2. Two flagella, one free, the other attaching posterior end of body or shell.

   Examples : *Bicosoeca, Poteriodendron.*

***Order. 3. Rhizomastigida***

1. Small and amoeboid forms.
2. Flagella one or more upto four.
3. Locomotion by pseudopodia and flagella.
4. Chiefly freshwater.

   Examples: *Mastigamoeba, Dimorpha*

***Order. 4. Kinetoplastida***

1. Small and more or less amoeboid in forms; no gullet.
2. Flagella basically one, but up to four.
3. Mostly parasitic forms, some are freshwater. Freshwater forms sessile, sometimes stalked.

   Examples: *Bodo, Trypanosoma, Leishmania.*

***Order. 5. Retortamonadina***

1. Gut parasites of insects or vertebrates.
2. One to four flagella. One flagellum associated with ventrally located cytostome.

   Examples: *Chilomonas.*

***Order. 6. Diplomonadina***

1. Small with a pellicle and often with a cytostome.
2. Bilaterally symmetrical flagellates, with two nuclei, each nucleus associated with four flagella.
3. Mostly parasitic.

   Examples: *Giardia, Hexamita.*

***Order-7. Oxymonadina***

1. Symbiotic flagellates having one to many nuclei. Each nucleus associated with four flagella, some of which are turned posteriorly and adhere to body surface.

   Examples: *Oxymonas, Pyrsonympha.*

***Order. 8. Trichomonadina***

1. Four to six flagella, one of which is trailing.

2. Parasites in genital passage.

   Examples: *Trichomonas*.

***Order. 9. Hypermastigida***

1. Highly specialized forms with numerous flagella.
2. Kinctosomes arranged in a circle, plate or longitudinal or spiral rows.
3. Uninucleate or multinucleate but never binucleate.
4. Food ingestion by pseudopodia.
5. Symbionts in the guts of termites, cockroaches and wood roaches.

   Examples: *Leptomonas, Trichonympha, Barbulanympha.*

***SUPER CLASS - B - OPALINATA***

1. Body covered by longitudinal, oblique rows of cilia arising from anterior subterminal rows.
2. Cytostome is absent.
3. Two or many monomorphic nuclei.
4. Binary fission generally symmetrogenic.
5. Sexual reproduction involves syngamy with flagellated gametes (anisogamous).
6. Parasitic forms found in gut of frogs and toads; less commonly in fishes, salamanders and reptiles.

   Examples: *Opalina, Zelleriella.*

***SUPER CLASS - C - SARCODINA***

1. Protozoa with pseudopodia as feeding and locomotor organelles.
2. Flagella, when present, only in development.
3. Asexual reproduction by binary fission.
4. Mostly solitary and free-living , some parasitic and colonial.
5. Skeletons of various forms and composition characteristic of some groups.

**Class-1- Rhizopodea**

1. The locomiotory organalles are lobopodia, filopodia or reticulopodia, but never axopodia.
2. Generally creeping forms.

**Subclass (i) Lobosia**

1. Pseudopodia usually lobopodia.

***Order-1. Amoebida***

1. Body is naked without any exoskeleton.

2. Typically uninucleate, ectoplasm is clearly differentiated.
3. Nucleus with honeycomb lattice.
4. Largely fresh water , some marine; many parasites.

   Examples: *Amoeba, Entamoeba, Pelomyxa.*

***Order-2. Arcellinida***

1. Body enclosed in a shell or test with an **aperture** through which **the** pseudopodia protrude out.
2. They are free living mostly fresh water.

   Examples: *Arcella, Difflugia, Euglypha.*

**Subclass (ii) Granuloreticulosia**

1. They have fine granular reticulopodia.

***Order-1. Foramaniferida***

1. Chiefly marine species with mostly **multi-chambered** shells.
2. Shells may be composed of tectin, **which may** incorporate **foreign** particles, but most commonly the shells are calcareous.
3. Shells with one or more openings through which reticulopodia emerge. The reticulopodia are fine and branched forming network.
4. Chiefly marine

   Example: *Elphidium, Globigerina, Orbulina, Spirillina.*

**Subclass (iii) Filosa.**

1. Pseudopodia as filopodia, tapering and branching.
2. Body naked with a shall with single aperture.
3. Marine and freshwater.

   Examples: *Allogromia, Gromia.*

**Subclass (iv) Mycetozoia**

1. Body large, amoeboid, multinucleate forming plasmodia.
2. Life-cycle complex and has sexual reproduction, usually sporangia are formed which liberate spores.
3. Pseudopodia numerous and blunt.

   Examples: *Arcyria, Plasmodiophora.*

**Class-2. Actiopodea**

1. The locomotory organelles are delicate axopodia with axial filaments.
2. They are primarily sessile or floating forms.

3. Test may be present or absent.
4. Reproduction both asexual and sexual.

**Subclass- (i) Radiolaria**

1. Body comparatively larger, central capsule is perforated by one or many pores.
2. Siliceous skeleton.
3. Axopodia or sometimes reticulopodia or filopodia.
4. Exclusively marine.

The radiolarians can be divided into two groups : The *Polycystina* has a central capsule with evenly distributed pores or a pore field. The *Phaeodaria* have three pore fields and the organic nature of the capsule wall is different from that of Polycystina.

Examples : *Collozoum, Acanthometra, Aulocantha.*

**Subclass (ii) Acantharia**

1. Imperforate, nonchitinoid central capsule without pores.
2. Anisotropic skeleton of strontium sulphate.
3. Axopodia present. Marine.

Examples: *Acanthometra.*

**Subclass (iii) Heliozoia**

1. Spherical animal without central capsule commonly called *Sun-animalcules.*
2. Axpodia radiating.
3. Body cytoplasm is differentiated into outer vacuolated ectoplasm and inner dense endoplasm.
4. Skeleton absent (naked); if present, it is made up of siliceous scales and spines.
5. Primarily in fresh water.

Examples: *Actinophrys, Actinosphaerium, Camptonema.*

**Subclass (iv) Proteomyxidia**

1. Largely marine and freshwater parasites of algae and higher plants.
2. Pseudopodia are filopodia and reticulopodia.
3. Reproduction by binary or multiple fission, multiple fission in cyst.

Examples: *Vampyrella, Pseudospora.*

**Class - 3 - Piroplasmea**

1. Small , round, rod-shaped or amoeboid.

2. Parasite in red blood corpuscles of vertebrates.
3. Do not produce spores.

Examples: *Babesia*

**SUBPHYLUM II SPOROZOA**

1. Exclusive endoparasites.
2. The locomotory organelles are absent in adults.
3. Cilia or flagella may be present in gametes.
4. Nutrition saprozoic.
5. Asexual reproduction by multiple fission and sexual by syngamy, followed by spore formation.
6. Sporozoites are infective stage; nucleus is of single type.

**Class - I. Telosporea**

1. Pseudopodia are absent and locomotion is by gliding or body flexion.
2. Spores without capsules or filaments, naked or encysted.
3. Adult trophozoites with one nucleus.
4. Reproduction is both asexual and sexual.

**Subclass (i) Gregarinia**

1. Mature trophozoites are large and extracellular.
2. Reproduction is sexual with sporogony; spores contains eight sporozoites.
3. These are parasite of gut and body cavity of invertebrates.

Examples: *Gregarina, Monocystis.*

**Subclass (ii) Coccidia**

1. Mature trophozoites are small and intracellular.
2. Female gametes hologamous.
3. Sporzoites multiply by schizogony in tissue cells.

***Order 1. Eucoccida.***

1. Asexual and sexual both the phases occur in life-cycle.
2. They are parasite in epithelial and blood cells of invertebrates and vertebrates.

*Suborder (a) Eimeriina*

1. Macrogametes and microgametes develop independently.
2. Syzygy is absent. Zygote is non - motile.

3. Sporozoites are encased in sporocyst.

   Examples: *Eimeria*

*Suborder (b) Haemosporina*

1. Macrogametes and microgametes develop independently.
2. Syzygy is absent. Zygote (ookinate) motile.
3. Sporozoites are naked.
4. Schizogony in vertebrate host and sporogony in invertebrate hosts.

   Examples: *Plasmodium.*

**Class - 2 - Toxoplasmea**

1. Spore formation is absent.
2. Reproduction asexual by binary fission.
3. Cysts are formed which have many naked sporozoites. No. flagella or pseudopodia.

   Examples: *Toxoplasma, Sarcocystis.*

**Class - 3 - Haplosporea**

1. Spores are present.
2. Pseudopodia may be present but flagella absent.
3. Reproduction only by asexual means.
4. Parasites of fish and invertebrates.

   Examples: *Ichthyosporidium, Haplosporidium.*

**SUBPHYLUM III. CNIDOSPORA**

1. Adult trophozoite has many nuclei.
2. Spore formation occurs throughout life.
3. Spores have several cells having one or more polar filaments which are coiled threads and can be shot out.
4. Zygote gives rise to one or more trophozoites without sporogony.

**Class - I Myxosporidea**

1. Spores develop from several nuclei and in two or three valves. Polar capsule present.

**Order. 1. Myxosporida**

1. Spores are large with a bivalved membrane.
2. Trophozoite amoeboid and not intracellular.
3. Polar capsule 1, 2 or 4 with filaments.

   Examples: *Myxidium.*

**Class - 2 - Microsporidea**

1. Spores small, with a univalved membrane.
2. Polar capsule absent or present with 1 or 2 filaments.

   Examples: *Nosema.*

**SUBPHYLUM IV CILIOPHORA**

1. The ciliates comprise the largest subphylum of Protozoa. Some 8000 species have been described and many groups are still not well known. They are also the most animal like and exhibit a very high level of organelle development.
2. The shape of body is definite. The body wall is a complex living pellicle, containing alveoli, trichocysts and other organelles, in addition to infraciliature.
3. The body surface is covered with uniform cilia which function in locomotion.
4. The cilia around the cytostome region have become specialized as compound ciliary organelles called membranelles and undulating membranes in many ciliates that employ filter feeing.
5. Most of them have two types of nuclei-larger *macronucleus* and smaller *micronucleus*.
6. One or more contractile vacuoles are present in fresh water forms.
7. Reproduction asexual by transverse binary fission.
8. Conjugation also takes place with fusion of nuclei.
9. The anal aperture or cytopyge is a permanent aperture.
10. No alternation of generation in the life - cycle.

**Class-1 Ciliata**

1. Cilia are present during the whole or a part of their life cycle.
2. Mouth or cytostome and cytophanynx are often present.
3. Nutrition is holozoic.
4. Reproduction takes place asexually by binary fission or budding and sexually by conjugation.
5. Mostly free living in fresh water, marine but parasites, colonial and sedentary forms also occur. Class Ciliata is divided into four subclasses:

**Subclass (i) Holotricha.**

1. Cilia usually short and uniform, arranged in longitudinal rows all over the body.

2. No adoral spiral of long cilia or membranelles.
3. Buccal ciliature either absent or, if present, usually inconspicuous. The subclass Holotricha is divided into following orders:

***Order -1- Gymnostomatida***

1. Chiefly large ciliates with no oral ciliature.
2. Cytostome opens directly to outside.

   Examples: *Didinium, Nassula, Coleps.*

***Order-2-Trichostomatida***

1. With vestibular but not buccal ciliature.
2. Free living and parasitic forms.

   Examples: *Colpoda, Balantidium.*

***Order-3-Chonotrichida***

1. Vase-shaped ciliates lacking body cilia.
2. A funnel at the free end of the body bears vestibular cilia.
3. Chiefly marine and ectocommensel on crustaceans.

   Examples: *Spirochona, Chilodochona.*

***Order-4-Apostomatida***

1. Body with spirally arranged ciliation.
2. Cytostome mid-ventral.
3. Marine parasite. Life cycle complex usually involving two hosts , one of which is crustacean.

   Examples: *Hyalophysa.*

***Order-5- Astomatida***

1. Body ciliation uniform.
2. Cytostome absent.
3. Commensals or endoparasites living in the gut and body cavity of oligochaete worms.

   Examples: *Hoplitophrya, Anoplophrya.*

***Order-6- Hymenostomatida***

1. Mouth usually lies on the side of the body at the end of peristome, remains permanently open and leads into cytopharnx.
2. Buccal cavity present with membranes.
3. Small body with uniform cilia.

   Example: *Paramecium*

***Order-7- Thigmotrichida***

1. A small group of fresh and marine water ciliates found in association with bivalve molluscs.
2. Anterior end of body bears a tuft of thigmotactic cilia.

   Example: *Boveria, Thigmophrya.*

**Subclass (ii) Peritricha**

1. Adults without body cilia.
2. The apical end of the body typically bears a conspicuous buccal ciliature.
3. Mostly sessile bearing a stalk.

   It bears a single order-Peritrichida.

   Examples: *Vorticella, Carchesium.*

**Subclass (iii) Suctoria**

1. Sessile and stalked body.
2. Young with cilia. Adult with few to many tentacles.
3. Adult completely devoid of any ciliature.

   This subclass bears only one order-Suctorida.

   Examples: *Ephelota, Podophrya, Acineta.*

**Subclass (iv) Spirotricha**

1. Reduced body cilia.
2. Well developed conspicuous buccal ciliature.

***Order-1- Heterotrichida***

1. With uniform body cilia or body encased in a lorica and body cilia absent.

   Examples: *Stentor, Nyctotherus, Spirostomum.*

***Order-2- Oligotrichida***

1. Body cilia reduced or absent.
2. Conspicuous buccal membranelles, commonly extending around apical end of body.

   Examples: *Halteria, Strombidium.*

***Order-3- Tintinnida***

1. Loricete, mostly free swimming ciliates with inconspicuous oral membranelles when extended.
2. Largely marine.

   Examples: *Favella, Codonella.*

***Order-4- Entodiniomorphida***

1. Ectocommensel in the digestive tract of herbivorous mammals.

2. Body cilia reduced or absent.
3. Prominent buccal ciliature often in separate anterior clumps.
4. Posterior end may be drawn out into spines.

Examples: *Entodinium, Cycloposthium.*

***Order-5- Odontostomatida***

1. Body laterally compressed and wedge-shaped, with carapace.
2. Reduced body and buccal cilia.

Example : *Saprodinum.*

***Order-6- Hypotrichida.***

1. Body dorsoventrally flattened.
2. Cillia on ventral side, forning cirri.

Examples: *Euplotes, Urostyla, Oxytricha.*

**CHILOMONAS**

| | | |
|---|---|---|
| Phylum | - | Protozoa |
| Subphylum | - | Sarcomastigophora |
| Superclass | - | Mastigophora |
| Class | - | Phytomastigophorea |
| Order | - | Cryptomonadida |
| Genus | - | *Chilomonas* |

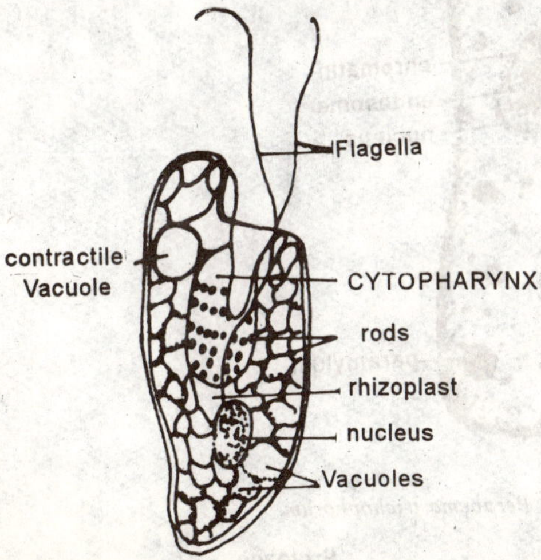

Fi.g. 1.1 *Chilomonas*

1. It occurs in abundance in stagnant water and is common in laboratory culture.
2. Body is oval, colourless, about 35m long.
3. Body with a deep anterior pit, the cytopharynx.
4. Two flagella projecting forwards from the cytopharynx.
5. Endoplasm is distinctly alveolar. Contractile vacuole lies on one side near the anterior end.
6. Nucleus spherical with endosome found at posterior end.
7. Mode of nutrition is saprozoic.

**PERANEMA**

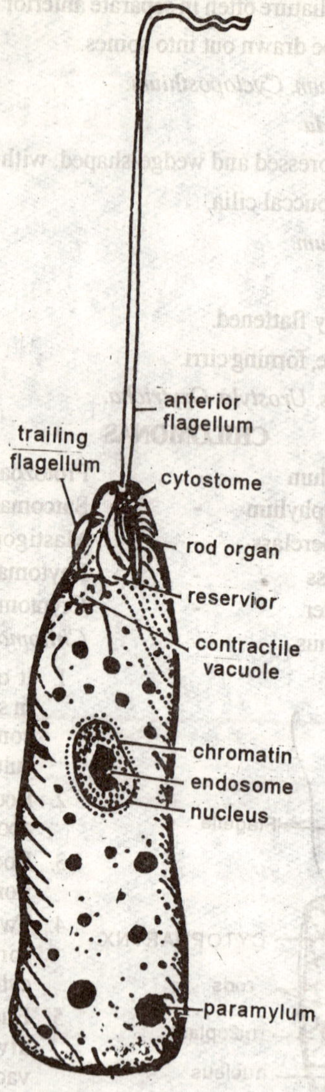


Fig. 1.2. *Peranema trichophorum.*

| | | |
|---|---|---|
| Phylum | - | Protozoa |
| Subphylum | - | Sarcomastigophora |
| Super Class | - | Mastigophora |

| | | |
|---|---|---|
| Class | - | Phytomastigophorea |
| Order | - | Euglenida |
| Genus | - | *Peranema* |

1. It is widely distributed in stagnant fresh water.
2. Body is covered over by an elastic and spirally striated pellicle.
3. Body is broad and truncated posteriorly and gradually tapers anteriorly.
4. Stigma, paraflagellar body and chromatophores are absent.
5. Paired, unequal flagella are present, each arising from its own blepharoplast in the wall of reservoir at anterior end.
6. A pair of hyaline rod-like structures forming the rod organ, running alongside the reservoir and supports the gullet which can be opened or closed.
7. Nucleus is centrally placed with a large; sometimes irregular endosome.
8. It feeds phagotrophically. The prey is seized by the rod organ and pushed into the cytostome.

## CHLAMYDOMONAS

| | | |
|---|---|---|
| Phylum | - | Protozoa |
| Subphylum | - | Sarcomastigophora |
| Super class | - | Mastigophora |
| Class | - | Phytomastigophorea |
| Order | - | Volvocida |
| Genus | - | *Chlamydomonas* |

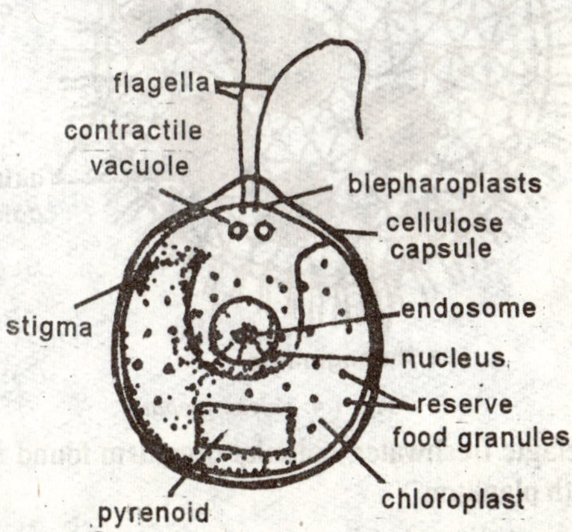


Fig. 1.3. *Chlamyaomonas.*

1. It is animal solitary commonly found in a stagnant freshwater cosmopolitan.
2. Body is ovoid or flattened surrounded by a thick cell-wall made of cellulose.
3. It has two flagella , originating from the blenharoplasts situated in endoplasm.
4. Body is green due to presence of a cup-shaped chlroplast. A pyrenoid is embedded in the basal part of choloroplast.
5. Nucleus is vesicular.
6. Red stigma and a pair of contractile vacuoles are found near the base of flagella.
7. Nutrition holophytic i.e. plant -like .
8. Reserve food as starch grain.
9. Asexual reproduction by binary fission ,often in the palmella stage .
10. The sexual reproduction by the fusion of flagellate isogamy.

## VOLVOX

The systematic position is same as that of *Chlamydomonas*

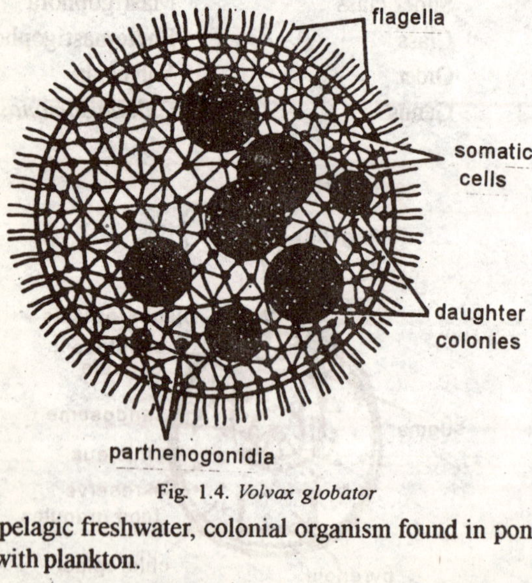


Fig. 1.4. *Volvax globator*

1. It is a pelagic freshwater, colonial organism found in ponds and lake; along with plankton.
2. It has cosmopolitan distribution.
3. The colony looks like a small green ,hollow sphere filled with a watery jelly.

4. Colony contains thousands of zooids (somatic cells) arranged in a single peripheral layer .the cells are connected by protoplasmic threads and not differing in size.
5. Each zooid has two flagella ,two or more contractile vacuoles, cup-like chloroplast, a single nucleus,a red stigma but no gullet .
6. Nutrition is holophytic.
7. The colones are polarised since they swim always in one region hence called anterior pole.locomotion by combined action of flagella.
8. The colony increases in size by binary fission. the zooids may be somatic and reproductive.the reproductive zooids found posteriorly take part in reproduction .
9. Some of these zooids are called parthenogonidia repeatedly divide to form daughter colonies which are liberated by the disintegration of parent colony .
10. During sexual reproduction some specialized cells ,called an antheridia give rise to several biflagellate microgametes.other cells ,called archegonia ,give-rise to non-flagellate macrogametes.Both fuse in water to form zygote that forms new colony after dormancy.

## NOCTILUCA

| | | |
|---|---|---|
| Phylum | - | Protozoa |
| Subphylum | - | Sarcomastigophora |
| Super class | - | Mastigophora |
| Class | - | Phytomastigophorea |
| Order | - | Dinoflagellida |
| Genus | - | *Noctiluca* |

1. *Noctiluca* is a marine and pelagic organism ,occurring near sea shores often in large numbers.
2. They are abundantly found in atlantic and pacific seas.
3. It has a large , melon-like, bilaterally symmetrical body measuring 1-15.mm. in diameter and enclosed in a firm cuticular pellicle.
4. The cytoplasm is highly vacuolated and differentiated into ectoplasm and endoplasm.
5. In a depression at the ventral pole is a longitudinal oral groove representing the sulcus.This groove has an oval mouth leading into the gullet and two flagella; a long, coarse transversely striated flagellum called *tentacle*,other in delicate *flagellum*.

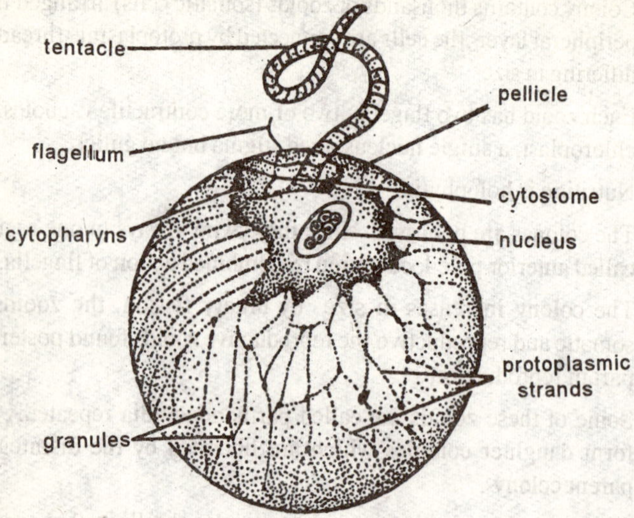


Fig. 1.5. *Noctiluca*

6. From the central sulcus and of protoplasm containing the nucleus, various delicate strands run through the watery interior to the periphery.
7. Nutrition is holozoic i.e. animal like.
8. Reproduction takes place asexually by binary fission, as well as by syngamy. The syngamy involves fusion of two organisms and is followed by formation of *Zoospores*.
9. When agitated, it glows with bluish or greenish light. The luminescence is caused by photogenic granules *luciferin*, which under the influence of enzyme, *luciferase*, emits light.

**CERATIUM**

The systematic position is as that of *Noctiluca.*

1. *Ceratium* includes marine as well as fresh water species.
2. It is common in U.S.A.
3. The body is usually covered with a cellulose plate grooved by a transverse and a longitudinal furrow each containing a flagellum, which projects out through a pore in theca.
4. The cellulose plates, which are produced into an anterior or apical and two posterio-lateral processes giving it a triangular shape.
5. The transverse groove is called as *annulus* running, around spiral girdle. Longitudinal groove is called as *sulcus*.

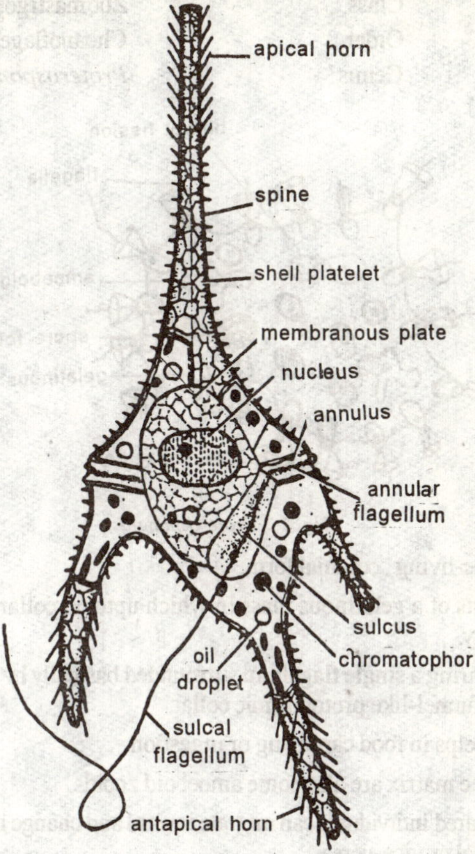


Fig. 1.6 *Ceratium hirrundinella*, semidiagrammatic, with portion of shell removed.

6. Cytoplasm is differentiated into ectoplasm and endoplasm.
7. Cytoplasm contains starch granules, chromoplast, nucleus and a complex system of contractile vacuoles.
8. It is mainly holophytic but some species are holozoic.
9. Reproduction is by oblique binary fission.
10. Cyst formation sometimes occurs.
11. Several species are known to be luminescent.

## PROTEROSPONGIA

| | | |
|---|---|---|
| Phylum | - | Protozoa |
| Subphylum | - | Sarcomastigophora |
| Super class | - | Mastigophora |

| | | |
|---|---|---|
| Class | - | Zoomastigophorea |
| Order | - | Choanoflagellida |
| Genus | - | *Proterospongia* |

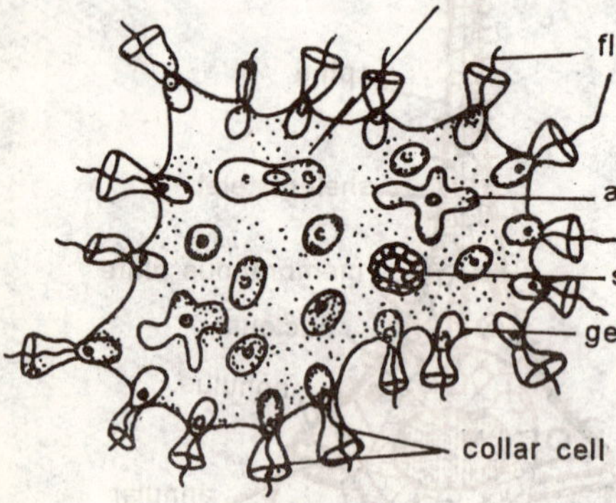


Fig. 1.7 *Proterospongia*

1. It is a free-living , colonial form.
2. It consists of a gelatinous mass in which upto 60 collared cells lying at periphery.
3. Each bearing a single flagellum surrounded basically by a thin, transparent and funnel-like protoplasmic collar.
4. Collar helps in food capturing or ingestion.
5. Inside the matrix are also some amoeboid zooids.
6. Any collared individuals can migrate inward and change into an amoeboid individual or vice-versa.
7. *Proterospongia* is often regarded as a connecting link between protozoan and sponges.

## MASTIGAMOEBA

| | | |
|---|---|---|
| Phylum | - | Protozoa |
| Subphylum | - | Sarcomastigophora. |
| Super class | - | Mastigophora. |
| Class | - | Zoomastigophorea |
| Order | - | Rhizomastigida. |
| Genus | - | *Mastigamoeba* |

1. It is a free-living flagellate found in fresh water ponds and lakes.
2. Body is irregular or amoeboid in form pushing out several finger like pseudopodia, which usually disappear when the organism swims. One flagellum is present.

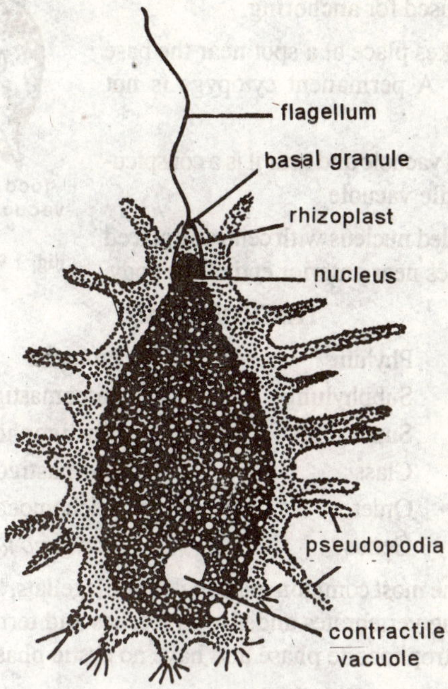


Fig. 1.8 *Mastigamoeba*

3. Protoplasm contains single nucleus and contractile vacuole.
4. Nutrition is holozoic. Pseudopodia help in ingestion of food.
5. It reproduces by a longitudinal binary fission.
6. It is considered to be a connecting link between Mastigophora and Rhizopoda.

## BODO

| | | |
|---|---|---|
| Phylum | - | Protozoa |
| Subphylum | - | Sarcomastigophora |
| Super class | - | Mastigophora |
| Class | - | Zoomastigophorea |
| Order | - | Kinetoplastida |
| Genus | - | *Bodo* |

1. It is a free-living animal found in the faecal matter.
2. Body is small, naked , more or less elongated, and with one surface more convex than the other.

3. A pair of unequal flagella arise from a depression at the anterior end, one flagellum trailing behind and used for anchoring.
4. Ingestion takes place at a spot near the base of flagella. A permanent cytopyge is not known.
5. Small rostral vacuole and near it is a conspicuous contractile vacuole.
6. Single rounded nucleus with centrally placed endosome lies near anterior end of the body.

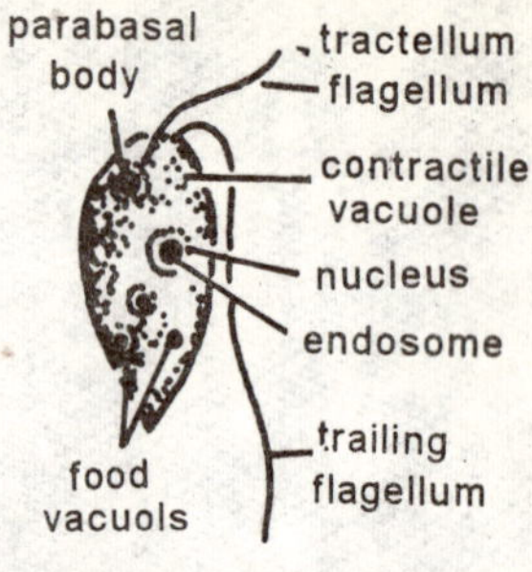


Fig. 1.9 *Bodo saltans.*

## TRICHOMONAS

| | | |
|---|---|---|
| Phylum | - | Protozoa |
| Subphylum | - | Sarcomastigophora |
| Super class | - | Mastigophora |
| Class | - | Zoomastigophorea |
| Order | - | Trichomonadina |
| Genus | - | *Trichomonas.* |

1. It is one of the most common intestinal multiflagellats, living in the intestine of all the vertebrates and also in leeches and termites. They exist only in the trophozoite phase and have no cystic phase.
2. It has cosmopolitan distribution.
3. The body is pear shaped tapering posteriorly, provided with four flagella of which one is directed backwards united to the body by an undulating membrane. Body measures 5-20 in length.
4. Cytostome is present anteriorly, used for ingestion of food.
5. Blepharoplast is minute situated at the anterior end. Extending posteriorly from the blepharoplast is club-shaped parabasal body.
6. A contractile fiber, the *costa* runs along the line of attachment of the undulating membrane.
7. The body is supported by a cylinder of microtubules, the *axostyle*, which extend backwards and projects from posterior end as *spike*. The *spike* helps to anchor the animal while feeding.
8. The endoplasm contains a large nucleus situated anteriorly. The endosome is prominant.
9. Feeding is saprozoic but most species ingest bacteria, yeast and other solid food.
10. Reproduction takes place exclusively by longitudinal binary fission.

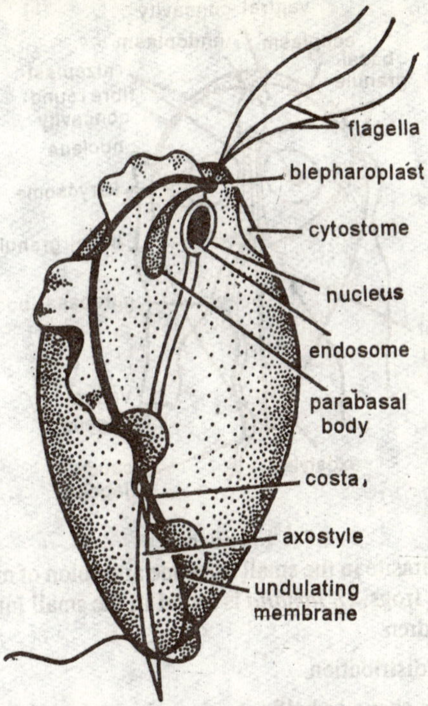


Fig. 1.10 *Trichomonas*

11. No cyst formation as transmission occurs directly in the active, free trophozoite stage.

*Pathogenecity* In man *Trichomonas buccalis* is found in mouth; *T. homonis* in colon and *T. vaginalis* in vagina of female (*T. vaginalis* is found in large number in the leucorrhoeic discharge of females. It causes inflammation of the vaginal mucosa.) Of other animals *T. foetus* of cattles, *T. gallinae* of birds are important. The disease caused by them is called *Trichomoniasis.*

## GIARDIA

| | | |
|---|---|---|
| Phylum | - | Protozoa |
| Subphylum | - | Sarcomastigophora |
| Super class | - | Mastigophora |
| Class | - | Zoomastigophorea |
| Order | - | Diplomonadida |
| Genus | - | *Giardia* |

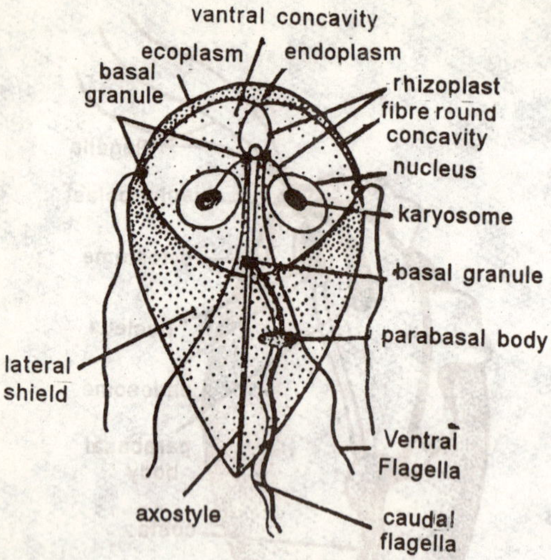


Fig. 1.11 *Giardia*

1. *Giardia* is a parasite in the small intestine and colon of man, rats, mice, dogs, cats and frogs. *G. lamblia* is found in the small intestine of man, especially children.
2. Cosmopolitan distribution.
3. Body is oval in shape or half pear-shaped with broad forward anterior end and narrow posterior end
4. The body is bilateral symmetrical.
5. The dorsal side is convex, but the ventral surface contains a slight concavity, forming a sucking disc to the wall of intestine.
6. There are two vesicular nuclei and four pairs of long flagella. The basal granules of foremost flagella are inter-connected and also connect with a pair of nuclei.
7. Axostyle forms the median longitudinal axis of the body.
8. Just behind the sucking disc is a deep staining parabasal body.
9. Cytopharynx is absent and the rhizoplast make complicated loops through the cytostome.
10. Reproduction takes place by longitudinal binary fission.
11. It forms oval thick-walled cysts. Division occurs in the cyst so that the cyst has four nuclei.
12. It prevents absorption of fats by the host, the unabsorbed fat causes diarrhoea.

*Pathogenecity:* It interferes with the absorption of food and causes *chronic enteritis, fever, anaemia, allergy* and *acute enterocolitis.*

## TRICHONYMPHA

| | | |
|---|---|---|
| Phylum | - | Protozoa |
| Subphylum | - | Sarcomastigophora |
| Super class | - | Mastigophora |
| Class | - | Zoomastigophorea |
| Order | - | Hypermastigida |
| Genus | - | *Trichonympha* |

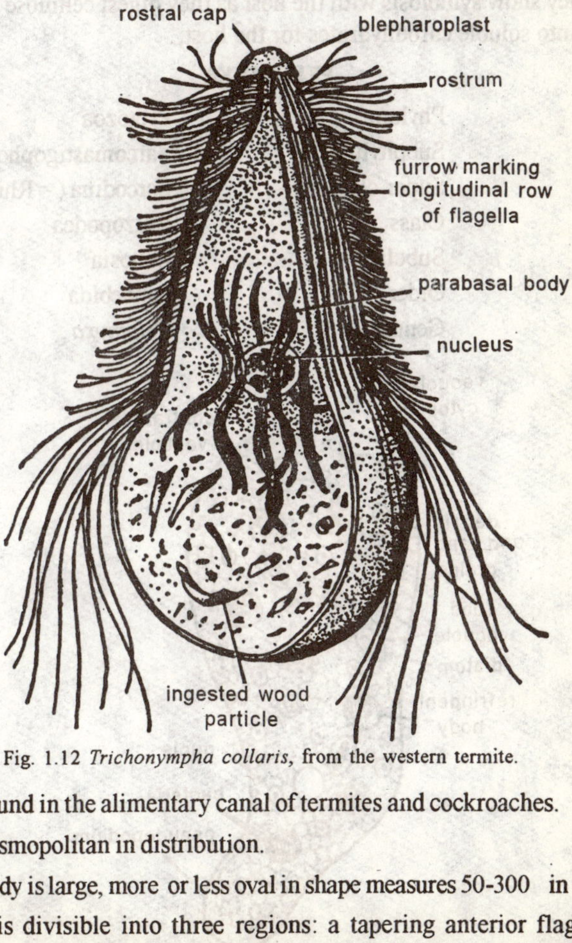


Fig. 1.12 *Trichonympha collaris,* from the western termite.

1. It is found in the alimentary canal of termites and cockroaches.
2. It is cosmopolitan in distribution.
3. The body is large, more or less oval in shape measures 50-300 in length.
4. Body is divisible into three regions: a tapering anterior flagellated rostrum, bearing apical cap; an intermediate bell-like portion, bearing numerous flagella; and a posterior rounded non flagellated region.

5. All the flagella are connected through fibrils with a hemispherical blepharoplast, lying at the base of the cap. Some flagella are arranged in longitudinal rows.
6. Several elongated parabasal bodies and axostylar filament run from rostral to hinder region of body.
7. In the ectoplasm are oblique fibres, an alveolar layer and tranverse myonemes. In the endoplasm are longitudinal myonemes and a nucleus.
8. Reproduction by longitudinal binary fission. Under unfavourable conditions it undergoes encystment.
9. They show symbiosis with the host as they digest cellulose and convert it into soluble carbohydrates for the host.

**PELOMYXA**

| | | |
|---|---|---|
| Phylum | - | Protozoa |
| Subphylum | - | Sarcomastigophora |
| Super class | - | Sarcodina ( = Rhizopoda) |
| Class | - | Rhizopodea |
| Subclass | - | Lobosia |
| Order | - | Amoebida |
| Genus | - | *Pelomyxa* |

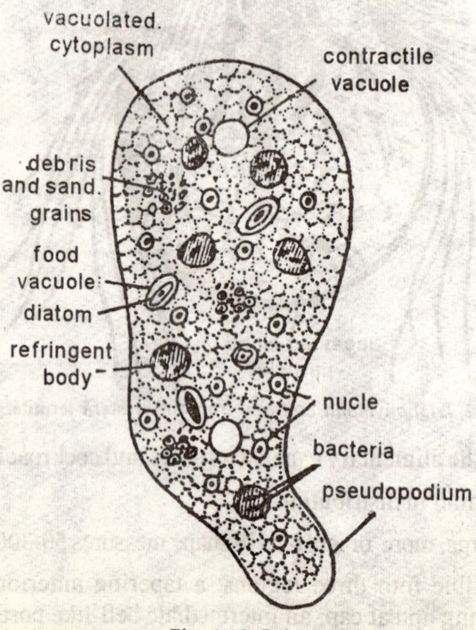


Fig. 1.13 *Pelomyxa.*

1. It is often known as *giant-amoeba* being about 25mm. in length. It is a fresh water from living in the mud of stagnant water.
2. The body is asymmetrical and the body shape changes constantly.
3. It has a single large, hyaline and blunt pseudopodium.
4. The cytoplasm is highly vacuolated and contains 100-1000 nuclei, all of which are of albuminous nature and which are alike.
5. It contains a number of peculiar, some-what rounded refringent bodies, which are of albuminous nature and which constantly harbour bacteria of *Cladothrix* sp. (symbiotic).
6. Nutrition is holozoic. Diatoms form the chief food. Numerous food vacuoles, and many fluid filled vaculoes but not contractile vaculoes. Reserve food material in the form of glycogen granules.
7. Asexual reproduction by *Plasmotomy*, in which the cytoplasm divides 2-6 times and each containing few nuclei.
8. Sexual reproduction by the formation of gametes which conjugate in pair. Encystment in some species.

## ARCELLA

| | | |
|---|---|---|
| Phylum | - | Protozoa |
| Subphylum | - | Sarcomastigophora |
| Super class | - | Sarcodina |
| Class | - | Rhizopodea |
| Subclass | - | Lobosia |
| Order | - | Arcellinida |
| Genus | - | *Arcella* |

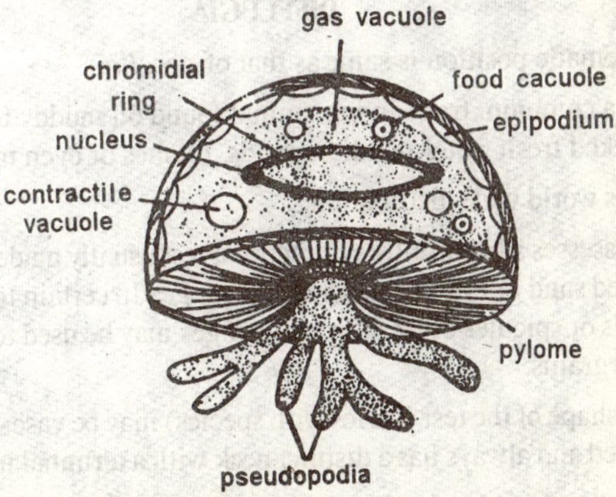


Fig. 1.14 *Arcella.*

1. It is commonly found in stagnant water attached to weeds growing in fresh water. It also occurs in moist forest soil, mosses etc.
2. It has cosmopolitan distribution.
3. The body is covered by a yellow to brown, thick, hard , transparent and disc-shaped shall or test secreted by animal itself.
4. The shall is made of pseudochitin, silica and iron.
5. The ventral surface has an inverted funnel-like depression, leading to central mouth or *pylome*.
6. 1-6 small, hyaline, finger-like, simple or branched lobopodia are extended through the pylome, which function as locomotory organelles and also help in feeding.
7. The cytoplasm is differentiated into nonvacuolated ectoplasm and endoplasm. The body is attached to the inner test by filamentous ectoplasmic strands known as *epipodia*.
8. Endoplasm contains food vacuoles, contractile vacuoles, reserve food material and two prominant nuclei in *A. vulgar* is and upto 15 in *A. podophrya.*
9. Nutrition is hologoic. Food includes mainly green and colourless flagellates.
10. A characteristic feature is the presence of a ring of granules called *chromidia.*
11. Asexual reproduction by binary fission. Two nuclei divide forming four nuclei, two of which extruded from pylome. This mass secretes a new shall. Multiple fission also occurs.

**DIFFLUGIA**

Systematic position is same as that of *Arcella*.

1. It is a common, free- living animal found on muddy floor of small leaf chacked fresh water puddies, ponds, ditches or even moist soil.
2. It has world wide distribution.
3. It possesses a beautifully symmetrical test usually made up of intricately placed sand grains of a definits size range. In certain localities, diatom, shells or spicules of fresh water sponges may be used to supplement the sand grains.
4. The shape of the test (varies with species) may be vase-shaped, or flask-shaped and always has a distinct neck with a terminal mouth or pylome.
5. About six slender, hyaline lobose pseudopodia may protrude out from pylome.

6. Cytoplasm is differentiated into ectoplasm and endoplasm. A nucleus surrounded by chromidia; numerous contractile vacuoles and food vacuoles are scattered around the periphery of the endoplasm.
7. Often the animals are a vivid green because of zoochlorellae in their cytoplasm as symbionts.
8. Mode of nutrition holozoic , it feeds upon micro-organisms and filamentous algae such as *Spirogyra.*

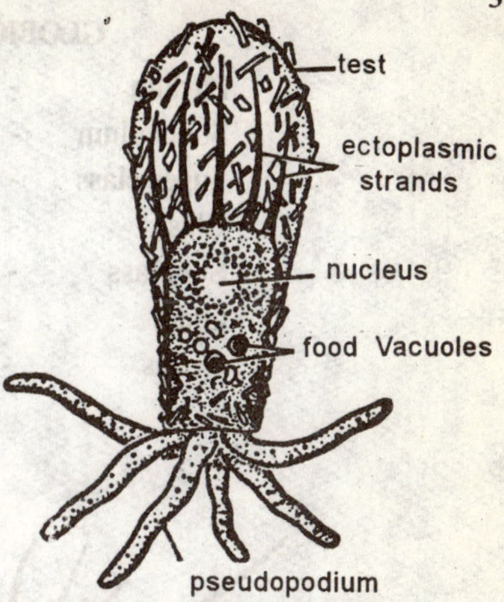

Fig. 1.15 *Difflugia oblonga*

9. It reproduces by binary fission, but multiple fission has also been reproted.
10. *Difflugia oblonga* is very active in the spring and most of the summer , but becomes encysted within its test in the late fall.

## EUGLYPHA

Systematic position is same as that of *Arcella.*

1. It is an aquatic animal found in freshwater, especially common in the sphagnum moss.
2. It has word wide distribution.
3. It has membraneous hyaline, oval, flask-shaped or elongated test composed of oblique rows of close-fitting, overlapping, rounded siliceous scale or plates.
4. The shall opens by denticulate mouth or pylome.
5. The pseudopodia are very fine, filiform, *filopodia*, which are often branched.

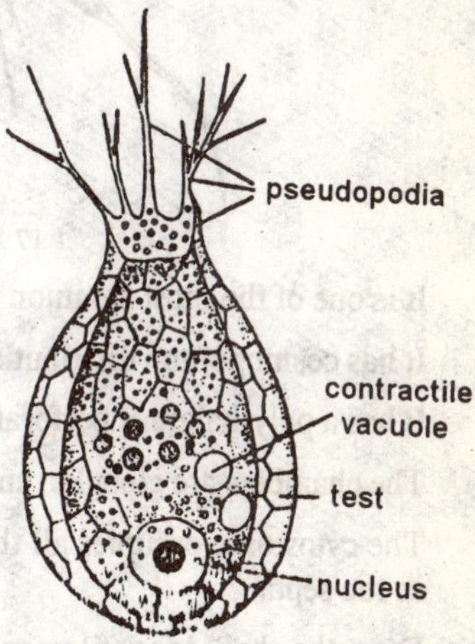

Fig. 1.16 *Euglypha*

6. Ectoplasm is indistinct. Endoplasm contains shall-plates, vacuoles, one or two large centrally placed nucleus surrounded by perinuclear chromidia.
7. It reproduces by encystment. Multiple fission is not evident.

## GLOBIGERINA

| | | |
|---|---|---|
| Phylum | - | Protozoa |
| Subphylum | - | Sarcomastigophora |
| Super class | - | Sarcodina |
| Class | - | Rhizopodea |
| Subclass | - | Granuloreticulosia |
| Order | - | Foraminiferida |
| Genus | - | *Globigerina* |

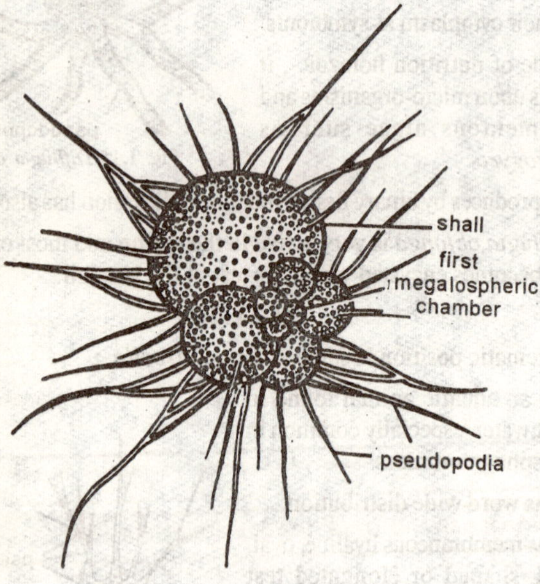


1.17 *Globigerina*

1. It is one of the most common and abundant *pelagic foraminiferan.*
2. It has cosmopolitan distribution.
3. It has a polythalamus perforated and calcareous shall.
4. The chambers are globular, and are arranged in a helicoid spiral.
5. The cytostome occupies all the chambers passing through the opening in the septa.
6. From the shall, long, filamentous, viscous and contractile pseudopodia extend all around. They form pseudopodial network. The pseudopodia are very important. They exibit streaming movement.
7. Shells of dead individuals sink to the bottom of the sea like rain, forming there a grey mud called the *Globigerina ooze*.

## ALLOGROMIA

| | | |
|---|---|---|
| Phylum | - | Protozoa |
| Subphylum | - | Sarcomastigophora |
| Super class | - | Sarcodina |
| Class | - | Rhizopodea |
| Subclass | - | Filosa |
| Genus | - | *Allogromia* |

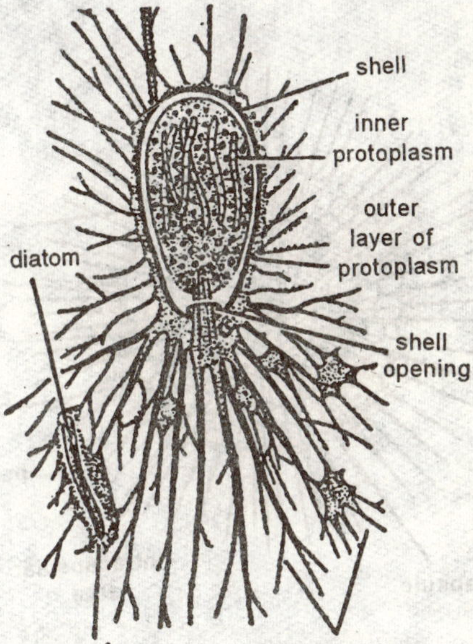


Fig. 1.18 *Allogromia oviformis.*

1. It is found in both fresh water and marine water.
2. Body of animal is covered by a shall which is one chambered.
3. The cytoplasm flows out from the terminal opening of shall and flows around the shall, in this way shall becomes internal.
4. Cytoplasm contains a single nucleus, a contractile vacuole and a food vacuole.
5. The pseudopodia are very long and delicate, and join to form a network, hence called *reticulopodia*. They help in capturing prey.
6. Reproduction by multiple fission.

## THALASSICOLA

| | | |
|---|---|---|
| Phylum | - | Protozoa |
| Subphylum | - | Sarcomastigophora |
| Super class | - | Sarcordina |
| Class | - | Actinopodea |
| Subclass | - | Radiolaria |
| Genus | - | *Thalassicola* |

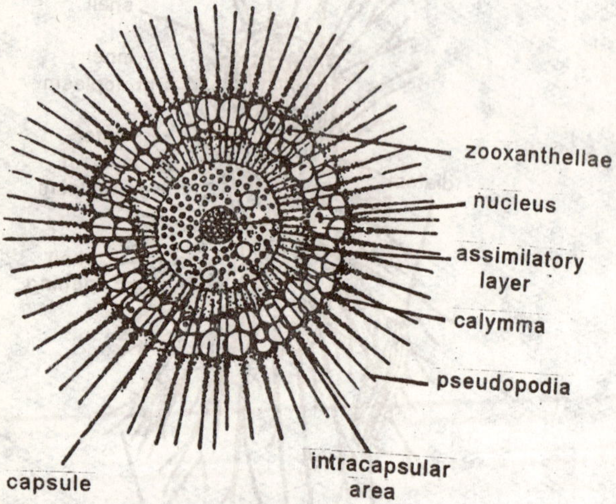


Fig. 1.19 *Thalassicola*

1. It is a marine, solitary and pelagic.
2. It has a perforated and membraneous central capsule.
3. The intercapsular cytoplasm contains a single large nucleus and a number of small vacuoles. The extracapsular cytoplasm has the usual three regions : inner thin assimlatory zone; middle thick vacuolar zone or calymma with large closely-packed, fluid-filled vacuoles lying in gelatinous material and often having yellow symbioitic dinoflagellates; and outer thin pseudopodial region.
4. Contractile vacuole is absent.
5. Nutrition is holozoic
6. Reproduction by binary fission and swarm-spore formation.

## COLLOZOUM

The systematic position is the same as that of *Thalassicola*.

1. It is a large, colonial form having a sub-spherical body measuring about 3-4 cm. in length. They are present in deep sea.
2. Protoplasm is divided into an intracapsular protoplasm by a perforated membrane. Skeleton is absent.
3. The vacuolated extra-capsular protoplasm is common to the entire colony having numerous central capsules embedded in it. Each central capsule indicates a zooid of the colony, which has a single nucleus.

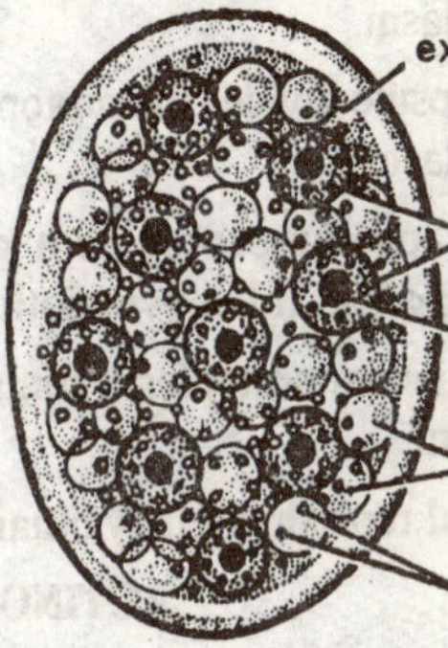

Fig. 1.20 *Collozoum*

4. The central capsule undergoes repeated division, while the extra -capsular protoplasm does not divide.
5. Contractile vacuole is absent.
6. Reproduction by *merogony*.

## LITHOCIRCUS

The systematic position is the same as that of *Thalassicola*.

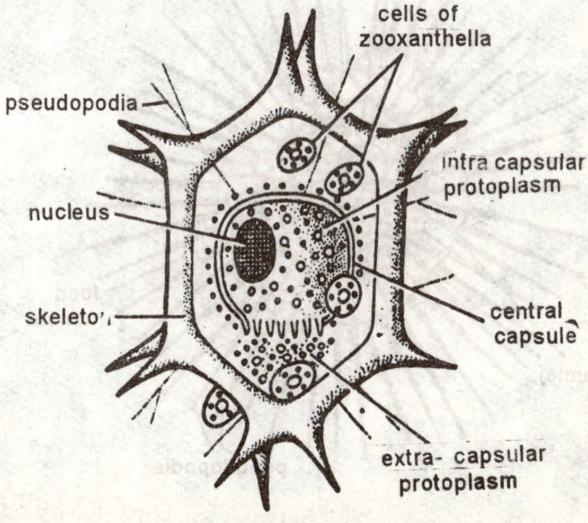

Fig. 1.21 *Lithocircus*.

1. It is a deep sea, solitary animal with a skelelton.
2. Body is covered by a shall of siliceous skeleton.
3. The animal is provided with a central capsule embedded in the protoplasm.
4. The capsule divides the protoplasm into extracapsular and intracapsular protoplasm.
5. Extracapsular protoplasm gives rise to the radiating thread-like pseudopodia. It has symbiotic yellow dinoflagellates.
6. Intracapsular protoplasm contains a large nucleus. Contractile vacuoles is absent.
7. Asexual reproduction by binary fission.

## ACTINOSPHAERIUM

| | | |
|---|---|---|
| Phylum | - | Protozoa |
| Subphylum | - | Sarcomastigophora |
| Superclass | - | Sarcodina |
| Class | - | Actinopodea |
| Subclass | - | Heliozoida |
| Genus | - | *Actinosphaerium* |

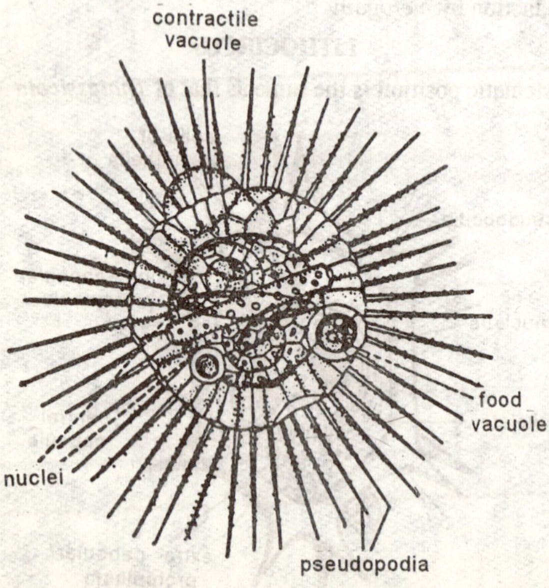


Fig. 1.22 *Actinosphaerium*

1. It is a fresh water from, abundantly found among floating vegetation. It is commonly called *Sun animalcule*.
2. Cosmopolitan distribution.
3. Several stiff needle-like axopodia radiate from the central spherical body of frothy protoplasm.
4. Protoplasm is divided into a highly vacuolated or frothy ectoplasm and a denser granular endoplasm.
5. Several contractile vacuoles in the ectoplasm.
6. Several nuclei are present in endoplasm.
7. The axial filament of pseudopodia do not reach up to nucleli, but terminate freely in the deeper granular layer of ectoplasm.
8. Nutrition is holozoic , voracious in feeding. Axopodia help in feeding rather locomotion.
9. The animal reproduce asexually by *plasmotomy*. The whole organism divides into two daughter spheres, each contains several nuclei.
10. Autogamy and encystment also observed.

## ACTINOPHRYS

The systematic position is same as that of *Actinosphaerium*.

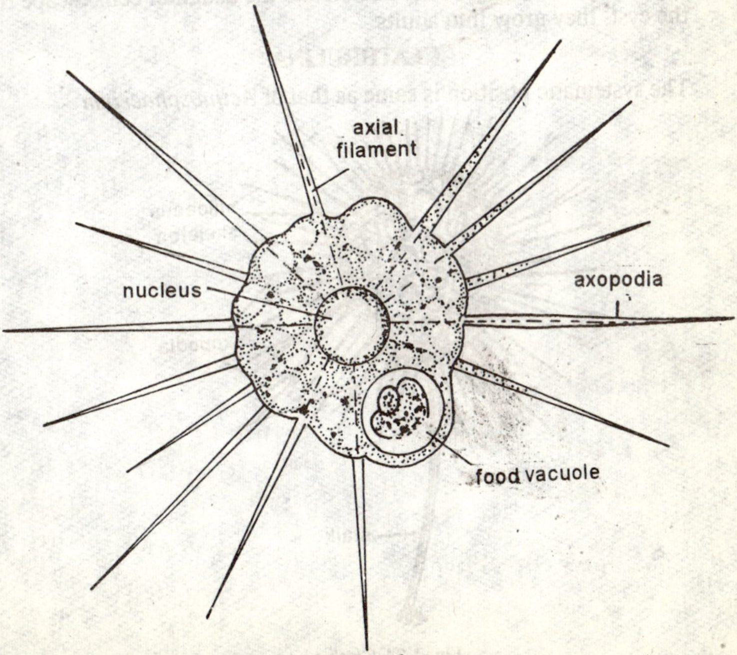


Fig. 1.23 *Actinophrys*

1. It is found in fresh and marine waters.
2. It has cosmopolitan distribution.
3. The body is round from which radiate thin, long pseudopodia, each with a central axial filament covered with an adhesive granular ectoplasm. They are called axopodia.
4. The axial filaments of axopodia are attached to the nuclear membrane.
5. The cytoplasm is not distinctly divided into ectoplasm and endoplasm.
6. The ectoplasm have many vacuoles.
7. Endoplasm is gelatinous and contains various vacuoles and food particles besides a single nucleus.
8. Nutrition is holozoic , whenever food or prey touches the axopodia, they attach to it. The axopodia paralyze the prey and at times shorten to convey the prey to central mass of cell.
9. Reproduction occurs by binary fission and also by *paedogamy*. During paedogamy the animal withdraws its pseudopodia gorges itself on flagellates and encysts. The cyst is bilayered outer gelatinous and membranous inside. It divides into many uninucleate cysts. Each secondary cyst divides into two cells enclosed in the cyst. The nucleus of each cell than divide twice to form four haploid nuclei. Two of the nuclei degenerate. The remaining two nuclei of cyst cells then fuse to form zygote. The zygote reproduces by binary fission and the daughter cells escape from the cyst, they grow into adults.

## CLATHRULINA

The systematic position is same as that of *Actinosphaerium*.

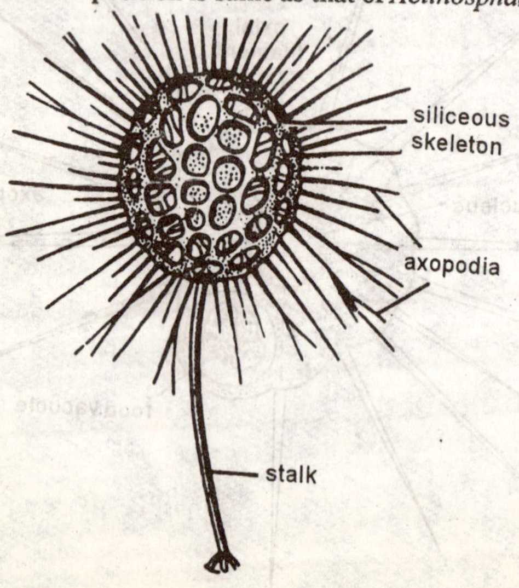


Fig. 1.24 *Clathrulina*

1. It is a fresh water protozoan.
2. The body is enclosed in a perforated skeletal sphere of silica, and is attached by a long, slender stalk.
3. The axopodia project through the skeletal investment.
4. Cytoplasm contains single nucleus.
5. It undergoes multiple fission in the active condition and produce biflagellate merozoites, called *flagellule*. The flagellule latter grow into adults.

## BABESIA

| | | |
|---|---|---|
| Phylum | - | Protozoa |
| Subphylum | - | Sarcomastigophora |
| Superclass | - | Sarcodina |
| Class | - | Piroplasmea |
| Genus | - | *Babesia* |

1. It is a minute, intracorpuscular blood parasite of cattles, dogs and monkeys.
2. The trophozoite lives in the R.B.C. of the host and has a rhizoplast but no flagellum and pigment.
3. Thick is the intermediate host and vector. Upon reaching in its gut, the trophozoites change into gametes and fuse in pairs to form zygote.

Fig. 1.25 *Babesia*

4. The zygotes are motile and bore the alimentary canal and reach the ovary and parasitize the ova.
5. In ova they divide and produce amoeboid spores which pass on to the next generation of host (tick).
6. Some of the spores reach the salivary glands cells where the spore becomes multinucleated and produce sporozoites.
7. Tick transfers the sporozoites into the blood of the primary host when it feeds upon its blood.
8. It causes the formidable *Texas Fever* or *Red Water Fever* in cattles in America and Australia.

## SARCOCYSTIS

| | | |
|---|---|---|
| Phylum | - | Protozoa |
| Subphylum | - | Sporozoa |
| Class | - | Toxoplasmea |
| Genus | - | *Sarcocystis* |

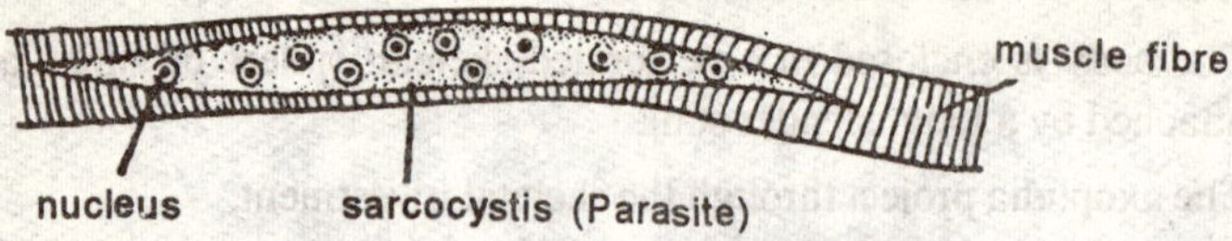


Fig. 1.26 *Sarcocystis*

1. It is a muscle parasite of vertebrates especially sheeps, rabbits, pigs, rats, cattles, horses and in reptiles and birds.
2. The parasite usually encysts within striped muscle fibres of the host, forming a long multinucleate spindle-like *cyst* which may be several mm. in length. It is known as *Miescher's or Rainey's tube*.
3. Within each tube a large number of sickle-shaped spores are formed. The spores are called *Rainey's corpuscles.*
4. Very little is known about the mode of transmission and life cyle. When ingested the spores hatches in the intestine of new host liberating a single sporozoite , which penetrates an epithelial cell to grown into a trophozoite. Through blood circulation it reaches to the striped muscle fibres, where it grows into plasmodium.
5. They produce a toxic substance *sarcocystin* and cause a disease known as *sarcosporidiosis.*

## NOSEMA

| | | |
|---|---|---|
| Phylum | - | Protozoa |
| Subphylum | - | Cnidospora |
| Class | - | Microsporidea |
| Genus | - | *Nosema* |

1. *Nosema bombycis* is a parasite in the gut of silk worm larvae where it causes *pebrine*-a destructive silkworm disease.
2. Asexual multiplication of the trophozoites and possibility of some spores developing in the same host cause enormous damage to the host's tissue, the result is fatal.
3. Its another species is *N. apis.* It is parasite in the epithelial tissue and the Malpighian tubules of honey-bees causing a fatal disease nosema.
4. The spores are very small, each

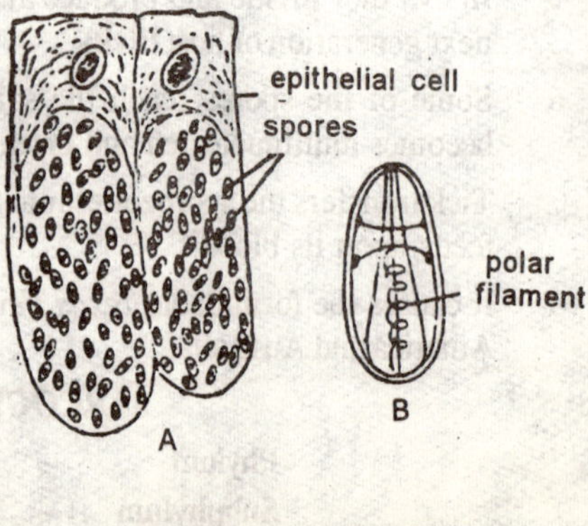


Fig. 1.27 *Nosema* A—Spares of *Nosema* apis in stomach cell. B—Spore of *N.bombycis*

containing a coiled thread, the polar filament, that represents a polar capsule without a wall.

## PRORODON

| | | |
|---|---|---|
| Phylum | - | Protozoa |
| Subphylum | - | Ciliophora |
| Class | - | Ciliata |
| Subclass | - | Holoricha |
| Order | - | Gymnostomatida |
| Genus | - | *Prorodon* |

1. It is a fresh water and marine ciliate.
2. The body is ovoid but sometimes variable in shape, but, when fully extended, it measures 165-200 in length and 40-60 in width.
3. Body is ellipsoidal, with the anterior end slightly wider then the posterior.
4. Body is covered over by cilia. The cilia of anterior and posterior ends are somewhat, longer than elsewhere.

Fig. 1.28 *Prorodon griseus.*

5. The terminal cytostome leads into a short oval cytopharynx.
6. The cytoplasm is differentiated into ectoplasm and endoplasm.
7. Ectoplasm of *P. griseus* contains no trichocysts or other inclusions except 20-30 deeply, embedded, large, parallel, rod-like trichite. These form a highly distensible oral basket around the subterminal cytostome.
8. Endoplasm is grayish and occasionally may become green because of numerous symbiotic *zoochlorellae.*
9. Numerous food vacuoles are present. Cytopyge at the posterior end.
10. A large contractile vacuole is present at extreme posterior end.
11. Two nuclei; the massive macronucleus located centrally, may be spherical or oval containing a prominent nuclelolus. A small pear-shaped micronucleus is found attached externally to the macronucleus.
12. When the pond water dry up, it develops a thick-walled protective cyst around itself. Almost immediately after excysting, the small animals conjugate followed by binary fission. Macronucleus also divides.

13. Macronuclear reorganization takes place only in the permanent protective cyst and requires at least four days.

## COLEPS HIRTUS

The systematic position is same as that of *Prorodon.*

1. It is a widely distributed fresh water ciliate with an ability to tolerate wide range of temperature, oxygen, salinity and acidity.
2. The body is barrel shaped, brown in colour and is slightly more convex dorsally. It measures about 40-65 in length and 20-30 in width.
3. In the alveolar layer of ectoplasm beneath the pellicle, the body is covered with armour plates which are modified into oral teeth at the anterior end, and caudal spines at posterior end.

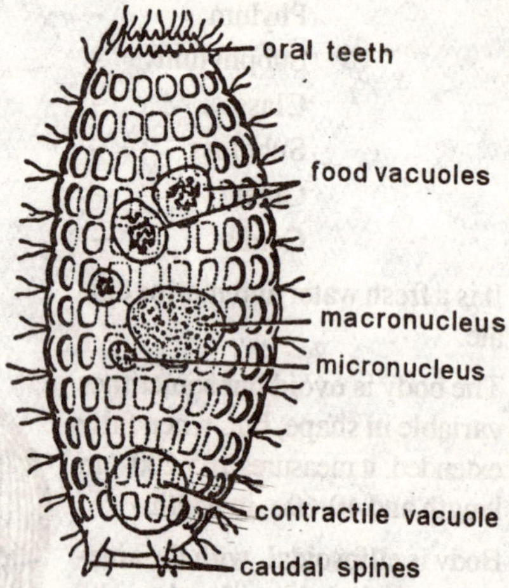

Fig. 1.29 *Coleps hirtus.*

4. The plates are organic in composition and are arranged in four anterioposterior girdles of 20 plates each which encircle the animal , like the staves of a barrel.
5. Each plate bears elevated rectangular areas that give the surface of the ciliate the appearance of being divided into more or less symmetrical squares.
6. Shorter cilia located between the longitudinal plates and longer cilia occur around the anterior cytostome. Longer cilia help in locomotion.
7. The animal is scavanger. It ingests large protozoans and green flagellates.
8. Endoplalsm contains many food vacuoles, a posterior contractile vacuole and two nuclei. The macronucleus is centrally placed and is spherical, the micronucleus is small.
9. A subterminal pygostyle is present.
10. Asexual reproduction by binary fission. Conjugation is not reported in this animal.

## DIDINIUM NASUTUM

The systematic position is same as that of *Prorodon.*

1. It is fresh water protozoan found in ponds and ditches where large protozoan like *Paramecium, Colpoda, Frontonia* etc. are in abundance.
2. It is commonly known as *'water bear'*.
3. Body is barrel-shaped, measuring 80-200 in length, and produced at the anterior end into a conical proboscis that bears cytostome at the tip.

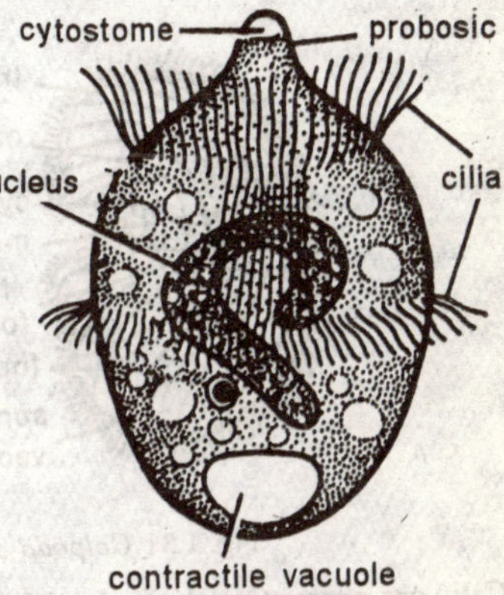

Fig. 1.30 *Didinium*

4. The cilia occur in two rings only, one ring of cilia is close to the base of proboscis and the other ring on the posterior end.
5. The trichocyst-like bodies are present only in proboscis region.
6. Within the endoplasm, a horse-shoe shaped macronucleus may be seen; a contractile vacuole and cytopyge are both in the extreme posterior end of the animal.
7. The animal is voracious. Its enormous appetite is satisfied by a diet of large ciliate like Paramecium etc.
8. Asexual reproduction by binary fission at a rate of 3-4 divisions per day.
9. Sexual reproduction by conjugation. It encysts under unfavourable conditions.

## COLPODA CUCULLUS

| | | |
|---|---|---|
| Phylum | - | Protozoa |
| Subphylum | - | Ciliphora |
| Class | - | Ciliata |
| Subclass | - | Holotricha |
| Order | - | Trichostomatida |
| Genus | - | *Colpoda* |
| Species | - | *cucullus* |

1. It is free-living, freshwater ciliata.
2. The shape of animal superficially resembles a flattened kidney. It measures 80-120 in length.

3. A clear defined, diagonal groove is located on the aboral surface extending from the right side of the posterior third of body to the middle where it forms a prominant notch. The groove continues for a short distance on the cytostome lying near the left side.

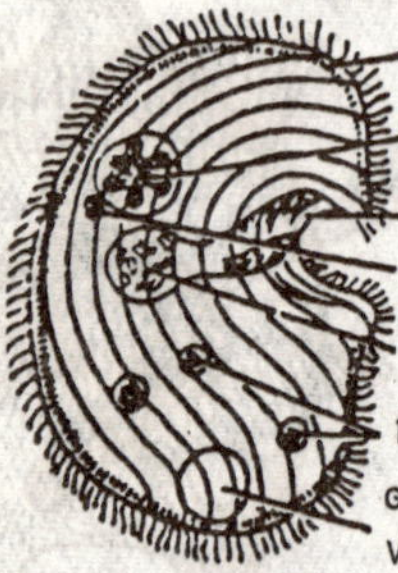


Fig. 1.31 *Colpoda cucullus*, oral view.

4. Cilia are present on the right side of cytopharynx and rows of cilia on the left side of same organelle give that area a cross-hatched appearance. Delicate cilia throughout the body.
5. Ectoplalsm possesses small clear trichocysts which have a doubtful function in this species.
6. In the endoplasm a spherical macronucleus, a small micronucleus are present in the middle region. A contractile vacuole present at the posterior end. Many food vacuoles are present.
7. Binary fission takes place in thin walled cyst.
8. Under unfavourable conditions thick walled cysts are formed. The encysted *Colpoda* remains in a dormant state for long periods of time and may become active again even after 38 years.

**BALANTIDIUM**

| | | |
|---|---|---|
| Phylum | - | Protozoa |
| Subphylum | - | Ciliophora |
| Class | - | Ciliata |
| Subclass | - | Holotricha |
| Order | - | Trichostomatida |
| Genus | - | *Balantidium.* |

1. It is commonly found in the intestine of man, pigs, sheeps, camels, cockroaches etc. It is abundantly found in the rectal contant of the frogs.
2. It has cosmopolitan distribution.

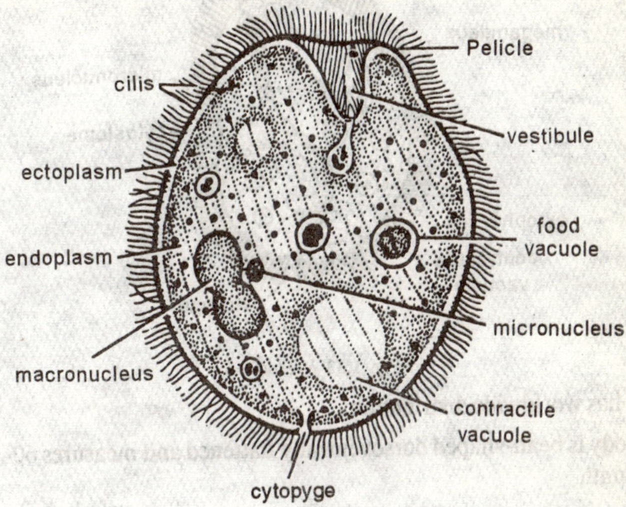


Fig. 1.32 *Balantidium coli*

3. It is spherical in shape measuring 60-70   in length and 40-60   in breadth.
4. The entire body is covered by longitudinal slightly spiral rows of cilia.
5. There is a large peristomal depression or vestibule leading to cytostome at the anterior end.
6. Broad posterior end bears a cytopyge.
7. Cytoplasm is differentiated into ectoplasm and endoplasm.
8. The endoplasm contains food-vacuoles, two contractile vacuoles and two nuclei. The macronucleus is large rod-like while micronucleus is small and spherical.
9. Nutrition is holozoic. Food consists of erythrocytes, tissue fragments, bacteria and faecal debris.
10. Reproduction takes place by binary fission and conjugation.
11. Cysts are formed, but not multiplication occurs inside the cysts.

*Pathogenicity:* It causes ulceration of the colon and invades the mucosa by virtue of cytolysin, which it is believed to secrete. It generally cause *diarrhoea*, but sometimes cause severe or fatal dysentery.

**NYCTOTHERUS**

The systematic position is same as that of *Balantidium.*

1. It is a parasitic ciliate in the rectum of frogs and intestine of cockroaches.

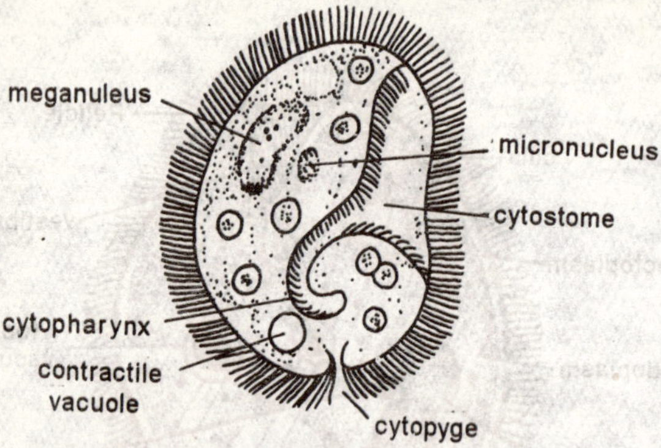


Fig. 1.33 *Nyctotherus*

2. It has worldwide distribution.
3. Body is bean-shaped dorsoventrally flattened and measures 60-120 in length.
4. The peristome starts from the anterior end and extends down to the middle of left side, where it leads by cytostome into a curved cytopharynx.
5. The entire body of animal is covered with short equal-sized cilia arranged in longitudinal rows. Peristome bears an adoral row of specially large cilia, that extended into the cytopharynx.
6. A permanent cytopyge at the posterior end of the body.
7. Cytoplasm is differentiated into ectoplasm and endoplasm.
8. Endoplasm encloses several food vacuoles, a large contractile vacuole (at the posterior end); a kidney shaped macronucleus and a small micronucleus infront of cytopharynx.
9. Nutrition is holozoic. Food consists of bacteria, undigested food particle etc.
10. Asexual reproduction by binary fission. Sexual by conjugation.
11. Encystment during breeding season of host. The cysts are swallowed by the tadpoles with vegetation.

## FRONTONIA

| | | |
|---|---|---|
| Phylum | - | Protozoa |
| Subphylum | - | Ciliophora |
| Class | - | Ciliata |
| Subclass | - | Holotricha |
| Order | - | Hymenostomatida |
| Genus | - | *Frontonia* |

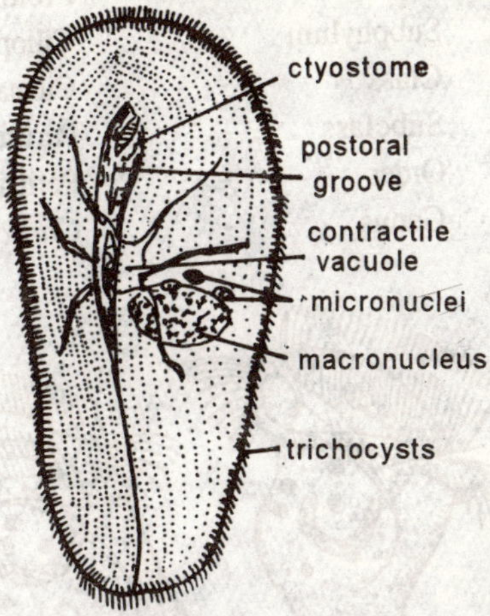


Fig. 1.34 *Frontonia leucas.*

1. It is a large fresh water animal.
2. The anterior end of body is somewhat flattened dorsoventrally, where as the posterior end is more cylindrical. It measures 150-300 in length and the width 70-150
3. The body is uniformly covered with cilia except the oral region.
4. Cytoplasm is differentiated into ectoplasm and endoplasm.
5. Embedded in the ectoplasm are numerous fusiform trichocysts, usually right angles to and in contact with pellicle.
6. Located ventrally in the anterior fourth or third of the body, the cytostome proper in more or less heart shaped with its point towards the anterior end. From this side a very narrow postoral groove extends almost to the posterior end of the body, then swings toward the opposite side at posterior end.
7. Endoplasm contain a very large contractile vacuole surrounded by 8-10 radial canals.
8. At the same level of the contractile vacuole is a large oval macronucleus. Many (3-5) micronuclei are found near macronucleus.
9. It is omnivorous feeds on bacteria , algae, amoebae, arcellae and even rotifers.
10. Symbiotic zoochlorellae sometimes occur in the cytoplasm.
11. Reproduction by binary fission and conjugation.

## CARACHESIUM

| | | |
|---|---|---|
| Phylum | - | Protozoa |
| Subphylum | - | Ciliophora |
| Class | - | Ciliata |
| Subclass | - | Peritricha |
| Order | - | Peritrichida |
| Genus | - | *Carachesium* |

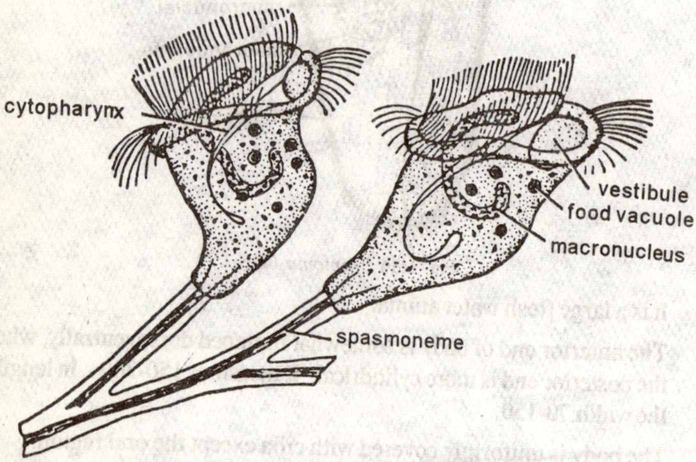


Fig. 1.35 Zooids of *Carchesium polypinum.*

1. It is fresh water as marine colonial ciliate with branching stalks.
2. The zooids are bell-shaped and largely resemble *Vorticella* in structure.
3. The stalk of each zooid has a separate spasmoneme so that it contract independently of the stalks of other zooids in the coloney.
4. The arrangement of cilia reaches its greatest complexity in the adoral or oral ciliature.
5. Endoplasm contains macronucleus and micronucleus, food vacuoles and contractile vacuole.
6. It reproduces by binary fission, conjugation and sometimes encystment occur.

## STENTOR

| | | |
|---|---|---|
| Phylum | - | Protozoa |
| Subphylum | - | Ciliophora |
| Class | - | Ciliata |

| | | |
|---|---|---|
| Subclass | - | Spiroptricha |
| Order | - | Heterotrichida |
| Genus | - | *Stentor* |

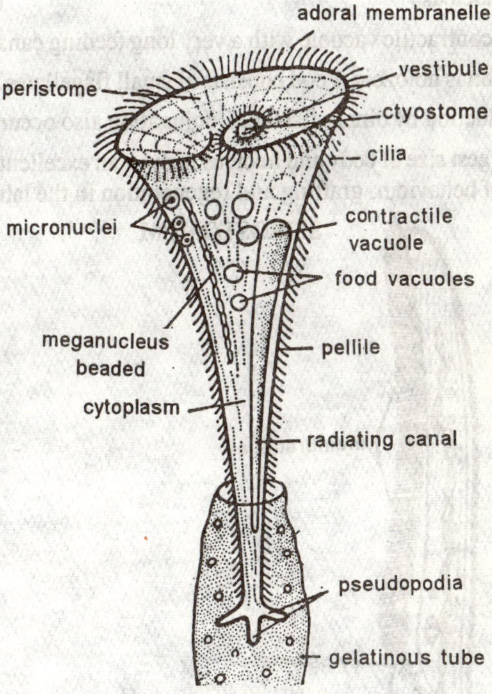


Fig. 1.36 *Stentor*

1. It is a large, sessile, fresh water ciliate, having a sky-blue pigment called as *stentorin.*
2. It has world-wide distribution.
3. It measures 1-4 mm. in length and is pyriform in shape.
4. The body is covered over by cilia, arranged in longitudinal rows.
5. The animal attaches itself to aquatic plants, animals and other solid objects by its lower narrower end, which gives out pseudopodia to serve as holdfast.
6. In some species the lower end in covered by a loose gelatinous tube.
7. The anterior ends bears a broad peristome encircled by a single row of adoral membranelles arranged spirally in a clock-wise manner.

8. The peristome spirals down to the cytosome that leads into a short tabular cytopharynx.
9. The nuclear apparatus consists of a long beaded macronucleus with 15-20 beads or lobes and about 80 micronuclei lying close to the macronucleus.
10. Single contractile vacuole with a very long feeding çanal.
11. Nutrition is holozoic. Food consists of small flagellates and rotifers.
12. Reproduction by binary fission. Conjugation also occurs in late spring.
13. The largest size of body and nucleus make it an excellent material for the study of behaviour, grafting and regeneration in the laboratory.

**SPIROSTOMUM**

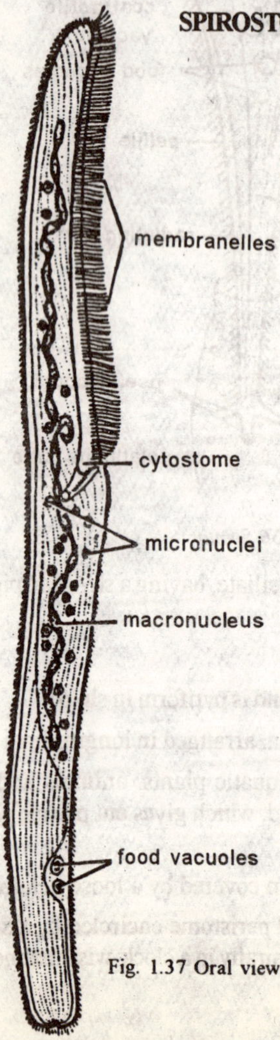


Fig. 1.37 Oral view of *Spirostomum*

The systematic position is same as that of *Stentor.*

1. It is fresh-water , free living animal.
2. It has a slim, long , cylindrical and worm-like body, measuring 3-4 mm. in length.
3. Body shows anticlock-wise spiral striations.
4. Highly developed contractile myonemes run longitudinally in the ectoplasm.
5. Peristome, occupying nearly 1/3 anterior ventral surface of the body.
6. The cilia are short, uniform except for some caudal cilia which are slightly larger, secrete mucous.
7. A large contractile vacuole and cytopyge lie at posterior end of the body.
8. Nuclear apparatus consists of a long beaded macronucleus and several micronuclei.
9. They undergo conjugation and a complicated process of binary fission.

## STYLONCHIA

| | | |
|---|---|---|
| Phylum | - | Protozoa |
| Subphylum | - | Ciliphora |
| Class | - | Ciliata |
| Subclass | - | Spiroptricha |
| Order | - | Hypotrichida |
| Genus | - | *Stylonchia.* |

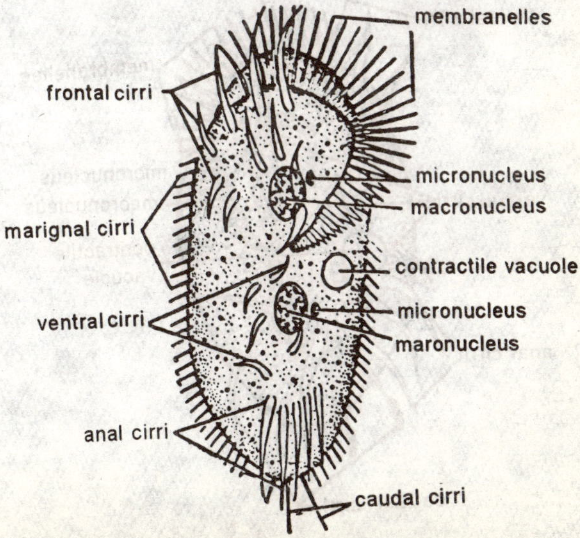


Fig. 1.38 Oral view of *Stylonychia*

1. It is one of the most prevalent ciliate found in infusion culture made from natural waters or soils.
2. The body is elongated, oval, rigid and depressed with an arched dorsal surface and flat ventral surface. It measures 150 in length.
3. The sides of the body are fringed with marginal cirri which are made up of fused cilia in bundles.
4. The ventral surface bears an eleborate system of cirri, including 8 frontals, 5 ventrals, 5 annals, and 3 caudals.
5. The cilia of dorsal surface are represented by few stiff bristles.
6. Peristome is triangular or semicircular. It begins at the anterior end and reaches upto middle of the body.
7. Cytostome remains permanently open.
8. Mode of nutrition holozoic. It takes bacteria, green flagellates. Cannibalism in the scarcity of foods.
9. Two large, oval macronuclei and two small micronuclei. Contractile vacuole is single.
10. Reproduction mainly by transerse binary fission. Conjugation also takes place.

## EUPLOTES PATELLA

The systematic position is same as that of *Stylonchia.*

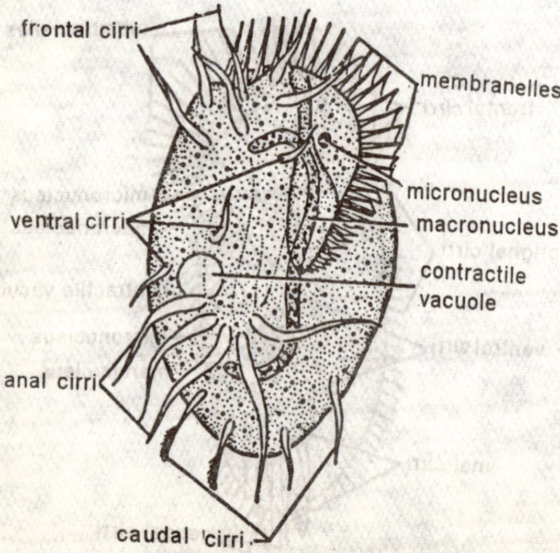


Fig. 1.39 Oral view of *Euplotes patella.*

1. It is found in fresh water and brackish water.
2. Body is subcircular to elliptical with dorsal side convex and the ventral side concave. It measures 90 in length and 50 in width.
3. On the dorsal side there are six distinctive ridges with corresponding rows of sensory bristles embedded in rosettes of granules.
4. The ventral surface bears the cirri, of which 6 are frontals, 3 ventral, 4 annals and 4 caudals.
5. A small peristomial field is surrounded by 34-35 memebranelles arranged in a spiral manner. On the right side there is somewhat raised peristomial lip.
6. Undulating membrane is absent.
7. Dense endoplasm contain a C-shaped macronucleus. Small micronucleus lies near the macronucleus.
8. A contractile vacuole lies dorsally.
9. An ill-defined cytopyge is present.
10. Mode of nutrition holozoic. It chiefly feeds upon *Chilomonas.*
11. It multiplies by binary fission.
12. Conjugation also takes place among the different mating types of a syngen.

## EPHELOTA

| | | |
|---|---|---|
| Phylum | - | Protozoa |
| Subphylum | - | Ciliophora |
| Class | - | Ciliata |
| Subclass | - | Suctoria |
| Genus | - | *Ephelota* |

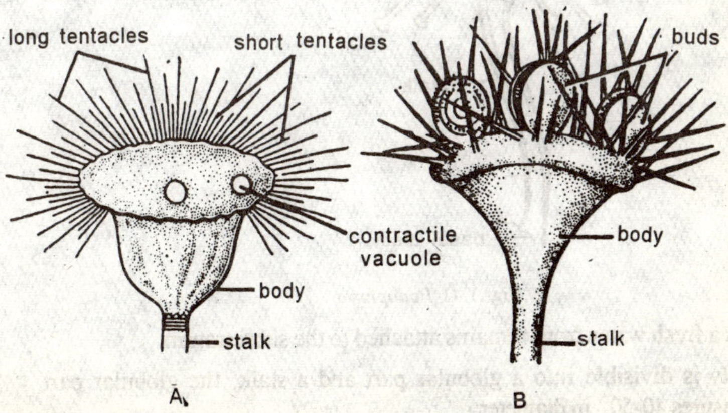


Fig. 1.40 *Ephelota*

1. It is a large marine form, found in the sea-water. It remains attached to various objects like algae etc.
2. The body is spherical and bears a stalk.
3. Two types of tentacles are present on body; long pointed for piercing the prey and short tentacles are cylindrical for sucking.
4. Cytoplasm contains few contractile vacuoles. The macronucleus is highly branched. Micronucleus is small.
5. The distal end bears a number of buds, each receiving a branch of macronucleus. The nuclei behave as in ordinary binary fission. Thus mode of reproduction is by exogeneous buds. The buds detach and acquire cilia on one surface. After a short active free swimming existence, they lose cilia and grow into tentaculate adults.

**PODOPHRYA**

The systematic position in same as that of *Ephelota.*

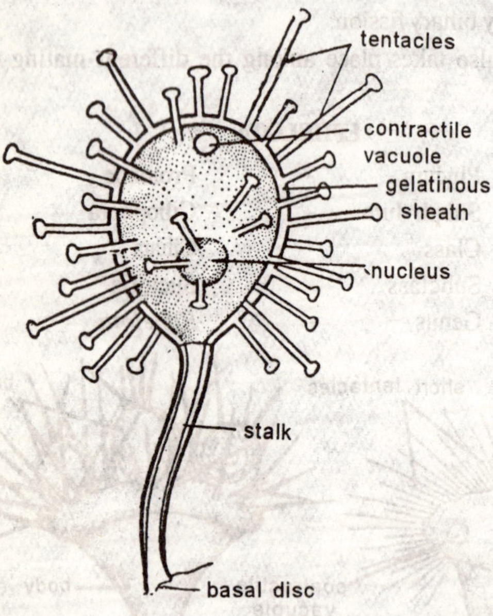


Fig. 1.41 *Podophrya*

1. It is a fresh water from remains attached to the subsetratum.
2. Body is divisible into a globular part and a stalk, the globular part measures 40-50 in diameter.

3. The body bears a number of tentacles that give a pin-cushion like appearances. They are evenly distributed. They consist of a core of fluid endoderm surrounded by firm ectoplasm.
4. The cytoplasm contains a large spherical macronucleus in the centre and several micronuclei.
5. A single contractile vacuole near the periphery. The entire body is covered by a gelatinous sheath.
6. Mode of nutrition is holozoic. It feeds other ciliates. The chief food in *Paramecium.* As soon as the tentacles come in contact with the prey, additional tentacles join in holding and eating it.
7. Reproduction by external budding. A bud is an out growth of the body. It receives parental cytoplasm, a part of parental macronucleus and micronuclei (derived from the parental micronuclei by regular division). The bud acquiring cilia and finally separates from parent. After free-swimming the bud comes to rest and attach itself by one end. It grows and given rise to an adult.
8. Conjugation takes place by the fusion of two adjacent stalked adults. Encystment during unfavourable conditions.

# 2

# EUGLENA

*Euglena* belongs to the Order-Euglenoidida (Class-Mastigophora), which comprises relatively large forms, usually of a conspicuously green colour. In the members of this order one or two flagella are present, arising from the invaginated anterior end of the body called *reservoir*. All chlorophyll bearing members possess a light-sensitive organelle called *stigma*. Several non - pigmented forms e.g. *Peranema* etc. are also placed in this order because of their basic structure.

**SYSTEMATIC POSITION**

| | | |
|---|---|---|
| Phylum | - | Protozoa |
| Sub phylum | - | Sarcomastigophora |
| Super class | - | Mastigophora |
| Class | - | Phytomastigophora |
| Order | - | Euglenoidida |
| Family | - | Euglenoidae |
| Genus | - | *Euglenma* |

**Habits and Habitat**

*Euglena* is a common, solitary and free-living flagellate found in stagnant water containing nitrogenous organic matters. Such water with high organic contents is called *polysaprobic*. Water with moderate organic contents is called *mesosaprobic,* and with low organic contents is *oligosaprobic*. The pure water is described as *kathrobic*. There are about 100 species of *Euglena* widely distributed all over world in polysaprobic fresh water environment. It is also know to occur in soil, mud and often has been found on the bark of trees. *Pringsheim* (1953) mentioned that some euglenae are marine and some are parasite. *Euglenomorpha* lives in the rectum of tadpoles, where they play an unknown symbiotic role. The most common species of *Euglena* is *E. viridis.* The common Indian species are *E. agilis, E. orientalis and E. fusiformis.* The present description is based on the biology of *E. viridis* (Gr. *eu* = true; *glene* = eyeball; and *L., viridis* = green). They are found in such a large number that they produce a green scum on the surface of ponds.

**Culture**

They can be cultured in laboratory. For preparing culture, some cow or hourse dung in boiled in water in a jar, and allowed to cool it for two days. Now pour some weeds from the pond containing euglenae into the jar, and place the jar near the light source. In a few days euglenae develop in the medium. Another method can be utilized for the preparation of culture. Boil the pond or tap water for some times and allow it to cool. Add 20 boiled wheat grains to this water and inoculate with euglenae. Keep the jar in sunny place. Wheat grains may be added monthly to maintain the culture.

**Morphology**

*Shape and size. Euglena viridis* is a minute elongated and spindle-shaped form. It is pointed at the posterior end and blunt at the anterior end. The posterior end is called *tail piece*. It measures 60 in length and 15-20 in breadth at the thickest part of the body.

*Pellicle.* The shape of the animal is maintained by the pellicle which presents parallel or spiral striations. The striations may or may not be present on the tail piece. According to *Chadfaud* (1937) the pellicle is two layered, having an outer thin *epicuticle* and deeper and thick inner *endocuticle*. The pellicle in soft and thin and enabling the animal to change its shape considerably. The pellicle is made of proteins but not of cellulose. Under electron-microscope the pellicle consists of interlocking longitudinal strips and usually associated with microtubular fibrils, the *myonemes* (*Leedale, 1964*).

In some species of *Euglena* just below the pellicle very small elongated or spherical bodies are present, these are believed to secrete mucilage and hence are called *mucous bodies* or *muciferous bodies*.

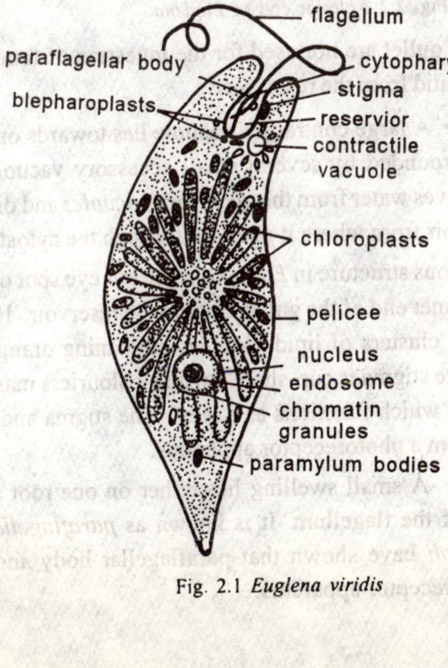


Fig. 2.1 *Euglena viridis*

*Cytostome and cytopharynx.* Shortly to one side of the middle line at the anterior end, is a small opening called *cytostome* or *cell mouth*. The cytostome leads into a tube called *gullet* or *cytopharynx* which in turn opens at its other end into a more or less spherical chamber called the *reservoir*. The pellicle invaginates into the gullet through the cytostome and lines the reservoir. The pellicle here contains only the epicuticle layer.

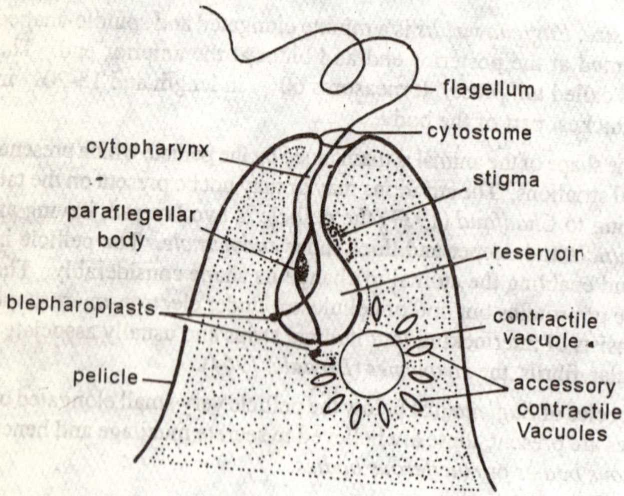


Fig. 2.2 Anterior end of *Euglena*.

The cytostome and gullet are not used for the ingestion of food but as a canal for the escape of fluid from the reservoir.

*Contractile vacuole.* A large contractile vacuole lies towards one side of the reservoir. It is surrounded by several small accessory vacuoles. The contractile vacuole receives water from the *accessory vacuoles* and discharges the fluid into the reservoir from where it goes out through the cytostome.

*Stigma.* A conspicuous structure in *Euglena* is the red eye spot or *stigma*. This is placed near the inner end of the gullet close to the reservoir. It consists of protoplasm in which clusters of lipid globules containing orange or red carotinoid pigments. The stigma is cup-shaped with a colourless mass of oily droplets in its concavity which functions as a lens. The stigma and paraflagellar body, together form a photoreceptor apparatus.

*Paraflagellar body.* A small swelling lies either on one root or at the junction of two roots of the flagellum. It is known as *paraflagellar body*. *Chadefaud* and *Provasoli* have shown that paraflagellar body and stigma together form the photoreceptor apparatus.

*Flagellum*. A single, long whip-like flagellum emerges out of the cytostome through cytopharynx. This whip-like process, by its lasting movements, causes the organism to move in water. The flagellum of *Euglena* is also called *tractellum* because it does not propel the body from behind but draws its forward. The flagellum arises from the bottom of the cytopharynx, is forked at the base and is connected by two *rhizoplasts* to the *basal granules* or *blepharoplasts*. Each blepharoplast is a compact granule that gives rise to a separate filament (*axoneme*) and in most of the species the two axonemes soon unite to form a definite flagellum. In *E. viridis* the flagellum is enlarged at the junction of two axonemes or just behind it to form the paraflagellar body. The paraflagellar body probably contains lactoflavin as *sensitiger*.

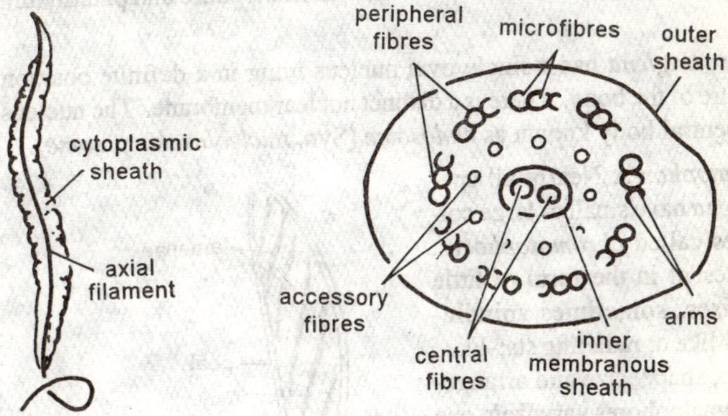


Fig. 2.3 *Euglena.* flagellum & T. S. of flagellum

According to some workers, there are two flagella in *E.spirogyra*, one long and the other is short . Each arising from the basal granule located at the base of reservoir. The short flagellum does not extend beyond the neck of the reservoir and is often adherent to the long flagellum giving the appearance of bifurcation.

Each flagellum is cylindrical structure about 0.25 in diameter. It is composed of a longitudinal bundle of microtubular fibres (axonemes) enclosed within a unit membrane which is continuous with the cell membrane. The microtubular fibrils of the axoneme are arranged in a precise pattern, best seen in a T.S. Two microtubules occupy the central position. They are called *central fibres* and remain surrounded by inner membranous sheath. The central fibres are surrounded by nine double microtubular fibrils called *peripheral fibres*. Each peripheral fibre is made of two microfibres. One of the microfibres bears a double row of short *arms* projections all pointing in the same direction. The distal portion of flagellum contains numerous minute fibres, known as *mastigonemes*, which project laterally (*Manton, 1959*). Such a flagellum that bears mastigonemes is called *stichonematic type*. Mastigonemes are believed to project rigidly from the flaglellum in two rows arranged in the plane of flagellar undulation.

The structure of blepharoplast is similar to that of the centriole found in animal cells. There are evidences which show that basal body contains DNA and may have some power of self-replication.

*Cytoplasm*. Two types of cytoplasm can be distinguished in *Euglena*, a dense outer layer, the *ectoplasm* and the central mass, the *endoplasm*, which is more fluid. The endoplasm contains nucleus, chromatophore and paramylum bodies.

*Nucleus*. *Euglena* has a single oval nucleus lying in a definite position near the centre of the body. There is a distinct nuclear membrane. The nucleus contains a central body known as *endosome* (Syn. *nucleolus, karyosome*).

*Chromatophores*. Nearly all species of *Euglena* have small or large coloured bodies called *chromatophores*. They are present in the form of little plates or discs, sometimes spindle-shaped band-like or radiating star-like. The number, shape, size and arrangement of chromatophores vary from one species to another. The colour of chromatophore is green due to the presence of green pigment called *chlorophyll*. The chlorophyll consists of two pigments, the green *chlorophylla a* and a *yellowish* carotene.

Under electron-microscope the chromatophore is swollen in the centre where it contains two non-pigmented hemispherical bodies called *pyrenophores*. Attached to one of each are protein bodies called *pyrenosome*. Each pyrenosome is covered with a *paramylum shell*. The two pyrenosomes constitute a *pyrenoid*.

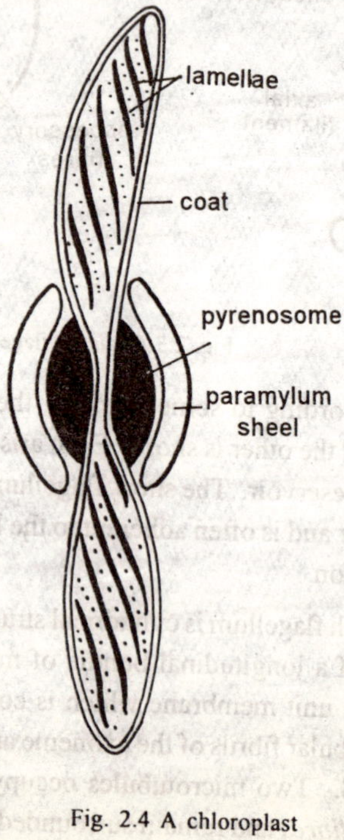


Fig. 2.4 A chloroplast

Pyrenoids are associated with the formation of a kind of animal carbohydrate called *paramylon*. Paramylon granules thus produced may be found scattered through the cytoplasm.

The chromatophore is surrounded by two membranes. The outer membrane is similar to plasma membrane. The inner membrane is similar to internal membrane system of the chloroplast. The *matrix* or *stroma* contains the chlorophyll-bearing bundles or *thylakoids*. The matrix contains the photosynthetic carbon cycle enzymes, ribosomes and DNA.

*Paramylon*. Here and there are scattered light bluish-green refractile granules called *paramylon*. It is a polysaccharide similar to starch but it does not give violet colour with iodine solution.

*Other endoplasmic organelles*. The endoplasm outside the nucleus clearly has a system of double-membranes forming the *endoplasmic reticulum*. Numerous small granules, termed *mitochondria*, are scattered throughout the cell. The mitochondria are important as a location of many of the biochemical reactions of the cell during respiration. *Golgi complex* are also found in abundance. The *ribosomes* also scattered and remain attached to the endoplasmic reticulum.

In some species of *Euglena*, a fine fibril called *rhizoplast*, extends from one of the blepharoplast to a small granule on the nucleus. This connection suggests that the flagellum is under the direct control of nucleus. The flagellum, blepharoplast, rhizoplast and the granule on the nucleus are said to form *infra-ciliary system* or *neuromotor system*.

## Physiology

### *Locomotion*

There are two methods of locomotion is *Euglena:*

1. Flagellar movement
2. Euglenoid movement

1. *Flagellar movement.* The process of the flagellar movement is not clearly understood because, except a few , its movements are not easy to observe in most animals. The whip-like movement of flagellum moves the animal forward and at the same time rotates it around the axis. Thus the animal moves forward in a spiral manner at a fair speed of 3.6 mm. per minute. Various theories have been put forward to explain the mechanism of flagellar movements in *Euglena*.

*Butschli* was first to describe the flagellar movement. He observed that the flagellum undergoes a series of lateral movement and in doing so, a pressure is exerted on the water at right angles to its surface. This pressure creates two forces: One directed parallel and other at right angles, to the main axis of body. The parallel force will drive the animal forward and the other force would rotate the animal on its own axis.

According to *Gray* (1928), the flagellum beats sideways in such a way that its movement proceeds from one end to other. These waves create two types of forces, one in the direction of the movement and the other in the circular direction with the main axis of body. The first force will drive the animal forward and the later would rotate the animal. The posterior end of body describes a smaller circle than the anterior end while the body is propelled forward through the water like a rotating plane which always remains inclined at an angle of $30^0$ to the axis of movement.

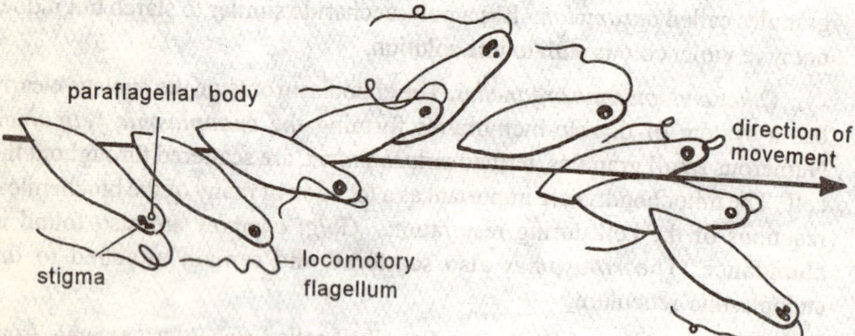


Fig. 2.5 *Euglena*. Successive stages in the glagellar movement.

*Lowends* (1941-43) also studied the locomotion in *Euglena*. His views are a little different from those of *Gray*. According to him, the flagellum is always held backwards during swimming and the movement in it occurs in a spiral manner from base to its tip. The flagellum beats, on an average, at the rate of 12 beats per second with increasing velocity and amplitude.

*Krijisman* (1952) observed the movement of flagellum under the dark-field microscope. According to him the normal movement of the flagellum is on the lateral side in a whip-like manner due to which it is called *effective stroke* in which the flagellum remains on outer side and moves quickly downwards. After this in *recovery stroke* the flagellum is brought up in a curved and relaxed state. As a result of these movements the animal moves forwards and rotate around its axis.

While the flagellum is moving it must be dissipating energy (*Brokaw*) as a result of work done against the surrounding water. The energy is supplied by ATP (adenosine triphosphate) present in the mitochondria to the blepharoplasts. The function of mastigonemes present on the flagellum is unknown.

*Euglena* is unable to swim backwards as the direction of motion can not be reversed.

2. *Euglenoid movements or Metaboly.* Sometimes a creeping movement is found in *Euglena*. It is brought about by the contraction and expansion of body. A peristaltic wave of contraction and expansion passes over the entire

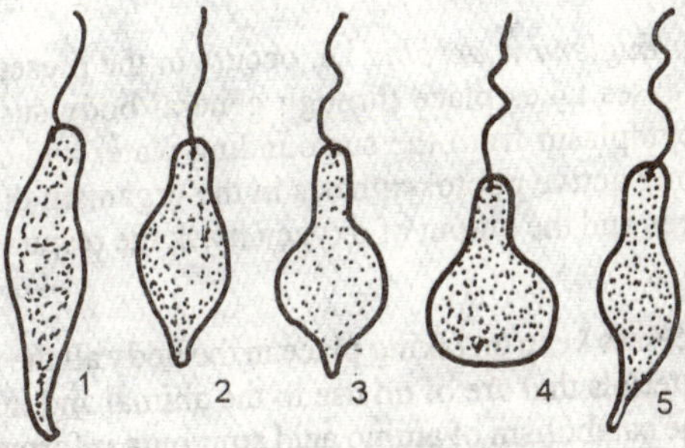


Fig. 2.6 Euglenoid movement in *Euglena*.

body from anterior to posterior end and the animal moves forward. The body becomes shorter and wider first at the anterior end, then in the middle and later at the posterior end. The contraction and expansion of the body is due to thin, elastic pellicle and the myonemes present beneath it. This is a very slow movement. *Lowends* found that an *Euglena* of (50-64 long) travelled at 168 .

**Nutrition**

The mode of nutrition in *Euglena* is *mixotrophic* i.e. the nutrition is accomplished either by *holoplytic* or *saprozoic* or by both the modes.

*Holophytic* or *Autotrophic nutrition.* Although *Euglena* has a mouth (cytostome) and gullet, it is doubtful if any food is ingested. The food is manufactured photosynthetically, as in plants, with the aid of light, carbon dioxide and the chlorophyll present in chromatophores. This mode of nutrition is known as *holophytic* or *autotrophic*. Photosynthesis enables the oxygen and a water molecule to be released , while hydrogen of the water molecule reduces the carbon dioxide present through a series of steps to form carbohydrates and more water. The carbohydrates are stored as a starch-like substance called *paramylum*.

*Saprozoic nutrition.* When the animal is kept in dark for a long time or the chlorophyll is destroyed by treatment with certain drugs, the colourless *Euglena* resulting can continue to live if provided with a medium rich in organic nutrients. This seems to indicate that organic substances in solution are absorbed through the surface of the body, that is, *saprophytic* nutrition.

*Myxotrophic nutrition*. The euglenae are sometimes holophytic and sometimes saprozoic depending upon the external conditions for maintaining life.

According to recent information, the *Euglena* is *amphitrophic*. It is holophytic in sun light and saprozoic in dark. In the sunlight, it is holophytic but not completely as it needs Vitamin $B_{12}$ or $B_1$. In the absence of these vitamins it will not multiply. This nutritional versatility is called *amphitrophy*.

**Respiration**

Respiration in *Euglena* is *aerobic* i.e. occurs in the presence of oxygen. The exchange of gases takes place through general body surface. Oxygen diffuses into the cytoplasm from the surrounding water and carbon dioxide diffused out. During active photosynthesis in the organism diffusion of carbon dioxide inwards and the output of oxygen mask the respiratory diffusion.

**Excretion**

Chemical processes keep on taking place in the body all the times, some of which produce materials that are of no use to the animal and may even prove harmful. During the catabolism of amino acid ammonia is formed. It is highly toxic and dissolved in water. This ammonia must be excreted as soon as it is formed. Some of the ammonia is diffused out through general body surface. Some of the ammonia is excreted out by contractile vacuole. The excretory role of contractile vacuole has gained support in the past few years. (*Chaudefaud*). The contractile vacuole is surrounded by specialized granular cytoplasm called excretory cytoplasm. Ammonia with large quantities of water first collect in the region of excretory cytoplasm as small droplets as accessory vacuoles which later on fuse to form the contractible vacuole. Periodically, the vacuole reaches its maximum size (*diastole*) and then burst (*Systole*) so as to discharge its contents in reservoir, from where waste passes out through the cytostome.

**Osmoregulation**

The primary function of contractile vacuole is *osmoregulation* or removal of excess of water from the body. The cytoplasm is denser than the surrounding water and thus has a relatively higher osmotic pressure. The surrounding

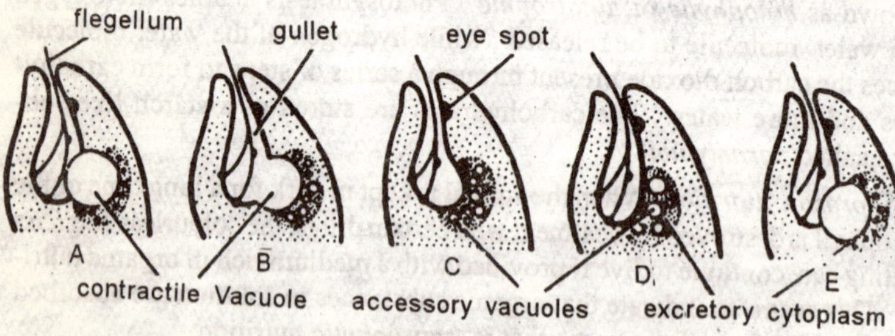


Fig. 2.7 Successive stages of contractile vacuole of *Euglena*.

water constantly enters the body by endosmosis through the pellicle. Moreover, water is also formed in the body during respiration. The excess of water from all over the body flows towards the excretory cytoplasm (described above) and collects in small accessory vacuoles. These water droplets unite to form the contractile vacuole (diastole) and during systole the water is discharged in the reservoir.

**Behaviour**

Like other protozoans, *Euglena* possesses a good degree of irritability i.e. the power of responding to external stimuli. This power enables it to avoid unfavourable environmental conditions. While swimming forward rotating and gyrating on its long axis *Euglena* may reach an unfavourable area. When this happens it turns violently usually heading into favourable area again. For this the gyration of the anterior end is suddenly widened and a new spiral path is taken. This behaviour gives the appearance of an animal trying a particular direction and correcting the error, if misled. Thus, such a behaviour is called *trail* and *error* behaviour, but it is better to call it *avoiding reaction*.

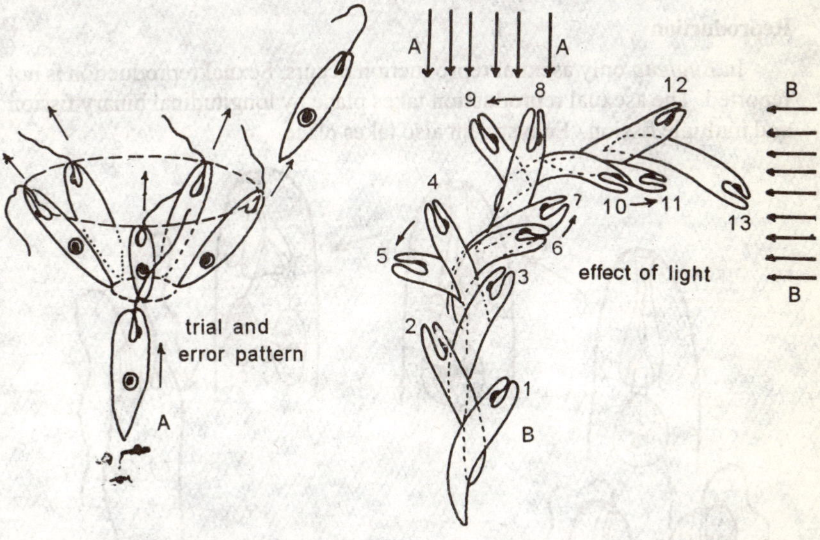


Fig. 2.8 A— Avoiding reaction of *Euglena*; and B—Effect of light on *Euglena*.

*Reaction to light or phototaxis*. *Euglena* is sensitive to light. It swims towards an ordinary light such at that from a window and avoids strong light. The light is preceived only at the anterior end by the specialized eye-spot. This specialization is important since they depend upon photosynthesis for their nourishment. *Euglena* orientates itself parallel to rays of light whenever the paraflagellar body is shaded by the eye spot or stigma. The animal adjusts its position to the direction of light moving either towards or away from it. It is now know after the experiments of *Tchakotine* (1936) that if the stigma is experimentally destroyed by ultraviolet light, the light-reactions disappear.

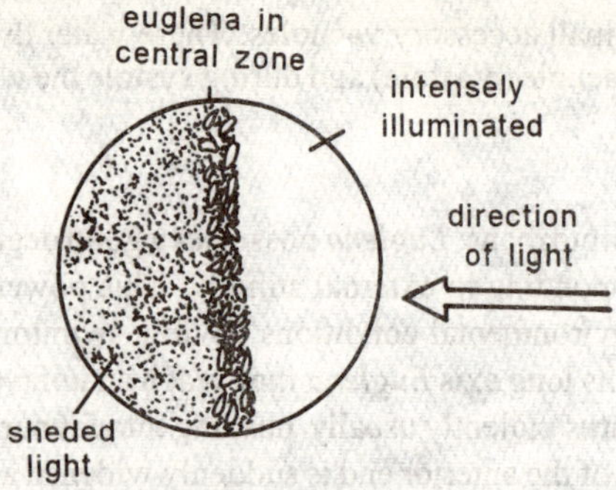


Fig. 2.9 Reaction to light in *Euglena.*

*Euglena* gives avoiding reaction to mechanical, thermal and chemical stimuli on a trial and error pattern.

**Reproduction**

In *Euglena* only asexual reproduction occurs. Sexual reproduction is not reported. The asexual reproduction takes place by longitudinal binary fission and multiple fission. Encystment also takes place.

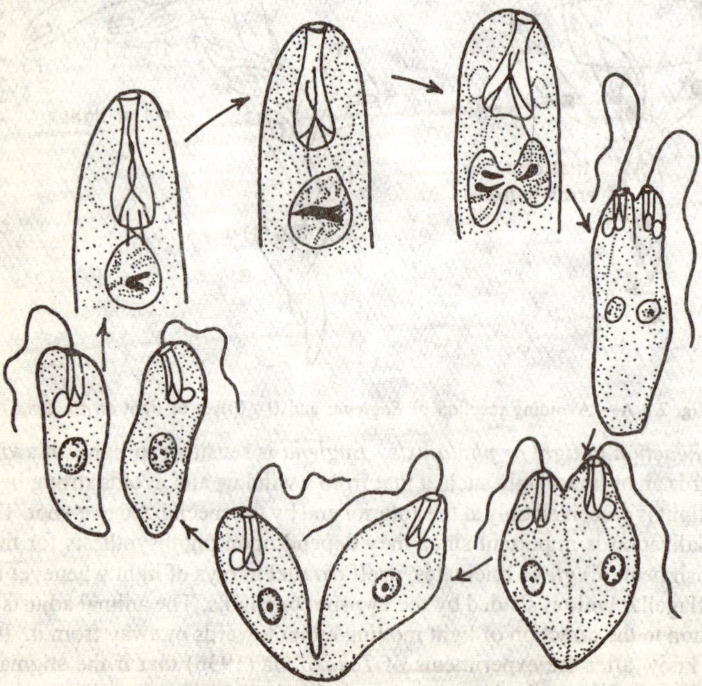

Fig. 2.10 *Euglena* showing binary fission.

*Londitudinal binary fission.* Longitudinal binary fission takes place during favorable conditions. Nuclear division takes place within the nuclear membrane. The chromomeres in the vegetative stage, forms pairs of chromosomes each of which divides longitudinally into two. The endosome becomes constructed into two approximately equal parts. The nucleus before its division migrates from the centre of body to lie just beneath the bases of the flagellum. During nuclear division, duplication of anterior extranuclear organelles such as flagella, blepharoptasts , cytostome, cytopharynx, reservoir, contractile vacuole, chromatophores etc. also take place. Soon after this, a longitudinal groove appears at the anterior end and gradually proceeds backwards as a result of which the body of animal is divided longitudinally into two halves. This division is *symmetrogenic* as one daughter is the plane mirror image of the other.

In some species, it has been observed that the paraflagellar body disappears before the onset of division. In others, the stigma breaks up into small pieces prior to binary fission. According to some workers the flagellum is completely lost and each daughter cell develops its own new flagellum. Often division takes place while the animals are in encysted condition.

*Multiple fission and Pamella Stage.* During unfavaourable conditions, multiple fission takes place in Euglena. It secretes very rapidly thick or thin cyst within which it divides into several (16-32) daughter euglenae. Sometimes, it loses its flagellum and round up into an alga-like cell in which metabolism continues and reproduction occurs by fission, thus forming extensive green scums on the surface of ponds. This condition is called palmella stage. Later flagella grow in these daughter euglenae which become free and develop into adult.

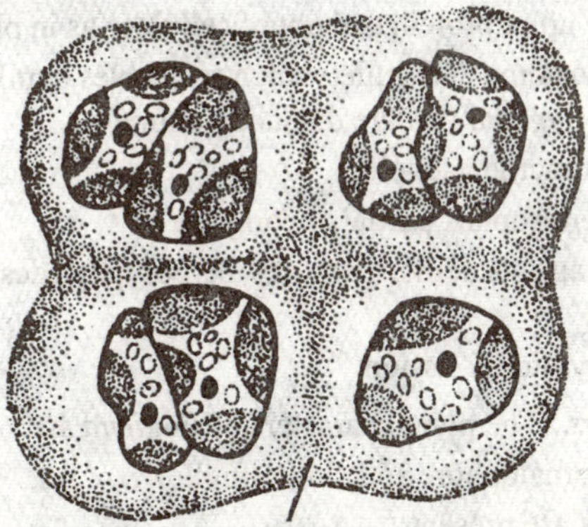


Fig. 2.11 *Euglena* snowing palmella stage.

*Encystment.* This is not a method of reproduction but is a mean to overcome unfavourable conditions. In the process of encystment, the cysts are formed by a special type of yellowish-brown carbohydrates. Their shape is generally rounded and their walls are formed of three concentric layers. During encystment binary fission may occur one or more times. On return of favourable conditions cyst wall breaks, the animal becomes active and emerge from the cyst to lead a normal life.

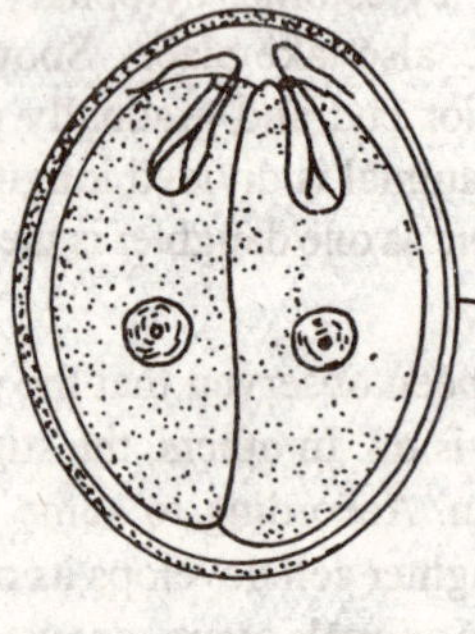


Fig.2.12 Binary fission in a cyst.

In different species of *Euglena*, the cyst may be thick, stalked or operculated and the organism lying centrally or eccentrically.

**Position of *Euglena***

*Euglena* is considered to be an animal by zoologists and plant by botanists.

***Animal Characters.*** The animal characters of *Euglena* include:

1. Pellicle is made up of protein and not of cellulose as in plants.
2. Presence of cytostome and gullet which resembles similar structures in the relatives of *Euglena* that are distinctly animals.
3. Presence of contractile vacuole.
4. Nutrition is heterotrophic, partial or complete.
5. Utalization of amino acids, pentoses or polypeptides as sources of nitrogen.
6. Reproduction by binary fission.

***Plant Characters.*** The plant characters of *Euglena* are:

1. Presence of chromatophore and chlorophyll.
2. Pyrenoid bodies are present.
3. Mode of nutrition is holophytic.
4. Existence of palmella stage.
5. Photosynthesis takes place in sun-light.

The main plant like character is presence of chromatophores and pyrenoids. But these are present only in some species and these can survice even without them. Moreover , the chlorophyll of this organism contains fewer pigments then the chlorophyll of plants. Only few species of *Euglena* are autotrophic and they too only partly. In the absence of light the mode of nutrition becomes saprozoic. During sun light the carbohydrate formed is paramylon which differs from the plant starch as it does not give violet colour with iodine solution. Palmella stage is an adaptation rather than affinity with plants.

Thus it appears that some of the species of *Euglena* are true animal like while few others behave as animal at one time and plant at other time. Its position suggest that we should regard *Euglena* as the intermediate stage in the evolution of plants and animals.

# 3

# ENTAMOEBA HISTOLYTICA

Most amoebae are free-living, but some are endoparasites. They have adapted themselves to live actively in the intestine of many invertebrates and vertebrates. Besides weakening the host body by obtaining nourishment from it, these many cause diseases and, hence, called *pathogenic parasites.* One of such pathogenic amoeba is *Entamoeba histolytica.* It was discovered by *Lable* (1859). *Losch* (1874) discovered its pathogenic nature. *Koch* (1887) first saw it in the intestinal section and *Kortulis* (1887) in the liver abscess. The name *E. histolytica* was given by *Schaudin* (1903). The structure of this parasite has been studied by *Nollar* (1922) and *Kofoid* (1927). *Elmassion* (1909) and, *Walker* and *Sellards* (1913) studied its precytic stage while *Hartmann* and *Prowazek* (1907) studied it cystic stages. *Dobell* (1909-28) studied its binary fission. *Cleveland* and *Sanders* (1939) studied the excystment. The ultrastructure has been studied by *Miller et al* (1961) and *Feltcher et al* (1962). It causes 'amoebic dysentery' or 'amoebiasis' that is why it is commonly called 'dysentery amoeba'. *E. histolytica* also occur in other mammals, lower vertebrates and some invertebrates, often without evident damage to these hosts. Laboratory turtles, snakes and frogs are common sources for study. The contents of lower intestine may be examined by diluting with physiological saline solution on slide. Iodine solutions disclose the nuclei in temporary stain, while iron hemotoxylin is preferred for permanent preparation.

**SYSTEMATIC POSITION**

| | | |
|---|---|---|
| Phylum | - | Protozoa |
| Sub-phylum | - | Sarocmastigophora |
| Superclass | - | Sarcodina |
| Class | - | Rhizopodea |
| Order | - | Amoebida |
| Genus | - | *Entamoeba* |
| Species | - | *histolytica* |

**Distribution**

*E. histolytica* is world-wide in distribution. It is known to be widely distributed in tropical, subtropical, temperate regions. The incidence is high in India, China, Mexico and parts of S. America. *Craig* (1962) estimated that about 10% of the human population is infected by this parasite.

**Habits and habitat**

As mentioned *E. histolytica* occurs in man and other mammals like dogs, cats, rats, monkeys, baboon, chimpanzee, gorilla etc. It lives in the upper part of large intestine (colon) of man. It lives in the mucous and submucous layers of colon. It secretes a toxic substance with dissolves and destroys the mucous lining. From here, certain individuals are carried by blood stream to liver, lungs, kidneys, gonads and even the brain. The parasite feeds on the dissolved tissues, bacteria, and blood corpuscles. It causes a serious and often fatal disease known as *amoebic dysentery*. Patients of this disease discharge mucous and blood in their stool.

**Morphology**

The active motile form of *E. histolytica* is called the *trophozoite* which usually ranges in long diameter from 15-25 . It resembles *Amoeba* in its structural details. It is bounded by a thin, flexible, transparent semipermeable membrane, the *plasmalemma*. The cytoplasm is differentiated into outer ectoplasm and inner endoplasm. The ectoplasm forms a thin, clear, relatively firm outer layer. The endoplasm is granular and more fluid part. A single pseudopodium is formed of ectoplasm only.

In the endoplasm a nucleus and several vacuoles are present. The nucleus is vesicular and rounded. It is bounded by a delicate nuclear membrane, whose inner surface is encrusted with a fine peripheral layer of *chromatin granules*. The *karyosome* or *endosome* is small and centrally placed in the nucleus. It is often surrounded by a clear area or *halo*. The size of nucleus is about 4-6 in diameter and it contains six chromosomes.

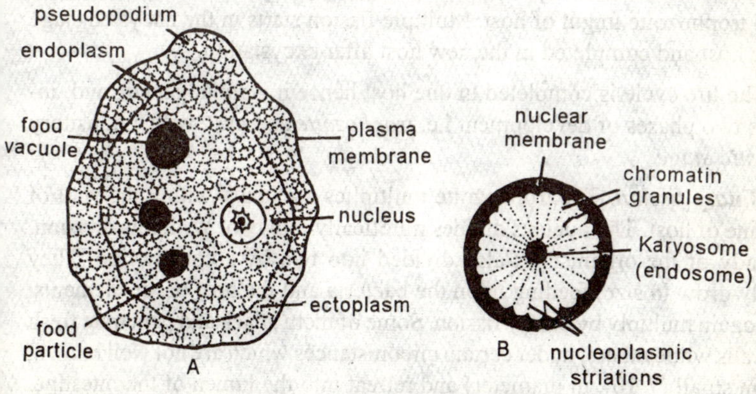


Fig. 3.1 A= Trophozoite of *Entamoeba histolytica*, B= nucleus

In the endoplasm, food vacuoles vary in size and number are present. They enclosing red blood corpuscles, white blood corpuscles and debris of epithetical cells and bacteria. The contractile vacuole is entirely absent.

## Physiology

### *Locomotion*

*E. histolytica* is a monopodial as it gives out a single, large and broad pseudopodium in the direction of movement. The pseudopodium is called *lobopodium*. In the lobopodium, the outer clear ectoplasm remains sharply differentiated from the inner granular endoplasm. The movement is slow and sometimes called limax *type* movement as it resembles the slow crawling of slug (*Limax*).

### *Nutrition*

The mode of nutrition is *holozoic*. It feeds mainly upon blood corpuscles, fragments of epithelial cells and bacteria. It is believed to secrete *cytolysin*, a protein digesting enzyme that dissolves the epithial mucous membrane and liberates the blood. It than capture blood corpuscles with pseudopodium. It also absorbs substances saprozoically from the surrounding medium.

### *Respiration*

*E. histolytica* is aerobic when oxygen is available but anaerobic when oxygen is not available. It requires very little oxygen hence *microaerobi.*

### *Excretion*

The nitrogenous waste product formed is ammonia which is diffused out in host's tissue.

## Reproduction and Life-Cycle

In *E. histolytica* only asexual reproduction occurs. It occurs in two ways : *binary fission* and *multiple fission.* The binary fission takes place in the active trophozoite in gut of host. Multiple fission starts in the encysted stage in one host and completed in the new host after excystation.

The life cycle is completed in one host hence it is monogenetic and includes two phases of development i.e. *trophozoite* and *cyst*, with a transitory *precystic stage.*

***Binary fission.*** The trophozoite multiplies asexually within the wall of intestine of host. The nucleus divides mitotically into two. With this division, the body of the organism is also divided into two by a constriction. They rapidly grow in size, feeding upon the bacteria and the host tissue elements. They again multiply by binary fission. Some of them attack and penetrate fresh host cells while others, under certain circumstances which are not well known, remain small (7-10 in diameter) and retreat into the lumen of the intestine. These small forms are rounded and sluggish and are known as *psecystic* or *minuta forms.* The minuta forms do not eat tissue cells and blood corpuscles but feed only on the bacteria. It is the non-motile and non feeding stage thus it is non-pathogenic to man.

***Encystment.*** The precystic forms undergo encystation only in the luman of the intestine. They round up and secrete a thin, flexible, refractile and resistant cyst wall around themselves. The cyst of *E. histolytica* is spherical body measuring 10-15 in diameter. Its cytoplasm is clear and hyaline containing two *chromatoid bodies* and many *glycogen vacuoles.* According to *Pitelka* (1963), the chromatoid bodies disappear as the cyst matures. At this stage the cyst is *uninucleate.* Soon the single nucleus divides mitotically into two and the cyst becomes *binucleate.* Next division results in four nuclei and the cyst now becomes *tetranucleate* or *quadrinucleate.* Now the cysts come out of the body of host with the faeces. The cysts can remain alive for many weeks outside the body of the host till they reach in the new host.

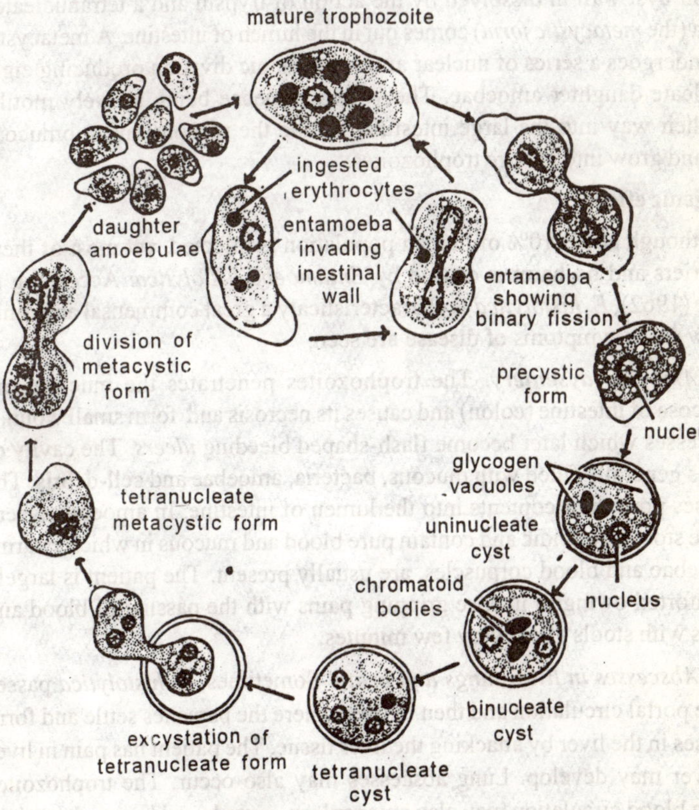


Fig. 3.2 Life Cycle of *Entamoeba histolytica*

*Transfer to new host.* The cysts reach the alimenatary canal of new host with contamination of food and water. The food can be contaminated by any of the following water :

1. By dirty feeding habit.
2. By coming in contact with contaminated water.
3. By collection of cysts on food substances by insects like house-fly, cockroach etc.
4. By eating uncooked vegetables and fruits.

***Excystation*** and ***metacystic forms.*** The excystation of cyst and metacystic development have been observed and studied by *Dobell* (1928) and, *Cleveland* and *Sanders* (1930) in cultures. Excystation is the process of transformation of cysts to the trophozoites. The cysts reach to alimentary canal of host. There is not effect of gastric juice in stomach on them. On reaching the intestine their cyst wall in dissolved by the action of trypsin and a tetranucleated amoeba (the *metacystic form*) comes out in the lumen of intestine. A metacystic form undergoes a series of nuclear and cytoplasmic division producing eight uninucleate daughter amoebae. The young amoebae being actively motile, make their way into the large intestine, invade the mucous and submucous lining and grow into mature trophozoites.

**Pathogenic effects**

Although about 10% of human population is infected, yet most of them are carriers and no harm is caused by *Entamoeba histolytica.* According to *Hoarse* (1962), *E. histolytica* is characteristicaly a great commensal and, only in a few cases symptoms of disease are seen.

1. *Amoebic dysentery.* The trophozoites penetrates the mucosa and submucosa of intestine (colon) and causes its necrosis and form small wounds or abscesses which later become flash-shaped bleeding *ulcers.* The cavity of ulcers is generally filled with mucous, bacteria, amoebae and cell-debris. The abscesses pour their contents into the lumen of intestine. In amoebic dysentery, the stools are acidic and contain pure blood and mucous in which swarms of amoebae and blood corpuscles, are usually present. The patient is largely discomforted owing to intense gripping pains with the passing of blood and mucous with stools after every few minutes.

2. *Abscesses in liver, lungs and brain.* Sometimes, *E. histolytica* passes into the portal circulation and then to liver where the parasites settle and form abscesses in the liver by attacking the liver tissue. The patient has pain in liver and fever may develop. Lung abscesses may also occur. The trophozoites through blood circulation may also enter spleen, gonads and brain where they also destroy the tissue causing specific abscesses. Abscesses elsewhere are rare. In acute cases, they might prove to be fatal.

***Treatment.*** The permanent cure is sometimes difficult. For prompt relief of acute dysentery the injections of *Emetin* are given. Certain antibiotics such as

*Fumgillin, Aureomycin, Erythromycin* and *Terramycin* are more effective for oral treatment. In chronic cases, certain arsenic compound like *Thiocarbarsome* and *Carbarsome* and iodine compounds such as *Yatren, Diodoquin* and *Vioform* are effective. For amoebiasis of liver and lungs *Chloquine* is quite effective

***Prevention.*** Following measures are essential in the prevention of the disease :

1. Washing of hand with antiseptic soaps and water before taking food and after toilet.
2. Sanitary disposal of faecal matters.
3. Perfect sanitation and protection of water and vegetable from pollution.
4. Protection of foods and drinks from contamination by house-flies, cock-roaches etc.
5. Use of boiled water and properly washing the vegetables and fruits.
6. Avoiding open field and road sides for passing out stools and chemicals treatment of human faecces, if used as fertilizers.

**Species of Entamoeba**

***Entamoeba coli***

It is the most common species to occur in about 50% of the human population world over. It lives in the lumen of colon and does not enter the tissues of the wall. It is a non-pathogenic species as it does not cause any disease to man. It feeds on bacteria, undigested food and other debris but never on the tissue. It is, therefore, called *endocommensel.*

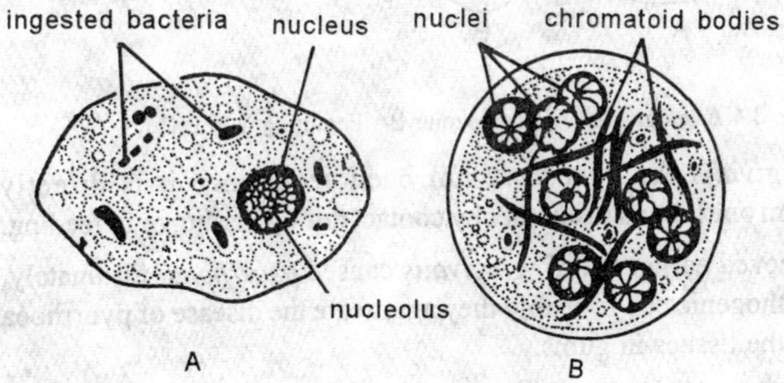


Fig. 3.3 *E. coli.* A = Tropozoite B = Cyst

Like *E. histolytica* its trophoites measure 15-40 in diameter. The cytoplasm is less distinct in ectoplasm and endoplasm. The nucleus has a comparatively larger nucleus or endosome not placed in the centre. The nucleus measure 5-8 in diameter. Numerous food vacuoles are seen but they contain bacteria, faecal debris etc. The cysts are spherical or often ovoid. Immature cysts contain 1, 2, 4 nuclei while the mature cyst has 8 nuclei. In young cyst besides 4 nuclei one or more glycogen bodies and small number of chromatoid bodies are present. The infection takes place through cysts in stool of infected man passed out and taken by new host by contaminated water or food. Its life-history is now known. According to *Hegner*, the cysts hatch as 8 nucleated amoebae.

***Entamoeba gingivalis***

The mouth amoeba, *E. ginglivalis* lives in the mouth usually between the teeth and gums. It is also found in abscesses of gums, tonsils etc. According to *Kofold* 75% people over 40 years of age harbour this amoeba. The trophozoite measures 10-20 in diameter. The cytoplasm in differentiated into an outer clear ectoplasm and a hyaline and vacuolated endoplasm. The nucleus is present in endoplasm measuring 2-4 in diamter. It centrally placed endosome or nucleolus. The trophozoite feeds on bacteria, cellular debries and white blood corpuscles. There are many pseudopodia all are with blunt ends.

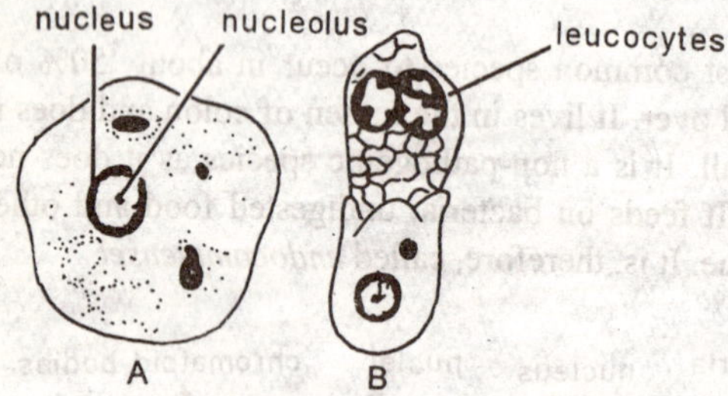


Fig. 3.4 *E. gingivalis*. A= Tropozoite B= Engulfing a W.B.C.

In *E. gingivalis* the cyst have not been seen since it is directly transmitted from one human to another by contact during kissing or in feeding.

It was believed earlier that *E. gingivalis* caused *pyorrhoea*. fortunately, they are not pathogenic but probably they aggravate the disease of pyorrhoea by destroying the tissues in gums.

***Entamoeba hartmanni***

*E. hartmanni* closely resembles the minute or precystic form of *E. histolytica*. It lives in the colon of man, invades the intestinal tissues and cause amoebic dysentery, but is less harmful. The trophozoites measure 9-14 in diameter. The nucleus is more compact. The cystic forms measure less than 10 in diameter.

# 4

# PARAMECIUM

*Paramecium* is the widely studied type of the class-Ciliata. The members of class Ciliata are much more developed than the other protozoans. Inspite of their being acellular, special organelles are found in them for ingestion, excretion etc. Thus division of labour is clearly seen in ciliates.

## SYSTEMATIC POSITION

| | | |
|---|---|---|
| Phylum | - | Protozoa |
| Subphylum | - | Ciliophora |
| Class | - | Ciliata |
| Subclass | - | Holotrichia |
| Order | - | Hymenostomatida |
| Family | - | Paramecidae |
| Genus | - | *Paramecium* |

## HABITS AND HABITAT

*Paramecium* is cosmopolitan in distribution. It is found in fresh water lakes, ponds, puddles, sewage pipes and rice fields. It is abundant in stagnant water containing decaying organic matter. It is omnivore in diet and reproduces in a variety of ways, which are asexual and sexual both.

There are 10 species of *Paramecium*. They differ in shape, size, number of micronuclei etc. The common species are *Paramecium caudatum, P.aurelia, P.bursaria* and *P. multimicronucleatum*.

### Culture

For general laboratory purpose, *P. caudatum* can be grown satisfactorily on a boiled-hay infusion medium to which wheat or rice grains have been added. It has also been maintained on several complex sterile media and simplest of which the yeast juice. Paramecia are easily grown in chalky medium (Nacl-80mg., $NaHCO_3$-4mg., $CaCl_2$-4mg, KCl-4mg and $CaH_4(PO_4)\ H_2O$ 1.6 mg. all are dissoslved in one litre of distilled water). Add 6-8 drops of skimmed milk weekly. The jar is kept away from direct light.

Other less uniform but very satisfactor method of growing paramecia include a monoflora suspension of bacterium, *Pseudomonas ovalis*, and a suspension of pure, living yeart.

## MOROPHOLOGY

### External Feature

***Shape and Size.*** The first description of *Paramecium* was given by *Christian Huygens* (1678). The term *Paramecium* was coined for the first time by *John Hill* (1752) from the Greek word *paramekos* meaning oblong. *Paramecium* , in general appearance appears like the sole of the slipper, hence it is popularly known as "*slipper-animalcule*." One species was named *P. caudatum* (*Joblot*) because of its resemblance to an imprint of human foot.

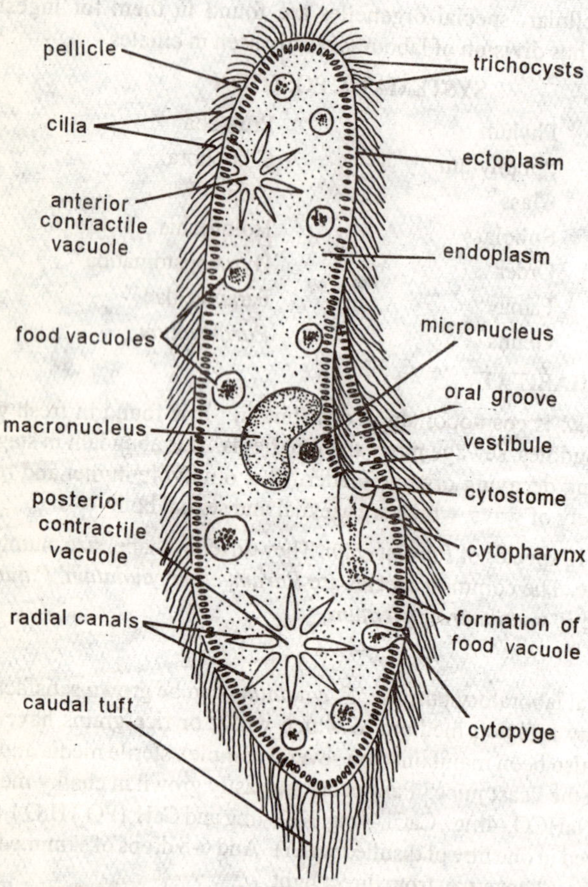


Fig. 4.1 *Paramecium caudatum.*

They vary in size. The largest species *P. caudatum* measures 180-300 in length; *P. aurelia* is 170-250 and *P. trichium* measures only 62 in length.

**Colour.** Some species are somewhat translucent and colourless , others are vary from light gray, white to pale yellow. *P. bursaria* is green because of the presence of *Zoochlorellae* in the endoplasm.

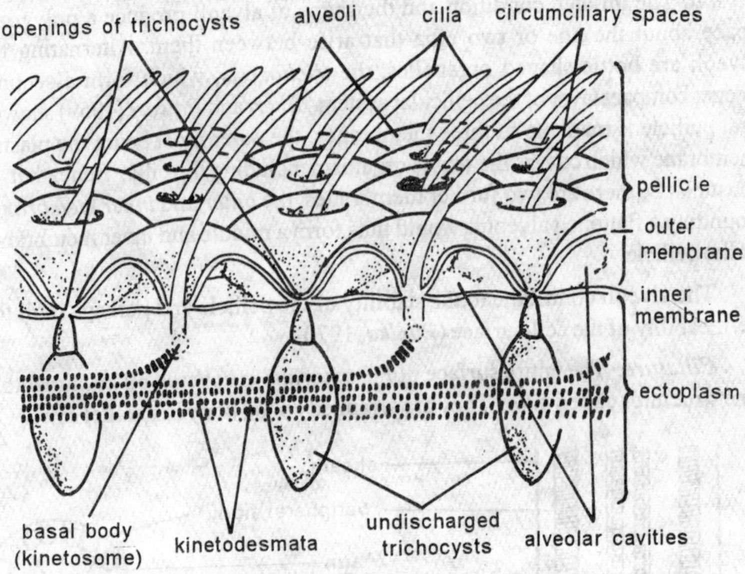


Fig. 4.2 Pellicular system in *Paramecium*

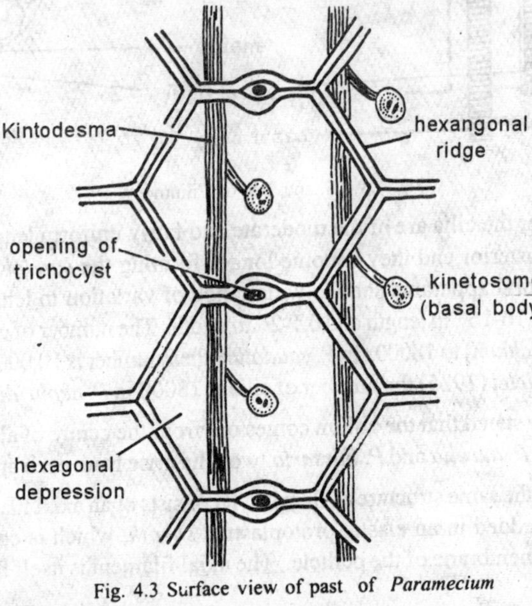


Fig. 4.3 Surface view of past of *Paramecium*

***Structure.*** The shape of the body is definite because the body of ciliates is typically covered by a complex, living *pellicle*, usually containing a number of different organelles. Under light microscope the pellicle appears homogenous. Under electron-microscope the pellicle has inflated kidney shaped *alveoli*. The inflated condition and the shape of alveoli produce a polygonal space about the one or two cilia that arise between them. Alternating the alveoli are bottle shaped organelles, the *trichocysts*, which form a second, deeper compact layer of the pellicular system. *Ehert* and *Powers* (1959) showed that pellicle is made up of three membranes, the outer most being the plasma membrane which covers the whole organism including the cilia. Beneath it, in most place, there are two further membranes, the *outer* and *inner membranes* bounding a flattened alveolus would thus form a middle and inner membranes of the pellicle.

The alveoli contribute to the stability of the pellicle and perhaps limit the permeability of the cell-surface (*Pitelka*, 1970).

***Ciliature***. The entire surface of *Paramecium* is covered by in-numerable fine structures called *cilia*. They are arranged on the body in parallel rows. On

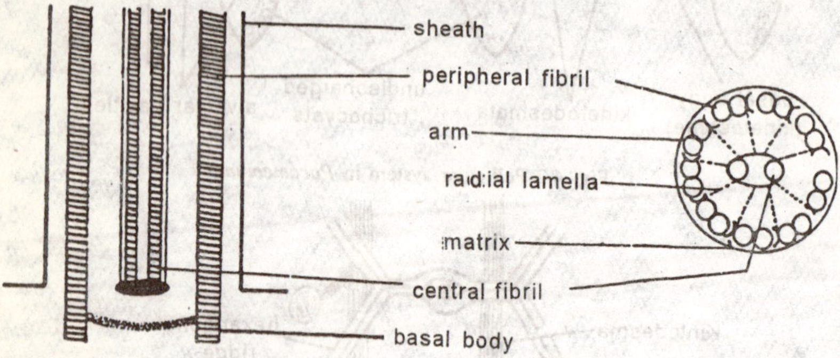


Fig. 4.4 L.S. and T.S. of Cilium

the body proper the cilia are of the moderate and fairly uniform length, but at the extreme posterior end they become longer forming the *caudal tuft*. The cilia found in oral apparatus show a wide range of variation in length. Each cilium measure 10-16 in length and 0.1-2 in width. The number of cilia varies from 350 (*P. trichium*) to 18000. In *P. caudatum* their number is 10,000 - 14,000. According to *Gelei* (1925) the number of cilia is 18000 in *P. nephridiatum*.

As already stated that the cilium comes out from the centre of alveolus of the pellicle. In *P. aurelia* and *P. bursaria* two cilia arise from each area.

Cilia have the same structure as flagella. It consists of an axial filament, the *axoneme*, embedded in an elastic protoplasmic *sheath*, which is continuous with the outer membrane of the pellicle. The axial filament is itself formed of

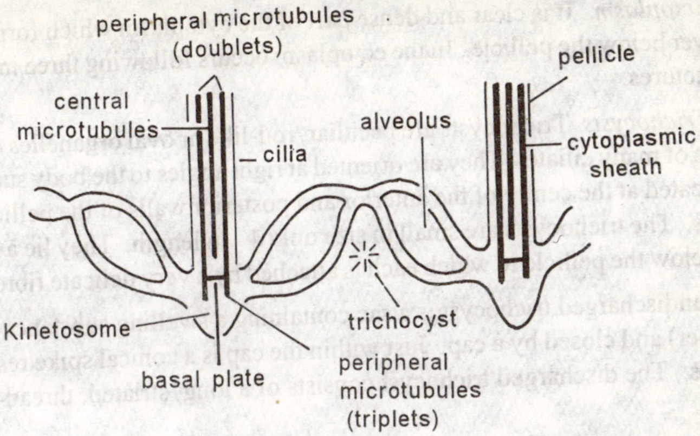


Fig. 4.5 V.S. through the surface of *Paramecium* to show position of cilia, kinetosome and Trichocyst

eleven fibrils that extends its full length. Nine of these fibrils are double and occur in a ring called *peripheral fibres*. Each is composed of two adhering subfibres, one of which is provided with a pair of arms all disposed in same direction. The remaining two are *central fibres* which are independent surrounded by a thin membraneous central *sheath*. Nine radially directed fibres are disposed between the peripheral fibres and the inner sheath like spokes of a wheel. The peripheral fibres continued below in to the basal body or *Kinetosome.*

**Internal Features**

*Cytoplasm.* The cytoplasm is differentiated into an outer ectoplasm and inner endoplasm.

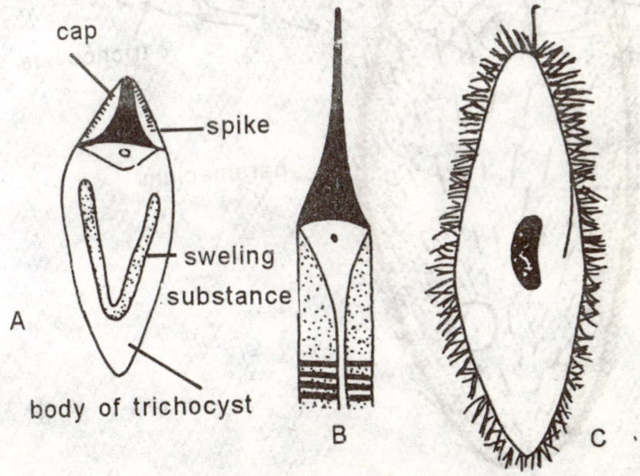


Fig. 4.6 Trichocyst A= entire, B = tip of discharged trichocyst, C=Trichocysts discharged after staining with per ink.

1. *Ectoplasm.* It is clear and dense part of the cytoplasm which forms its outer layer below the pellicle. In the ectoplasm occurs following three important structures.

(i) *Trichocysts.* Trichocysts are peculiar, rod-like or oval organelles characteristic of many ciliates. They are oriented at right angles to the body surface being located at the centre of the anterior and posterior walls of the pellicular polygons. The trichocysts are small in size only 4 in length. They lie a little (1-2 ) below the pellicle to which each is attached by a very delicate fibre.

An undischarged trichocyst is a sac containing a swelling substance (the quelkorper) and closed by a cap. Just within the cap is a conical spike resting on a basis. The discharged trichocyst consists of a long, striated, thread-like

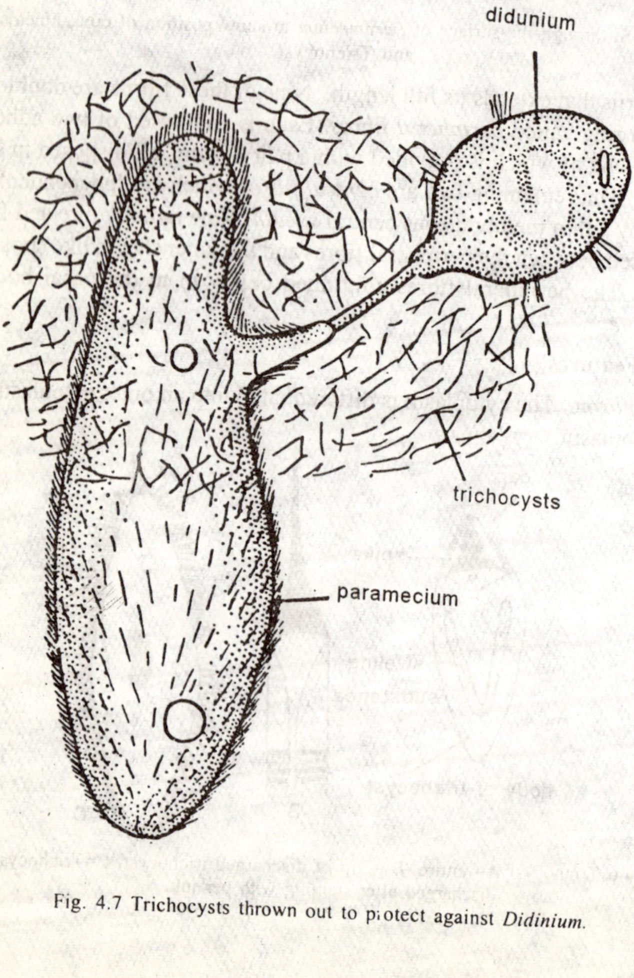


Fig. 4.7 Trichocysts thrown out to piotect against *Didinium.*

shaft surmounted by a barb. The shaft is not evident in the undischarged trichocysts and is probably polymerized in the process of discharge. The trichocysts are discharged on mechanical, chemical or electric stimuli. They are discharged very rapidly(within several milliseconds).

The trichocysts are non-toxic type in *Paramecium* thus they are not defensive organelle. Probably they may be used in anchoring the animal when feeding.

Area in the neighbouring of the nucleus is believed to the centre of origin of trichocysts. Recently, it has been established that trichocysts are produced by the kinetosomes. A kinetosome first ejects a trichocyst granule or tricystosome which forms a budding that develops into a trichocyst.

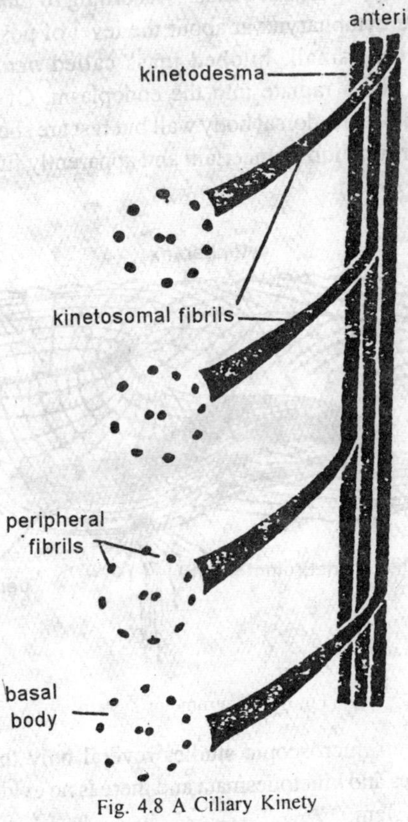

Fig. 4.8 A Ciliary Kinety

(ii) *Infra-ciliary system.* As mentioned earlier, each cilium arises from a basal body or kinetosome, located in the alveolar layer. The kinetosomes that form a particular longitudinal row are connected by means of fine, striated fibrils, called *kinetodesma*. The kinetosomes plus the fibrils of that row make up a kinety. The longitudinal bundle of fibrils run to the right side of the row of kinetosomes, and each kinetosome gives rise to one kinetodesmos (fibril), which joins the longitudinal bundle and extends anteriorly. A single kinetodesmos is tapered and extends for varying distance, as a part of bundle. At the kinetosome, the kinetodesmos is connected to certain of the kinetosome triplets.

A kinety system is characteristic of all ciliates although there are variations in details of the pattern. The kineties with their cilia constitute the neurolocomotor organelle. The kineties control and coordinate the beating of cilia.

(iii) *The fibrillar system or Neuromotor apparatus.* The fibrillar system is visible with the help of nuclear stains. According to *Lund* (1933) on the left dorsal wall of the cytopharynx at about the level of posterior margin of the cytostome is a very small, bilobed mass called *neuromotorium*. From neuromotorium, fibrils radiate into the endoplasm. Of these four or more usually pass almost to the dorsal body wall but rest are shorter and not definite in position. Their function is uncertain and apparently function as part of the cilium anchorage system.

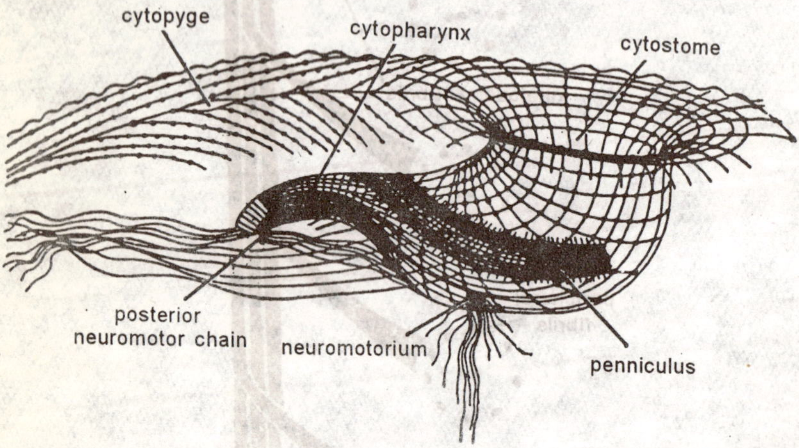


Fig. 4.9 Neuromotor system

But the electron microscopic studies reveral only the presence of the alveoli, kinetosomes and kinetodesmata and there is no evidence regarding the existence of this system.

***2. Endoplasm*** : The endoplasm of Paramecium is semi-fluid and less granular. It contains following structures :

(i) ***Nucleus.*** *Paramecium* is *hereokaryotic* having two types of nuclei. The smaller one or the *micronucleus* is lodged in a depression at one side of the larger nucleus called *meganucleus* or *macronucleus*. The meganucleus is bean or kidney shaped body lying approximately in the centre. It is a compact structure , containing fine threads and tightly packed discrete chromatin granules of variable size. The meganucleus controls the vegetative activities of the animal. The micronuclei vary in number in different species. These are smaller structure and may be compact (*P.caudatum*) or vesicular (*P.aurelia*).*P.caudatum* and *P. bursaria* each has one micronucleus. *P. aurelia* has two and *P.multimicronucleatum* has 3-9 micronuclei. The micronuclei are concerned with reproduction.

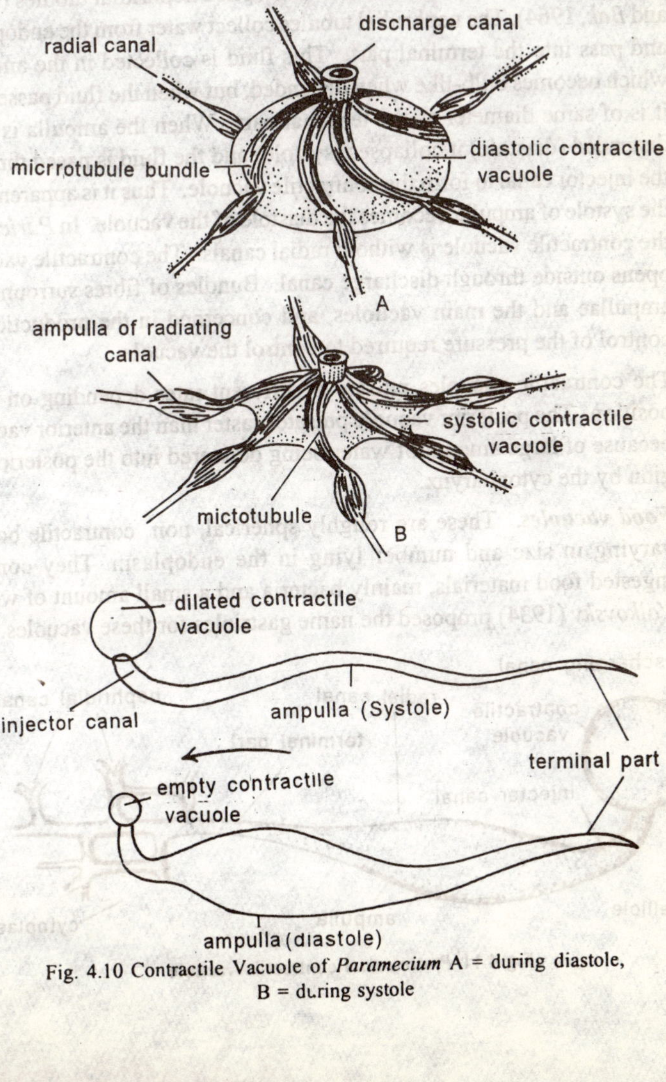


Fig. 4.10 Contractile Vacuole of *Paramecium* A = during diastole, B = during systole

(ii) ***Contractile Vacuoles :*** In *Paramecium* there are two contractile vacuoles occupying somewhat fixed positions in the endoplasm. One vacuole is located at the posterior, and other vacuole at the anterior end of the body.(In *P. multimicronucleatum* there are 2-7 contractile vacuoles). The anterior vacuole is sometimes called the nuclear vacuole being near the nucleus and the posterior one is called the peristomial vacuole being in vicinity of peristome. Each contractile vacuole is surrounded by 5-12 radiating canals. Each radiating canal consists of (a) a tabular terminal part (b) an ampulla and (c) an injector canal. The injector canal opens into the vacuole. Electron microcopy reveals that the outer or terminal part is surrounded by a network of fine coiled nephridial tubules (*Elliot* and *Bak*, 1964). The nephridial tubules collect water from the endoplasm and pass into the terminal part. This fluid is collected in the ampulla which becomes bulb-like when distended, but when the fluid passed out it is of same diameter as the terminal part. When the ampulla is fully distended (diastole) it collapses (systole) and the fluid is pased through the injector canal to form the contractile vacuole. Thus it is apparent that the systole of ampulla becomes the diastole of the vacuole. In *P.trichium* the contractile vacuole is without radial canals. The contractile vacuole opens outside through discharge canal. Bundles of fibres surround the ampullae and the main vacuoles, and concerned in the production or control of the pressure required to control the vacuole.

The contractile vacuoles pulsate at different rates depending on their position. The posterior vacuole pulsates faster than the anterior vacuole because of large amount of water being delivered into the posterior region by the cytopharynx.

(iii) ***Food vacuoles.*** These are roughly spherical, non contractile bodies varying in size and number lying in the endoplasm. They contain ingested food materials, mainly bacteria and a small amount of water. *Volkovsky* (1934) proposed the name gastrioles for these vacuoles.

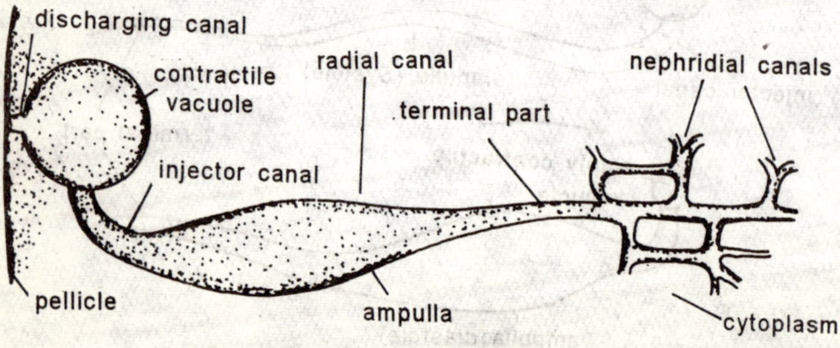


Fig 4.11 *Paramecium* contractile vacuole.

**Oral groove and associated structures.** The oral groove is located on the ventral surface extending back-wards into a funnel shaped depression known as *vestibule* consisting of inveginated body pellicle. The vestibule leads into buccal cavity which directly leads into the cytostome or mouth which is fixed oral opening. The cytostome directly opens into the wide cytopharynx which extends toward the centre of body. The cytopharynx forms a food vacuole at the posterior end.

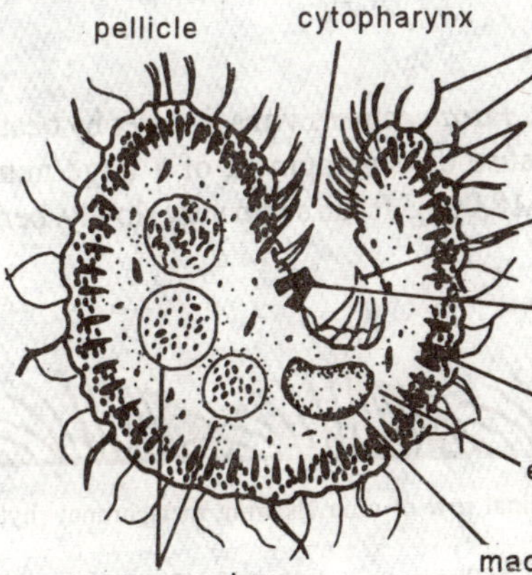


Fig. 4. 12 T.S. of *Paramecium* through the region of cytopharynx to show ciliation.

Large and hard cilia are present in these structures. These are specialized cilia and form four organelles which help in food ingestion.

(a) *Endoral membrane.* It is formed by the fusion of one row of cilia. It runs transversely along the right wall and partly encircles the opening of vestibule into cytostome.

(b) *Dorsal peniculus.* This is formed by four rows of densly arranged cilia. which are placed in a spiral manner on the left side of buccal cavity. They end behind the cytostome.

(c) *Ventral peniculus.* It is also formed by four rows of cilia. It extends backwards to a distance on the left side of buccal cavity and ends near the cytostome.

(d) *Quadrulus.* It is also formed of four rows of cilia which are less densly arranged. They are placed in a spiral manner along the dorsal wall of the buccal cavity. It ends near the dorsal peniculus.

The quadrulus and peniculi control food passage while endoral membrane guards the entrance of buccal cavity. A group of long cilia, *post buccal cilia* hang behind the lateral wall of cytopharynx.

***Cytopyge.*** Just behind the cytostome on the ventro lateral surface there is a minute pore called *cytopyge*. Through this cytopyge the undigested food particles are eleminated out.

## PHYSIOLOGY

**Locomotion.** The locomotion in Paramecium takes place by two methods:

1. Ciliary locomotion

2. Metaboly

***1. Ciliary Locomotion.*** *Paramecium* swims in water by beating of its cilia and has put on speed because of development of a large number of these locomotary organs (10,000-18,000). It can swim very slowly, perhaps 5 meters per hour.

Fig. 4.13 A longitudinal row of cilia showing metachronal rhythm.

A cilium performs whip-lash movement by which water is propolled. In its beat each cilium performs an *effective* and a *recovery stroke*. During the effective stroke the cilium is outstratched and moves from a forward to a backward position. In the recovery stroke the cilium is bent over to the right against the body and is brought back to the forward position in a counterclock-wise movement. The recovery position offers less water resistance and is somewhat analagous to feathering an oar. The direction of effective stroke is oblique to the long axis of the body. This causes the animal to swim in a spiral path and at the same time to rotate on its longitudinal axis. The ciliary beat can be reversed, and the animal can move backward. This backward-movement is associated with the so called avoiding reaction (behaviour). The direction and intensity of beat is controlled by levels of $Ca^{++}$ and $K^{+}$ions (*Eckert*, 1972).

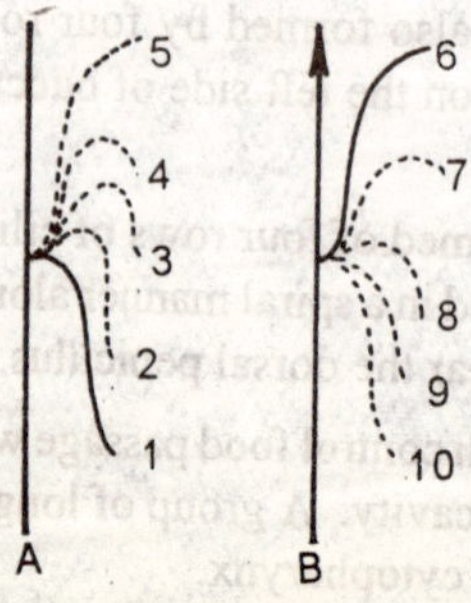


Fig. 4.14 A = recovery stroke, B=effec.ive stroke

All the cilia of body can neither move simultaneously nor are they capable of independent movement . The cilia of a longitudinal row oscillate in such a way that the front cilium moves a little earlier than the one behind it. Thus the cilia move one after the other from front backwards. This movement of cilia is called *metachronal rhythm*. However, the cilia of a transverse row move simultaneously. This movement is called *synchronous rhythm*. The movement of cilia is controlled by the infra ciliary system and ATP plays an important role (*Serawin*, 1967).

***2. Metaboly.*** The body of *Paramecium* possesses elasticity. It can pass through a passage narrower than its body, after which it regains its normal size. such type of movement is called metaboly. This type of body movement is caused by the cytoplasm.

***Nutrition.***

The free-living ciliates are almost entirely holozic. It feeds on bacteria, small protozoan, yeast, algae etc. but its favourite food is *Bacterium coli*. It can devour 2-5 millian *Bacterium* in 24-hours. Thus *Paramecium* is a selective feeder, showing considerable choice of food, and often rejecting undigestible non-nutritive particles.

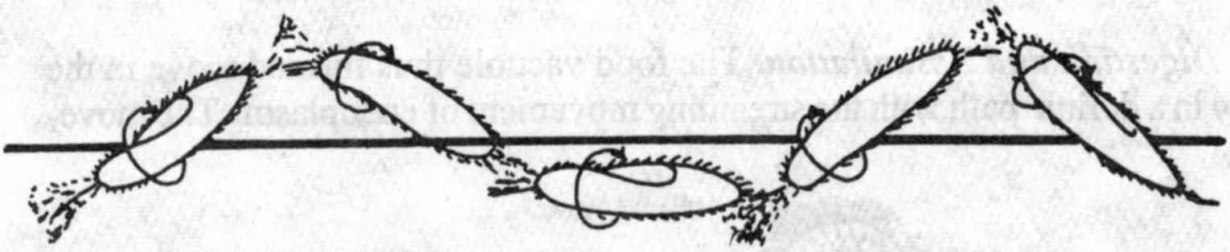

Fig. 4.15 Locomotion in *Paramecium*.

***Feeding mechanism.*** It feeds while stationary or swimming slowly. In feeding , the cilia of the oral groove produce a current of water that sweeps in an arc-like manner down the side of body and over the oral region. The food particles are sucked with water current into the oral groove and then to the vestibule. Ciliary tract of vestibule direct the food particles into the buccal cavity. Some of these particles are discarded and left out. Quadrulus and peniculi regulate the passage of food into the cytopharynx. At the bottom of cytopharynx the food particles are gathered and form a rounded or ball-like mass, the *food vacuole*. It grows rapidly and when the vacuole attains a maximum size it is pinched off and takes its course in endoplasm. The food vacuole besides food particles also contains a little amount of water. The exect mechanism by which the food vacuole is liberated is not properly known probably post-oral fibres help in this process. The food vacuoles are formed in about 1-5 minutes depending upon the abundance of material and rate of feeding. The size of vacuoles depend upon the quantity and quality of food, viscocity of surrounding medium and the physiological state of individual.

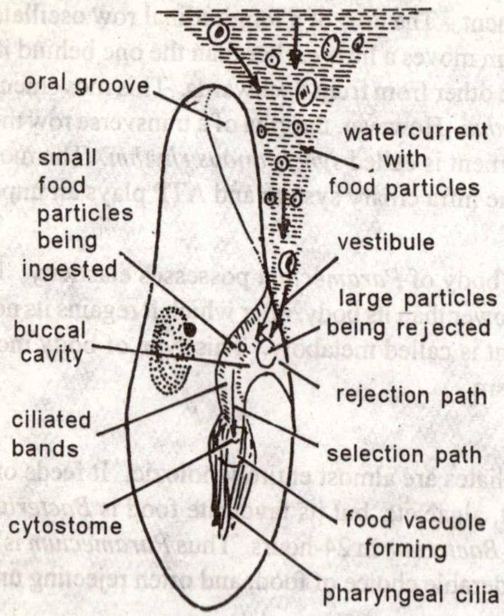


Fig. 4.16 *Paramecium* drawing water current to obtain food.

***Digestion and Assimilation.*** The food vacuole thus formed move in the body in a definit path with the streaming movement of endoplasm. The move-

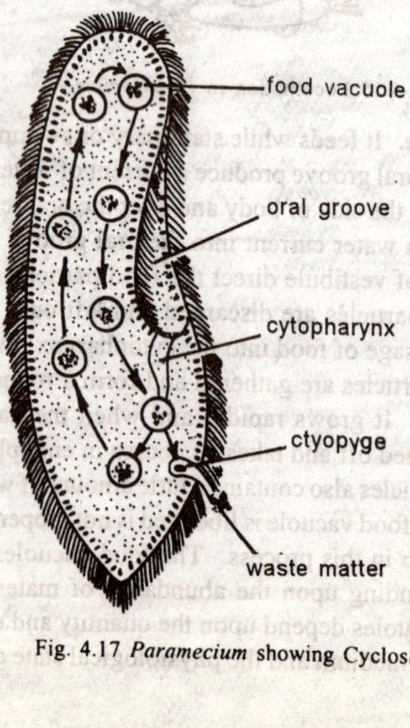


Fig. 4.17 *Paramecium* showing Cyclosis

ment of food vacuoles is called *cyclosis*. Soon after it enters the endoplasm the food vacuole moves, posteriorly up to the end where it stops momentarily and them turns dorsally and turning anteriorly it moves forwards. On reaching the anterior end of the body it moves back again along the oral side casting off unabsorbed remains. The food vacuole remains in the body from one to three hours.

On leaving the cytopharynx the vacuole first decreases in size and finally increased. The decrease is due to osmotic concentration of endoplasm as it is higher than the vacuole, the water is lost from the vacuole and it becomes small.

The digestion of food takes place in the food vacuoles in which the medium is first acidic and later alkaline. As a result of digestion the proteins are hydrolysed into amino acids by *proteases*, carbohydrates into glucose by *carbohydrases*, and probably fat is also digested. The process of cyclosis can be demonstrated experimentally in *Paramecium*. If milk, coloured with *Congo Red* stain, is given to *Paramecium*, the fat particles of milk first become red due to the acidic reaction of the enzymes in food vacuoles, then they gradually turn purple and finally blue due to alkaline medium in the food vacuoles. Complete digestion takes place mainly during the later alkaline phase of food vacuole. The digested food materials are throughly distributed to all parts of body during cyclosis. Reserve food material is glycogen and fat-droplets which remain scattered throughout the endoplasm.

If *Paramecium* does not get food, it feeds upon the reserve food materials in the endoplasm. When even these are exhausted the macro-nucleus becomes smaller, the animal becomes transparent, vacuoles appear in the cytoplasm, cilia disappear and degeneration of specialized structures set in. This phenomenon is called *inanition*. It restores its normal shape when supplied with rich food such as bacteria.

***Egestion.*** The vacuoles become smaller and finally the undigested food particles are released from the cytoproct or cytopyge into surrounding medium.

*Paramecium bursaria* displays symbiotic relationship with algae. The endoplasm is filled with green zoochlorellae. It takes its food material manufactured by zoochlorellae.

**Respiration.**

Respiration takes place by diffusion through the general body surface. Oxygen is taken in and used for burning of food to obtain energy. As a result carbon dioxide is formed which is diffused out in surrounding water.

**Excretion.**

Most of the nitrogenous excretion in *Paramecium* is carried by diffusion through general body surface. Contractile vacuoles also serve for excretion as urea has been detected in them. The chief nitrogenous product is ammonia which is diffused out.

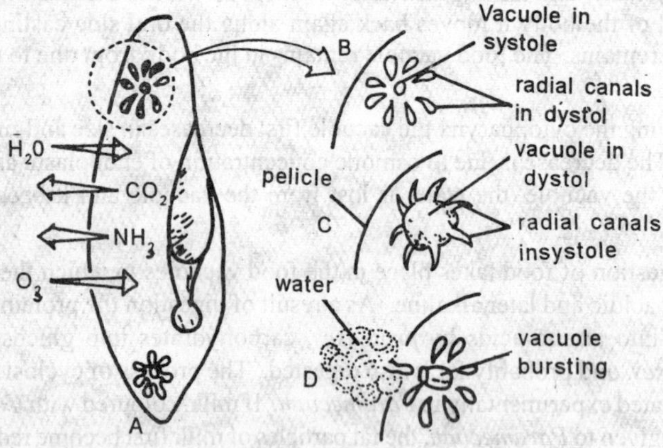


Fig. 4.18 ***Paramecium*** **diagrametic representation of respiratio, excretion and osmoregulation.**

**Osmoregulation.**

Two contractile vacuoles regulate the water content of the body. They occupy rather fixed position, one being anterior and the other posterior. Each vacuole is surrounded by 5-12 radiating canals. Both vacuoles derive energy to work from the mitochondria crowded around them.

**Behaviour.**

*Paramecium* is extremely sensitive to different types of stimuli and shows its reaction to them. The response is positive when the animal moves towards a stimulus and negative when it moves away. To an adverse stimulus the animal continuous to give the avoiding reaction until it escapes. Various stimuli and their response given by *Paramecium* are:

1. *Reaction to mechanical stimuli* (*Thigmotaxis or Thigmotropism*). Response to contact is varied. If the anterior end lightly touches a fine point, the animal shows strong avoiding reaction. A slow moving individual responds positively to contact with an object by coming to rest upon it.

2. *Reaction to temperature* (*Thermotaxis or Thermotropism*). The optimum range of temperature for paramecia is 24-280°C, but they can withstand considerably lower and higher temperature than the optimum range. Extreme cool and hot water is avoided.

3. *Reaction to chemical (Chemotaxis or Chemotropisin)*. It shows avoid-

ing reaction to chemical like base. If a drop of weak salt solution(5%) is introduced in a *Paramecium* culture on a microslide, the animal show negative response. However, they show a positive response to the weak acids.

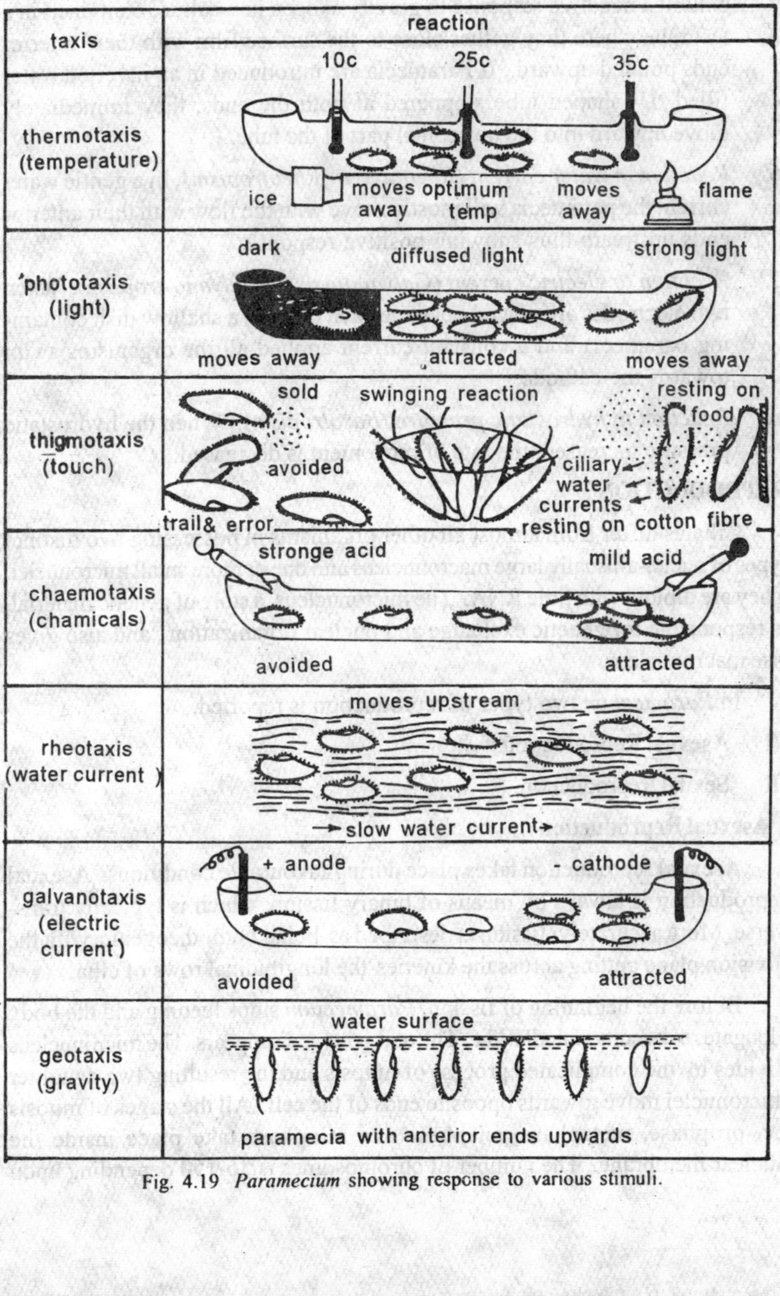


Fig. 4.19 *Paramecium* showing response to various stimuli.

4. *Reaction to light (Phototaxis or Phototropism).* It shows negative response to bright light as well as dark. It responds to the diffused light. However, P. bursaria is positively phototactic.

5. *Reaction to gravity (Geotaxis or Geotropism). Paramecium* generally exhibit a negative response to gravity as seen in a culture contained in a test tube where they gather close to the surface film with their anterior ends pointed upward. If Paramecia are introduced in an inverted water filled 'U' shaped tube stoppered at both the ends, they immediately move upward into the horizontal part of the tube.

6. *Reaction to water current (Rheotaxis or Rheotropism).* In a gentle water current the paramecia will mostly move with the flow with their anterior ends upstream thus showing positive response.

7. *Reaction to electric current (Galvanotaxis or Galvano-tropism).* When two electrodes are placed opposite each other in a shallow dish containing paramecia and a constant current applied all the organisms swim towards the cathode.

8. *Reaction to hydrostatic pressure (Barotropism).* When the hydrostatic pressure increased, the rate of movement is decreased.

## REPRODUCTION

Ciliates differ from almost all other organisms in possessing two distinct types of nuclei-a-usually large macronucleus and one or more small micronuclei. They are diploid with little RNA. The micronucleus, a store of genetic material, is responsible for genetic exchange and nuclear organization , and also gives rise macronucleus.

In *Paramecium* two types of reproduction is reported.

I Asexual Reproduction

II Sexual Reproduction.

### I Asexual Reproduction

Asexual reproduction takes place during favourable conditions. Asexual reproduction is always by means of binary fission, which is typically transverse. More accurately, fission is described as being homotheogenic, with the division plane cutting across the kineties-the longitudinal rows of cilia.

Before the beginning of fission, *Paramecium* stops feeding and the body elongates to become spindle-like. The oral groove disappears. The micronucleus divides by the complicated process of mitosis and the resulting two daughter micronuclei move towards opposite ends of the cell. All the stages of mitosis i.e. prophase, metaphase, anaphase and telophase take place inside the nuclear membrane. The number of chromosomes is 36-150 depending upon the species. When this is happening, the macronucleus also elongates, divide transversely by amitotic process in which the number of chromosomes may differs. A transverse constriction appears in the middle. The constriction deepens gradually. In the mean time two oral grooves begin to form one in the anterior half and another in the posterior half. Two original contractile vacuoles of parent remain, one in each half of the dividing parent individual. Two contractile vacuole are later formed. Two new buccal structures also appear. Now the constriction completely divides the parent into two daughter paramecia of equal size and with a complete set of cell-organelles. These grow to full size before dividing again.

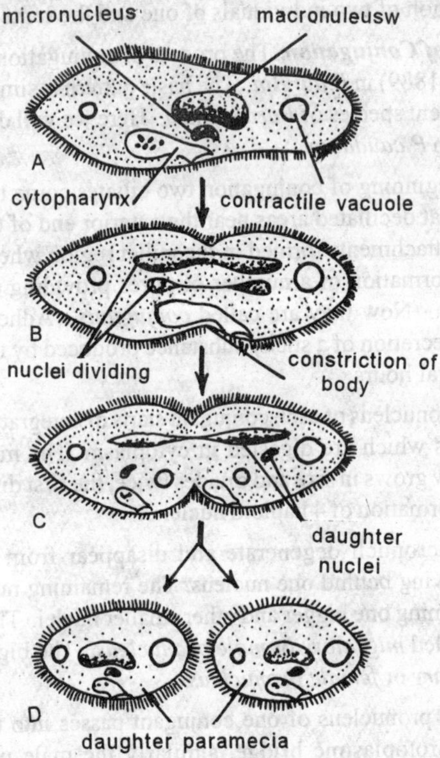


Fig. 4.20 *Paramecium caudatum showing binary fission.*

The newly formed daughter paramecia neither resemble each other nor the typical parent, so that for a while they can be distinguished as the anterior daughter (*Proter*) and the posterior daughter (*Opisthe*). Soon they assume the typical form.

The entire process of binary fission is completed with 30-120 minutes depending upon the availability of food and temperature. In *P. caudatum* it may occur 2-3 times a day and in *P. aurelia* about 5 times a day, so that a large number of them can be obtained in a short time. All the individuals which are produced from a single parent, collectively, form a clone and all the members of a clone are genetically alike.

The binary fission cannot go on indefinitely. After several generations, the rate of fission declines.

**II Sexual reproduction**

*Paramecium* reproduces sexually in which nuclear material of two individuals is interchanged. This method is called *conjugation*, but it is simply a temporary union of two individuals of one and the same species.

***Process of Conjugation.*** The process of conjugation is given by *Hertwig* and *Maupas* (1889) in *P.aurelia*. The basic pattern is similar with slight variations in different species of *Paramecium*. Here we will discuss the process of conjugation in *P.caudatum*.

At the beginning of conjugation two ciliates come together and become attached first at deciliated areas near the anterior end of their ventral surface, and later by attachments formed in the gullet region where membrane fusion leads to the formation of a number of pores providing cytoplasmic bridges between them. Now they are called *conjugants*. Adhesion probably takes place by the secretion of a sticky substance produced by the cilia. Attachment lasts for several hours.

The macronucleus of each conjugant starts disintegrating. It simply breaks up into pieces which are digested in cytoplasm. The micronucleus of each conjugant now grows in size and divides twice, the first division being meiotic result in the formation of 4 haploid nuclei.

Three micronucli degenerate and disappear from cytoplasm of each conjugant leaving behind one nucleus. The remaining nucleus divides again unequally forming one bigger and other smaller nuclei. The smaller nucleus is active and called *migratory* or *male pronucleus*. The bigger one is inert and called *stationary* or *female pronucleus*.

The male pronucleus of one conjugant passes into the other conjugant through the protoplasmic bridge, similarly the male pronucleus of other conjugant passes into the first conjugant. Now the male and female pronuclei fuse results in the formation of zygote nucleus. The complete fusion of two nuclei from two different forming a zygote nucleus is called *amphimixis*. The process cannot be called as fertilization because there is no fusion of gametes and their cytoplasm.

After about 12 hours or so the conjugants are separated and are now termed as *exconjugants*. The zygote nucleus divides thrice in each exconjugant resulting 8 nuclei. The divisions are mitotic. Out of these 8 nuclei 4 assume large size and become macronuclei. Remaining 4 are micronuclei. Out of these 4 micronuclei 3 disappear leaving behind only 1 micronucleus. It divides again forming two and at the same time the exconjugant divides by binary fission into two cells. Each cell having two macronuclei and one micronucleus. The

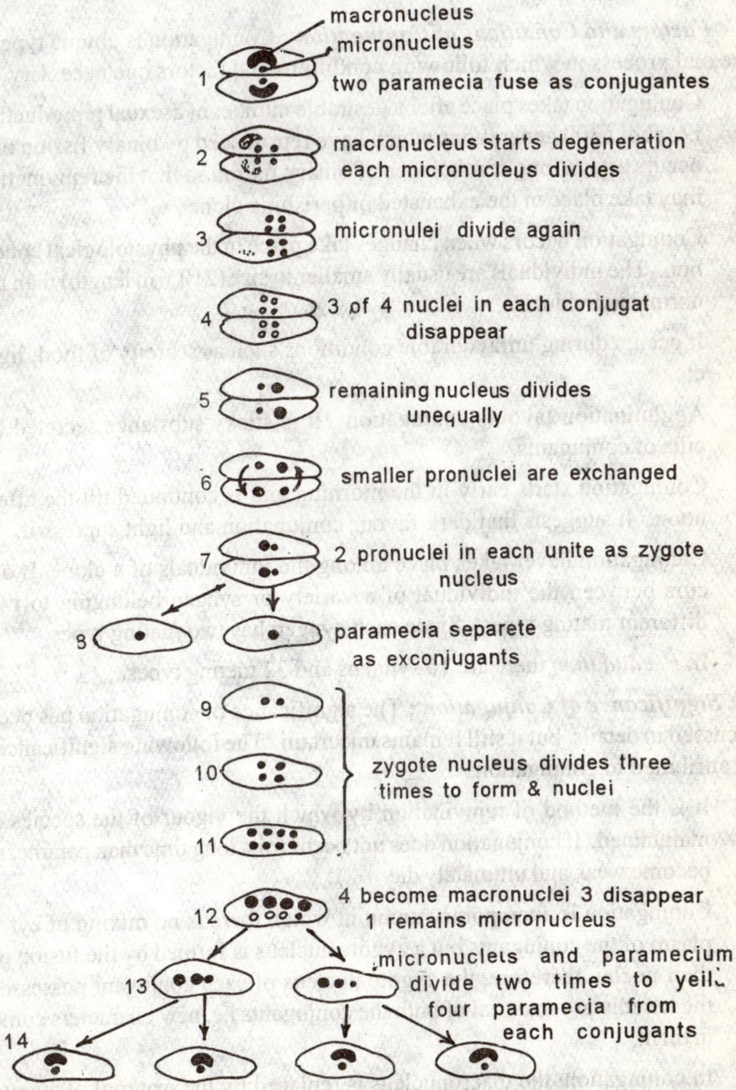


Fig. 4.21 *Paramecium caudatum* showing stage of conjugation.

micronucleus again divides into two micronuclei and at the same time the cell divides into two daughter paramecia by binary fission. Each daughter having one macronucleus and one micronucleus.

Thus each conjugant produces four daughter individuals in the end of conjugation. The whole process of conjugation takes about 96 hours to complete.

***Factors and Conditions of Conjugation :*** Conjugation is unique type of a sexual process in which following conditions and factors one necessary.

1. Conjugation takes place after a desirable number of asexual reproduction i.e. about 300 generations which have reproduced by binary fission or it occurs after a long continuance of binary fission so that the rejuvination may take place in the exhausted or perishing clone.
2. Conjugation occurs when changes take place in the physiological condition. The individuals are usually smaller in size (210 in length) than the normal individuals.
3. It occurs during unfavourable conditions such as scarcity of food, light etc.
4. Agglutination favours conjugation. It is sticky substance secreted by cilia of conjugants.
5. Conjugation starts early in the morning and is continued till the afternoon. It suggests that dark favour conjugation and light supress it.
6. Conjugation never takes place among the individuals of a clone. It occurs between the individual of a variety or syngen belonging to two different mating types. Since each syngen has two mating types.

   In *P. caudatum* there are 16 syngens and 32 meting types.

***Significance of Conjugation*** : The significance of conjugation has been discussed in details, but it still remains uncertain. The following significances are attributed to conjugation.

1. It is the method of rejuvination by which the vigour of the species is maintained. If conjugation does not occur for a long time than paramecia become weak and ultimately die.
2. Conjugation is a temporary union in which there is no mixing of cytoplasm of the conjugants but a zygote nucleus is formed by the fusion of their nuclei, therefore, the zygote nucleus of each conjugant possesses the herediatary material of both the conjugants i.e. new characters come in them.
3. In conjugation, the macronucleus is replaced by the material of zygote nucleus. The number of chromosomes is restored in macronucleus by conjugation due to which the vigour of daughter paramecia is renewed.

4. The function of micronucleus is to maintain balanced chromosomes and gene complex.

According to *Sonneborn* (1957) in *P. aurelia*, there are 14 syngens and 28 mating types. These are as follows,

**Table. Showing the mating types of *P. aurelia* complex.**

| Syngen number | Mating type | Other mating types with which conjugation occurs (and %) | Group | + or - type |
|---|---|---|---|---|
| 1 | I | II (95%), X (40%) | A | - |
| | II | I (95%), IX (40%), XIII (10%), V (1%) | | + |
| 2 | III | IV (95%) | B | |
| | IV | III (90%) | | |
| 3 | V | VI (95%), XVI (40%), II (1%) | A | - |
| | VI | V (95%) | | + |
| 4 | VII | VIII (95%), XVI (95%) | B | - |
| | VIII | VII (95%), XV (60%) | | + |
| 5 | IX | X (95%), II (40%) | A | - |
| | X | IX (95%), I (40%) | | + |
| 6 | XI | XII (95%) | B | - |
| | XII | XI (95%) | | - |
| 7 | XIII | XIV (95%) II (10%) | B | - |
| | XIV | XIII (95%) | | + |
| 8 | XV | XVI (95%), VIII (60%) | B | - |
| | XVI | XV (95%), VIII (95%), V (40%) | + | |
| 9 | XVII | XVII (95%) | A | - |
| | XVII | XVII (95%) | | - |
| 10 | XIX | XX (95%) | B | - |
| | XX | XIX (95%) | A | - |
| 11 | XXI | XXII (95%) | A | |
| | XXII | XXI (95%) | | |
| 12 | XXIII | XXIV (95%) | B | - |
| | XXIV | XXIII (95%) | | + |
| 13 | XXV | XXVI (60%) | C | - |
| | XXVI | XXV (60%) | C | - |
| 14 | XVII | XXVIII (95%) | A | - |
| | XXVIII | XXVII (95%) | | - |

Fig. 4.22 *P. caudatum* there are 16 syngrns and 32 mating types

## Aberrant behaviour in reproduction

*Paramecium* shows certain variations in its nuclear behaviour during fission and conjugation these deviations are Autogamy, Endomixis, cytogamy and Hemixis.

**1.** ***Autogamy.*** *Diller* (1934,36) described a process of self-fertilization or autogamy occuring in a single individual of *P. aurelia.* About once a month *P. aurelia* undergoes this process.

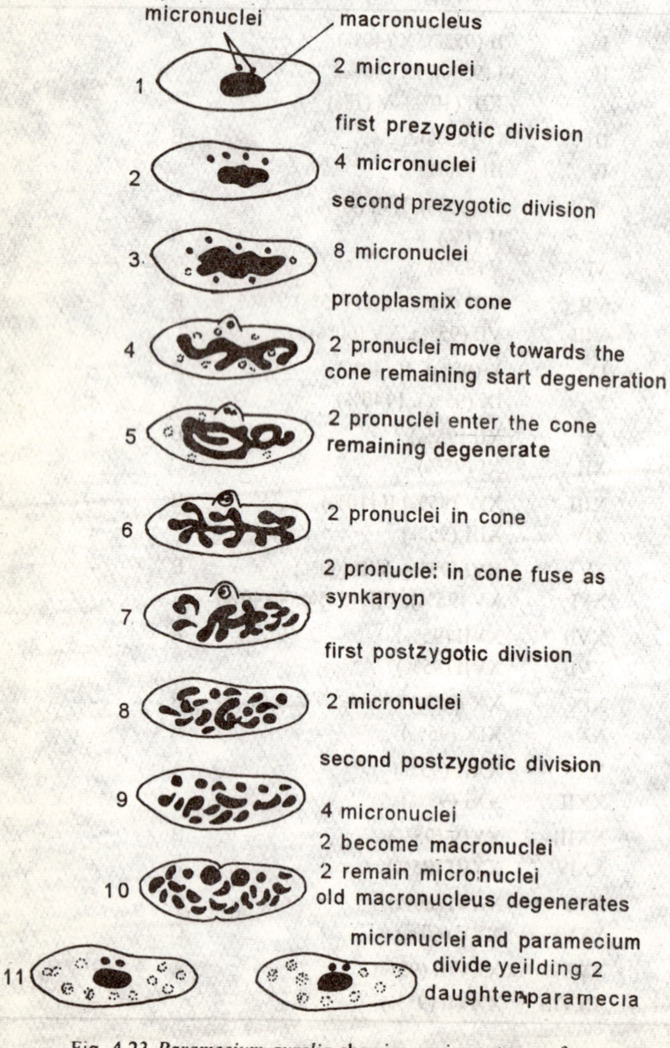


Fig. 4.23 *Paramecium aurelia* showing various stages of autogamy

During autogamy the macronucleus increase in size to form an irregular structure and later breaks up into small fragments are absorbed in the cytoplasm. The two micronucli divide twice forming eight nuclei. As a rule the first division is meiotic thus all the eight nuclei are haploid. Out of these eight nuclei, seven disintigrate and only one nucleus is left. It again divides to form two nuclei. These nuclei are called *gametic nuclei.* By this time a peculiar cone-like bulge appears on the surface of autogamous animal, slightly to the right and anterior to the cytostome. Because of its position it is called the *oral cone* or *protoplasmic cone.* The gametic nuclei are at first elongated and spindle like bodies. Often there is slight size variation between them so that they may be termed to as male (smaller) and female (larger) nuclei. They approach each other and then move into the cone and fuse to form zygote nucleus. Later it comes out of cone and the cone disappears. The zygote nucleus divides twice forming four nuclei. Two of these nuclei increase in size to become macronuclei and remaining two as micronuclei. The cell body and micronuclei divide by binary fission to form two daughter paramecia each with a macronucleus and two micronuclei.

**Significance of Autogamy**

1. Rejuvination in the species is accomplished by autogamy.
2. A new macronucleus is formed from the material of micronuclei.
3. Paramecium regains its lost vigour by this process.

**2. *Endomixis.*** Endomixis was first observed by *Woodruff* and *Erdmann* (1914) in *P. aurelia*. In endomixis only the nuclear reorganization occurs after which the body of *Paramecium* divides into daughter paramecia. It occurs as follows:

When endomixis starts, the macronucleus disintigrates and disappears. The two micronuclei divide twice resulting in the formation of eight nuclei( all diploid ). Out of these 8 nuclei six disintigrate and only two are left. Now the *Paramecium* divides by fission into two daughters each with one nucleus. The nucleus of each daughter divides twice forming four nuclei. Two nuclei enlarge to form macronuclei and remaining two micronuclei. The two micronuclei, again divide to form four micronuclei in the mean time the cell divides again by fission into two daughters each containing one macronucleus and two micronuclei. In the end four daughter paramecia are formed from one individual.

According to *Diller* (1936), *Sonneborn* (1947) and *Whichter-man* (1953) the endomixis does not occur in *Paramecium*.

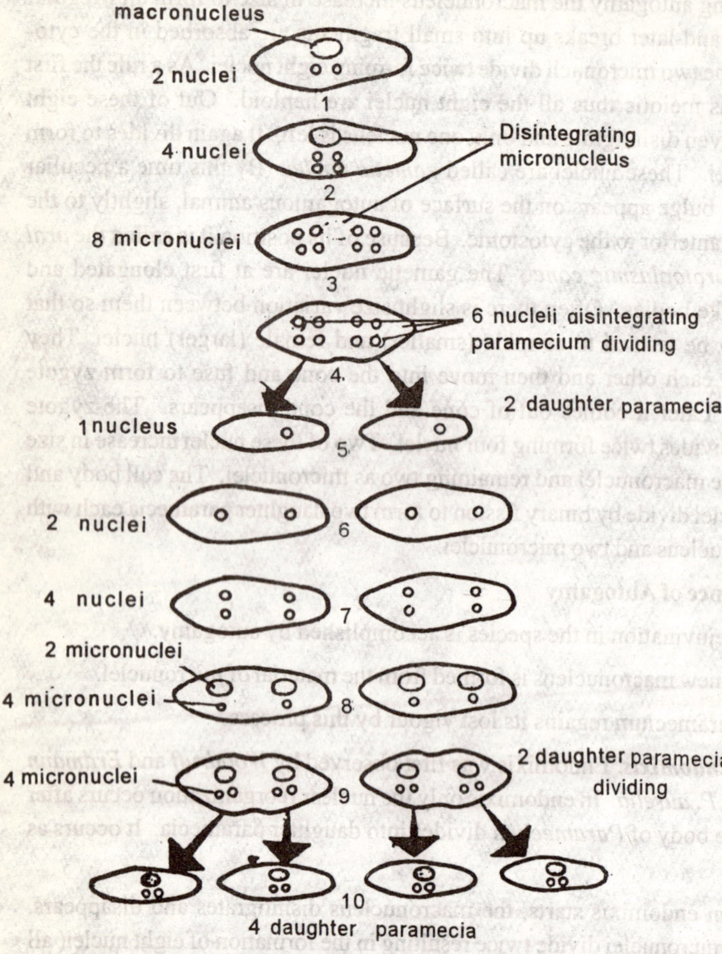


Fig. 4.24 *Paramecium aurelia* showing endomixis.

**3. *Cytogamy.*** *Wichterman* (1940) observed the process of cytogamy in *P. caudatum*. The process is something intermediate between conjugation and autogamy. Two individuals stick by their oral surfaces, the micronucleus divides twice (first division in meiotic) of these three disintegrate into the cytoplasm, the remaining one divides once producing two gametic nucleli. Self-fertilization takes place. The conjugants now separate and called exconjugants. Exconjugants behave and multiply as in conjugation. The process is completed in about 13 hours.

The genetical consequence of cytogamy is always homozygosity.

**4. *Hemixis.*** This method of nuclear reorganization was observed by *Hartman* in *P. aurelia* and was supported by *Diller* (1936). During hemixis the micronucli do not participate. The macronucleus throws out two or more equal or unequal fragments to be absorbed in the cytoplasm. The remainder of macronucleus and micronuclei behave normally at the cell division.

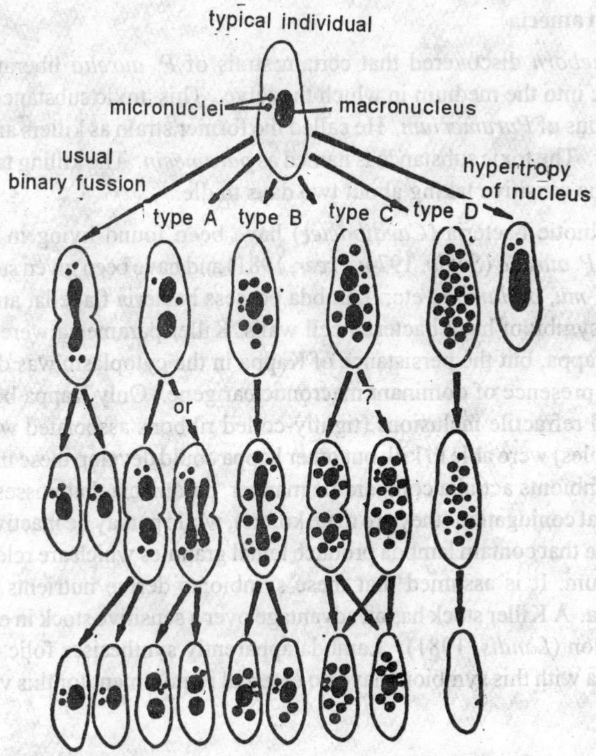


Fig. 4.25 *Paramecium aurelia* showing hemixis.

*Diller* classified hemixis into four types.

1. *Type A.* It is the simplest form of hemixis characterized by a division of the macronucleus into two or more parts in the cytoplasm.

2. *Type B.* It is characterized by the extrusion of 1-20 or more chromatin balls from the macronucleus.

3. *Type C.* It is characterized by the simultaneous splitting of macronucleus into 2 or more major portions and extrusion of balls into cytoplasm.

4. *Type D.* It is considered to be the pathogenic conditions in which micronucleus undergoes complete fragmentation into chromatin balls

that disappear from the cell later on. Micronuclei generally disappear before the dissolution of macronucleus.

The hemixis is considered as an act of purification on the part of the macronucleus and the process of chromatin material cast off by it are held in contain waste and harmful substances.

**Killer Paramecia**

*Sonneborn* discovered that certain strais of *P. aurelia* liberate a toxic substance into the medium in which they live. This toxic substance kills the other strains of *Paramecium*. He called the former strain as killers and later as *sensitives*. The toxic substance is named as *paramecin*. The killing takes place slowly, the sensitive taking about two days to die.

Symbiotic bacteria (*Caedibacter*) have been found living in the cytoplasm of *P. aurelia* (*Soldo,* 1974; *Preer*, 1981) and have been given such names as *kappa, mu, pi, lambda* etc;. Lambda possess bacteria flagella, and several types of symbiont have bacterial cell walls, Killer paramecia were found to contain kappa, but the persistance of Kappa in the cytoplasm was dependent upon the presence of dominant macronuclear gene. Only Kappa bodies that contained refractile inclusions (tightly-coiled ribbons associated with virus-like granules) were able to kill, but other kappa could develop these inclusions. Other symbionts act in a comparable manner, paramecia that possess mu kill sensitive at conjugation (they are mate killers), while pi may be inactive Kappa. Paramecia that contain lambda produce lethal granules which are released into the medium. It is assumed that these symbionts derive nutrients from the paramecia. A Killer stock has an advantage over a sensitive stock in ecological competition (*Landis*, 1981). Lambda apparently synthesises folic acid, and paramecia with this symbiont have no external requirement for this vitamin.

5

# GENERAL CHARACTERS & CLASSIFICATION OF PORIFERA

Sponges which constitute phylum Porifera, are the most primitive of the multicellular animals. Neither true tissues nor organs are present, and the cells display a considerable degree of independence. All the members of the phylum are sessile and exhibit little detectable movement. This combination of characteristic convinced *Aristotle, Pliny,* and other ancient naturalists that sponges were plants. In fact it was not until 1765, when internal water current were first observed by *Ellis*, that the animal nature of sponges became clearly established. As a result, *Linnaeus, Lamarck* and *Cuvier* placed the sponges under Zoophytes to Polypes in their systems, regarding them as allied to anthozoan coelenterates, and many investigators sought to find in sponges the polyps they thought must be there. *de Blaiville* (1816) recognized the lack of affinities of sponges with coelenterates and separated them into a group Spongiaria allied to Protozoa, but this idea gained little attention, and, through much of the 19th century, sponges were placed with coelenterates, usually under the name Coelenterata or Coelentera or Radiata. Morphology and Physiology of sponges were first understood by *Robert E.Grant* (1825), who coined the term Porifera (L. *Porous* = Pore; *ferro* = to bear). *Huxley* (1875) and *Sollas* (1884) proposed the complete separation of Sponges from other Metazoa on the ground of many peculiarities, but this view did not find its present general acceptance until about the beginning of 20th century. Sponges are now recognized as constituting a separated isolated branch of the Metazoa, named Parazoa, after *Sollas.*

**General Characters**

1. All sponges are aquatic, mostly marine but a few are fresh-water.
2. They are multicellular organisms with *cellular grade of body organization* without forming tissues or organs.
3. They are sessile, solitary or colonial.
4. They have asymmetrical or radially symmetrical bodies.
5. The body is cylindrical, tubular or vase-like.

6. The body wall is diploblastic with outer dermal epithelium and inner gastral epithelium with a gelatinous non-cellular mesenchyme in between.
7. The body surface bears numerous minute pores i.e. ,ostia for the ingression of water.
8. They possess peculiar canal system through which water current flows drawing food and oxygen inside the body and carrying away excretory and reproductive products.
9. Internal skeleton is present in the mesenchyme in the form of calcareous or siliceous spicules or of proteinous spongin fibres.
10. Digestion is entirely intracellular.
11. Nervous and sensory cells are lacking.
12. All sponges possess great power of regeneration.
13. They are mostly bisexual but asexual form also exist.
14. Reproduction takes place both by asexual and sexual methods.
15. Asexual reproduction takes place by budding, fission or gemmule formation and sexual reproduction by sperms and ova.
16. Fertilization is internal and at the same time it is cross fertilization.
17. Cleavage is holoblastic and development is indirect followed by a free swimming ciliated larva, the *amphiblastula* or *parenchymula*.

**CLASSIFICATION**

The classification of sponges presented great difficulty, and no one scheme has been unanimously accepted by specialists on sponges. The classification is based on the type of skeleton, and this serves to mark off rather sharply two groups of sponges, the calcareous and the glass sponges. The remaining silicons and horny sponges are, however, less clearly delimited, and they are variously subdivided by different workers. The classification here adopted is that given by *Topsent, Wilson* and *de Laubenfels*.

Phylum Porifera is divided into three classes:

**Class-1. Calcarea**

The members of this class possess their skeleton composed of separate calcareous spicules made of calcium carbonate (L.; *Caleis*= lime; Gr; *sponges* = sponge).

1. Members of this class are smaller in size, extending upto 15 cm. in length; may be solitary or colonial. All are marine.
2. Shape of body usually cylindrical or vase-like.

3. Osculum is narrow, terminal and fringed with spicules.
4. The colour of the body is white or dull-brown.
5. Skeleton consists of spicules which may be monaxon, triradiate or tetraradiate.
6. Collar cells or choanocytes are comparatively large.
7. The canal system is ascon, sycon or leucon type.

***Order-(i) Homocoela***

1. Body is small, cylindrical and thin walled without any foldings.
2. Canal system ascon type.
3. Spongocoel is lined throughout by choanocytes.

   Examples: *Leucosolenia, Clathrina.*

***Order-(ii) Heterocoela***

1. Body is comparatively bigger, vase-like, thick walled with internal foldings.
2. Canal system sycon or leucon type.
3. Choanocytes are restricted to the flagellated chambers.
4. Spongocoel is lined by dermal epithelium.

   Examples: *Scypha* or *Sycon, Grantia.*

**Class 2. Hexactinellida or Triaxonida**

The members of the class are called glass-sponges having triaxon or hexactinal spicules made up of silica.

1. Members of this class are of moderate size, extending up to one meter in length and are usually solitary.
2. The body shape is usually cylindrical, cup-shaped or funnel-shaped.
3. The osculum is wide and usually covered over by a siliceous sieve plate.
4. The skeleton consists of triaxon or hexactinal siliceous spicules, often fused resembling spun glass.
5. Dermal epithelium is wanting.
6. The canal system is simple rhagon type.
7. The choanocyte cells are restricted to finger-like flagellated chambers.
8. They are exclusively marine and occur chiefly in tropical deep sea water.

   Class Hexactinellida is divided into two orders :

***Order (i) - Hexasterophora***

1. The beautiful glass-like body is cylindrical, very light and attached to the substratum directly.

2. The spicules are hexasters having star-like appearance.
3. The flagellated chambers are radially and regularly arranged.

Example: *Euplectella* (Venus's flower basket).

***Order (ii) - Amphidiscophora***

1. The body is usually rounded or oval and attached to the substratum by root tufts.
2. The spicules are amphidiscs having a convex disc, bearing a crown of backwardly directed marginal teeth at both ends.
3. The flagellated chambers are irregularly arranged.

Example: *Hyalonema* (Glass rope sponge).

**Class 3. Demospongiae**

This class comprises diverse and complicated sponges with skeleton of siliceous spicules or spongin fibres or a combination of both or totally absent. (Gr., *demos* = frame; *spongos* = sponge).

1. Members of this class are highly organised, varying from small to large size and may be solitary or colonial.
2. The body shape is variable being rounded, oval, cup-like or funnel-like.
3. The skeleton is made up of silicious spicules or spongin fibres or both. The spicules are monaxon or tetraxon.
4. Canal system is complicated and leucon type.
5. Choanocytes are restricted to small flagellated chambers.
6. They are mostly marine but few sponges are fresh water.

Class Demospongiae is divided into three sub-classes:

***Subclass-I. Tetractinellida***

1. Body is rounded or flattened.
2. Siliceous spicules, if present are tetraxon, or absent.
3. Spongin fibres are absent.

***Order (i) Myxospongida***

1. Spicules are absent.
2. Body surface is velvet-like

Example: *Oscarella.*

***Order (ii) Carnosa***

1. Homogenous leuconoid structure.
2. Spicules are of equal-size.

Examples: *Chondrilla, Plakina, Halina.*

***Order (iii) Choristida***

1. Body form is usually elaborate:
2. Both small (microscleres) and large (megascleres) spicules are present.

Example: *Geodia, Ancoria.*

***Subclass-II, Monaxonida***

1. Body form variable, occurring in shallow and deep water.
2. The skeleton consists of monaxon spicules distinguished into megascleres only.
3. Spongin fibres may or may not be present.

Sub-class Monaxonida is divided into four orders:

***Order (i) Hadromerina***

(i) Megascleres are mostly tylostyles i.e. the broad end is knobbed.

(ii) Microscleres are usually absent and if present, are star-like.

(iii) Spongin fibres are absent.

Examples: *Cliona* (Boring sponge), ***Suberites, Poterion.***

***Order (ii) Halichondrina***

(i) Megascleres are always of more than one kind.

(ii) Microscleres are usually absent.

(iii) Spongin fibres are present but they are very scanty.

Example: *Halichondria.*

***Order (iii) - Poecilosclerina***

1. Megascleres are usually of two or more kinds and united by spongin fibres into a regular network.
2. Microscleres may be curved, bow-shaped or C-shaped.

Examples: *Myxilla, Microciona.*

***Order (iv) - Haploscerina***

(i) Megascleres are always diactinal and pointed at both ends.

(ii) Microscleres may or may not be present.

(iii) Spongin fibres are usually present.

Examples: *Chalina, Spongilla* (fresh-water).

***Sub-Class 3, Keratosa***

(i) The body form is usually rounded and massive with a leathery surface dark colour, occuring generally in warm shallow water.

(ii) The skelelton is exclusively composed of spongin fibres.

(iii) The spicules are totally absent.

Examples: *Euspongia* (Bath sponge), *Hippospongia* (Horse sponge).

**CLATHRINA**

| | | |
|---|---|---|
| Phylum | - | Porifera |
| Class | - | Calcarea |
| Order | - | Homocoela |
| Genus | - | *Clathrina* |

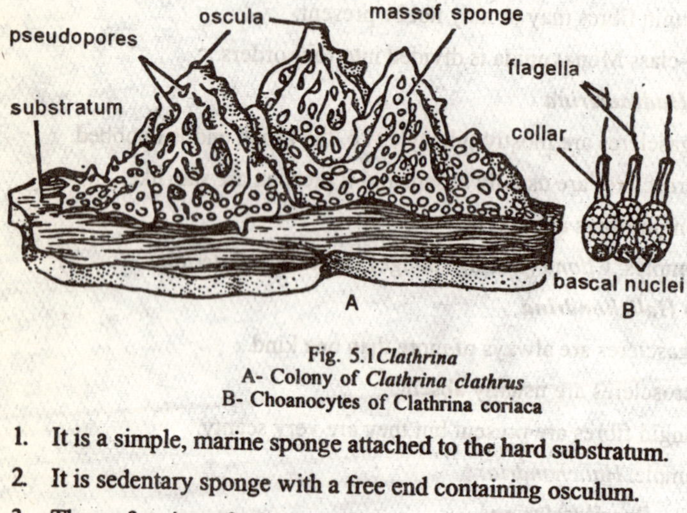


Fig. 5.1 *Clathrina*
A- Colony of *Clathrina clathrus*
B- Choanocytes of Clathrina coriaca

1. It is a simple, marine sponge attached to the hard substratum.
2. It is sedentary sponge with a free end containing osculum.
3. The surface is perforated by minute apertures called ostia, through which the water enters the body. Canal system asconoid type.
4. Skeleton consists of calcareous spicules which are triradiate with equal rays.
5. Choanocytes lines the spongocoel.
6. Dermal membrane or cortex is lacking.
7. Development includes a *parenchymula* larva and the so called olynthus stage.

**GRANTIA**

| | | |
|---|---|---|
| Phylum | - | Porifera |
| Class | - | Calcarea |
| Order | - | Heterocoela |
| Genus | - | *Grantia* |

Fig. 5.2 *Grantia*

1. It is a small, solitary or colonial, marine sponge.
2. It is found in the Gulf of St. Lawrence.
3. The body is complex vase-like.
4. Each cylinder bulges in the centre and opens to the exterior by an osculum.
5. The osculum is encircled by a fringe of large giant monaxon spicules forming oscular fringe.
6. The body surface is covered by the dermal membrane forming a cortex of varying thickness having special cortical spicules. The spicules are tangential radiate.
7. Body wall is composed of outer dermal epidermis, middle mesenchyme and inner flattened epithelium lining the spongocoel.
8. Canal system is syconoid.
9. Choanocytes are restricted to the radial canal, and the flagella arise directly from nuclei.
10. Bisexual. Reproduction by asexual and sexual methods. Asexual reproduction by budding and regeneration. Sexual by gametes. Larva is an amphiblastula.

## LEUCILLA

The classification is the same as that of *Grantia.*

1. It is a small, marine and solitary sponge.
2. It is found in Europe and U.S.A.
3. The body is somewhat globose in shape as the anterior region is narrow and the posterior region is broad.
4. The osculum is situated at the anterior most end and is surrounded by oscular fringe of monaxon spicules.

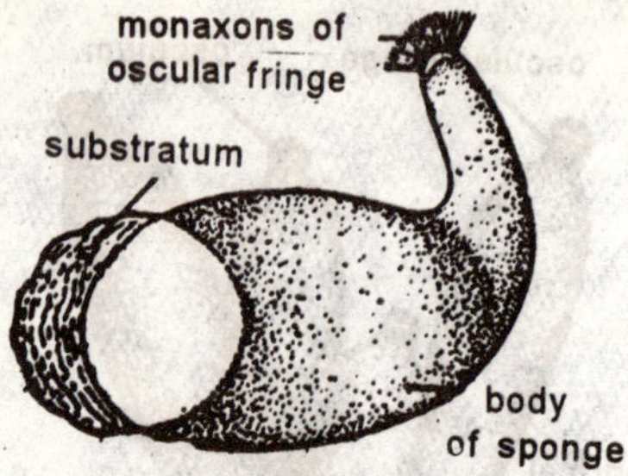


Fig. 5.3 *Leucilla*

5. The spongocoel is narrow while the body wall is quite thick.
6. Canal system is primitive leuconoid type. Each radial canal is subdivided into elongated flagellated chambers grouped around a common excurrent canal.
7. The skeleton is of inarticulate type.

## LEUCANDRA

The systematic position is the same as the of *Grantia*.

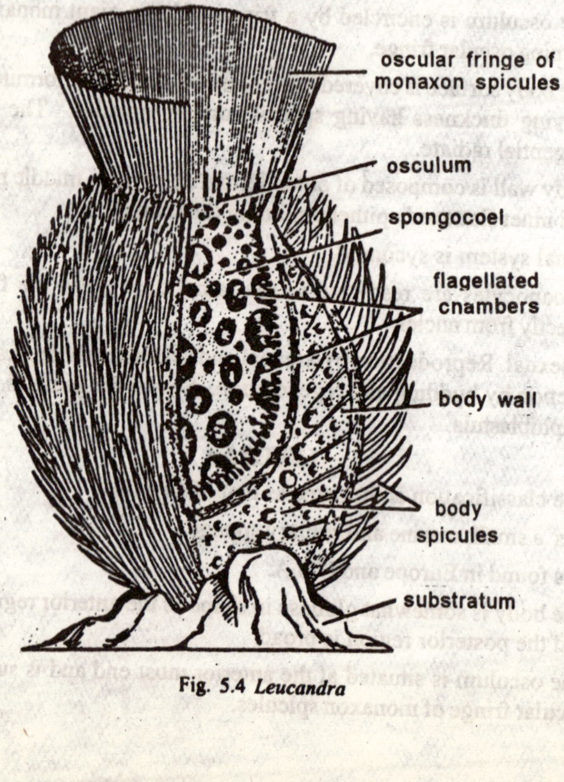


Fig. 5.4 *Leucandra*

1. It is a small, marine and solitary sponge.
2. Body is globose in shape.
3. Body wall is thick. Osculum is situated at the anterior end and is surrounded by oscular fringe of monaxon spicules.
4. Dermal skeleton is formed of tangentially arranged triradiate spicules which may sometimes develop an apical ray.
5. The flagellated chambers are small and spheroidal.

**OSCARELLA**

| | | |
|---|---|---|
| Phylum | - | Porifera |
| Class | - | Demospongiae |
| Subclass | - | Tetractinellida |
| Order | - | Myxospongida |
| Genus | - | *Oscarella* |

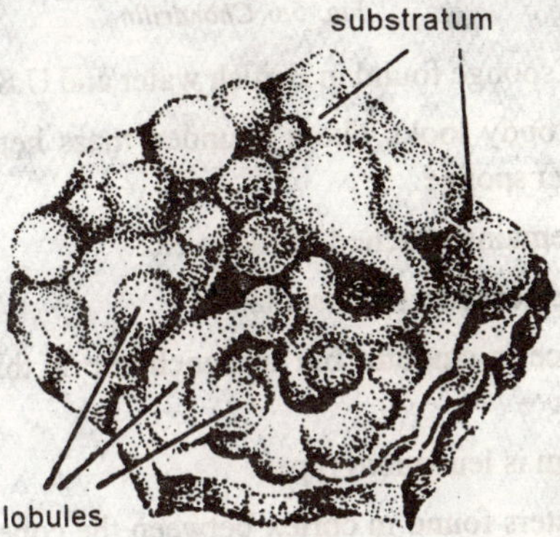


Fig. 5.5 *Oscarella*

1. It is found in shallow water forming encrustation over the rocks and submerged objects.
2. It occurs in British water.
3. It is a bright coloured small sponge whose upper surface is lobulated.
4. The canal system is leuconoid type.
5. The lobules are covered with thin dermal layer. Each lobule contains excurrent canal surrounded by flagellated chambers and the depression between the lobules house the incurrent canal.
6. Skeleton is absent i.e. the spicules are completely absent.

**CHONDRILLA**

Phylum Porifera

| | | |
|---|---|---|
| Class | - | Demospongiae |
| Subclass | - | Tetractinellida |
| Order | - | Carnosa |
| Genus | - | *Chondrilla* |

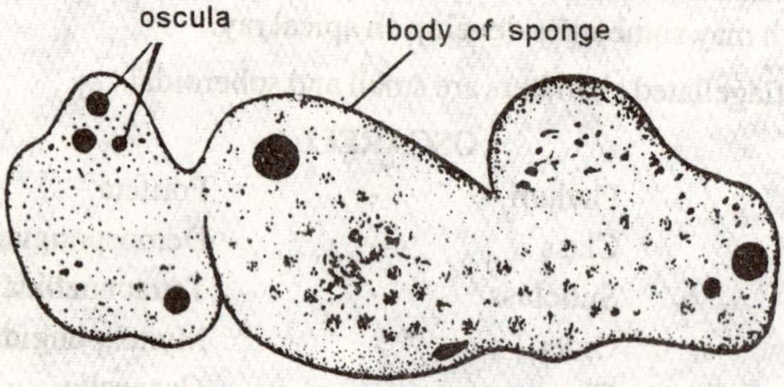


Fig. 5.6 *Chondrilla*

1. It is marine sponge found in British water and U.S.A.
2. The entire body looks like a rounded mass hence commonly called chicken-liver sponge.
3. The body remains attached to rocks etc.
4. Free surface contains few osculae.
5. Body surface is smooth, but mesenchyme is thick and stony due to spherasters.
6. Canal system is leuconoid type.
7. The spherasters found in cortex between the cones and are well developed while other spicules are absent.

## CLIONA

| | | |
|---|---|---|
| Phylum | - | Porifera |
| Class | - | Demospongiae |
| Subclass | - | Monaxonida |
| Order | - | Hadromeria |
| Genus | - | *Cliona* |

1. It forms low encrustation on rocks, mollusc shells, coral skeleton and other calcareous objects.
2. It is cosmopolitan in distribution.
3. It is commonly called 'Boring sponge' or 'Sulphur sponge'.
4. It is light yellow in colour but may be green or purple , and forms a mass upto 20cm. in diameter.

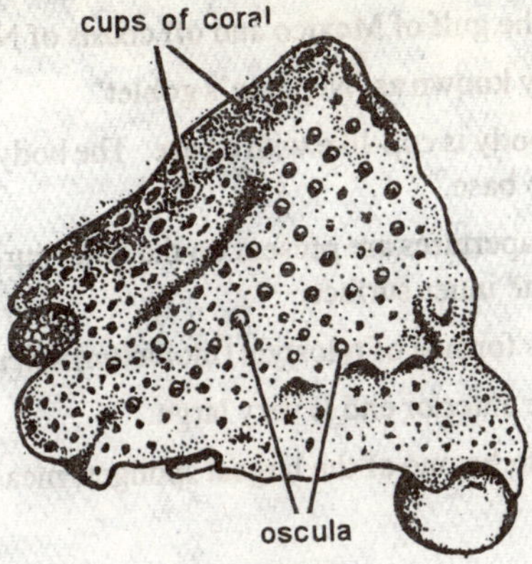


Fig. 5.7 *Cliona*

5. It enters the interior of above animals and lives permanently there, forming extensive burrows and tunnels.
6. Each tunnel has a raised osculum.
7. Monaxon spicules form the skeleton of the individual.
8. Sexual reproduction by the formation of gametes. Development indirect through a larval stage. The larva grows into a compact mass.
9. It has sulphurous odour hence commonly called sulphur sponge.

## POTERION

The systematic position is the same as that of *Cliona*.

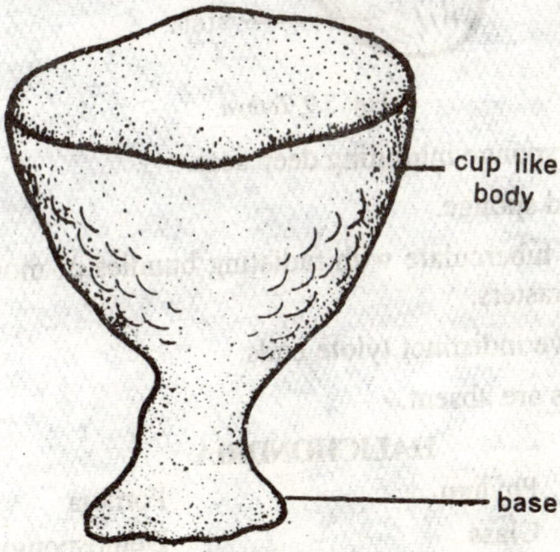


Fig. 5.8 *Poterion*

1. It is a solitary, marine and boring sponge.

2. It is found in the gulf of Mexico and off coasts of North Carolina.
3. It is commonly known as 'Neptune's goblet'.
4. The shape of body is cup-like with a base. The body remains fixed to the substratum by base.
5. The incurrent apertures are present on the outer surface while excurrent apertures on the inner surface.
6. The skeleton is formed of tylostyle (monaxon) spicules.
7. Osculum at the anterior end is very large.
8. It is supposed to be one of the largest sponges, measuring 1-2 metres in hight.

## TETHYA

The systematic position is the same as that of *Cliona.*

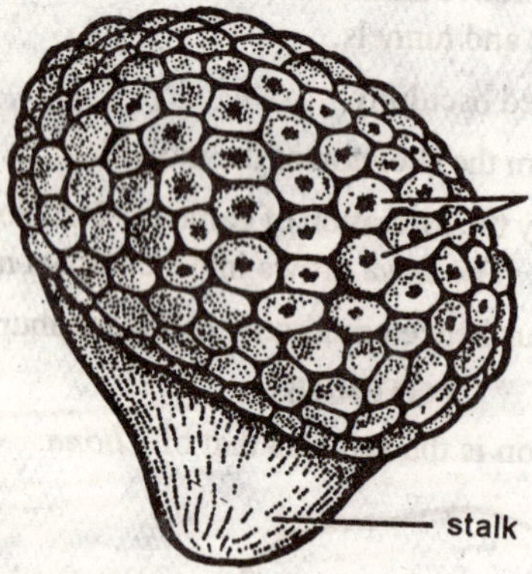


Fig. 5.9 *Tethya*

1. It is a marine sponge inhibiting deep seas.
2. It is a rounded sponge.
3. Body surface tuberculate with radiating bundles of monaxon spicules and several exasters.
4. The styles have indistinct tylote ends.
5. Spongin fibres are absent.

## HALICHONDRIA

| | | |
|---|---|---|
| Phylum | - | Porifera |
| Class | - | Demospongiae |
| Subclass | - | Monaxonida |
| Order | - | Halichondrina |
| Genus | - | *Halichondria* |

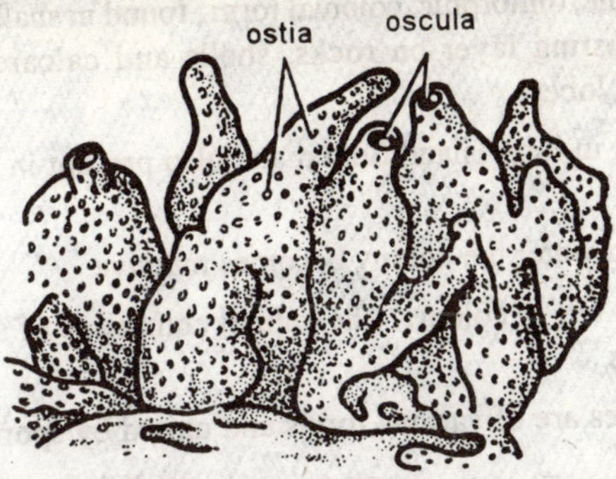


Fig. 5.10 *Helichondria*

1. It is a colonial sponge. It forms low encrustations on rocks in shallow waters.
2. It is cosmopolitan in distribution.
3. It is commonly called the crumb of 'bread sponge'.
4. It is soft and yellowish or green or brown in colour.
5. The surface presents one or more large rounded openings called oscula.
6. Skeleton consists of oxeas lying without any arrangement.

## MICROCIONA

| | | |
|---|---|---|
| Phylum | - | Porifera |
| Class | - | Demospongia |
| Subclass | - | Monaxonida |
| Order | - | Poecilosclerina |
| Genus | - | *Microciona* |

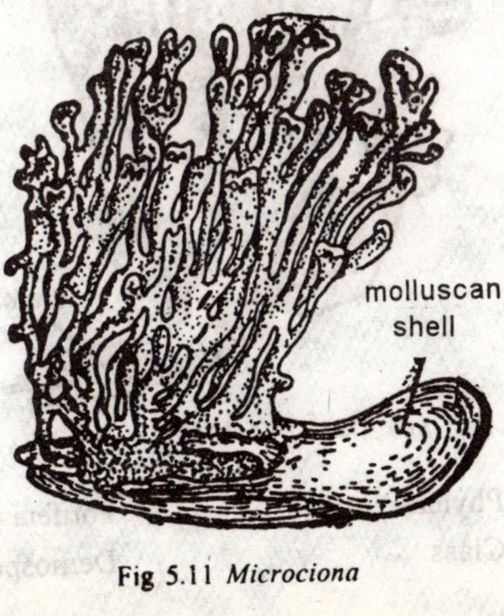


Fig 5.11 *Microciona*

1. It is a marine, dimorphic, colonial form , found in shallow water as a soft, thin encrusting layer on rocks, shells and calcareous objects with ascending lobes.
2. It is found in long Island Sound and also present in South Carolina to Cape Cod.
3. In deep sea water the colony becomes massive.
4. It is bright red in colour. The animal comprises of several finger-like projections.
5. The spicules are of various forms and united by spongin into a regular network.
6. The animal is used in regeneration and reassociation experiments. This animal has tremendrous capacity to reassociate and regenerate.
7. Sexual reproduction by gametes. Development through a free swimming larval stage in early August.

## CHALINA

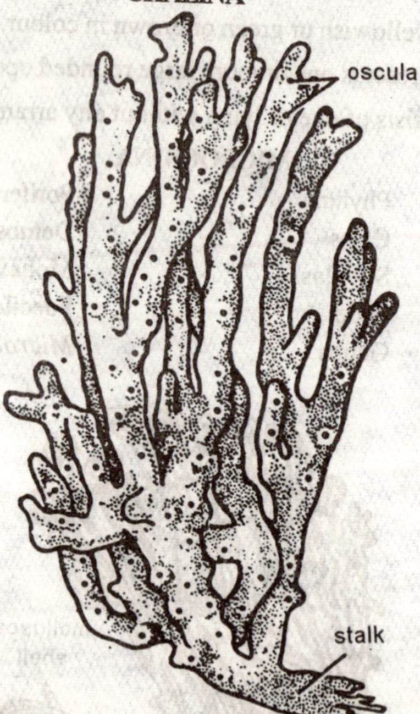


Fig. 5.12 *Chalina*

| | |
|---|---|
| Phylum | Porifera |
| Class | Demospongiae |

| | | |
|---|---|---|
| Subclass | - | Moxaxonida |
| Order | - | Haplosclerina |
| Genus | - | *Chalina* |

1. Chalina is commonly called Dead man's finger found in deep water, orange-brown or red in colour.
2. It is found in Europe, common from Rhode Island to Labrador. and North Carolina to Cape Cod.
3. It is much branched and each branch looks like a finger, so also called mermaid's glove.
4. Each branch is perforated by numerous openings or oscula.
5. Canal system is leucon type.
6. Monaxon siliceous spicules embedded in spongin fibres constitute the skeleton of this animal.
7. Asexual reproduction by budding and sexual reproduction includes a free-swimming larva.

## SPONGILLA

The classification is the same as that of *Chalina*.

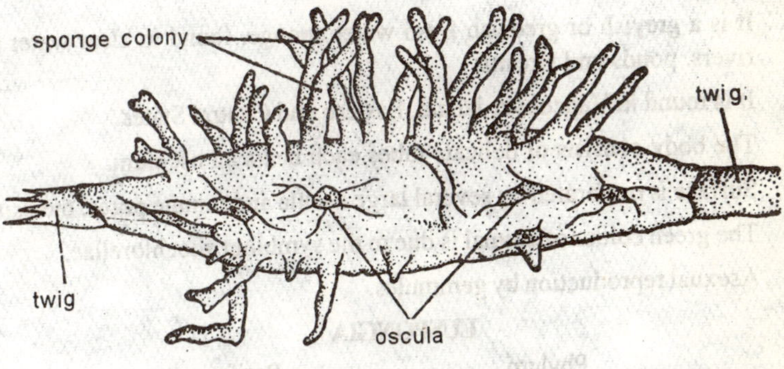


Fig. 5.13 *Spongilla*

1. Spongilla is a colonial fresh - water sponge, found attached to submerged sticks and plants of ponds, lakes and streams.
2. It is cosmopolitan in its distribution. It is found mainly in Atlantic, European and American waters.
3. Siliceous spicules form the skeleton. The spicules are oxeas, embedded in spongin.
4. Canal system rhagon type.

5. Highly branched colony with numerous oscula.
6. Asexual reproduction by budding and gemmule formation, sexual reproduction also occurs.
7. Gemmules are protected by the amphidisc spicules.

## EPHYDATIA

The classification is the same as that of *Chalina*.

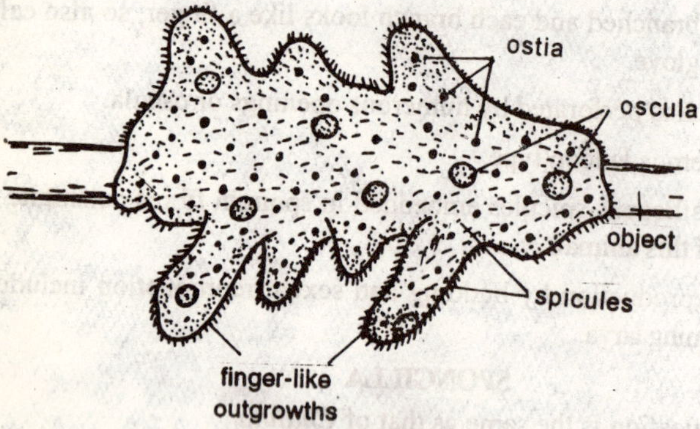


Fig 5.14 *Ephydatia*

1. It is a greyish or greenish fresh water sponge, found in clear water in rivers. ponds and streams.
2. It is found in Vancouver Island, Eastern and Central States.
3. The body consists of delicate tubes each bears an osculum.
4. Surface is perforated by several large oscula and minute pores, the osita
5. The green colour of animal is due to the symbiont zoochlorellae.
6. Asexual reproduction by gemmules.

## EUSPONGIA

| | | |
|---|---|---|
| Phylum | - | Porifera |
| Class | - | Demospongiae |
| Subclass | - | Keratosa |
| Genus | - | *Euspongia* |

1. It is found in shallow water on rocky bottom.
2. It has cosmopolitan distribution.
3. It is commonly called 'bath sponge'.
4. It forms a large brownish more or less spherical mass covered by a dark, non-living leathery membrane of spongin.

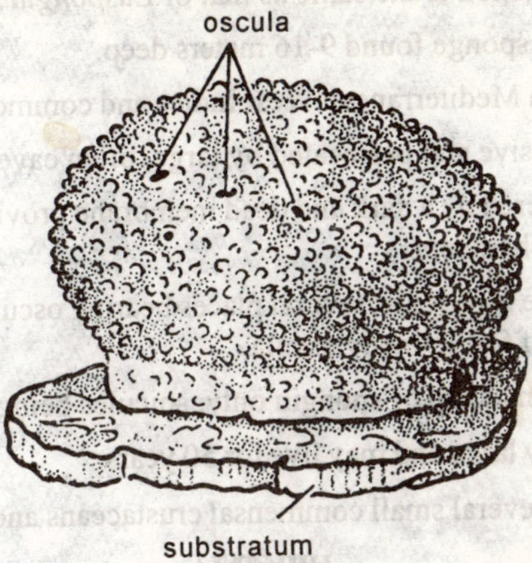


Fig 5.15 *Euspongia*

5. The surface bears small projects, the conules, supported by stiff spongin streams.
6. The body surface contains large openings called as oscula, and small openings called inhalant canals.
7. Internally the sponge resembles a liver in both consistency and colour.
8. It is unisexual and males are rare.
9. Skeleton is made of only spongin fibres. Spicules are absent. Many small worms and crustaceans live inside it as commensals.
10. It can live upto 50 years.
11. It is commonly used in bathing, washing automobiles, furnitures, applying face creams etc.

## HIPPOSPONGIA

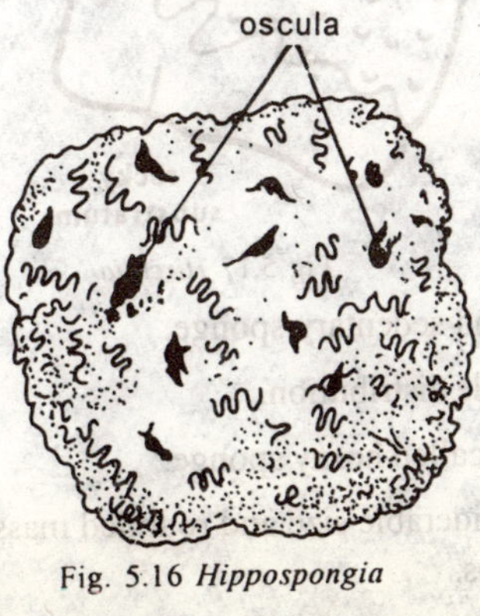


Fig. 5.16 *Hippospongia*

The classification is the same as that of *Euspongia*.

1. It is a horny sponge found 9-16 meters deep.
2. It is found in Mediterranean West Indies and common in Florida.
3. Body is massive and permeated by large, often cavernonals.
4. Body is covered by a dark and dead membrane provided with numerous flagellated chambers.
5. Body surface is raised at several places bearing oscula. A single sponge is made up of few to many oscula.
6. Skeleton is made up of spongin only, spicules being absent.
7. It grows very large and may live for 50 years.
8. It harbours several small commensal crustaceans and worms.

## HIRCINIA

The classification is the same as that of *Euspongia*.

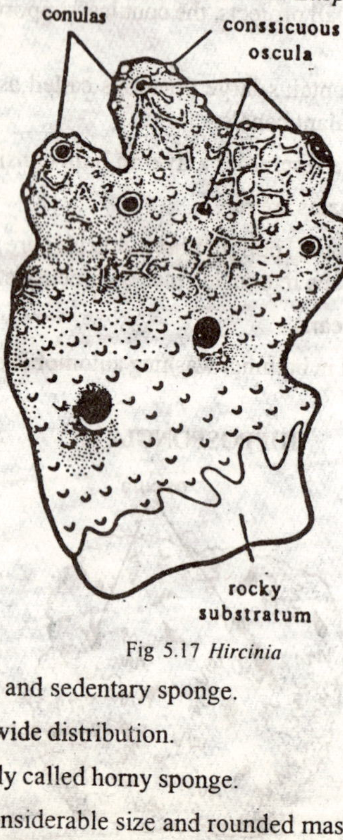


Fig 5.17 *Hircinia*

1. It is a marine and sedentary sponge.
2. It has world-wide distribution.
3. It is commonly called horny sponge.
4. It attains a considerable size and rounded massive form. Body surface contains conules.

5. The body colour is due to presence of symbiotic blue-green algae.
6. The canal system is diplodal leucon type.
7. Skeleton consists of spongin fibres only.

**EUPLECTELLA**

| | | |
|---|---|---|
| Phylum | - | Porifera |
| Class | - | Hexactinellida |
| Order | - | Hexasterophora |
| Genus | - | *Euplectella* |
| Species | - | *Aspergillum* |

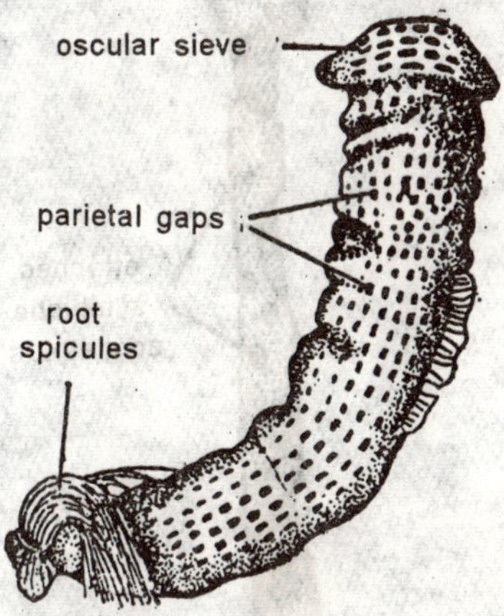


Fig. 5.18 *Euplectella*

1. It is a deep sea form found near Philipine islands and has become curved due to slow water current.
2. It is commonly called Venus' flower basket and is a beautiful glass sponge.
3. It is made of six-rayed siliceous spicules which are held together by siliceous cement resulting in a beautiful basket-like structure.
4. The curved upper end of the body is closed by an oscular sieve.
5. From the base arise root spicules which are long siliceous threads which fasten the sponge to its unstable slimy substratum.
6. The spongocoel contains a pair of crustacea(Spongicola) which had commensal life throughout their life. For this reason Euplectella is a beautiful presentation for newly married couple in some countries.

**HYALONEMA**

Phylum Porifera

| | | |
|---|---|---|
| Class | - | Hexactinellida |
| Order | - | Amphidiscophora |
| Genus | - | *Hyalonema* |

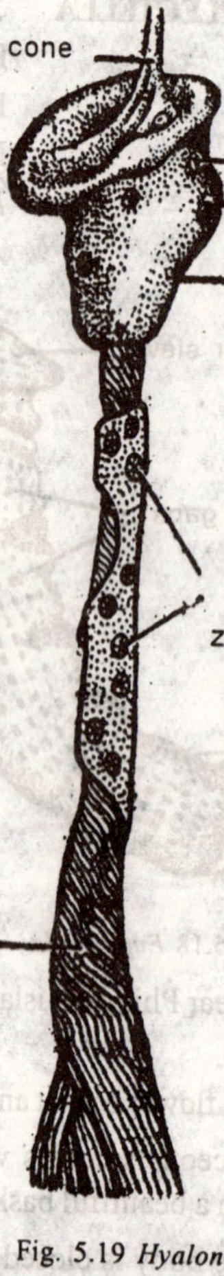


Fig. 5.19 *Hyalonema*

1. It is marine sponge found along new England coast.
2. It is commonly called as 'glassrope sponge'.
3. It has a spherical or oval about 30cm. long body fastened in the mud by a root-tuft of long twisted spicules.
4. The spicules continue through the body as an axis or columella and often project above it as the gastral cone.
5. The gastral cone bears openings of the excurrent canals.
6. The surface may be depressed to enclose a cavity, the spongocoel, or

may be flat in which case there is no spongocoel and the excurrent canals open out directly.

7. It possess large amphidisc spicules like fresh-water sponges. Entire body contains small, branching, six-rayed spicules.
8. The flagellated chambers are arranged radially and in parallel planes in the sponge wall.
9. Some anemones live attached to the root-tuft and certain crustaceans live inside the body as commensals.

## PHERONEMA

The classification is the same as that of *Hyalonema.*

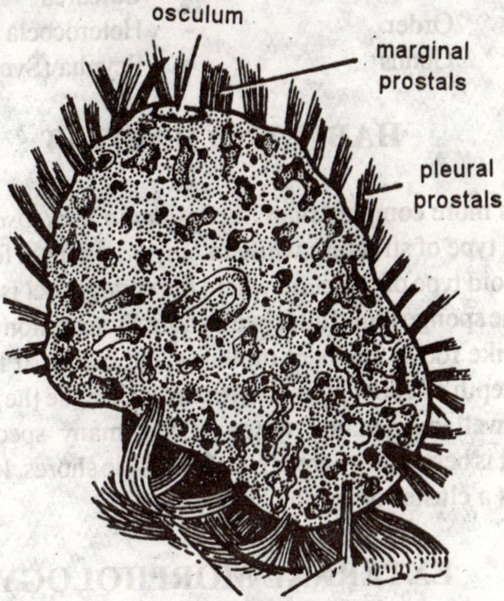


Fig 5.20 *Phercnema*

1. It is found on the bottom of sea. It occurs in U.K., U.S.A. and Europe.
2. It has a cup or bowl-shaped body.
3. Body thick-walled with a large spongocoel, which opens above by an osculum.
4. Osculum is guarded by spicules or marginal prostals.
5. Glassy spicules project from the body surface, often in tufts called the pleural prostals, giving the appearence of glass wool.
6. Skeleton consists of amphidisc microsclers.
7. Canal system is of eurypylour leuconoid type.
8. Broad, ill-defined root-tufts made of silica, which are joined and serve for attachment to the substratum.

# 6

# SCYPHA (SYCON)

SYSTEMMAIC POSITION

| | | |
|---|---|---|
| Phylum | - | Porifera |
| Class | - | Calcarea |
| Order | - | Heterocoela |
| Genus | - | Scypha (Sycon) |

## HABIT AND HABITAT

*Scypha* is more complex than *Luecosolenia. Leucosolenia* shows primitive asconoid type of structure while in *Scypha*, there is folding of body wall to show syconoid type of structure where the spongocoel is reduced. *Scypha* is a small, marine sponge found attached by a sticky secretion to some submerged solid objects like rocks, shells of molluscs and corals. It is found in shallow water upto a depth of 100 meters where waves provide the animal with plenty of food and well oxygenated water. It has many species and is widely distributed but is best, known from North Atlantic shores. It may be free-living but also forms a cluster or colony by budding.

## EXTERNAL MORPHOLOGY

*Scypha* or *sycon* possesses a vase-shaped and radially symmertrical body which measures from 1 to 3 cm in length. The colour is not specific but varies from grey to light brown. Near the attached end are found a few small tubular projections, or the *buds*, which are formed not so extensively as a *Leucosolenia* or other asconoid spongs. The free end of the vase-shaped individual bears an opening, the *osculum* (*exhalent* or *excurrent* pore), which is fringed with long, straight, needle-like monoaxon spicules. This *oscular fringe* checks small animals from entering into the body. The body surface is thrown into regularly arranged polygonal elevations from which project spear-like spicules, called *oxeotes* (monaxon spicules) that impart a bristly appearance to the body. The polygonal elevations are separated by deep grooves, bearing minutes pores, i.e., the *ostia*, which lead into the central body cavity, the *spongocoel* or *paragastric cavity*, through a system of canals.

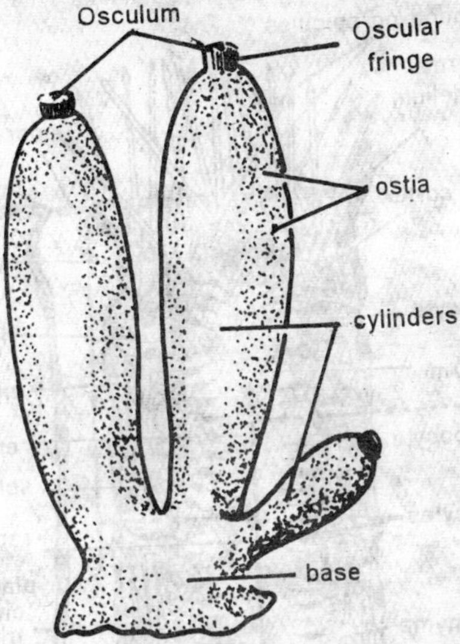

Fig. 3.1 Scypha (Sycon)

# CANAL SYSTEM

As in other sponges, *Scypha* has the peculiarity of having canal system which consists of complex system of pores and canals. The body is traversed by numerous canals of several types. The body wall is thick and so folded as to form regularly arranged alternating invaginations and evaginations, forming sycon type of canal system. The different components of canal system of sycon are given below:

1. *Ostia or dermal pores.* The external large openings of the incurrent canals are stretched over by a thin membrane, called the *poremembrane*. It bears two or more openings for the ingress of outside water into the body of sponge. These pores are known as *ostia* (L., *ostium*, door) or *dermal pores*. Because of the presence of contractile cells the *myocytes*, around them the ostia can reduce in diameter and thus regulate the amount of ingressing water.
2. *Incurrent canal.* The incurrent canals are narrow and almost squarish in cross-section. They end blindly at their inner extremities a little distance away from the spongocoel. Externally and become somewhat dilated. The dilated part of each canal is closed externally by a thin *pore-membrane* the bears 3 or 4 minute openings, the *dermal ostia* already referred to. The pore membrane also contains myocytes, some of which surround the dermal ostia and serve to close them when necessary. The incurrent canals are lined by flat pinacocytes similar to htose lining the spongocoel.

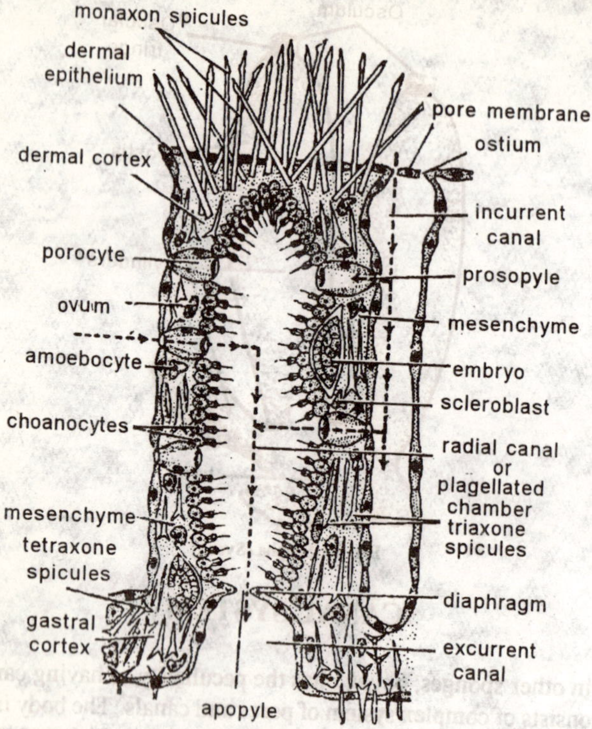


Fig. 3.2 A part of vertical section of Scypha showing histology and canal system.

3. *Radial canals.* The radial canals are relatively wide and are polygonal in cross-section. They are closed at their outer ends which lie some distance inside the polygonal surface elevations. Internally they open into the excurrent canals by apertures called *apopyles*. An apopyle is a large circular opening in the centre of a thin partition, the *diaphragm*, that stretches between the adjoining radial and excurrent canals. It can be contracted or dialted by myocytes that border it. The radial canals are lined by a layer of loosely arranged, flagellated, collared cells known as the *choanocytes*. The latter resemble the choanoflagellate Protozoa. A choanocyte is a somewhat oval or rounded cell that contains a nucleus and one or more vacuoles and bears at its inner end a long, whiplike, vibratile flagellum, partly surrounded by a delicate, transparent, contractile collar. The collar is formed by side-to-side fusion of 20-30 processes from the cell. The flagellum arises from a prominent organelle, the *centroblepharoplast,* situated within the cell. It is a combination of the centriole and the blepharoplast. It is connected by a thread, the *rhizoplast* with another organnelle, the *parabasal body,* lying on the nucleus.

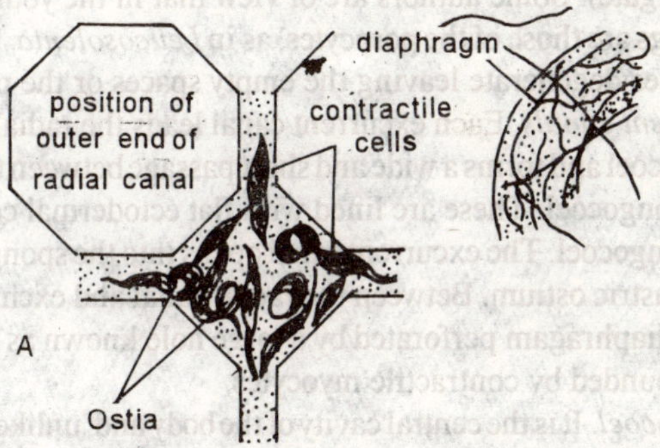


**Fig. 3.3 Apopyle lined by myocytes in *Scypha*.**

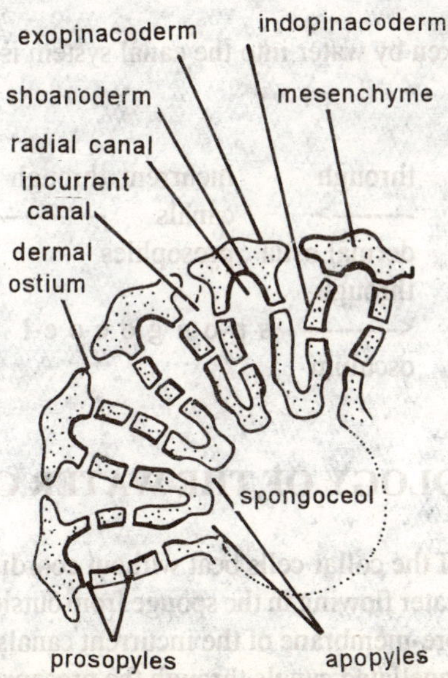


**Fig. 3.4. A portion of a T.S. of Scypha**

The incurrent canals communicate with the radial canal by short narrow channels called the *prosopyles.*

4. *Prosopyles.* The incurrent canals are communicated with the radial canals through intercellular spaces, called *prosopyles* (Gr., *pros,* near + *pyle,* gate). Some authors are of view that in the young *Sycon* these openings are those of the porocytes, as in *Leucosolenia,* but in adult the porocytes degenerate leaving the empty spaces or the prosopyles.
5. *Excurrent canals.* Each excurrent canal leads the radial canal into the spongocoel and forms a wide and short passage between the radial canal and spongocoel. These are lined with flat ectodermal cells like that of the spongocoel. The excurrent canal opens into the spongocoel through wide gastric ostium. Between the radial canal and excurrent canal lies a thin diaphragam perforated by a large hole known as *apopyle* which is surrounded by contractile myocytes.
6. *Spongocoel.* It is the central cavity of the body and, unlike *Leucosolenia,* is not lined by the flagellated collar cells or *choanocytes,* which have departed to line the radial canals, while the epidermal pinacocytes have immigrated into the spongocoel (Gr., *spongos,* sponge + *koilos,* hollow) to form its lining. This cavity is comparatively a narrow cavity.
7. *Osculum.* The spongocoel leads to outside through a terminal opening, the *osculum.* The oscula are provided with *sphincters* are lined by special contractile pinacocytes called myocytes (Gr., *Myos,* muscle + *kytos,* cell).

The course taken by water into the canal system is as under -

| | | | | |
|---|---|---|---|---|
| Water | through | incurrent | through | radial |
| from | --------> | canals | ------------> | canals |
| outside | dermal ostia | prosophles | | |
| | through | | | through |
| outside | <--------- | s p o n g o c o e l | | <--------- |
| | osculum | | | epopyles |

## PHYSIOLOGY OF THE WATER CURRENT

The flagella of the collar cells beat without coordination yet maintain a steady current of water flowing in the sponge from outside. The current enters the pores in the pore-membrane of the incurrent canals and fills the latter, it then passes into flagellated canals through the prosopyles. From here it goes into the excurrent canals by way of apopyles. Finally, the current is thrown into the spongocoel by the large gastric ostia from where it escapes through the osculum. The osculum is also surrounded by myocytes serving to form an *oscular sphincter*, which can close the osculum.

The water current flowing through the canal system brings with it food in the form of micro-organisms and oxygen. The food, chiefly unicellular algae, bacteria and organic microfragments, is collected by the collar cells just the same way as described for *Leucosolenia*. The outgoing current carries with it undigestible wastes and nitrogenous excreta.

## CELLULAR ORGANIZATION

Studies on structure and morphogenesis of *Scypha* have clearly revealed the presence of two types of cellular layers, the *pinacoderm* and the *choanoderm*. The former controls the interrelations between the mesenchyme and the external medium, while the later controls mainly the nutrition of the animal.

### 1. Pinacoderm

The pinacoderm comprises (i) the *exopinacoderm* (dermal epithelium) covering the entire body surface except the dermal ostia abd osculum and (ii) the *endopinacoderm* which includes the epithelial lining of the incurrent canals and spongocoel. The pinacoderm is composed of large, flattened, polygonal cells, the *pinacocytes* (Gr., *pinako,* plank + *kytos,* cell). In profile, these cells present a central bulging containing the nucleus. The margins of adjacent cells are closely cemented together. Pinacocytes are highly contractile and they can greatly increase or reduce the surface area of the sponge body.

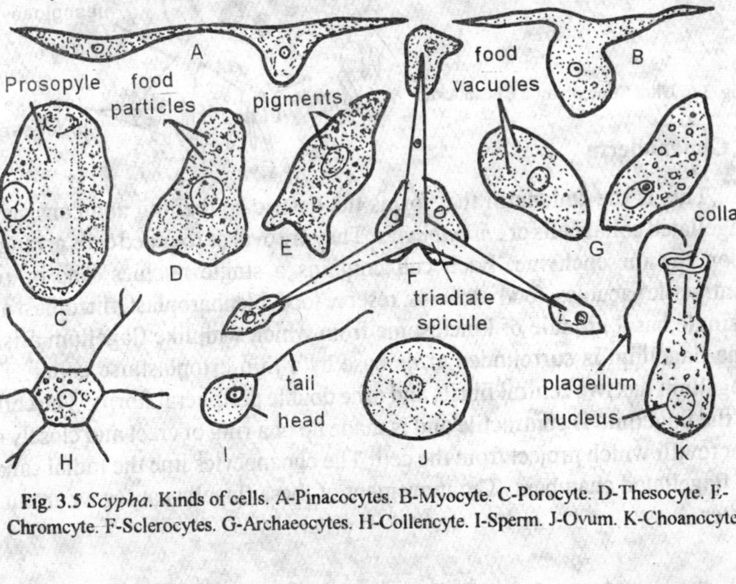


Fig. 3.5 *Scypha*. Kinds of cells. A-Pinacocytes. B-Myocyte. C-Porocyte. D-Thesocyte. E-Chromcyte. F-Sclerocytes. G-Archaeocytes. H-Collencyte. I-Sperm. J-Ovum. K-Choanocyte.

In the lining of the incurrent canals, some of the pinacocytes are modified to form tubular cells, the *porocytes,* connecting the incurrent canals with the radial canals through their intracelluar channels, the *prosopyles.* The porocytes are thin-walled cells, open at both ends and with the nucleus present in the peripheral cytoplasm. As has already been noted, the porocytes in the adult sponge are lacking altogether. Pinacocytes surrounding the osculum, outer or dermal ostia and inner ostia or apopyles are elongated and contractile and act as muscle cells, called the *myocytes.* They form the sphincters around them to regulate these openings.

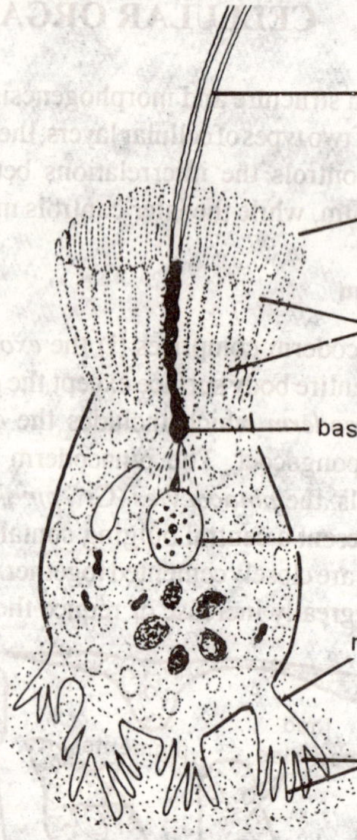


Fig. 3.6. Ultra Structure of a Choanocyte.

## 2. Choanoderm

This is the endoderm that forms the gastric epithelium and consists of flagellated collar cells or *choanocytes.* They are oval or rounded cells arranged upon the mesenchyme. Each cell contains a single nuclues, one or two contractile vacuoles, food vacuoles, reserve food, blepharoplast, rhizoplast and a single basal granule or kinetosome from which whiplike flagellum arises. The flagellum is surrounded at its base by a thin cytoplasmisc collar. The flagellum has two central fibrils and nine double peripheral fibrils. The collar of the flagellum is contractile and is made up of a ring of erect and closely set microvilli which project from the cell. The choanocytes line the radial canals or flagellated chambers. The movement of these flagella causes a current of water.

**3. Mesenchyme**

The mesenchyme which fills the space between the two body layers is composed of a thin gelatinous matrix which contains man kinds of amoeboid cells such as:

(i) *Collencytes* - Star-shaped cells which are sometimes united to form a network.

(ii) *Chromocytes* - Pigmented cells with lobose pseudopodia. They probably impart colour to the sponge.

(iii) *Thesocytes* - Cells with lobose pseudopodia and are laden with reserve food material.

(iv) *Scleroblasts* - These secrete the spicules.

(v) *Trophocytes* - They are loaded partly or completely with digested food. They serve to transfer food from one place to another.

(vi) *Phagocytes* - These cells collect food from the choanocytes through their pseudopodia and also engulf excreta and damaged tissues.

(vii) *Archaeocytes* - or undifferentiaed embryonic cells that give rise to sex cells.

(viii) *Lophocytes* - These are amoeboid cells with a tuft of fibrils at one or both ends and occurs in connection with the pinacocyte epithelium. Its function is unknown but it may secrete fibrillar bundler.

(ix) *Desmocytes* - They are long slender cells lying at the basement of the canals.

(x) *Gland cells* - They are derived from amoebocytes and secrete mucous.

(xi) *Germ cells* - Some of the amoebocytes (archaeocytes) differentiate into sex cells i.e. ova or sperms. However in some sponges, they are said to be modified choanocytes.

(xii) *Myocytes* - Around the apopyles, osculum, and ostia are also present a number or narrow elongated highly contractile cells, the *myocytes*.

The mesenchyme also contains the spicules which are monaxonic and triradiate in form. These serve to strengthen the body and hold it up right.

## SKELETON

The skeleton of *Sycon* consists of calcareous spicules having crystalline appearance which supports and protects the soft parts. They have an axis of organic material around which is deposited the inorganic substance i.e. calcium carbonate or hydrated silica. The spicules are secreted by sclerocyte cells and each ray of the spicule is said to be secreted by two such cells. The sclerocytes are in their turn produced by scleroblast cells. Spicules or *Sycon* show a variety of shapes and sizes and have a particular arrangement with regard to their canal system. The following are the main types of spicules:

(i) Large single rayed, needle-shaped *monaxon spicules* arranged in a circlet around the osculum, (ii) simple spear-like monaxon spicules projecting from dermal cortex opposite the outer ends of the radial canals, (iii) *Triaxon* or three rayed spicules are present along the radial canal with their one end pointing towards the distial ends of the canals, (iv) Four rayed, *tetraxon spicules* are present in the thick gastral cortex alongwith triaxon spicules and surround the spongcoel. Thus the long monaxons surround the osculum in the form of oscular fringe while the short monaxons lie parallel to the radial canals. The triradiates mostly lie along the radial canals with one ray facing externally. The tetraradiate spicules also lie in the same region.

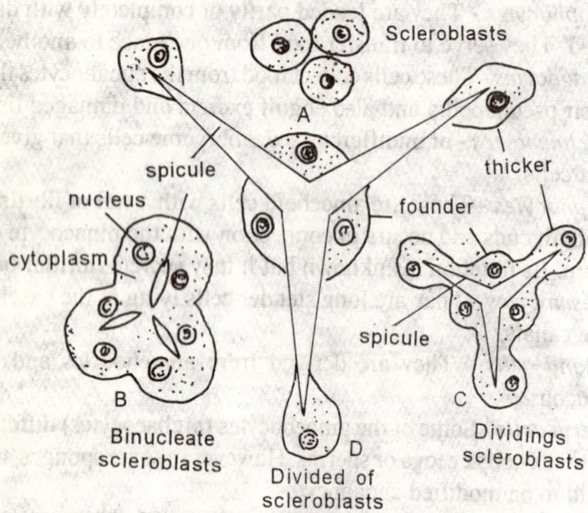


Fig. 3.7 Secretion of a triradiate spicule.

## NUTRITION

Nutritionally, *Scypha* is both holozoic and saporozoic. Its food consists of microscopic organisms, organic debris and *detritus.*

*Ingestion* takes place in an interesting way. The ceaseless, though uncoordinated, beating of the flagella of the choanocytes set up a water current that draws the food particles into the radial canals through the dermal ostia, incurrent canals and prosopyles. The flagellum undulates from the base to the tip and this creates a current that starts from the base of the flagellum and passes towards its tip. This in turn draws the food particles towards the collar, because it surrounds the base of the flagellum and here they adhere to its outer sticky surface. These particles are then engulfed by pseudopodia arising from the choanocyte at the base of its collar.

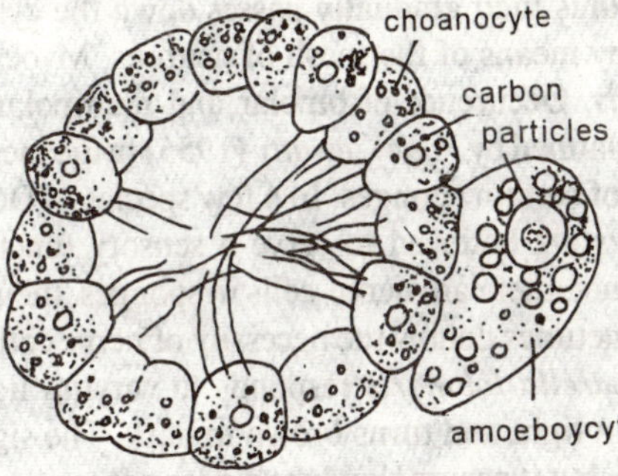


Fig. 3.8. Amoebocytes receiving carbon-grains from choanocytes.

*Digestion* in *Scypha* is completely intracellular and occurs in food vaculoes in much the same way as in Protozoa. Enzymes screted in the food-vacuoles by the surrounding cytoplasm gradually digest the food. Contents of the vacuoles are first acidic and later become alkaline. This corresponds to the acidic phase of digestion in the stomach and alkaline phase in the intestine of higher animals. All the 3 types of enzymes, *viz., proteases, carbohydrases* and *lipases,* have been isolated, but it is not certain whether they are secreted by *Scypha* itself or by bactria that about in it.

*Distribution* of digested food to other cells (pinacocytes, myocytes, porocytes, etc.) is affected in two ways: by cell to cell diffusion and by amoebocytes, called *trophocytes*, that take up wholly or partly digested food from the choanocytes and move about in the mesogloea, delivering nutrient to various cells.

*Storage* of food occurs in certain amoebocytes, called *thesocytes,* in fthe form of glycogen, fat, glycoprotein or lipoprotein.

Egestion is brought about by choanocytes and amoebocytes that throw off indigestible pieces into the canals, from where they are ultimately driven out through the osculum by the outgoing current of water.

## SENSITIVITY

Sponges do not have a nervous system and their responses are mainly local. The osculum is the most sesitive part, it collapses when the incoming current is polluted, or a stronge beam of light is focussed over it or, when the sponge is partly submerged or is touched by some object. It seems that under the influence of a stinulus mycocytes contract causing the closure of the osculum. The stimulus then gradually passes down the cylinder for a short distance probably by means of the mesh of neurons. Myocytes are probably *neuromuscular cells*. Occurence of bipolar and multipolar nerve-cells, the *neurons*, has been claimed by *P. de Cewatty* (1955) in the mesohyl beneath the dermal membrane of certain sponges. In a few species of Demospongiae and Calcarea *lophocytes* are believed to serve a sensory function (*W.C. Jones* 1962). Whether or not, there are nerve-cells in sponges, their responses are so simple that one sometimes doubts the necessity of nerve cells. For instance, a sponge named *Oscarella lobularis* responds to varying light intensities; it becomes deep red in light and turns brown in dark. The significance of this change is unknown, but it occurs slowly and does not seem to be controlled by nerve cells. Similarly many sponges secrete lot of slime when lifted out of water. It is perhaps a direct response of pinacocytes at the surface to a change in their environment. It does not call for a nervous system.

### Respiration

Respiration is aerobic and is accomplished externally by the general body surface or pinacocytes and internally by choanocytes and lining cells. Each cell respires for itself, oxygen entering the cell and carbon dioxide leaving the cell by simple diffusion. Amoebocytes distribute oxygen within the mesenchyme and carry away carbon dioxide. The entire process is again *intracellular*.

### Excretion

The nitrogenous wastes is ammonia for most part and in all probability it leaves the cell by diffusion. It has also been claimed some investigators that the nitrogenous wastes first are taken up by the amoebocytes which are then discharged into the spongocoel, from where they pass out in the water current through osculum.

### Reproduction

*Scypha* reproduces asexually as well as sexullay as given below :

#### 1. Asexual Reproduction

(i) *Budding* - It is a very common method of asexual reproduction. Masses of archalocytes migrate to the tips of spicules clumps that project from the surface in the form of bud. These bud either remain attached and grows in size, thus increasing the bulk of the sponge or it may finally breaks up and form new sponge.

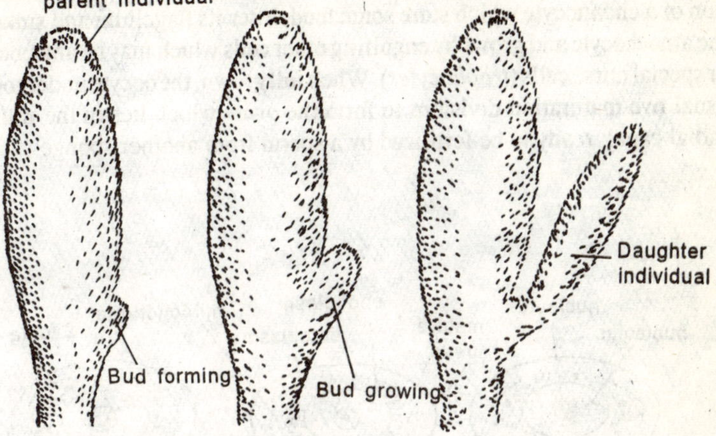


Fig 3.9. Bud formation in *Scypha*.

(ii) *Regeneration* - Regeneration is the power to replace the parts lost by injury. The power of regeneration has developmed to such an extent that if a sponge is thoroughly teared or squeezed in the sea water so as to separate the cells from one another, it is found that the cells slowly come together and gradually form a young sponge again.

**2. Sexual Rreprodution**

*Scypha* is a *monocious* sponge but due to *protogyny*, only cross-fertilization occurs. Special sex organs are lacking. Male sex cells, the *sperms* and the female sex cells, the *ova,* are found in the mesenchyme. They develop from undifferentiated amoebocytes, called *archaeocytes.*

(a) *Spermatogenesis.* The sperm mother cell or *spermatogonuim* is said to be an enlarged archaeocyte. However, it is supposed to be a modified choanocyte by *Gatenby,* who described the transformation of an entire flagellated chamber into spermatozoa. Soon the spermatogonium is surrounded by one or more flattened *cover cells* to form the *spermatocyst.* The cover cells are derived either by the division of the mother cell or from other amoebocytes. The spermatogonium undergoes two or three divisions to form the *spermatocytes* which give rise to *spermatozoa.* A mature sperm or *spermatozoon* consists of a rounded, nucleated head and vibratile tail, by the lashing movements of which it moves in water to reach other sponges.

(b) *Oogenesis.* The egg mother cell or *oocyte* is derived from a large archaeocyte with a distinct nucleus. This may sometimes arise by transformation of a choanocyte which store some food, loses its flagellum and sinks into the amoebocyte and grows by engulfing other cells which may be amoebocytes or special nurse cells (*trophocytes*). When full grown, the oocyte undergoes the usual two maturation divisions to form the *ovum* which lies in the wall of a radial canal, ready to be fertilized by a sperm from another sponge.

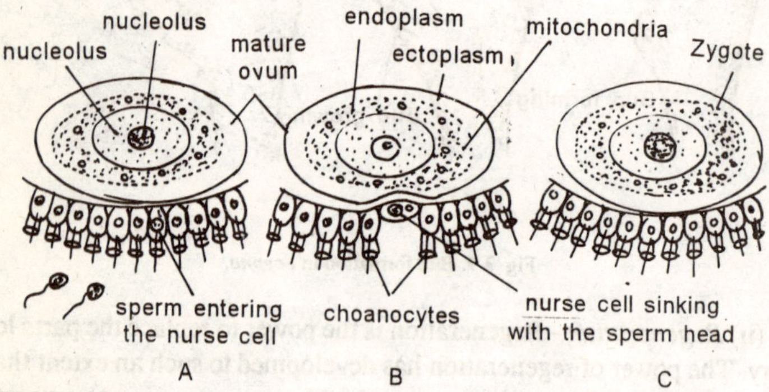


Fig. 3.10. Stages in fertilization *in situ* in a calcarean sponge.

Fertilization is internal. When the sperms mature, the spermatocysts rupture, discharging the sperms into the radial canals. From here, they are carried by the outgoing current of water to the exterior through the osculum. With the incoming water current, the sperms enter another sponge through this dermal ostia and reach the radial canals. Here, each sperm enters a wandering amoebocyte or a choanocyte. The invaded chaonocyte loses the flagellum, withdraws the collar, becomes amoeboid and migarates into the mesogloea. The amoebocytes and choanocytes penetrated by the sperms are known as the *sperm-transit cells*, as they serve to carry the sperms to the ova through the mesogloea. A sperm-transit cell adheres to a nearby ovum, which forms a conical depression to receive it. The sperm loses its tail and its head swells up. The swollen head gets surrounded by a capsule. The capsule with the sperm head enters the ovum. This completes the act of fertilization and results in the formation of a *zygote*. The sperm transit cell now departs. The zygote gets enclosed in a *brood capsule* formed by lining up of the neighbouring amoebocytes. Early development of the zygote occurs within the brood capsule of the parent.

## DEVELOPMENT

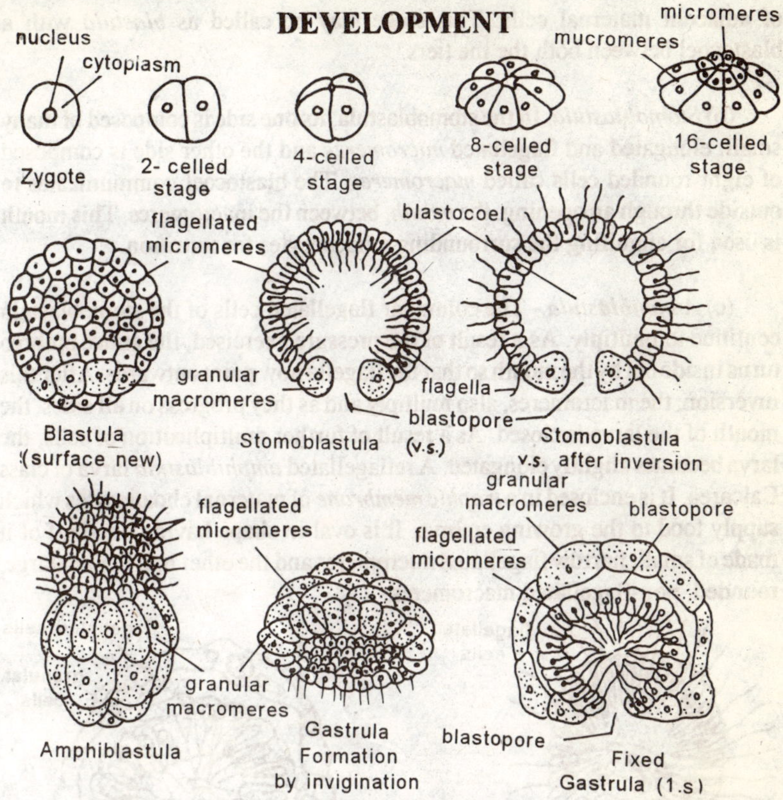


Fig. 3.11 Development of *Scypha*.

The development of *Sycon* includes the early embryonic period and larval period followed by metamorphosis.

### I. Embryonic Period

(a) *Cleavage.* The fertilized egg undergoes maturation and starts development within the mesenchyme of the parent. It undergoes equal and *holoblastic cleavage*. The first three cleavages are vertical forming 8 blastomeres. The horizontal division then divides the blastomeres unequally and of the 16 blastomeres produced, 8 are large cells or macromeres lying next to the parent choanoderm. They are designed to form future epidermis. Eight smaller cells or *micromeres* give rise to future choanocytes. At this stage, the embryo lies just beneath the maternal choanocyte layer is a flattened disc-shaped body. The micromeres undergo rapid mitosis to form several micromeres, elongate and each develops a flagellum on its inner end facing the blastocoel. The macromeres do not divide for sometime, become rounded and granular and an opeining is formed in their middle which helps in the ingestion of adjacent maternal cells. This stage may be called as *blastula* with a blastocoel between both the the tiers.

(b) *Stomoblastula*. In the stomoblastula, its one side is composed of many small, elongated and flagellated *micromeres* and the other side is composed of eight rounded cells called *macromeres*. The blastocoel communicates to outside through an opening, the *mouth*, between the macromeres. This mouth is used for engulfing the surrounding amoebocytes for nutrition.

(c) *Amphiblastula* - The columnar flagellated cells of the stomoblastula continue to multiply. As a result of the pressure exercised, the entire embryo turns inside out by the mouth so that the flagella now point outwards. After this inversion, the macromeres, also multiply and as they progress on all sides, the mouth of the larva is closed. As a result of further multiplication of cells, the larva becomes slightly elongated. A reflagellated *amphiblastula* larva of class Calcarea. It is enclosed in a *trophic membrane* of maternal choanocytes which supply food to the growing embryo. It is oval in shape having one-half of it made of small, narrow flagellated micromeres and the other half having large, rounded, non-flagellated macromeres.

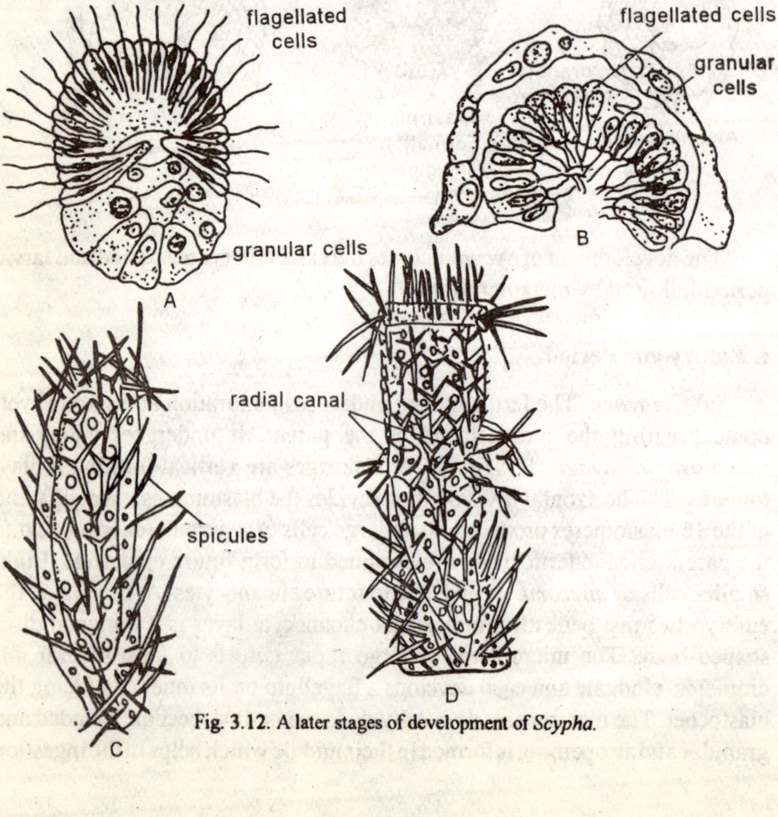


Fig. 3.12. A later stages of development of *Scypha*.

**2. Gastrula.**

Ultimately the amphiblastula settles down and undergoes gastrulation. The macromeres multiply more rapidly than the micromeres, so that the flagellated half of the larva is invaginaged into and overgrown by the granular non-flagellated half. The larva now becomes a typical double-walled *gastrula* with a *blastopore* at the invaginated side.

## METAMORPHOSIS

After swimming for some time, the flagellated half of the larva invaginates into or is overgrown by the large non-flagellated cells. With this process, the single-layered amphiblastula larva changes into a two layered *gastrula* with an opening, the *blastopore,* at the invaginated end. Outer layer of the gastrula consists of the non-flagellated granular cells and the inner of the flagellated cells. Gastrulation, thus in a way reverses the inversion that took place at the stomoblastula stage.

The gastrula soon fixes itself to some substratum by its blastoporal end and grows into a cylindrical structure. Its blastopore closes by the growth of the both the layers of cells along the substratum. At its free end develops a perforation which becomes the osculum. The outer non-flagellated cells flatten to form the dermal layer of pinacocytes. The inner flagellated cells for a time line the entire cavity which is the future spongocoel. Pores appear in the sides of the cylinder and start functioning as the dermal ostia. Through them water entres the spongocoel and leaves it through the osculum. At this stage the young sponge remebles adjust as coined sponge *Leucosolenia*. This is known as the *olynthus stage.* The wall of the cylinder gradually thickness due to the formation of mesoglea between the two layers of cells. Evagination from the dermal layer from the incurrent canals. The radial canals alone retain the flagellated cells, which by now become the choanocytes. In the meantime, spicules and various types of amoebocytes are differentiated in the mesoglea. The scleroblasts and myocytes are derived from the non-flagellated layer, while the archaeocytes and other amoebocytes develop from the flagellated layer of the larva. The mesenchyme cells are, thus, differentiated from the both the germ layers. With all this, a young *Scypha* comes into being.

# 7

# PHYLUM-COELENTERATA

Coelenterata was formerly regarded as Phylum divided into two subphyla : Subphylum Cnidaria including *Hydra, Obelia, Aurelia* Corals and Sea-anemones, and Subphylum Acnidaria including Ctenophora (*Pleurobranchia, Beroe* etc.) But during recent years *Hyman* (1980) and others regard the Coelenterata (Cnidaria) and Acnidaria as two distinct phyla on the basis of a number of dissimilarities between the two.

The coelenterates are regarded as primitive Metazoa in which the cells are organized to reach the tissue stage. In fact, all the basic types of tissues of higher animals are found in them viz. epithelial tissue for covering the body, muscular tissue for support, nervous tissue for conduction of stimuli and reproductive tissue for the reproduction.

**Derivation of Name**

All the coelenterates possess a single large cavity, the *gastrovascular cavity.* This cavity performs the function of digestion and distribution of digested food materials. It has one exit, the mouth. Term Coelenterata was first used by *Leuckart* (1847) for those animals in which the enteric carity form the body cavity. The name Coelenterata has been derived from two Greek words, *Koilos* = hollow, *enteron* = intestine. Thus the name of this phylum literally means hollow intestine but hollow bodies is considered more appropriate because actual intestine is not found in coelenterates.

**General Characters**

1. They are simplest Metazoa showing the cell-tissue grade of organization.
2. All the members of this phylum are aquatic, most of them are marine others are found in fresh water, none is parasitic.
3. They are either colonial or solitary. They are sedentary or free-swimming.
4. Symmetry is usually *radial* about an oral-aboral axis. in some forms, however, it is *biradial.*

5. Head and segmentation is absent.
6. They exhibit two different body forms, the *medusa* which is adopted for a pelagic existence and the *polyp*, which is adapted for an attachment benthic existence. Colonial forms have evolved in many polypoid forms.
7. One or more whorls of tentacles encircling the mouth at one end of the body. These are used for food-capturing, ingestion and defence.
8. The soft and delicate body may be supported by horny or calcareous exoskeleton or endoskeleton.
9. The body-wall is composed of two layers of cells (*diploblastic*), the outer epidermis and inner endodermis or gastrodermis, and an intervening *mesogloea*. The mesogloea may be thin or thick, cellular or acellular and is secreted by the epidermis and endodermis.
10. There are special stinging cells, the *cnidoblasts* which produce in them the peculiar *nematocysts* the organelles of offense and defence. About 17 types of such nematocysts have been identified.
11. Characteristic undifferentiated *interstitial cells* are found among the epidermial cells.
12. They are primitive in their lack of organs, their lack of fully differentiated epithelial and muscle cells. A muscular system consisting of contractile processes of epithelial cells.
13. Only one cavity lined with endodermal cells, is found in the body called *gastrovascular cavity*. It performs the function of digestion of food and distribution of digested food. It opens out through the mouth which opens into *stomodaeum*.
14. Mouth serves for ingestion of food as well as egestion of undigested food.
15. Digestion intracellular and intercellular.
16. There is no separate body cavity or coelom because coelom separate from alimentary canal, as found in higher Metazoa; is absent. Thus these are acoleomate.
17. There is no special organelles for respiration, excretion and circulation. Respiration and excretion through general body surface by diffusion.
18. Nervous system is primitive consisting of neurons. The neurons are usually arranged as a nerve net at the base of epidermal and gastrodermal layers, and the impulse transmission tends to be radiating.

19. Sense organs may be simple or complex. In some eye-spots or statscysts are found.
20. Asexual and sexual both types of reproduction is seen. Asexual reproduction by budding. In sexual reproduction gametes are formed in gonads. Gonads are simple without any duct.
21. Cleavage holoblastic. Development indirect with a ciliated free-swimming stereogastrula, called the *planula larva.*
22. The life-history usually exhibits *metagenesis* in which free swimming sexual generation (medusa) and sedentary asexual generation (polyp) alternate with each other.

**Classification**

Phylum-Coelenterata includes about 10,000 known species. About 5000 are known as fossils. They are grouped into three classes.

Class-I Hydrozoa.

Class-II Scyphozoa.

Class-III Anthozoa.

**Class-I-Hydrozoa**

(Gr. *Hydro* = Water, *zoon* = animal).

1. Mostly colonial and marine, a few solitary and fresh-water, sessile or free-living.
2. They are *tetramerous symmetrical* or *polymerous symmetrical.*
3. Members of this class are medusoid of polypoid or show both forms in their life-cycle.
4. Mesoglea is acellular.
5. Nematocysts occur only in the epidermis.
6. Gametes develop in the epidermis.
7. Hydromedusae are usually small and planktonic.
8. The asexual polypoid form arises from sexual medusoid form and the latter arises from the former, thus exhibiting metagenesis in the life-cycle of some hydrozoan e.g., *Obelia.*
9. Naked solitary species e.g., hydras probably stem from early polypoid forms which were not colonial.
10. Associated with colonial organization have been the evolution of a skeleton (support) and division of labour (Polymorphism).

11. Zygote undergoes complete cleavage and a hollow blastula is formed which changes into a solid *stereogastrula* whose ectodermal cells acquire cilia and then this stereogastrula changes into free-swimming *planula*. Planula settles down, attaches by its anterior end to a substratum and by budding forms the branched colonial polypoid stage.

The class Hydrozoa is divided into five orders :

***Order (i) Hydroida***

1. Solitary or colonial.
2. The polyps are more developed.
3. Medusoid stage present or absent.
4. Sense organs of medusae are exclusively ectodermal.

It is divided in following sub-orders :

*Suborder (a) Anthomedusae or Athec*

1. Skeletal covering, when present, does not surround hydrotheca. Freshwater or marine.
2. Polyps are naked.
3. Medusae are bud or bell-shaped. The gonads are situated on the manubrium. Statocysts absent.

Examples : *Hydra, Protohydra, Bougainvillea Tubularia, Hydractinia* etc.

*Suborder (b) Leptomedusae or thecate*

1. Usually marine.
2. Hydranth surrounded by a theca.
3. Medusa flattened, bowl shaped. Gonads are present on the radial canals. Eyespot and statocysts are present.

Examples : *Obelia, Sertularia, Plumularia, Aglaophania* etc.

*Suborder (c) Limnomedusae*

1. Polypoid stage without perisarcal skeleton.
2. Some only with medusoid stage

Example : *Graspeda custa.*

***Order (ii) Hydrocorallina***

1. Polypoid generation forming a colony which secretes a massive calcareous skeleton having minutes pores at the surface through which the polyps protuade.
2. Polyps dimorphic, nutritive gastrozooids and dactylozooids.

*Suborder (a) Milleporina*

1. Skeleton covered by only a thin epidermal layer.
2. Dectyiozooids are long, hollow and with tentacles.
3. Mature medusae lead an independent life.

   Example : *Millepora.*

*Suborder (b) Stylasterina*

1. Dactylozooids are small, solid and without tentacles.
2. Polyps are symmetrically arranged.
3. Medusa develop in special spore-sac and never free.

   Example : *Stylaster*

***Order (iii) Trachylina***

1. Medusoid forms only, polypoid forms are absent or reduced.
2. Medusa with a velum beneath the margin of the bell and the tentacles inserted above the margin.
3. Statocyst and tentaculocysts present.
4. The planula give rise to the adult through an *actinula larva.*

   Order Trachylina is divided into two suborders :

*Suborder (a) Trachymedusae*

1. Senseory tentacles in pits or vesicles.
2. Gonads on radial canals.
3. Margins of bell are smooth.

   Example : *Gonionemus. Petasus.*

*Suborder (b) Norcomedusae*

1. Sensory tentacles not enclosed.
2. Gonads on the floor of stomach.
3. Margins of bell scalloped by tentacle bases.

   Examples : *Cunina, Polycopa.*

*Order (iv) Siphonophora*

1. Free floating or fee swimming colonies, showing maximum polymorphism.
2. Individuals attached to a linear stem or a circular disc.
3. A *pneumatophore* or float filled with air is present at the top in some cases.

4. Oral tentacles absent.
5. Nematocysts large and powerful.
6. Medusae incomplete, never free.

This order is divided into two suborders :

*Suborder (a) Calycophora*

1. Pneumatophore is absent.
2. One or more swimming zooids are present in the upper part of the colony.

Examples : *Abyla, Diphyses, Praya.*

*Suborder (b) Physophorida*

1. A large pneumatophore is present at the upper end of the colony.
2. Zooids are polypoid and medusoid.

Examples : *Physalia, Halistemma.*

***Order (V) Chondrophora***

1. Pelagic, polymorphic, polypoid colony with a chitinous, circular gas-filled float.
2. A central gastrozooid surrounded by many gonozooids producing medusae.
3. Dactylozooid in a marginal circlet.

Examples : *Porpita, Velella.*

**Class-II Scyphozoa**

1. Class Scyphozoa include large jelly fishes.
2. Polyp satge is usually reduced or absent.
3. Medusae are large, umbrella-shaped without velum. Perisarc is absent.
4. The sense or organs are in the form of hollow *tentaculocysts* having *statolith.*
5. The gastral tentacles are endodermal.
6. Mesogloea enlarged and usually cellular.
7. The gonads are endodermal and release the gametes in stomach.
8. Stomodaeum is absent in the gastro-vascular system but gastric tentacles are present and the cavity is divided into interradial pockets by four ridges or speta.
9. Alternation of generation is seen in some.

10. They are exclusivelymarine.

The class is divided into 5 orders.

***Order (i) Lucernarida***

1. Mostly found in cold littoral waters.
2. Sessile and sedentary, attached to the substratum by the aboral stalk.
3. Body is goblet or trumpet shaped.
4. Mouth is four cornered with small oral lobes and a short manubrium.
5. Tentacles are present in some, they are perradial and interradial.
6. The umbrella is cup-shaped.
7. Gonads are long band-shaped lying on faces of septa.
8. Larva is planula with cilia.

Examples : *Lucernaria, Haliclystus.*

***Order (ii) Coronatae***

1. Free swimming forms living in deep sea.
2. Body conical or dome-shaped divided into an upper cone and lower crown by a coronary groove.
3. Solid tentacles are perradial and adradial, 4 interradial *rhopalina* present.
4. Four to sixteen tentaculocysts.

Examples : *Pericolpa, Atolla.*

***Order (iii) Cubomedusae***

1. Free-swimming found in warm and shallow waters.
2. The umbrella is four-sided and cup-shaped.
3. Four hollow inter-radial tentacles and four perradial tentaculocysts are present. Each tentaculocyst with one or more ocelli and a lithocyst.
4. A true velum is absent.
5. Gonads are leaf-like and there is no alternation of generation in life cycle.

Examples : *Chiropsalmus, Tamoya.*

***Order (iv) Semaeostomae***

1. Common free-swimming animals found all over the world.
2. The umbrella is disc-like.

3. Mouth surrounded by four oral arms.
4. Eight or more tentaculocysts are present.
5. Gastric pouches and septa are absent.

   Examples : *Aurelia, Cyanea.*

***Order (v) Rhizostomae***

1. Free-swimming forms found is shallow waters of tropical and subtropical oceans.
2. The mouth is obliterated by the growth across it of the oral arms.
3. The stomach is continued into the canals opening by funnel shaped apertures on the edges of the arms.
4. Umbrella is saucer or bowl-shaped.
5. Marginal tentacles absent.
6. Tentaculocysts are 8 or more in number.

   Examples : *Rhizostoma, Cassiopeia*

**Class-III Anthozoa or Actinozoa**

1. Exist only in the polyp form, no medusa stage.
2. Differ from hydrozoan and scyphozoan polyp in possessing a stomodaeum, paired mesenteries (their free ends bear coiled mesenteric filaments like the gastric filaments of scyphozoan but are partially ectodermal in origin).
3. Body wall consists of ectoderm and endoderm separated by a strong mesogloea having fibres and cells.
4. Stomodaeum consists of the same layers reversed i.e., its lining membrane is ectodermal. The mesenteries are formed by a double layer of endoderm with a supporting plate of mesogloea.
5. More complex nematocysts than in Hydrozoa and Scyphozoa are found in the tentacles, body-wall, stomodaeum and mesenteric filaments.
6. Muscular system is well-developed containing both ectodermal and endodermal fibres and endodermal muscule processes.
7. Nervous system is a typical nerve net.
8. Gonads develop in the mesenteries, sex cells are located in the endoderm; sperms and ova are discharged into the coelenteron.
9. The zygote develops into a planula which after swimming freely for sometime settles down and metamorphoses into the adult form.

10. Except in one doubtful instance there is no alternation of generations.

The class-Anthozoa is divided into 2 sub-classes :

**Subclass-1. Hexacorallina or Zoantharia**

1. Solitary or colonial marine forms.
2. Tentacles and mesenteries very numerous, arranged in multiples of six.
3. Tentacles are simple, unbranched, hollow cones.
4. Two siphonoglyphs are present and two pairs of directive mesenteries; the remaining mesenteries are generally arranged in couples with the longitudinal muscles of each couple facing one another.

**Order (i) Actiniaria**

1. The animals are commonly called sea-anemones. They are solitary or colonial, brightly coloured.
2. Tentacles and mesenteries are numerous.
3. Skeleton is absent.
4. One or more siphonoglyphs.

Examples : *Metridium, Adamsia, Edwardsia.*

**Order (ii) Madreporaria**

1. The animals are usually colonial.
2. Skeleton is external and calcareous.
3. Siphonoglyph is absent.
4. Polyps are very small.
5. These are stony corals.

Examples : *Corallium, Madrepora, Fungia, Meandra, Favia.*

**Order (iii) Zoanthidea**

1. Mostly colonial and some are solitary forms.
2. The tentacles are unbranched.
3. The skeleton is usually of a calcareous nature but in few cases there is a horny axial skeleton.
4. Polyps are small and often united by basal stolons.
5. Mesentries paired. Each pair with a complete and an incomplete mesentery.
6. They have single siphonoglyph.

Example : *Zoanthus, Polythoa.*

***Order (iv) Antipatharia***

1. These are compound tree-like animals.
2. The tentacles and mesenteries are comparatively few (6-24) in number.
3. The skeleton is present in the form of a branched chitinoid axis.
4. Two siphonoglyphs.

   Example : *Antipathes.*

***Order (v) Cerianthatia***

1. The individuals are long and solitary.
2. Skeleton is absent.
3. The mesenteries are incomplete.
4. Numerous, simple tentacles are arranged in two wholrs oral and marginal.
5. There is a single dorsal siphonoglyph.

   Example : *Cerianthus, Pachycerianthus.*

***Subclass-2 Octocorallina***

1. Colonial marine forms.
2. Only polyps no medusae.
3. Tentacles and mesenteries always eight in number.
4. The tentacles pinnate; i.e., produced into symmetrical branchlets.
5. Never more than one siphonoglyph, which is ventral in position, i.e., faces the proximal end of the colony.
6. The arrangement of mesenteries is not always in couples (or pairs) and all their longitudinal muscles are directed ventrally i.e., towards the same side as the siphonoglyph.

   This subclass is divided into following orders :

***Order (i) Stolonifera***

1. Found in shallow waters.
2. Polyps arising form a creeping stolon.
3. Skeleton either absent or of calcareous tubes or separate calcareous spicules.
4. Found on coral-reefs in Old and New World.

   Examples : *Tubipora, Clavularia.*

***Order (ii) Telestacea***

1. Colony formed of simple or branched stems arising from a creeping base.
2. Skeleton of calcareous spicules.

Example : *Telesto.*

***Order (iii) Alcyonacea***

1. The skeleton usually consists of calcareous spicules or the spicules become aggregated so as to form a coherent skeleton.
2. Some forms may be dimorphic.
3. Body is branched with a central axis.
4. Polyps embedded in fleshy coenochyme.

Examples : *Alcyonium, Gersemia.*

***Order (iv) Coenothecalia***

1. Blue or brown corals found on coral reefs in the Indo-Pacific waters.
2. Skeleton massive, calcareous and blue-green from iron salts.

Example : *Heliopora.*

*Order (v) Gorgonacea*

1. The individuals are compound tree-like.
2. Skeleton is calcareous or horny.
3. Spicules are present in mesogloea.
4. Siphonoglyph is absent.
5. A central skeletal axis composed of a horn-like substance, the *gorgonin.*

Examples : *Gorgonia, Corallium.*

*Order (vi) Pennatulacea*

1. The colony is usually elongated.
2. One end of the colony remains embeded in the mud at the sea bottom while other end bears the polyps.
3. The rachis is the axial polyp bearing numerous dimorphic polyps laterally.
4. The skeleton is calcareous or horny.

Examples : *Pennatula, Renilla, Pteroeides.*

## SERTULARIA

| | | |
|---|---|---|
| Phylum | - | Coelenterata |
| Class | - | Hydrozoa |
| Order | - | Hydroida |
| Sub order | - | Leptomedusae |
| Genus | - | *Sertularia* |

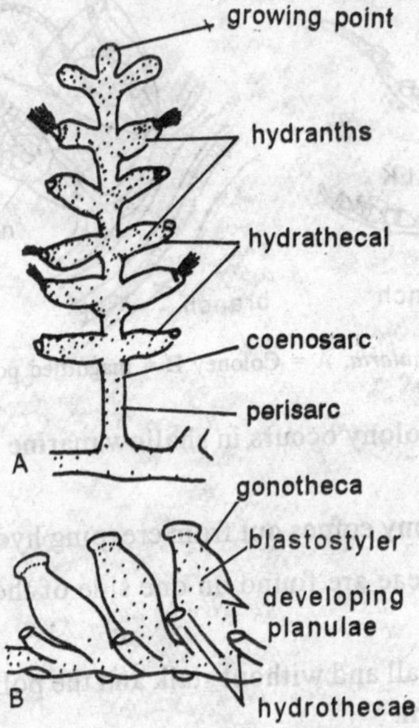


Fig. 7.1 *Sertularia* A = Branch with hydranths B = Branch with gonandis

1. It is a small branching colony occurs in shallow marine water on submerged objects or rocks.
2. Hydrotheca sessile i.e., not stalked, with *opercula* in pair along the stem being exactly opposite to each other.
3. Hydrothecae are large, polyps retract within themselves.
4. *Gonangia* are of simple form, much larger than hydrothecae, a few in a colony and present only in a certain period of three years.
5. Blastostyles produce *planulae* and no free medusae.

## PLUMULARIA

The classification is the same as that of *Sertularia.*

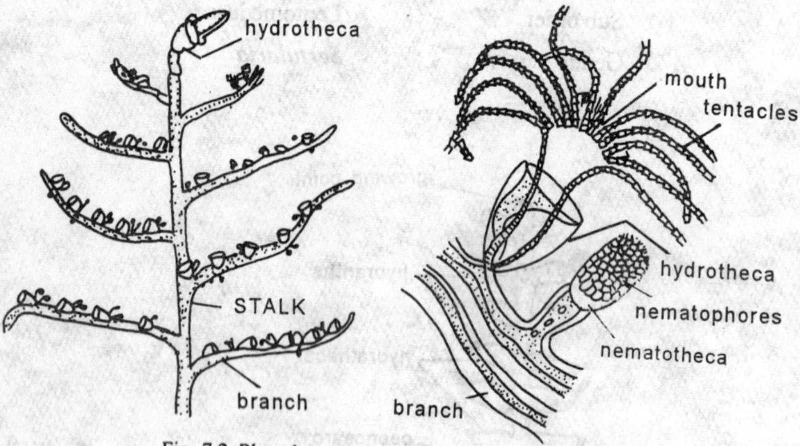


Fig. 7.2 *Plumularia.* A = Coloney B = magnified polpy.

1. It is a feather-like colony occurs in shallow marine water attached to some substratum.
2. The plume-like colony comes out from creeping hydrorhiza.
3. A series of hydrotheae are found on one side of the branches of the colony.
4. Hydrothecae are small and without stalk and the polyps can be partly retracted within them.
5. Special mouthless polyps known as *nematophore* are found which have long amoeboid projections.
6. The gonangia are long and superficially resembling the spadix inflorescence.

## CAMPANULARIA

| | | |
|---|---|---|
| Phylum | - | Coelenterata |
| Class | - | Hydrozoa |
| Order | - | Hydroida |
| Genus | - | *Campanularia* |

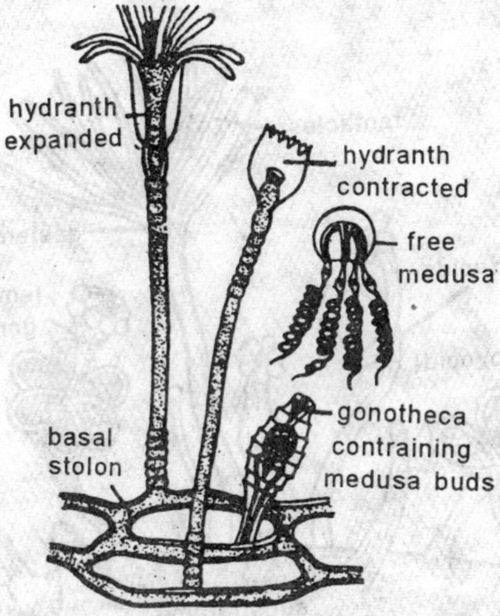


Fig. 7.3 *Campanularia*

1. It is a simple branched colony found attached to submerged marine objects in shallow water.
2. Hydrothecae, enclosing polyps, are bell-shaped usually stalked, without operculum and with or without marginal teeth.
3. Annulated stem at the base of the branch.
4. *Blastostyle* produces *planulae* and not medusae.
5. *Gonangium* long and slender.
6. Hypostome of hydranth or polyp trumpet-shaped surrounded by tentacles.

## HYDRACTINIA

| | | |
|---|---|---|
| Phylum | - | Coelenterata |
| Class | - | Hydrozoa |
| Order | - | Hydroida |
| Suborder | - | Anthomedusae |
| Genus | - | *Hydractinia* |

1. It is a small, marine hydroid found in shallow water attached to stones, sea-weeds and shells.
2. It is commonly found on Atlantic coast, U.S.A., Europe and also found in Sanjuan Island.

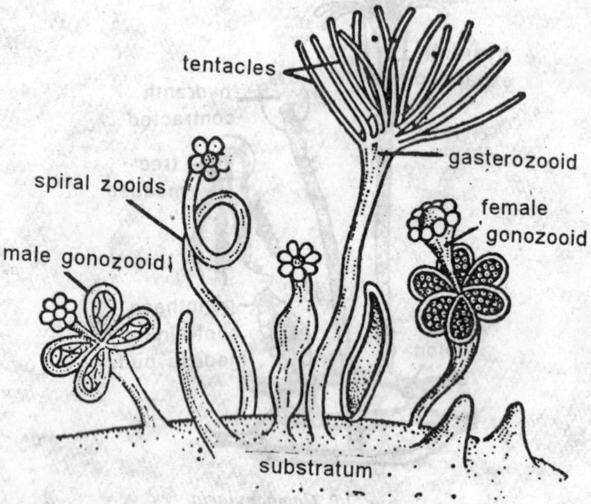


Fig. 7.4 *Hydractinia*

3. Zooids are of four types *Gastrozooid* (feeding polyp), *gonozooid* (reproductive zooid), *dactylozooid* (protective zooid) and *tentaculozooid* (sensory polyp).
4. The hydranths are simply mouthless and are called as dactylozooids.
5. The dectylozooids have very short tentacles abundantly supplied with nematocysts.
6. The dactylozooids are capable of very active movements.
7. With the help of massive coenosarc consisting of a number of branches, *Hydractinia* form a firm brownish crust on the surface of dead gatropod shells inhabited by hermit-crabs.
8. The association between *Hydractinia* and hermit carbs is known as *commensalism.*

## CERATELLA

Classification is the same as that of *Hydractinia.*

1. It is a colonial, sedentary hydroid coelenterate.
2. It is tree-like highly branched animal.
3. It consists of a branching axis composed of many intertwisting and anastomosing tubes.

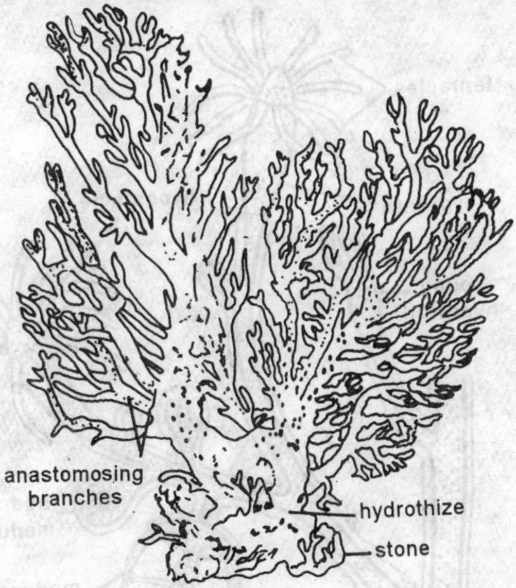


Fig. 7.5 *Ceratella*

4. Hydrotheca covers polyp while gonotheca covers medusa.
5. The zooids have scattered capitate tentacles.
6. The medusae bear gonads on the manubrium and devoid of lithocytes.
7. Eye spots are present.

## BOUGAINVILLEA

The classification is the same as that of *Hydractinia*.

1. It is a plant-like hydroid found in sea waters.
2. Two types of zooids are present the *polyp* or *hydranth* and *medusa*.
3. Polyps are vegetative responsible for nutrition and food catching, while medusa is reproductive in function.
4. Hydrotheca and gonotheca coverings are absent on polyps and medusa respectively.
5. Polyp is beset with a single circlet of filiform tentacles which surround the hypostome.
6. Medusa bears four radial canals and a wide velum.
7. Gonads are developed on the manubrium of medusa.

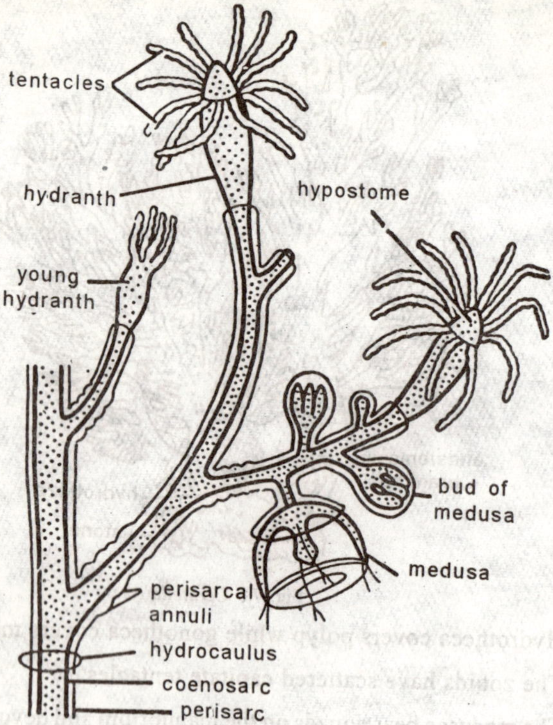


Fig. 7.6 *Bougainvillea*

## MILLEPORA

| | | |
|---|---|---|
| Phylum | - | Coelenerata |
| Class | - | Hydrozoa |
| Order | - | Hydrocorallina |
| Suborder | - | Milliporina |
| Genus | - | *Millepora* |

1. Marine colonial form found in tropical seas and associated with other corals.
2. Ectoderm secretes a calcareous yellowish or white skeleton of perisarc.
3. The dried coloney is perforated by numerous pores. These are of two different sizes; large *gastropores* and small *dactylopores* which surround the gastropores.
4. In living specimens two types of zooids- *gastrozoods* and *dactylozooids* which protude out of these pores.

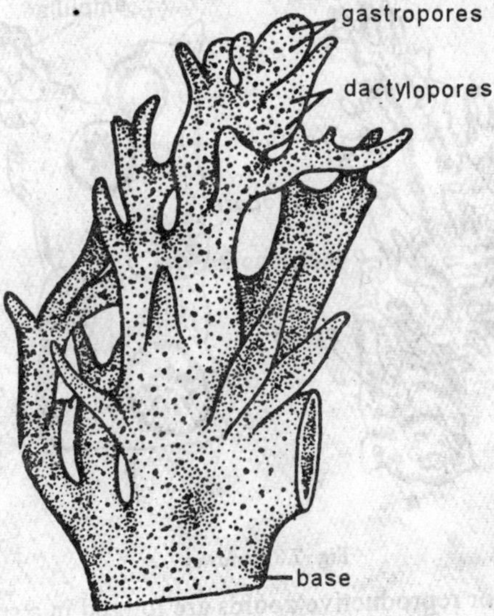


Fig. 7.7 *Millepora*

5. Gastrozooids are feeding zooids, having 4-5 short knobbed tentacles. The dactylozooids are protective zooids with capitate tentacles having nematocysts.
6. Pores lead into canals which form network in coenosarc.
7. Medusae with 4 or 5 rudimentary tentacles. These are free, simple and originate from coenosarc.

## STYLASTER

| | | |
|---|---|---|
| Phylum | - | Coelenterata |
| Class | - | Hydrozoa |
| Order | - | Hydrocorallina |
| Suborder | - | Stylasterina |
| Genus | - | *Stylaster.* |

1. It is highly branched, tree-like, deep pink colony found in tropical and subtropical seas.
2. Numerous upright branches of calcium carbonate are present.
3. On the branches are cup-shaped projections formed of serveral zooids.
4. *Gastrozooids* situated in the centre, while *dactylozooids* are in the periphery.

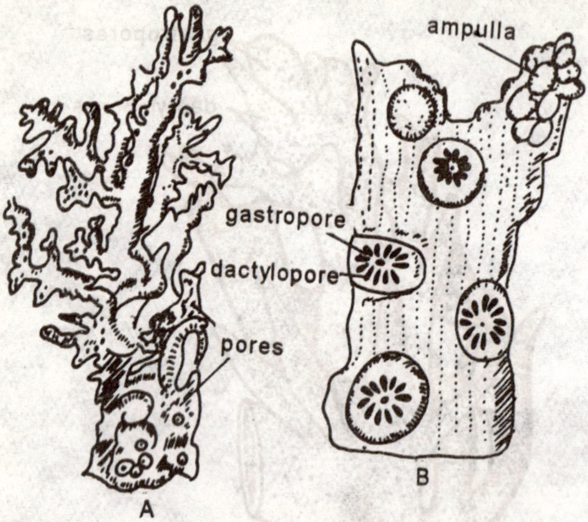


Fig. 7.8 *Stylaster*

5. *Gonophores* or reproductive zooids are lodged in a special chamber or *ampulla* of the coral.
6. Form the horizontal partition at the bottom of each cup projects a calcareous projection callled *style*, hence the generic name *Stylaster* is given.
7. The young sets free in the planula stage which later on metamorphoses into a new colony.

**GONIONEMUS**

| | | |
|---|---|---|
| Phylum | - | Coelenterata |
| Class | - | Hydrozoa |
| Order | - | Trachylina |
| Suborder | - | Trachymedusae |
| Genus | - | *Gonionemus.* |

1. It is a medusa found in bottom of sea water. It is cosmopolitan in distribution.
2. It is like the medusa of *Obelia* and generally bell-like or saucer shaped.
3. The convex outer surface of the bell is known as *exumbrella* while the concave inner surface is *subumbrella*.
4. From the centre of the concave or subumbrella surface hangs down a

short, hollow and quardrangular process known as the *manubrium* bearing four brief frilled oral lobes surrounding the mouth.

5. The rim of the margin of the medusa bears numerous 16 to 18 highly contractile tentacles provided with nematocysts and adhesive pads.
6. The adhesive pads are generally situated at the bending on each tentacle and helps in anchoring to marine plants at rest.
7. The gonads are situated on four radial canals and medusae are unisexual.

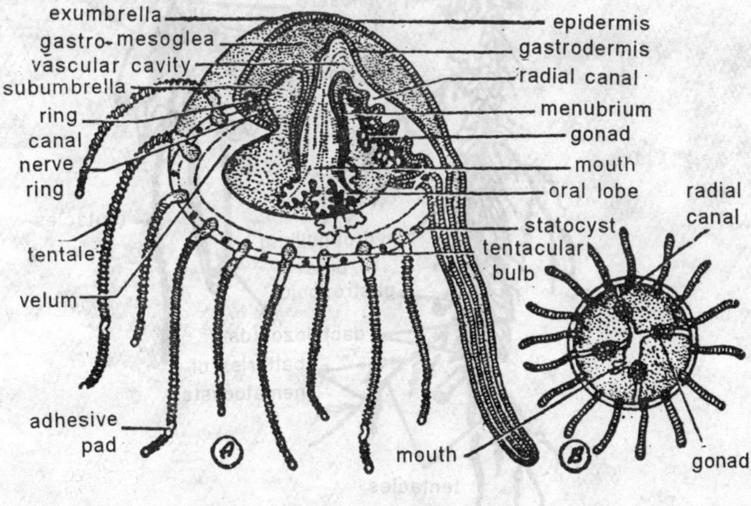


Fig. 7.9 *Goneonemus* A=Section to show internal structure B=Sub umbrellar side

8. Planula develops into small polyp which has mouth, tentacles and is known as *haleremite.*
9. The haleremite reproduces asexually and gives rise to planula-like buds which are nonciliated and are called *frustules.*
10. The frustules develop into new polyps which from external side bud-off gonophores. These naked gonophores give rise to medusae.

## HALISTEMMA

| | | |
|---|---|---|
| Phylum | - | Coelenterata |
| Class | - | Hydrozoa |
| Order | - | Siphonophora |
| Suborder | - | Physophorida |
| Genus | - | *Halistemma* |

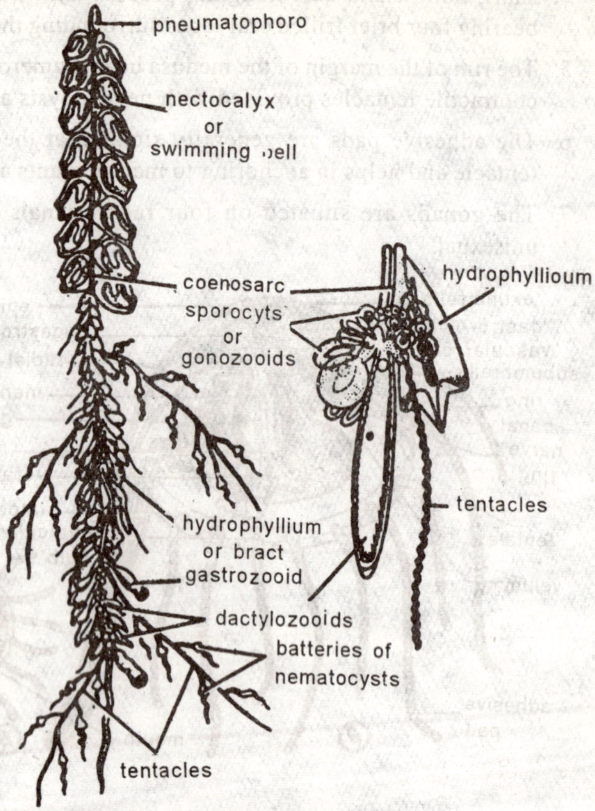


Fig. 7.10 *Halistemms* A=Coloney, B = Cormidium

1. It is a polymorphic colony found floating on the surface of sea in Mediterranean seas.
2. Uppermost end of the stem possesses an ovoid bubble-like body, containing air, known as *float or pneumatophore*.
3. Below the pneumatophore, there is a series of unsymmetrical medusae, called *nectocalyces* each having a deep bell-like body with a velum and without manubrium.
4. *Swimming bells* or *nectocalyces* contract, drawing water into their cavities and pumping it out to propel the whole organism through water.
5. Stem underneath divides into nodes and internodes.
6. Below certain nodes arise *polyps* with a long, branched tentacle and bearing batteries of *stinging capsule.*

7. In the remaining nodes, *dactylozooids* or *feelers* take the place of polyps.
8. Below these dactylozooids are *sporosacs*, some are males and other females.
9. Delicate leaf-like, transparent bodies, the bracts or *hydrophylia* spring from internodes.

## PHYSALIA

| | | |
|---|---|---|
| Phylum | - | Coelenterata |
| Class | - | Hydrozoa |
| Order | - | Siphonophora |
| Suborder | - | Physophorida. |
| Genus | - | *Physalia* |

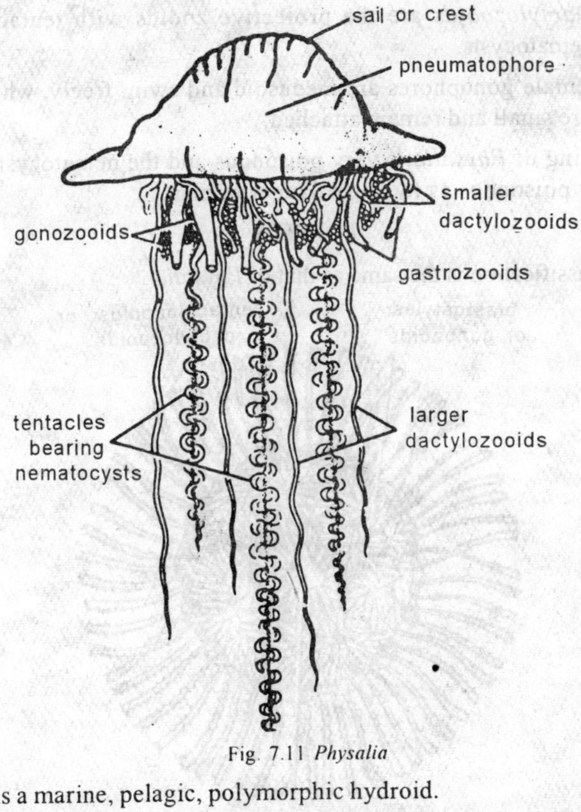


Fig. 7.11 *Physalia*

1. It is a marine, pelagic, polymorphic hydroid.
2. It is found in the gulf stream from Florida to Vineyard Sound and

occassionally to the Bay of Funday.

3. It is commonly called 'portuguese man-of-war', because it has a large, brilliant coloured pneumatophore or float which is like the cap of Great Napolean.
4. The dorsal side of float is called crest or sail.
5. The float is filled with a gas. The composition of the gas is 85-91% nitrogen, 1.5% argon and 7.5-13.5% oxygen.
6. The float or pneumetophore has a pore called *pneumatopore.*
7. The swimming bells or nectocalyces are absent.
8. Hanging from the underside of pneumatophore are three types of zooids :
   (i) *Gastrozooids* are nutrative zooids without tentacles.
   (ii) *Blastostyles* are reproductive zooids containing served medusae.
   (iii) *Dactylozooids* are the protective zooids with tentacles and nematocysts.
9. The female gonophores are medusoid and swim freely, while male ones are small and remain attached.
10. The sting of *Physalia* is very poisonous and the nematocysts are so highly poisonous as to cause danger to man.

**PORPITA**

The classification is the same as that of *Physalia.*

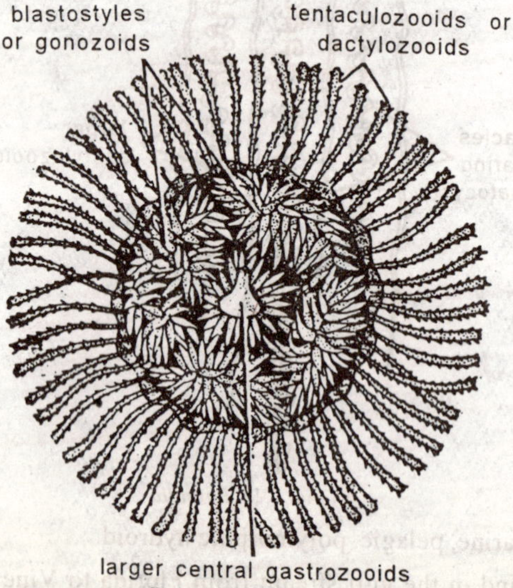


Fig. 7.12 *Porpita*

1. *Porpita* is a marine coelenterate found usually on South Atlantic Coast and occassionally near U.K. Coast.
2. The float is very large, circular and disc like (like the float of *Physalia*).
3. The colony consists of disc-like body enclosing a chambered chitinoid shell.
4. It exhibits the remarkable phenomenon of *Polymorphism*.
5. Long dactylozooids or tentacles are present around the disc.
6. The under-surface is beset with gonozooids or blastostyles provided with mouth bearing the medusae.
7. The single gastrozooid is present in the centre of disc on ventral side as in *Velella*.
8. Crest is wanting.

## VELELLA

The classification is the same as that of *Physalia*.

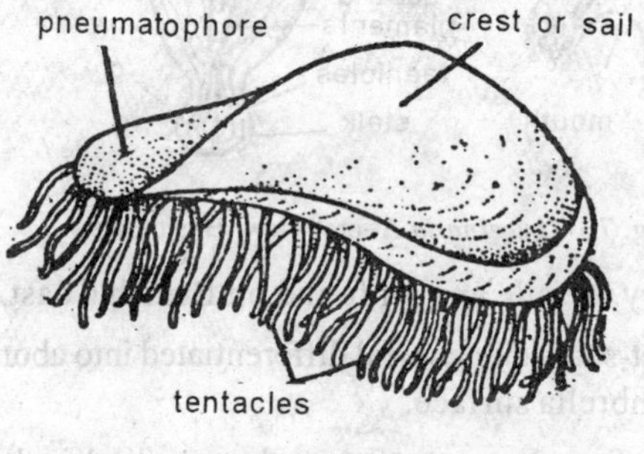


Fig. 7.13 *Velella*

1. It is a beautifully coloured pelagic form found in warm water. It is common is pacific coast, South Atlantic Coast.
2. Body rhomboidal with a medusa-like appearence.
3. Pneumatophore dorsal, in the form of a chambered chitinous disc and bears a vertical crest like ridges.
4. Hanging from the centre of the under surface is a single large gastrozooid which is surrounded by numerous gonozooids which give off medusa-buds.
5. The gonozooids are devoid of tentacles.
6. Dactylozooids hang down from the rim of the disc forming a circular fringe.

7. It is incapable of sinking down by altering the gas contents of the float.

## LUCERNARIA

| | | |
|---|---|---|
| Phylum | - | Coelenterata |
| Class | - | Scyphozoa |
| Order | - | Lucernarida |
| Genus | - | *Lucernaria* |

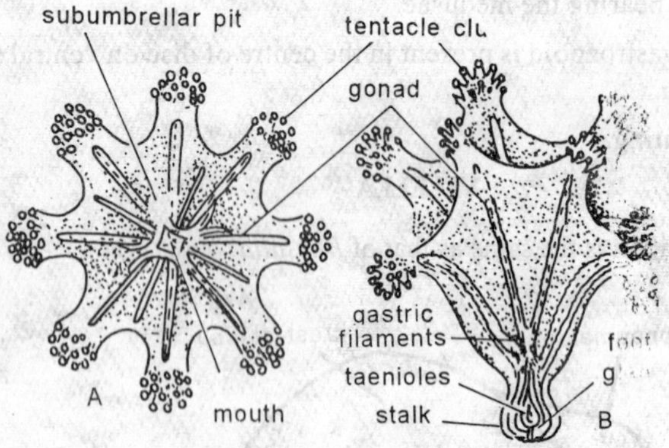


Fig. 7.14 *Lucernaria* A=oral view, B=side view

1. It is a sedentary, sessile animal found on British Coast.
2. Body is trumpet-shaped or conical differentiated into aboral exumbrella and oral subumbrella surface.
3. Exumbrella surface drawn out into a short, cylindrical stalk used for attachment to sea-weeds.
4. Margin of umbrella drawn into 8 short and hollow adhesive lobes or adradial arms.
5. Each arm bears a cluster of knobbed adhesive tentacles.
6. Mouth is cruciform (four cornered) with small oral lobes and a short manubrium.
7. Gastrovascular system consists of central stomach and four perradial pouches divided by four interradial septa.
8. Gastric filaments numerous.
9. Gonads are band-like borne on septa.
10. Marginal adhesive gastric pits and velum are absent.

## CHARYBDEA

| | | |
|---|---|---|
| Phylum | - | Coelenterata |
| Class | - | Scyphozoa |
| Order | - | Cubomedusae |
| Genus | - | *Charybdea* |

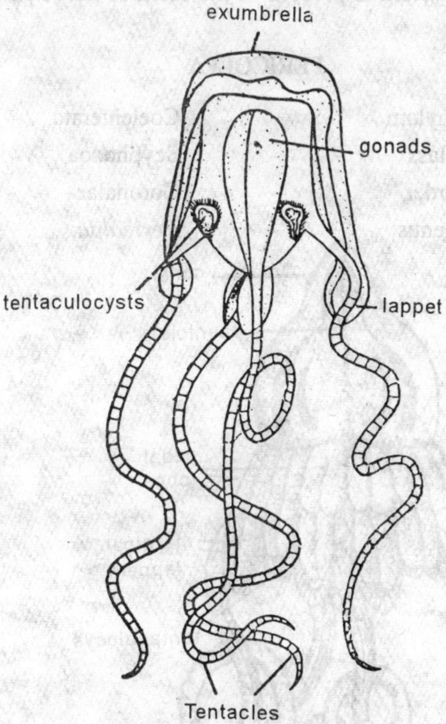


Fig. 7.15 *Charybdea*

1. It is a marine animal. It is commonly found in the warm shallow waters of tropical and subtropical region.
2. It is an active swimmer and swims through the water appearing beautiful.
3. These resemble a deep bell with somewhat flattened top and square in transverse section.
4. The margin of the umbrella bears 4 tentacles and 4 tentaculocysts.
5. The tentacles are inter-radial and tentaculocysts are perradial.

6. The tentaculocysts are set in deep marginal notches and the tentacles arise from gelatinous lobes.
7. The tentaculocysts are very complex, each bearing a lithocysts and several eye-spots.
8. The margin of the umbrella is produced into a false velum like the velarium of *Aurelia*.
9. The nervous system is present in the form of nerve ring around the margin of bell.

**PERICOLPA**

| | | |
|---|---|---|
| Phylum | - | Coelenterata |
| Class | - | Scyphozoa |
| Order | - | Coronatae |
| Genus | - | *Pericolpa* |

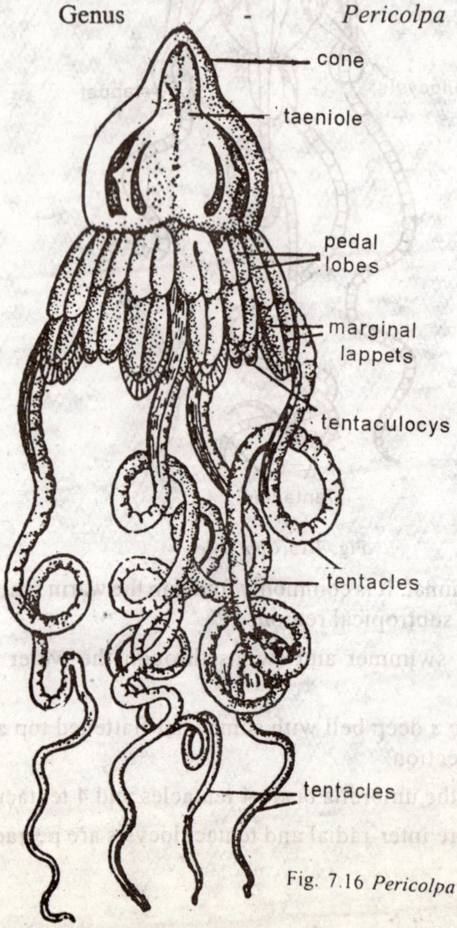


Fig. 7.16 *Pericolpa*

1. It is a beautiful marine solitary medusoid animal.
2. It is cosmopolitan in distribution, but abundent found in Greenland.
3. Umbrella is divided into apical cone and marignal crown by a furrow.
4. Marginal crown is divided into a series of pedal lobes and series of marginal lappets by a second horizontal furrow.
5. Marginal lappets and pedal lobes are present in same radii.
6. Tentaculocysts are present on four interradial pedal lobes.
7. Large mouth opens into the stomach by manubrium.

## CYANEA

| | | |
|---|---|---|
| Phylum | - | Coelenterata |
| Class | - | Scyphozoa |
| Order | - | Samaeostomae |
| Genus | - | *Cyanea.* |

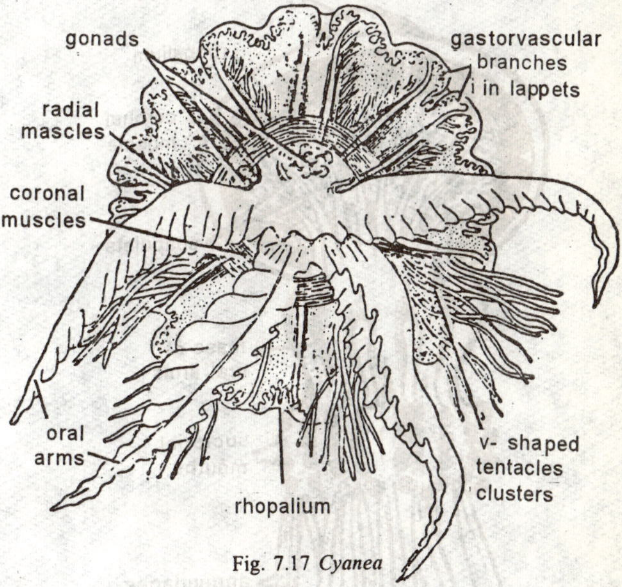


Fig. 7.17 *Cyanea*

1. It is the largest, solitary, bioluminescent and marine medusoid form.
2. It is found in the costal waters of America, Pacific coast and Polar regions.
3. It is commonly called as 'sun-jelly' or 'sea-blubber'.
4. Umbrella dome-shaped or saucer-shaped with margin scalloped into eight lappets having eight rhopalia.

5. Subumbrella carries four long oral arms and eight V-shaped bundles of tentacles, the marginal tentacles.
6. Four gonads, placed between oral arms and tentacles.
7. Circular canal absent. Radial canals branch profusely and extend into rhopalia and tentacles.
8. It produces burring sensation.

## RHIZOSTOMA

| | | |
|---|---|---|
| Phylum | - | Coelenterata |
| Class | - | Scyphozoa |
| Order | - | Rhizostomae |
| Genus | - | *Rhizostoma* |

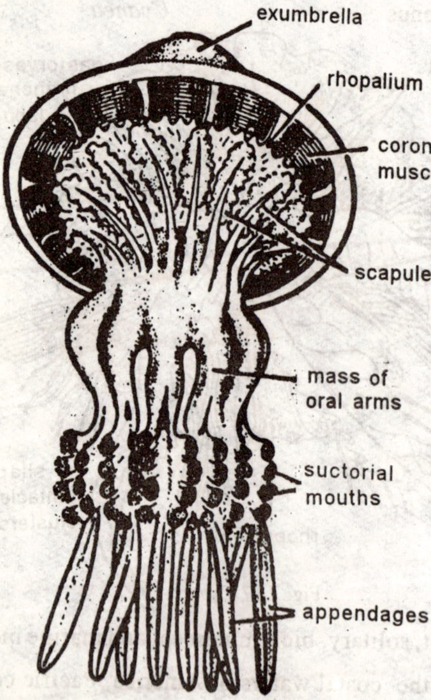


Fig. 7.18 *Rhizostoma*

1. It is inhabitant of shallow water in tropical and subtropical zones, sometimes, in temperate zones. These are also found in Indo-Pacific region and North Carolina.

2. The colour varies. The colour of umbrella is pale green with a deep reddish margin, arms bright blue.
3. The marginal tentacles are absent.
4. The orignal four arms become divided longitudinally into eight.
5. The mouth is one is young but it is replaced by numerous small "sucking mouths" in adult which lie along the lips.
6. Lips act as organs for external digestion.
7. It feeds upon small animals like fishes etc.
8. The mesogloea of the medusae of *Rhizostoma* is neither gelatinous nor mucilaginous. The entire medusa contains 91 to 96% water content.
9. The umbrella is about 2 inches in length and specimen is about 4 inches

## EDWARDSIA

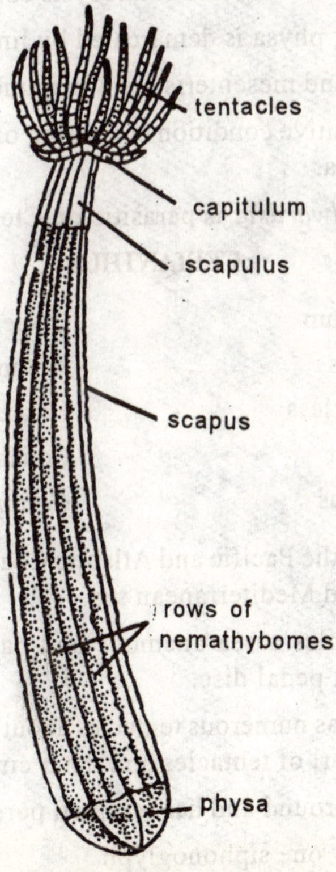


Fig. 7.19 *Edwardsia*

| | | |
|---|---|---|
| Phylum | - | Coelenterata |
| Class | - | Anthozoa |
| Subclass | - | Hexacorallia |
| Order | - | Actiniaria. |
| Genus | - | *Edwardsia.* |

1. It is a small, solitary, marine animal, burrows in sand. It is found in U.S.A., Southern California and North of Cape Cod.
2. Body is elongated and is diffrentiated into oral disc and column.
3. Oral disc is short and carries centrally placed mouth and a circlet of 16 tentacles arranged in two rows.
4. The cylindrical column has three parts, the *capitulum*, *scapulus* and *scapus*.
5. The body surface has 8 longitudinal ridges and the posterior half of the scapus contains rows of nematocysts called as *nemathybomes*.
6. The basal part or physa is demarcated by limbus from scapus.
7. Siphonoglyphs and mesenteris are 8 in number are present.
8. Septa are in primitive condition consisting of 8 macrosepta and more than 4 microsepta.
9. The young of *Edwardsia* is parasitic on Ctenophore.

## CERIANTHUS

| | | |
|---|---|---|
| Phylum | - | Coelenterata |
| Class | - | Anthozoa |
| Subclass | - | Hexacorallia |
| Order | - | Ceriantharia |
| Genus | - | *Cerianthus* |

1. It is found along the Pacific and Atlantic Coast, Cape Cod to Florida, Bay of Fundy and Mediterranean sea.
2. The form resembles a sea anemone and has cylindrical elongated body but without pedal disc.
3. The oral region has numerous tentacles about 130 arranged in 2 rows, the marginal whorl of tentacles and inner circlet of tentacles.
4. The lower end is round and has terminal pore.
5. Pharynx has only one siphonoglyph.
6. The ectoderm secretes in long tube of mucus in which the animal

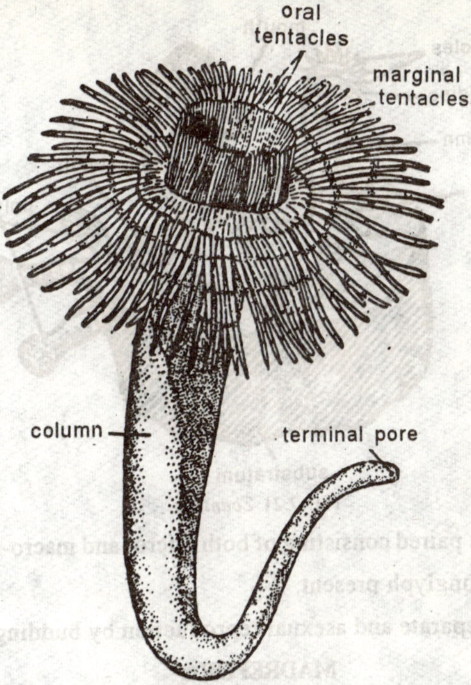


Fig. 7.20 *Cerianthus*

lives and the tubes is fixed in soft, sandy or mud sea bottom.

7. On the tube sand particles are deposited so it becomes hard.

## ZOANTHUS

| | | |
|---|---|---|
| Phylum | - | Coelenterata |
| Class | - | Anthozoa |
| Subsclass | - | Hexacorallia |
| Order | - | Zoantheria |
| Genus | - | *Zoanthus.* |

1. It is a colonial form found attached to rocks or corals, sponges, shells etc.
2. It is found is West Indies.
3. Body is differentiated into column and oral disc having mouth and marginal tentacles.
4. Cylindrical, small polyps arise singly from a network of stolon.
5. Tentacles are numerous, 48-60 in number arranged in one or two rows.

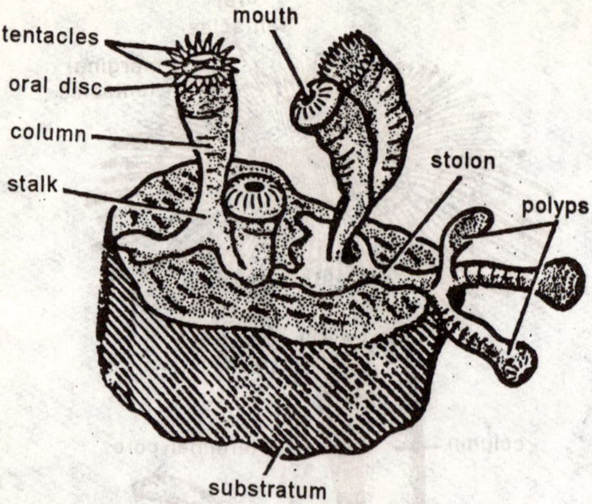


Fig. 7.21 *Zoanthus*

6. Mesenteries paired consisting of both micro- and macro-mesenteries.
7. Single siphonglyph present.
8. Sexes are separate and asexual reproduction by budding.

## MADREPORA

| | | |
|---|---|---|
| Phylum | - | Coelenterata |
| Class | - | Anthozoa |
| Subclass | - | Hexacorallia |
| Order | - | Madreporaria |
| Genus | - | *Madrepora* |

1. It is a colonial, symbiotic and marine coral found in West Indies and Florida.
2. It is commonly called as 'horn coral' and plays an important role in coral-reef formation.
3. Colony is branched with numerous, small crowded polyps in elevated cylindrical cups separated by coenostem.
4. Terminal polyps contain 6 tentacles and lateral polyps with 12 tentacles.
5. Internally mesentries are bilaterally arranged and coenosarc contains a network of canals.

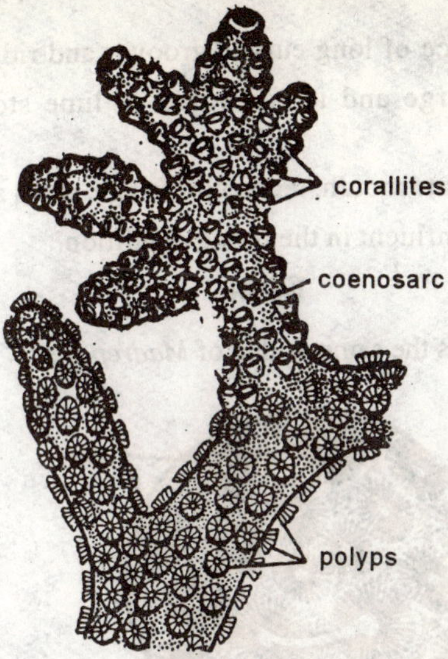


Fig. 7.22 *Madrepora*

6. Sometimes small crustaceans are found in association with horny corals.

## MEANDRINA

The classification is the same as that of *Madrepora.*

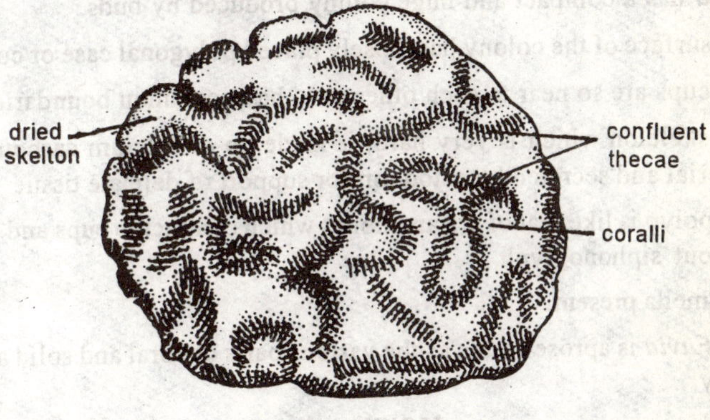


Fig. 7.23 *Meandrina*

1. Colonial form inhibiting the coastal waters of West India, Asia and America.
2. It is popularly known as *brain-coral* since resembles the human brain

due to the presence of long curved grooves and ridges.

3. The colony is large and is composed of lime stone secreted by ectoderm.
4. Polyps bear separate mouth, tentacular fringe, septa and mesenteries.
5. Polyps remain confluent in the living condition.

## FAVIA

The classification is the same as that of *Madrepora*.

Fig. 7.24 *Favia*

1. *Favia* is a massive stony reef building coral.
2. It is commonly found in Florida and West Indies.
3. *Favia* has a compact and huge colony produced by buds.
4. The surface of the colony has closely placed polygonal case or cups.
5. The cups are so near to each other as to have common boundaries.
6. The skeleton which is very hard, is made up of calcium carbonate material and secreted by ectoderm for support of delicate tissue.
7. The polyp is like that of sea anemones which contract in cups and are without siphonoglyph.
8. Columella present.
9. The *Favia* is aprose corals as the various parts of coral and solid and stony.

## FUNGIA

The calssification is the same as that of *Madrepora*.

1. It is a solitary and marine coral found in warm sea, generally in Gulf of California.
2. It is commonly called 'mushroom coral'.

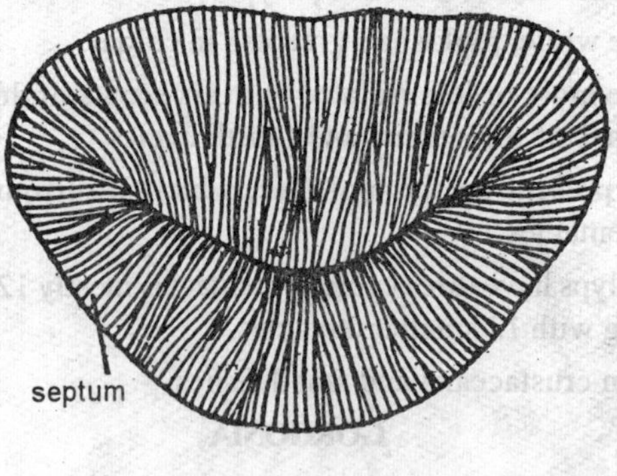


Fig. 7.25 *Fungia*

3. The coral is flat and discoidal, the theca is confined to the lower surface and small calcareous rods and synapticula, which connect septa with one another.
4. Adult animal contains a single large polyp with numerous tentacles.
5. Siphonoglyphs are absent.
6. The life-history includes planula larva which metamorphoses into adult.
7. Reproduction by monodisc strobilation that is plate-like structures are separated off one by one from the top of corallite. The broken disc are called *anthocyanthus* and develop its adults. The remaining broken part is called *anthocaulus* which grows again into new disc and is set off.

## ASTRANGIA

The classification is the same as that of *Madrepora.*

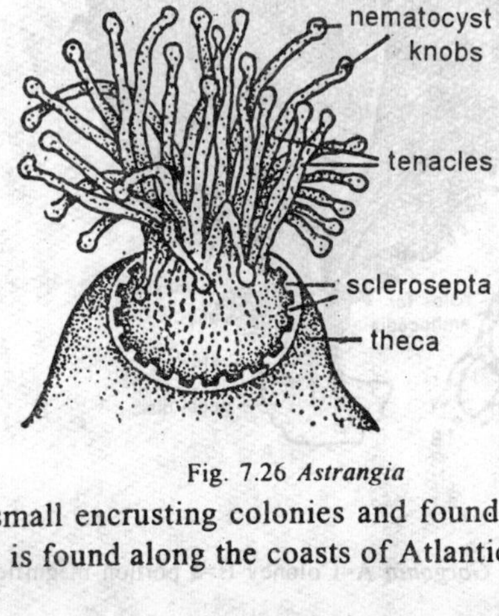


Fig. 7.26 *Astrangia*

1. It forms small encrusting colonies and found on rocks and shells- *Astrangia* is found along the coasts of Atlantic and America.

2. Colony consists of theca and polyps.
3. Polyps are white, pinkish or greenish in colour.
4. Zooids are more or less isolated with six sepata of the first cycle, six smaller ones of the second and fourth cycles.
5. Mouth is present on the oral disc, which is surrounded by double rings of tentacles.
6. Bigger polyps have three cycles of tentacles mostly 12 larger tentacles alternating with 12 smaller tentacles.
7. It feeds on crustaceans and small fishes.

## GORGONIA

| | | |
|---|---|---|
| Phylum | - | Coelenterata |
| Class | - | Actinozoa |
| Subclass | - | Octocorallia |
| Order | - | Gorgonacea |
| Genus | - | *Gorgonia.* |

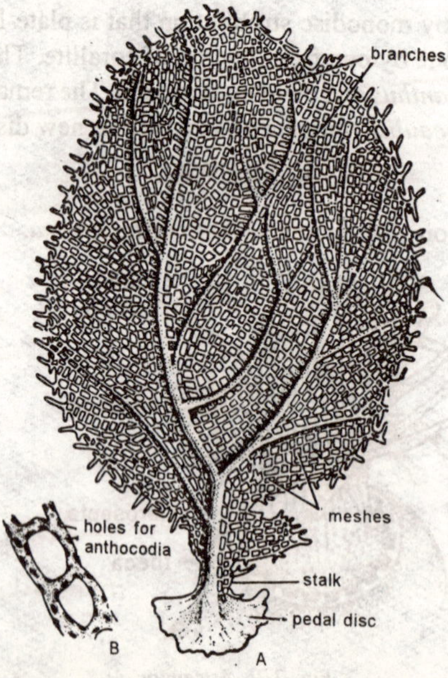


Fig. 7.27 *Gorgonia* A=Coloney B=a portion megnified.

1. These are present in all the seas attached to rocks and stones in shallow waters.
2. It is commonly known as 'sea-fan'.
3. It is yellowish or brownish colony with upright branches consisting of an axial rod extending throughout the colony along all the branches.
4. The skelton is made up of a flexible substance gorgonin and calcareous spicules embedded in mesogloea.
5. Polyps or anthocodia are retractile and present in rows on both the sides of the branches.
6. Sexes are separate.
7. The dried skeletons of the colony are used in decorative art.

## TUBIPORA

| | | |
|---|---|---|
| Phylum | - | Coelenterata |
| Class | - | Anthozoa |
| Subclass | - | Octocorallia |
| Order | - | Alcyonacea |
| Genums | - | *Tubipora* |

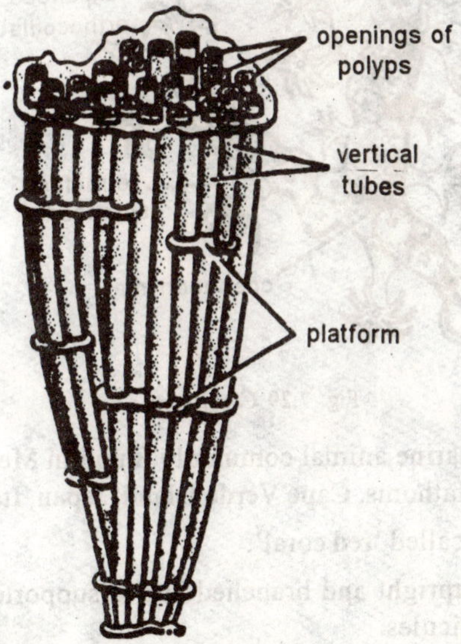


Fig. 7.28 *Tubibora*

1. It is marine, colonial animal found in shallow waters of tropical and temperate regions.

2. It is found abudantly in shallow water of Atlantic, Indian and Pacific oceans.
3. It is commonly called 'organ pipe coral'.
4. Colony consists of long, parallel and upright tubes closely fitted, and joined together at definite interval by horizontal calcareous tubes.
5. The polyps are bright green and secrete these tubes.
6. The skeleton is internal and is covered by ectoderm in living condition.
7. The mesogloeal spicules become closely fitted together and form a continuous tube for each polyp.
8. Asexual reproduction by peculiar budding. The base of original polyp expands from which new polyps originate.

## COMRALLIUM

The classification is the same as that of *Tubipora*.

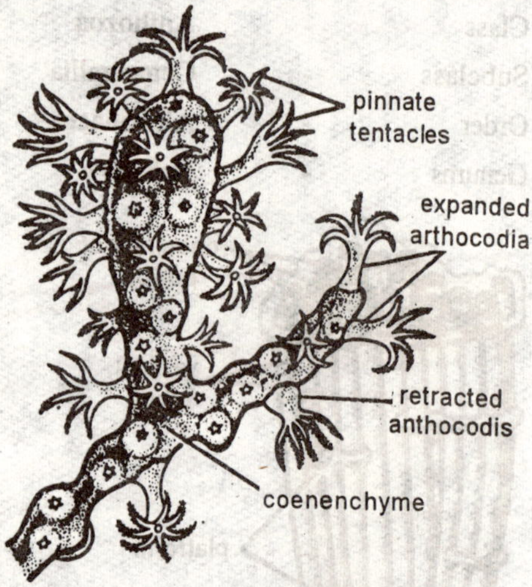


Fig. 7.29 *Corallium*

1. It is a clonial marine animal commonly found in Mediterranean at a depth of 10-30 fathoms, Cape Verde Island, Japan, Italy etc.
2. It is commonly called 'red coral'.
3. The colony is upright and branched and is supported by calareous axis of fused spicules.
4. The colony is dimorphic as there are two type of zooids (i) nutritive zooids, the *anthozooids* and (ii) for circulation of water in the colony, the *siphonozooids*.

5. The zooids are white in colour and the colony is dark red due to the prepesence of red calcareous spicules.
6. Gonads are borne by siphonozooids.
7. Planula develops inside the zooids and so the viviparity is found in *Corallium*.
8. The *Corallium* is of great economic importance as the precious red coral of commerce known as red moonga in N. India which is obtained from this animal.

## ALCYONIUM

The classification is the same as that of *Tubipora.*

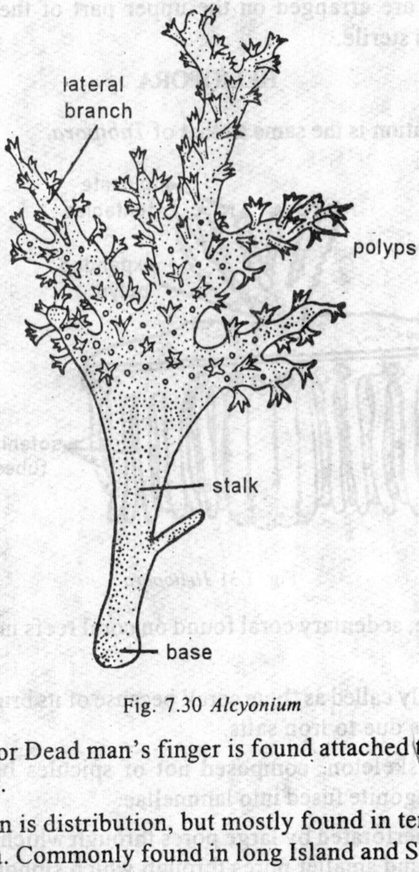


Fig. 7.30 *Alcyonium*

1. *Alcyonium* or Dead man's finger is found attached to stones or rocks in sea water.
2. Cosmpolitan is distribution, but mostly found in temperate and cold water of sea. Commonly found in long Island and St. Lawrence.
3. Colony of short and thick leathery lobes attached to stones.

4. Long polyps with tentacles except those of the outer end, polyps with tentacles can be contracted and everted.
5. *Coenechcyme* (mesogloea or soft parts of an alcyonarian colony), flesh and spicules present.
6. Colour of the colony is reddish or Yellowish; length is 4 to 10 cm.
7. The skeleton consists of calcareous spicules.
8. The food in *A. digitatum* consists of various fine fragments of muscles of fish but they may reject many kinds of fish ova (according to *Pratt*).
9. Gland cells occur in stomodaeum and it is probable that they secrete a fluid for digestion.
10. The polyps are arranged on the upper part of the colony, and the lower part is sterile.

**HELIOPORA**

The classification is the same as that of *Tubipora.*

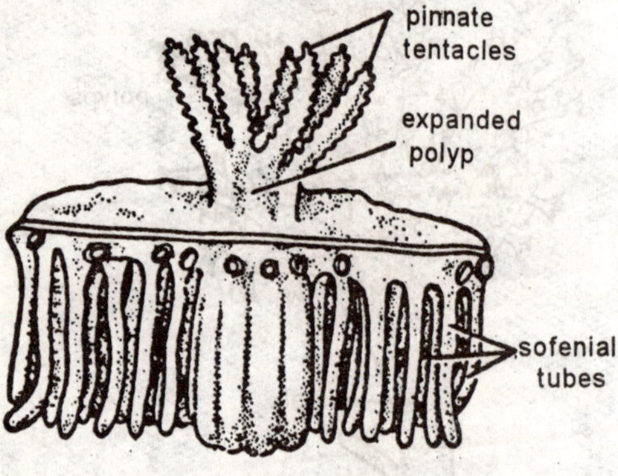


Fig. 7.31 *Heliopora*

1. It is a marine, sedentary coral found on coral reefs in the Indo-Pacific ocean.
2. It is commonly called as 'blue coral' because of its bright blue skeleton. The colour is due to iron salts.
3. Its massive skeleton, composed not of spicules but of crystalline fibres of arragonite fused into lammellae.
4. Skeleton is perforated by large pores through which ordinary polyps project out, and smaller pores through which siphonozooids project out.

5. It is traversed by tubular cavities of two sizes, larger cavities-few in number open out through large pores. Smaller cavities open out through small pores.
6. Surface contains flat coenechyme, which contazins solanial network connected to middle region of the polyp and also with erect solanial tubes.
7. Polyps dimorphic, large ordinary zooids and small siphonozooids.

## PENNATULA

| | | |
|---|---|---|
| Phylum | - | Coelenterata |
| Class | - | Actinozoa |
| Subclass | - | Octocorallina |
| Order | - | Pennatulacea |
| Genus | - | *Pennatula* |

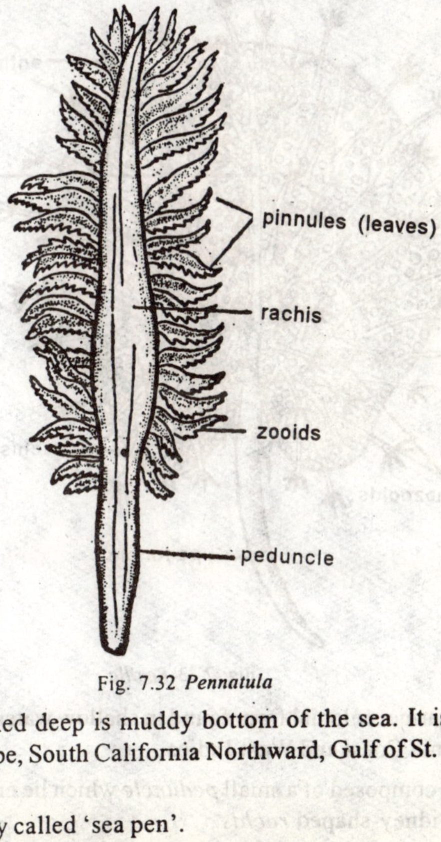


Fig. 7.32 *Pennatula*

1. It is found fixed deep is muddy bottom of the sea. It is commonly found in Europe, South California Northward, Gulf of St. Lawrence to Carolina.
2. It is commonly called 'sea pen'.

3. Colony is divided into two parts : (a) stalk, which is embedded in sand or mud and (b) an upper part, called the *rachis* or *horny axis*.
4. *Rachis* is elongated with paired lateral leaves or *pinnulae*.
5. Lateral leaves or pinnulae long, 20 to 25 in number on each side.
6. Polyps, arranged on horny axis, are of two types :
   (a) *Autozooids*, whose function is nutritive, are situated side-by-side as regular lateral branches giving an appearance of a feather.
   (b) *Siphonozooids*, which maintain circulation of water in the colony, are situated on the back of the axis.

## RENILLA

The classification is the same as that of *Pennatula*.

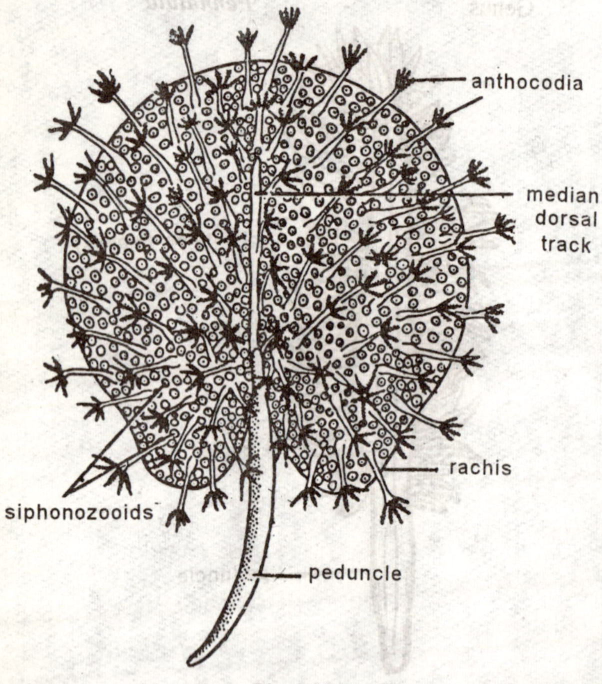


Fig. 7.33 *Renilla*

1. It is a marine, colonial form found in shallow waters of Carolina Coast, California Coast and West Indies.
2. Colony composed of a small *peduncle* which lie embedded in the mud and a kidney-shaped *rachis*.

3. *Rachis* has a broad ventral surface devoid of polyps and a dorsal surface covered with authozooids and siphonozooids.
4. *Autozooids* or *anthocodia* lie scattered on the dorsal surface are nutrive in function.
5. *Siphonozooids* are arranged in groups and maintain the circulation of water within the colony.
6. A median bare 'track' devoid of polyps extends from the peduncle to about the middle of rachis, where it terminates at a special exhalent siphonozooid.
7. Axial skeleton absent.

## VIRGULARIA

The classification is the same as that of *Pennatula*.

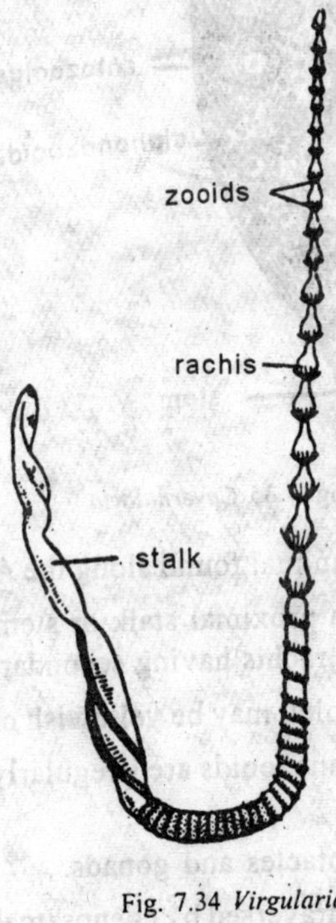


Fig. 7.34 *Virgularia*

1. *Virgularia* or "walking stick" is found partly embedded in warmer water having a muddy and soft bottom. It is found along Pacific Coast.
2. The rachis is long and elongated like a walking stick.

3. The elongated body is divisible into two parts, *a stalk* and the *rachis*.
4. The polyps arranged in transverse rows at regular intervals and are slightly fused and arise from the *rachis*.
5. The stalk is without polyps.
6. The stalk is thrust into the muddy bottom.
7. The distal rachis stand straight outside water.

## CAVERNULARIA

The classification is the same as that of *Pennatula*.

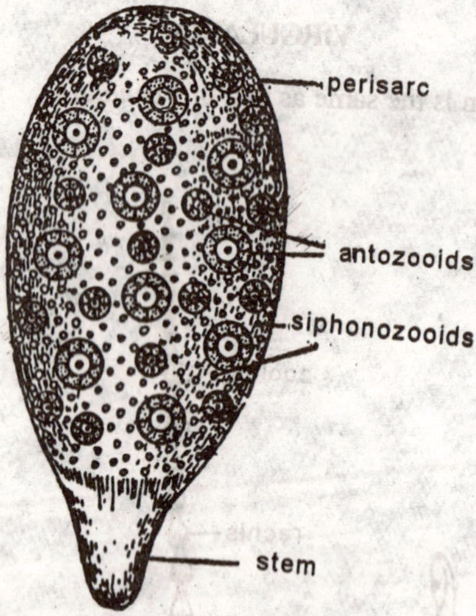


Fig. 7.35 *Cavernularia*

1. It is a colonial, marine animal found along the Atlantic Coast.
2. Body is divisible into a proximal stalk or stem which is devoid of anthocodia and a distal rachis having secondary polyps.
3. The colour is blue or violet, may be yellowish or greenish.
4. Anothozooids and siphonozooids are irregularly distributed all over the rachis.
5. Anthozooids having tentacles and gonads.
6. Coenosarc is fleshy and traversed by coenosarcal canals and contains long spicules.
7. It is a bioluminescent, emits light.

# 8

# OBELIA

The structure of other coelenterates is more complex than that of *Hydra*, therefore, its structure can not be considered as typical of coelenterates. *Hydra* is a fresh water animal where as most coelenterates are marine and colonial. Their life cycle is more complex then that of *Hydra*. *Obelia* is commonly studied as a typical example of marine and colonial coelenterates. Many species of *Obelia* are found but the most common is *Obelia geniculata*.

## SYSTEMATIC POSITION

| | | |
|---|---|---|
| Phylum | - | Coelenterata |
| Sub-phylum | - | Cnidaria |
| Class | - | Hydrozoa |
| Order | - | Hydroida |
| Sub-order | - | Leptomedusae |
| Genus | - | *Obelia* |

### Habits and Habitat

*Obelia* is a tree-like branched, marine colonial hydroid coelenterate growing on the surface of sea-weeds, rocks and pilings along the sea-coasts. Colonies appear like whitish or light-brown branched threads, about 30 mm. to several cms in height. The colony arises by budding from a single hydra-like individual, the buds fail to separate and after repeated budding there results a tree-like growth permanently fixed to some object and consisting of numerous members joined to the main stem. The members are known as *zooids*.

### External Morphology

If a colony of *Obelia* is examined under low power of microscope, it is found to be made up of a root-like horizontal part called *hydrorhiza*. From hydrorhiza arises a vertical branching stem about 2.5 cm. long known as *hydrocaulus*. Hydrocaulus and hydrorhiza consist of two distinct layers.

(*a*) ***Perisarc.*** It is a transparent, tough outer layer, yellow in colour, of

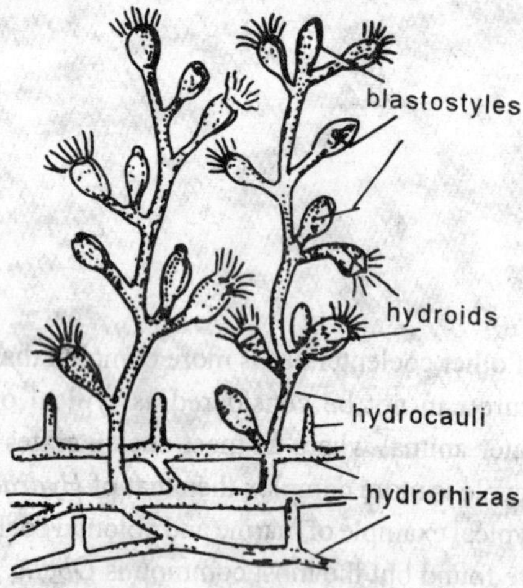


Fig. 8.1 *Obelia* colony

horny consistency. It is not composed of cells but is a cuticular secretion of thee ectodermal cells. The *perisarc* is tubular in shape and is at first in contact with the ectoderm but afterwards becomes separated from it by a small space, except at regular intervals. This condition is acquited secondarily, as during its formation it is in contact with ectoderm but, when it thickens, ectoderm withdraws except at some places. In some parts, perisarc is ringed. Perisarc serves as a protective layer and gives support to the branches having zooids.

(*b*) ***Coenosarc.*** Fibrous processes connect the perisarc with coenosarc. It is an inner, granular layer. It is hollow and its tubular cavity is continued into cavities of polyps, thus, forming a part of *gastrovascular cavity.* There is a fluid, which shows flickering movement similar to that of cilia. Coenosarc consists of an outer layer, the *ectoderm* and an inner layer, the *endoderm.*

From the upright stem, called *hydrocaulus,* several side-branches arise. These side-branches are ringed and their end is expanded into a hydra-like structure known as *hydranth* or *polyp*. Hydranth is a kind of *zooid.* Towards the proximal region of the colony are found cylindrical bodies known as *blastostyles.* These give rise to several small lateral offshoots, called *medusa buds*.

Thus, *Obelia* colony is *trimorphic*, having following types of zooids :

(a) Polyp or hydranth;

(b) Blastostyle, and

(c) Medusa.

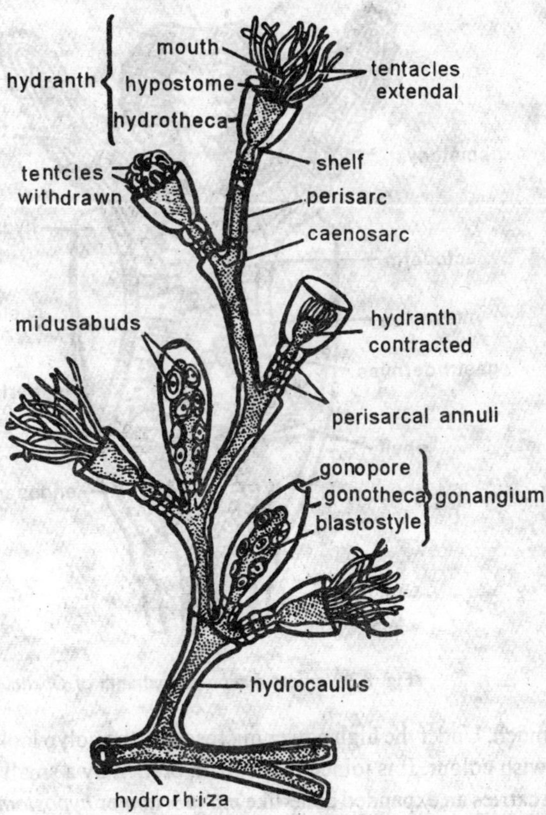


Fig. 8.2 *Obelia*. A part of colony under the microscope.

## Polymorphism

Due to the presence of three types of zooids or the individuals the *Obelia* colony (trimorphic) is said to exhibit the phenomenon of polymorphism which is correlated with the colonial organization of *Obelia*. The phenomenon of polymorphism is essentially one of division of labour i.e., different functions are assigned to different individuals rather than to parts of organs of one individual. It may be recalled here that in *Obelia* the feeding zooids (Hydranths), asexually reproducing zooids (Blastostyles), and sexually reproducing zooids (Medusae) perform their respective functions for *Obelia* and thus exhibit division of labour.

## Hydranth or Polyp

It is a nutrient polyp which in its essential structure resembles *Hydra*

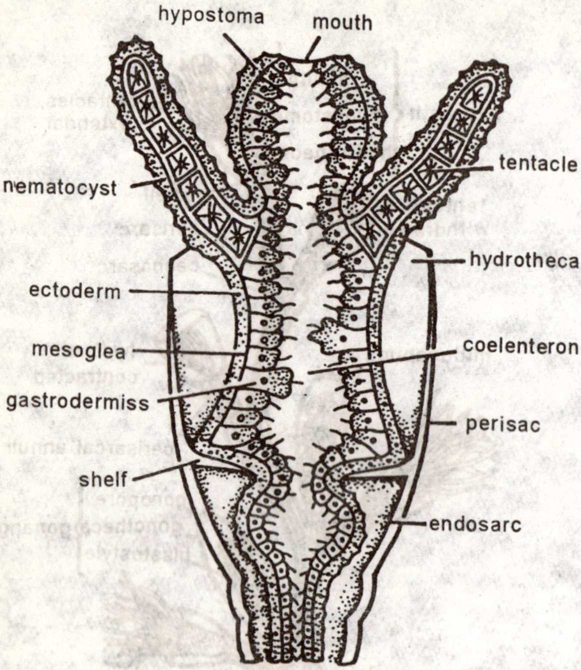


Fig. 8.3 Vertical section of hydranth of *Obelia.*

very much. Under the high power microscope, the polyp looks a hollow case of yellowish colour. It is joined to the main branch by a small stalk. At its distal end it carries an expanded cone-like *manubrium* or *hypostome* which measures about one third of the polyp. The *mouth* is situated at the tip of the manubrium. At the base of the manubrium are twenty four solid *tentacles* which are arranged in a circle. The manubrium is hollow; its cavity opens out by mouth while on the other side, it is continuous with the gastrovascular cavity of the stem. The mouth is elastic and capable of great dilation and contraction. It serves both for ingestion and egestion, there being no anus in this animal.

The perisarc, covering the polyp, forms a cup-like structure called the *hydrotheca* which is transparent and perfecty colourless. The hydrotheca forms a sort of circular *shelf* near its proximal end on which rests the polyp. Normally the polyp remains extended out of the manubrium contract rapidly and are almost completely withdrawn into the hydrotheca.

***Histology.*** As seen under light microscope in longitudinal section, the body wall of the hydranth is seen to consist of two layers of cells :

(a) the outer *ectoderm* or *epidermis*.

(b) the inner *endoderm* or *gastrodermis*,

and separating these layers is a delicate membrane called *mesogloea* or supporting lamella. The hypostome also consists of these two layers of cells. The layers of polyps are continued as coenosarcs. The other zooids are also made up of the same layers decribed above.

Ectoderm is thin and is made up of columnar *epithelio-muscular cell*. The cells are conical in shape. Their narrow inner ends are prolonged into involuntary muscle fibres, arranged longitudinally between epidermis and mesogloea.

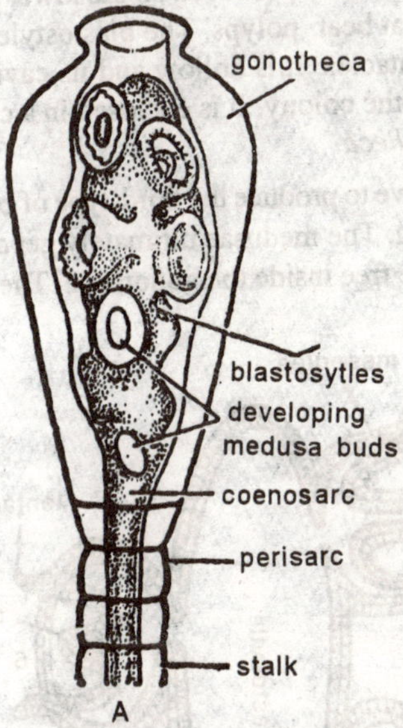


Fig. 8.4 *Obelia*. Gonangium

Interstitial cells may be present in the space between the narrow ends of ectodermal cells. But according to *Grove* and *Newell* the interstitial cells are completely absent. *Nematocysts* are also present in the ectoderm which act as organs of offence and defence. The nematocysts are specially abundantly present in the tentacles where they constitute annular batteries. The nematocysts are *basitrichous isorhizas* type. These are characterized by an oval capsule, a long theread open at the tip and bearing spines but without a butt.

Endodermal cells are large and glandular and their free ends have pseudopodia or flagella, the movement of the latter cause the entry of food particles into the coelenteron (enteron). These are known as *musculo-nutritive cells*. *Gland cells* with granular protoplasm are also present in endoderm. They produce digestive enzymes. Tentacles and hypostome also have a layer of unstriated muscle fibres lying between ectoderm and mesogloea which helps in rapid contraction of tentacles. In hypostome, muscle lying transversely serves to contract the cavity and act antagonistically to the action of longitudinal muscle fibres.

A nerve net is present on both the sides of the mesogloea. It is formed by large branched *nerve cells*.

**Blastostyle.**

The blastostyles are developed when the hydrocaulus has attained its full length. They are generally produced on the lower part of the colony in the axil of the branches that bear polyps. The blastostyle is a club-shaped zooid without mouth and tentacles. It is hollow and its cavity is continuous below with that of the rest of the colony. It is enclosed in a cylindrical expansion of the perisarc, the *gonotheca*.

The blastostyle serve to produce the third type of zooids, *i.e.*, the medusae. This occurs by budding. The medusae ultimately get constricted off from the blastostyle and become free inside the gonotheca. The gonotheca develops at

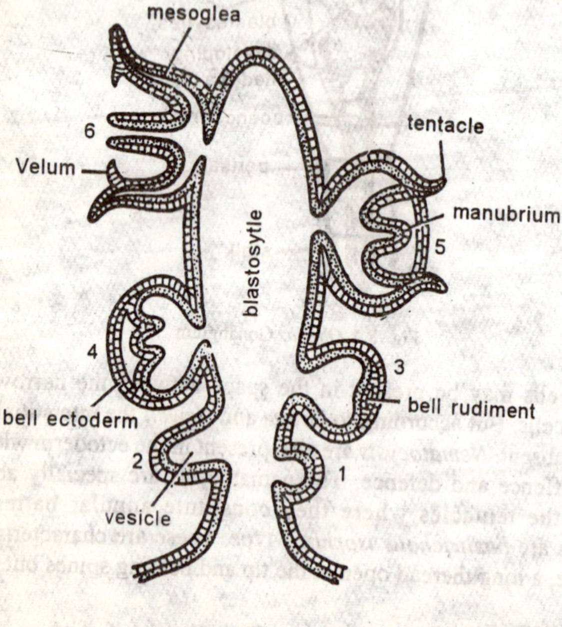


Fig. 8.5 Development of medusa from a blastostyle

its tip an aperture, the *gonopore*, and medusae escape into the sea through it. They are probably expelled through the gonopore by currents produced by rhythmic contractions of the polyps.

The blastostyle, gonotheca and the medusa buds are collectively called the *gonangium* or *gonanth*.

**Physiology of Polyps**

***Nutrition.*** The polyps are the feeding zooids. They supply food to the entire colony. The food consists of small living organisms. They are captured with the help of nematocysts and brought to the mouth by tentacles as in *Hydra*. The annuli round the stalks of the polyps give them sufficient flexibility to move *to and fro* in water in pursuit of food. A part of the food is digested in the gastrovascular cavity of the polyps by the action of the digestive juices secreted by the gland cells of the gastrodermis. This is called the *intercellular* or *extracellular* digestion. The partly digested food circulates in the gastrovascular cavity of entire colony by beating of the flagella of the gastrodermal cells and by convulsive peristaltic contractions of the polyps. It is ingested from the gastrovascular cavity by the pseudopodia of the gastrodermal cells in which it is digested in much the same way as in *Amoeba*. This is known as the *intracellular digestion*. From the gastrodermis, some of the digested food is passed on to the epidermis by diffusion. The undigested food is egestedout through the mouth of the polyp. Starch, cellulose and chitin are not digested. The digested food is supplied to the entire colony including blastostyle.

***Respiration.*** Respiration is aerobic and occurs through the general surface of the body. The cavity of the entire colony contains water, which enters through the mouth of the polyps. This water is kept circulating by the contractions of the polyps and beating of gastrodermal flagella. The epidermis is also in contact with water which diffuses through the permeable perisarc. Thus, almost all the cells are in contact with water which has oxygen dissolved in it. This oxygen diffuses into the cells and the carbon dioxide likewise diffuses out of them.

***Movements.*** Colony is incapable of locomotion. Its zooids, particularly the polyps, show movements. They can contract and expand their body and can bend their tentacles.

***Growth.*** Growth of the colony occurs more or less like a plant. It involves spreading of the branches of hydrorhiza along the substratum, development of new hydrocauli from the hydrorhiza and formation of new zooids on the branches arising from the hydrocauli. It is to be noted that untill a polyp is developed at the end of a branch, the latter does not produce the next branch. Blastostyles grow by budding in the axils of the older branches of a full-grown hydrocaulus. The stimulus leading to blastostyle formation is not clear.

***Excretion.*** Elimination of nitrogenous waste materials takes place by diffusion through the general body surface.

***Behaviour.*** The polyps are sensitive to stimuli. They respond to touch and unfavourable chemicals by contraction.

***Osmoregulation.*** The colony has no problem of osmoregulation as the sea-water is almost as dense as the cell contents. As the sea provides almost a uniform environment throughout the year and there is no dessication, *Obelia* colony does not experience unfavourable periods and is capable of indefinite survival.

**Medusa**

The medusae are the reproductive zooids, some produce eggs and others produce sperms, *i.e.*, the sexes are separate. Each medusa, when fully mature is a saucer-shaped structure measuring 6 or 7 mm in diameter. The outer convex side is called the *ex-umbrella* while the concave inner side is known as the *sub-umbrella*. The attachment of the medusa with the blastostyle is by the exumbrellar surface. From the centre of the sub-umbrella hungs a process called *manubrium*, which bears foursided *mouth* at its top.

The *mouth* leads into the enteric cavity which is divided into four *radial canals*. Near the edge of umbrella all the radial canals open into a *circular canal*. The cavity of the manubrium, the radial canals and the circular canal are all lined by the gastrodermal cells. The food is ingested through the mouth and its digestion takes place in the radial and the circular canals.

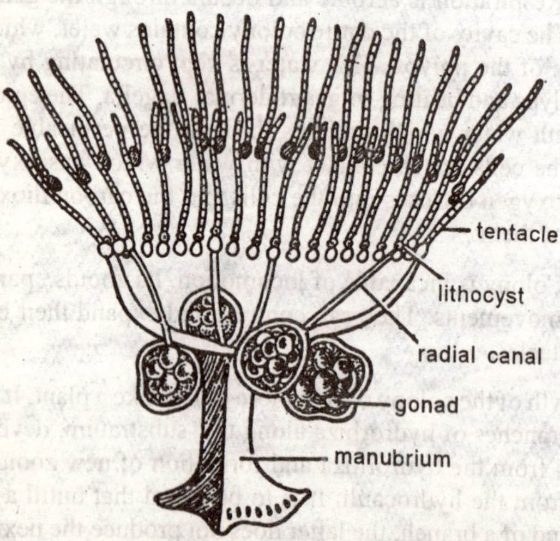


Fig. 8.6 *Obelia*. A medusa in oral view.

The peripheral edge of the umbrella is produced into a narrow inward fold called the *velum*. From the margin of the velum arise a large number of tentacles. In a young medusa there are sixteen tentacles but their number increases in a medusa. These tentacles, like the tentacles of the polyp, are solid structures having a core of endodermal cells surrounded by ectoderm. There is a definite plan of arrangement of these tentacles around the velum. In the young

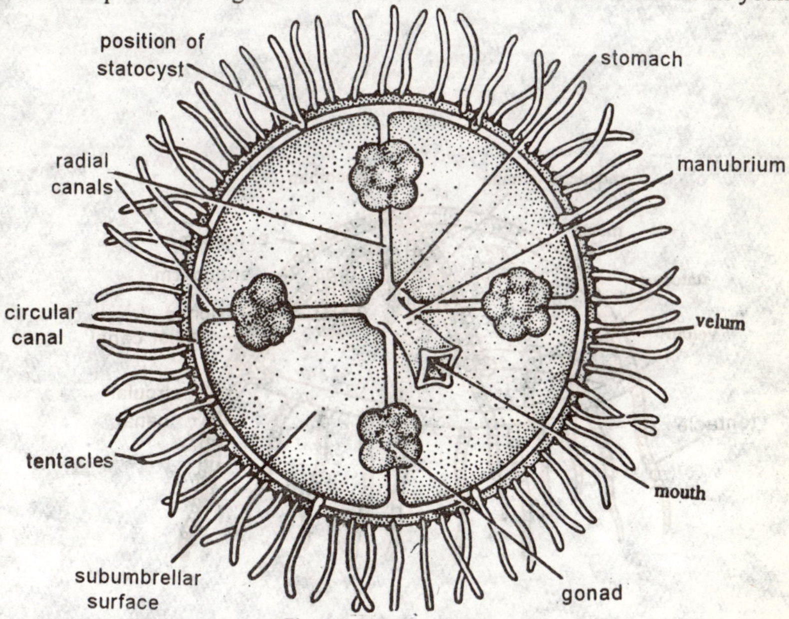


Fig. 8.7 An everted medusa

medusa (when there are only 16 tentacles), they are arranged in four groups, each group having 4 tentacles. The four tentacles placed against the four radial canals are the *per-radial tentacles*. Bisecting the four angles between the per-radials, are the four *inter-radii* having the *inter-radial tentacles*. The radii bisecting the inter-radii are the *ad-radial tentacles*. Thus there are four per-radial, four inter-radial and eight ad-radial tentacles.

At the base of eight ad-radial tentacles, a few interstitial cells get collected. These cells form a sort of reservoir of cells which form the nematoblasts. The receptor organs are present at the bases of these ad-radial tentacles. Each consists of a minute water-filled ectodermal sac called *statocyst*. Its cavity contains a movable calcareous particles, *otoliths* or *otocysts*. The cells are also provided with hair-like processes of sensory cells. These organs were previously supposed to be auditory organs. Now they are thought to be co-ordinating or balancing organs which help to maintain the equilibrium.

***Histology.*** Microscopically the structure of medusa resembles that of the hydranth polyp. It consists of the outer epidermis and inner gastrodermis with mesogloea in between the two. At the tip of the manubrium the gastrodermis is continuous with the epidermis. The whole of the ex-umbrellar and sub-umbrellar surfaces are covered by the epidermis while the cavity of the manubrium and the inner radial canals and circular canal are lined by the gastrodermis. The velum is made up of two layers of epidermis with a core of mesogloea. The

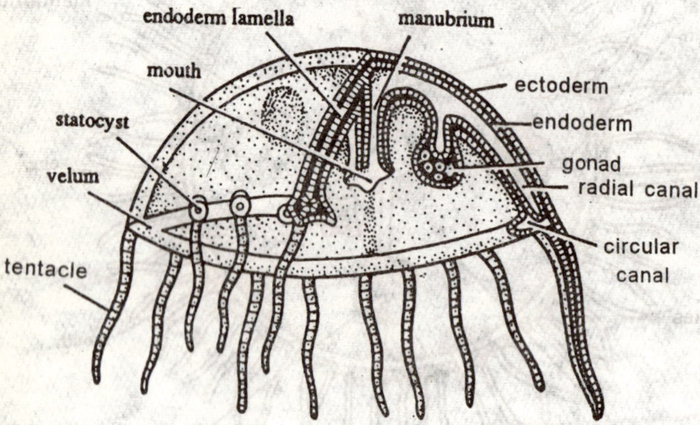


Fig. 8.8 Medusa in section

epidermis has a large number of stinging capsules. On the radial canals are usually present the gonads, one on each canal.

***Development of Medusa from Balastostyle.*** This small sac is formed with a cavity and encloses two layers of ectoderm. It is called *bell-rudiment*. It grows in size and assumes the form of sub-umbrellar cavity with the manubrium lying in the centre. Later on the bell-ectoderm (outer ectoderm) breaks up and the remaining part of it constitutes a circular shelf, termed as *velum*, projecting inwards from the margin of the umbrella. It is quite rudimentary here. Medusae develop as buds on the blastostyles. In initial stage of the formation of medusa it is just like a small diverticulum or projection of the cavity of the blastostyle. This diverticulum increases in dimension at its extremity and form a small vesicle enclosed by ectoderm, mesoglea, and endoderm. It also has enteron of blastostyle. The medusa remains in contact with the blastostyle by a narrow stalk. Subsequently, a cavity appears below the subumbrellar surface by the separation of distal region of ectoderm into two layers : (a) an outer ectoderm and (b) an inner ectoderm. There is a cavity between these two layers of ectoderm. When the medusa acquires its final shape, mouth breaks through the apex of the manubrium. The stalk also ruptures later on and the medusa becomes free and comes out through the opening at the distal end of gonotheca, and the medusa swims freely in sea-water.

*Radial symmetry.* The medusa, like polyp, is also radially symmetrical. The four radial canals mark the four principal radii, called *perradii* and the tentacles lying against the four radial canals are *perradial tentacles.* Between any two per radii is the *interradius* with *interradial tentacles.* Bisecting between a per-radius and adjacent interradius is the *adradius* and then the *subradius.* Thus there are four per-radii, four inter-radii, eight ad-radii and sixteen sub-radii. In *Obelia*, the radial canals, the angles of the mouth and four of the sixteen tentacles of the young medusa are perradial, four other tentacles are inter-radial and the remaining eight tentacles bearing the lithocysts are ad-radial.

**Physiology of medusa**

***Movement.*** The *Obelia* colony is fixed and does not show any bodily movement from one place to another. The polyps can control and expand their body and can bend their tentacles.

The medusae, on the other hand, float passively in water and are drifted along the water currents. They also swim actively by muscular contraction. The impulses, originating in the nerves of the umbrella bring about muscular contraction. By rhythmic contraction and expansion of umbrella, the water in the sub-umbrellar cavity is propelled behind and the animal moves ahead with a series of jerks. The contraction of the bell is brought about by the contraction of the ectodermal muscle-tail which are better developed along the sub-umbrellar surface, specially along the surface where they form a muscle ring. The waves of contraction start from the sub-umbrellar surface and fade towards the exumbrellar surface where they are completely absent. The bell opens mainly by the elastic mesogloea and to some extent by the contraction of the muscle-tails in the middle of the upper surface. During swimming, the body of the medusa may be tilted, thrown aside or more often turned inside out so that the sub-umbrellar surface becomes outer and convex with the manubrium arising from its apex.

***Nutrition.*** The medusa is carnivorous, feeding on nematodes, insects, crustanceans, small worms etc. The prey is captured by the tentacles with the help of nematocysts and ingested by the highly contractile mouth. The digestion takes place in stomach. It is both extracellular and intracellular. The digested food is distributed to the different parts of the medusa through radial and circular canals.

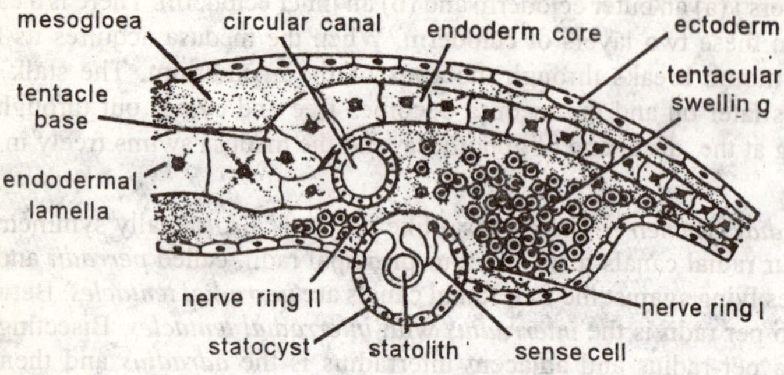


Fig. 8.9 *Obelia.* L.S. of the base of an adradial tentacle of medusa.

***Nervous system.*** The nerve cells of the epidermis and gastrodermis are present on either side of mesogloea to form nerve nets. However, the nerve

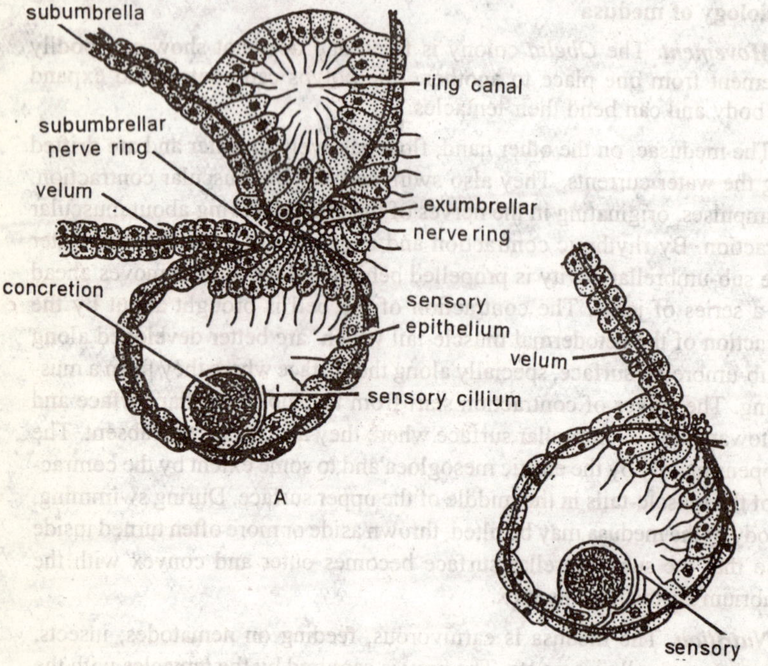


Fig. 8.10 Statocysts : A. in horizontal position; B. in tilted position

cells are specially concentrated along the margin of the umbrella to form two circular nerve rings on either side just below and above the base of the velum. The upper ex-umbrellar nerve ring supplies the tentacles while the lower nerve ring supplies the sub-umbrellar musculature and the statocysts.

***Sense organs.*** The medusa has eight marginal receptor organs situated at the bases of the eight adradial tentacles on the sub-umbrellar surface, just inside the bell-margin. Each consists of a minute and fluid-filled ectodermal sac, called the *statocyst* or *marginal vesicle.* Its cavity contains a movable round particle of calcium carbonate, called *statolith* or *otolith,* secreted by a large cell, termed *lithocyte.* The wall of the bell-margin, being connected basally with nerve cells. The free inner ends of these sensory cells bear fine protoplasmic sensory processes which arch over the statolith.

The statocysts are considered to be organs of *equilibrium* and muscular *coordination*. Their presence in medusae is associated with their active free-swimming habit. During swimming if the body becomes tilted, the statolith falls over the tilted side against the processes of sensory cells which become stimulated. In this way, nerve impulse in created and transmitted to the nerve ring. As a reesult, the muscle tails of the stimulated side contract more rapidly and the medusa is brought again into the normal position.

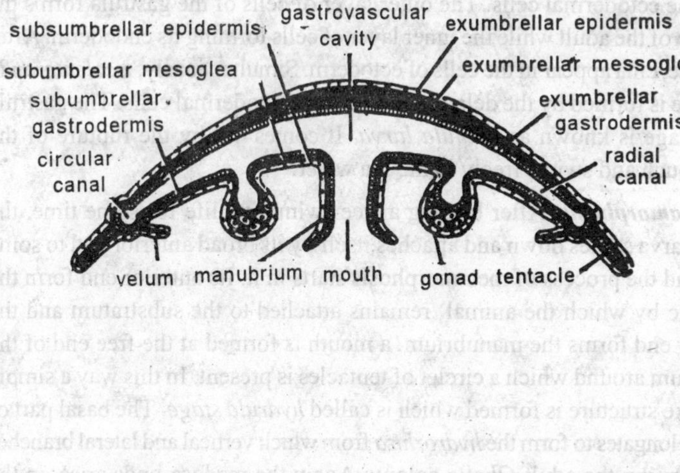


Fig. 8.11 *Obella.* A vertical section of medusa showing gonads.

**Reproduction**

In *Obelia*, the asexual reproduction takes place by budding as a result of which *hydranths* and *blastostyles* are formed. The blastostyles bud-off medusae which are the sexual zooids of *Obelia* because the gonads are found in four groups in each of them. The gonads of a medusa are of the same sex.

***Gonads.*** The gonads are situated perradially on the sub-umbrellar surface. Each gonad is an oval structure which is a group of undifferentiated interstitial cells situated in the mesogloea between the ectoderm and the endoderm. The ectoderm over the gonad is continued with that of the sub-umbrellar surface and the endoderm with the radial canal. The interstitial cells of the gonad form ova in female medusa and sperms in the male medusa. These cells are formed by the ectodermal cells of the manubrium. When the gametes mature, the ectoderm over the gonad ruptures to release them into the sea water.

***Fertilization.*** In *Obelia*, always cross fertilization takes place. The oval are fertilized in the body of the female medusa or in the sea water. The sperms come towards the ova along with the water current.

***Development.*** The zygote undergoes holoblastic and repeated equal cleavages forming a *blastula* which is hollow. The cavity of blastula is called *blastocoel* which is surrounded by a single layer of cells. Soon the cells start separating from the wall of the blastostyle and collect in the blastocoel due to which it becomes completely filled up by them. Thus, a solid gastrula or *stereogastrula* is formed as a result of *delamination* and multipolar immigration of the ectodermal cells. The outer layer of cells of the gastrula forms the ectoderm of the adult while the inner layer of cells forming its endoderm. After some time, cilia appear in the cells of ectoderm. Simultaneously, a *colenteron* in the centre is formed by the delamination of the endodermal cells. The gastrula in this stage is known as *planula larva.* It comes out by the rupture of the brood pouch and swims freely in the sea water.

***Metamorphosis.*** After leading a free swimming life for some time, the planula larva settles down and attaches itself by its broad anterior end to some object and the process of metamorphosis starts in it. Its anterior end form the basal disc by which the animal remains attached to the substratum and the posterior end forms the manubrium. a mouth is formed at the free end of the manubrium around which a circlet of tentacles is present. In this way a simple polyp-like structure is formed which is called *hydrula stage.* The basal part of hydrula elongates to form the *hydrorhiza* from which vertical and lateral branches arise forming the adult *Obelia* colony. Again the medusa buds arise on the blastostyles of this colony and start the sexual generation.

The free swimming medusae are very important in a fixed animal like *Obelia* because they help in its far and wide distribution so that more facilities may be available to it.

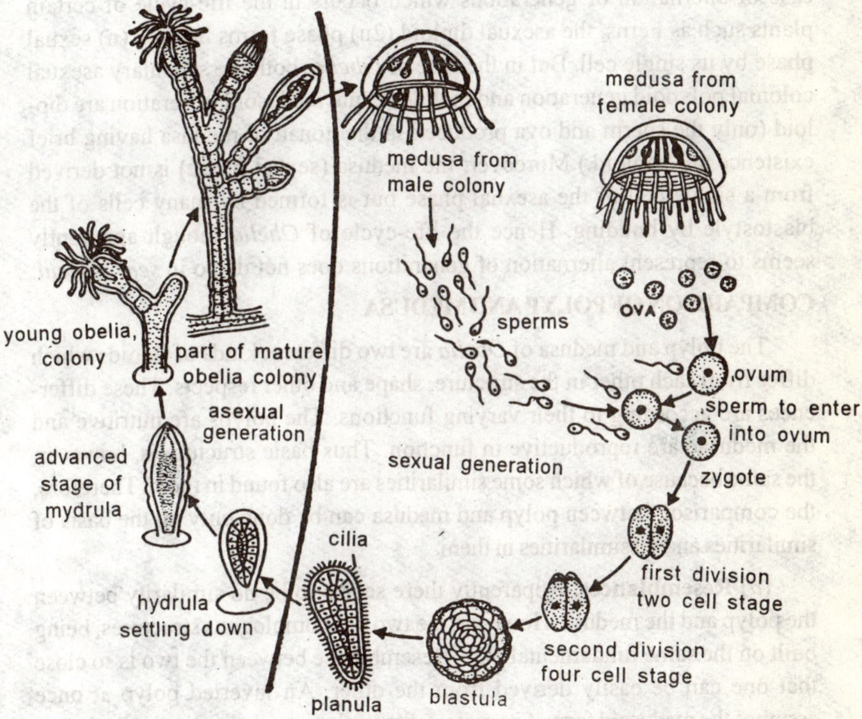


Fig. 8.12 Life-cycle of *Obelia*

## Alternation of Generations or Metagenesis

The life-cycle of *Obelia* colony described above gives the impression that alternation of generations or metagenesis takes place. The sedentary colonial asexual hydroid phase gives rise by budding to medusae which represent the sexual generation. The gonads of medusa produce sperms and ova. The fertilized ovum ultimately gives rise to planula larva which gives rise to *Obelia* colony. In other words the sexual medusa gives rise to asexual colony by sexual method. This alternation of generations or metagenesis confers upon the species not only genetical advantage but also ecological advantage. When each asexual and sexual form has a large reproductive capacity, one sexual, the other asexual, the species has the less chance of being eliminated due to predation or competition. However, in the strict sense the life-cycle of *Obelia* does not represent a true case of alternation of generations because in the true case of alternation of generations which occurs in the life-cycle of certain plants such as Ferns, the asexual diploid (2n) phase forms haploid (n) sexual phase by its single cell. But in the case of *Obelia*, both the sedentary asexual colonial polypoid generation and mobile sexual medusoid generation are diploid (only the sperm and ova produced in the gonad of medusa having brief existence are haploid.) Moreover, the medusa (sexual phase) is not derived from a single cell of the asexual phase but is formed by many cells of the blastostyle by budding. Hence the life-cycle of *Obelia* though apparently seems to represent alternation of generations does not do so in *sensustrictu.*

## COMPARISON OF POLYP AND MEDUSA

The polyp and medusa of *Obelia* are two different kinds of zooids which differ from each other in the structure, shape and other respects. These differences are according to their varying functions. The polyps are nutritive and the medusae are reproductive in function. Thus basic structure is, however, the same because of which some similarities are also found in them. Therefore, the comparison between polyp and medusa can be done only on the basis of similarities and dissimilarities in them.

(*a*) **Resemblances.** Apparently there seems to be no similarity between the polyp and the medusa. However, the two are homologous structures, being built on the same fundamental plan. Resemblance between the two is so close that one can be easily derived from the other. An inverted polyp at once assumes the medusoid form if its region of tentacles is pulled out and manubrium is pushed up. This, together with the thickening of mesogloea, especially on the upper side, brings the walls of the polyp together so that they meet everywhere to form the *gastrodermal lamella* except along the radial and circular canals. Thus, we find that the menubrium, subumbrellar surface, exumbrellar surface and tentacles of the medusa correspond to the manubrium, oral side, aboral side or foot, and tentacles of the polyp respectively. Moreover, the polyp and the medusa have the same layers, which occupy the same relative positions in both, the epidermis covering the outer surface, the gastrodermis lining the cavities and the mesogloea lying between the two. In both, the gastrovascular cavity has a single outlet and both have solid tentacles bearing cnidoblasts. The nature of food and modes of capturing and digesting it are also similar in the two. Finally, both arise as buds on the colony and show radial symmetry.

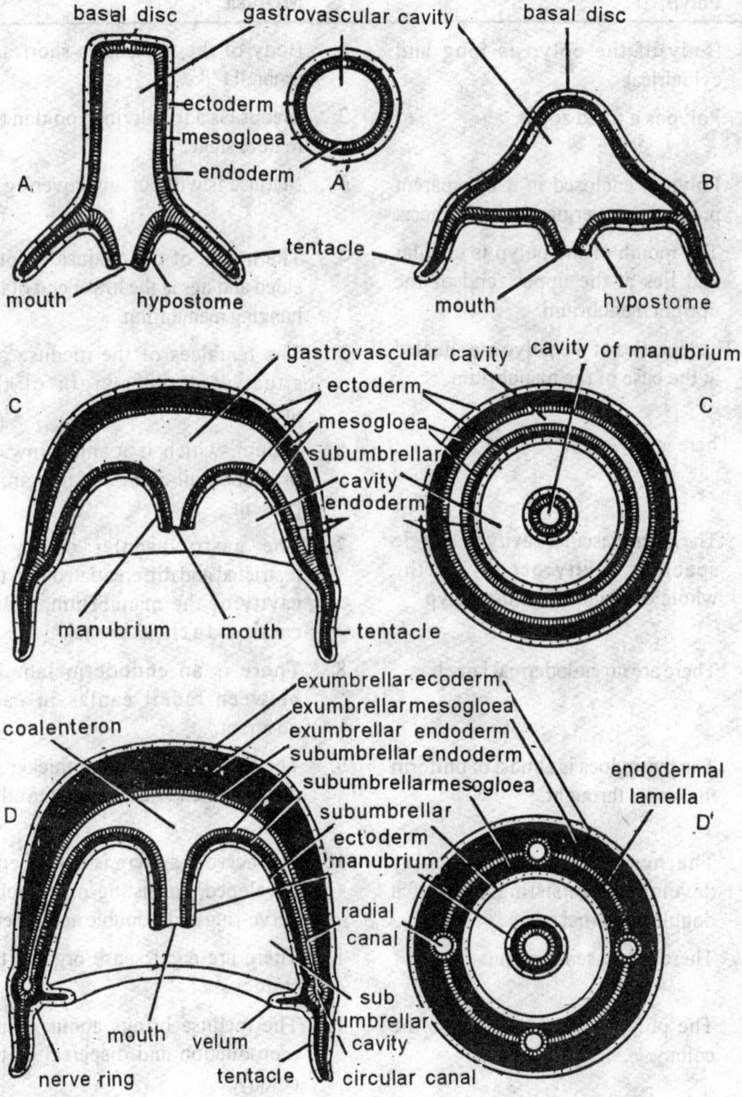


Fig. 8.13 Diagram illustrating the derivation of the medusa from the polyp. A-Polyp in L.S., A'-Polyp from with extended tenacular region, C-Vertical and C' transverse section of form with tentacular region extended into the form of a bell; D-Vertical and D' transverse seection of medusa.

## DIFFERENCES BETWEEN POLYP AND MEDUSA

| Polyp | Medusa |
|---|---|
| 1. Body of the polyp is long and cylindrical. | 1. Body of the medusa is short and umbrella-like. |
| 2. Polyp is a fixed zooid. | 2. Medusa is a free-living zooid in the mature stage. |
| 3. Polyp is enclosed in a transparent protective covering, the hydrotheca. | 3. Medusa is without any covering. |
| 4. The mouth of the polyp is circular and lies at the upper end of the upright manubrium. | 4. The mouth of the medusa is four-sided and lies at the lower end of the hanging manubrium. |
| 5. The tentacles of the polyp are situated at the base of the manubrium. | 5. The tentalces of the medusa are situated at the margin of the umbrella. |
| 6. There is no velum | 6. Velum, which is a small inward projection all along the margin, is present. |
| 7. The gastrovascular cavity is a single spacious cavity occupying the whole of the body of the polyp. | 7. The gastrovascular cavity is restricted and differentiated into the cavity of the manubrium, radial canals and a circular canal. |
| 8. There are no endodermal lamellae. | 8. There is an endoderm lamella between radail canlas in each quadrant. |
| 9. The mesogloea is almost of uniform thickness throught. | 9. The mesogloea is much thicker on the exumbrella side than on the other side. |
| 10. The nervous system is poorly developed, consisting only of a double nerve-net. | 10. The nervous system is much better developed, consisting of a double nerve-ring and a double nerve-net |
| 11. There are no sense organs. | 11. There are eight sense organs, the statocysts. |
| 12. The polyp feeds and protects the colony. | 12. The medusa brings about sexual reproduction and dispersal of the colony. |
| 13. Polyps are produced directly on the hydrocauli. | 13. Medusae are produced on the blastostyles and not directly on hydrocauli. |
| 14. Polyp belongs to the asexual generation. | 14. Medusa belongs to the sexual generation. |

Medusa shows a distinct morphological advancement over the polyp in having sense organs and nerve-ring in the nervous system; almost an independent muscular tissue; differentiation of gastrovascular cavity into region like gullet, stomach, radial canals and circular canal; gonads for sexual reproduction; and free-swimming life.

## Polyp Versus Hydra

Polyp of *Obelia* resembles *Hydra* in form, structure and mode of feeding. The two, however, differ in many respects. The more obvious differences are tabulated below -

### Camparision of the Structure of *Hydra* and *Obelia*

| *Hydra* | *Obelia* |
|---|---|
| 1. *Hydra* is a solitary animal but because of the presence of buds, sometimes it may also be branched. Later these buds separate from the parent and lead independent life. | 1. *Obelia* is a colonial animal. It appears like a branched tree because the lateral branches do not separate from it. |
| 2. It is semi-sedentary and remains temporarily attached on aquatic plants, stones or the wall of the pond. It can be occasionally seen floating in water. The *basal disc* helps in attaching it to the substratum. | 2. It is fully sedentary and ramins permanently attached to marine plants and rocks. It is attached to the substratum by means of *hydrorhiza.* |
| 3. Polymorphism is absent in it. Only polyp or hydranth stage is found. | 3. It is polymorphic animal which is found in three forms.<br>(a) Polyp or hydranth<br>(b) Blastostyle<br>(c) Medusa. |
| 4. There is no covering around it | 4. The polyp is enclosed in a conical cup-shaped *hydrotheca.* |
| 5. Its body is long and cylindrical, and its lower end closes to form the basal disc by which it remains attached to the substratum. | 5. The polyp is long and wine glass-shaped. Its both the ends are open and lower end is connected with the hydrocaulus through a circular shelf. |
| 6. The free distal end of *Hydra* is conical and called *hypostome.* There is a mouth at its apex. | 6. The hypostome of polyp forms its 1/3 part and a mouth is present at its apex. |

| | |
|---|---|
| 7. There are 6 to 10 tentacles around the mouth on the hypostome. These tentacles are hollow. | 7. The number of tentalce around mouth is about 25 and they are solid. |
| 8. A thin transparent membranous cuticle forms the outer covering of the body of *Hydra.* The perisarc is absent. | 8. The cuticle is absent in *Obelia* and the whole colony is covered by a thick and tough *perisarc.* |
| 9. The body wall of *Hydra* is made up of two layers the outer *ectoderm* and the inner *endoderm.* | 9. The same two layers are found in *Obelia.* |
| 10. A non-cellular *mesoglea* is present between these two layers. | 10. The mesoglea is also non-cellular. |
| 11. It is not so in *Hydra.* | 11. The two layers of the polyp are united with the two similar layers of the hydrocaulus. |
| 12. A gastrovascular cavity is present but it is closed at the lower or basal end. | 12. The gastrovascular cavity of the polyp is in communication behind with the coenosarc of the hydrocaulus and the hydrorhiza. |
| 13. The following seven types of cells are found in the outer layer (ectoderm).<br>(a) Epithelial cells<br>(b) Interstitial cells<br>(c) Nematocysts<br>(d) Sensory cells<br>(e) Nerve cells<br>(f) Gland cells<br>(g) Reproductive cells. | 13. Only five types of cells are found in the outer layer (ectoderm) of the polyp. The gland cells and reproductive cells are absent in it. |
| 14. The nutritive and reproductive activities both are done by the *Hydra.* The blastostyles and medusae are absent in it. | 14. In *Obelia,* the polyp is only nutritive form which supplies food to the colony. It has not reproductive function. The blastostyles (asexual) and medusae (sexual) are the reproductive forms. |

*Hydra* and *Obelia* both are the members of the Order Hydroida, Class Hydrozoa and Phylum Coelenterata, by which it is appeerent that there should be many similarities between them, however, there are also many dissimilarities in them. The following differences are met within their life histories.

## Differences between the reproduction and life histories of *Hydra* and *Obelia*

| *Hydra* | *Obelia* |
|---|---|
| **(A) Reproduction** | |
| 1. The reproduction is *Hydra* takes Place by two methods : (a) Asexual (b) Sexual. | 1. Only sexual reproduction occurs in *Obelia* although the whole colony grows by asexual method of budding. |
| 2. The asexual reproduction mostly occurs by *budding*. *Hydra* is also capable of fission and regeneration. The bud separates from the parent body after sometime and leads an independent life. | 2. In asexual reproduction, the colony increases in size and new *zooids* are formed which do not separate from parent colony. |
| 3. The sexual reproduction takes place by the formation of gametes. | 3. The gametes are also formed in the *Obelia*. |
| 4. The gonads are formed in the polyp condition in *Hydra*. Therefore, the polyp also acts as a reproductive zooids. | 4. The gonads are formed in the medusa form in *Obelia*. |
| 5. *Hydra* is mostly bisexual but unisexual individuals also occur. | 5. The Medusa of *Obelia* is always unisexual. |
| 6. The gonads of *Hydra* are temporary structures which are formed only during breeding season. | 6. The gonads of *Obelia* are permanent structures. |
| 7. The gonads are formed from the interstitial cells of the ectoderm and project out on the body wall. There is only a layer of ectoderm on their outer side. | 7. The gonads are formed on the subumbrellar surface from the ectoderm and are situated in the mesogloea between the ecto-and endoderm. |
| 8. The number of gonads is not definite. | 8. Only four gonads are present is a medusa. |
| 9. The male gonads or testes are found in the anterior region of the body while the female gonads or ovaries are situated in the posterior region near the base. | 9. The gonads in medusa are of one sex only i.e. either male or female and are situated on the radial canals |
| 10. The reproductive cells are formed by the division of interstitial cells and collect together to form gonads. | 10. The reproductive cells are formed in the ectoderm of the manubrium and migrate to mesogloea to form gonads. |

| | |
|---|---|
| 11. The mature female gametes do not come out in water from the parent body. | 11. The male and female gametes both separate from the parent body after maturation and swim freely in water. |
| **B. Life History** | |
| 12. The fertilization is external and completed in the ovum while attached to the parent body. | 12. The fertilization is external and takes place in the ovum swimming in the water. |
| 13. The development of the zygote takes place in the parent body. | 13. The development of zygote takes place in water. |
| 14. The cleavage of the zygote is holoblastic and equal. | 14. The cleavage of the zygote is also holoblastic and equal. |
| 15. A solid ball-like morula is formed as a result of division. Later with the formation of blastocoel, a blastula having eight blastomeres is formed. | 15. A solid morula is formed by division. It later changes into a single layered hollow blastula. |
| 16. As a result of delamination the blastula changes into a two-layered gastrula. | 16. By the process of delamination, the blastula changes into a solid gastrula. |
| 17. A double layered cyst is secreted by the outer layer of the gastrula around the embryo. The outer cyst wall is thick, hard and spiny but the inner cyst wall is thin and gelatinous. | 17. Cilia are formed on the ectodermal cells of the gastrula as a result of which an oval, ciliated *planula* larva is formed. |
| 18. In the encysted condition, the embryo detaches from the parent body and lies in the mud of the pond and remains in dormant condition. | 18. The planula larva leads a free swimming life and does not rest. |
| 19. The cyst ruptures with the onset of favourable conditions in spring season and a small *Hydra* comes out of it which grows into the adult. | 19. After leading a free swimming life, the planula larva attaches to some substratum and transforms by growing into an adult colony. The planula first changes into *hydrula* whcih forms the different zooids of the colony by budding. |
| 20. Alternation of generations is not found in the life history of *Hydra* because only one form of zooids is found in it. | 20. Alternation of generations or metagenesis occurs in the life history of *Obelia*. A regular alternation of asexual and sexual stages is found in it. |

♦♦♦

# 9

# CORAL AND CORAL REEFS

Corals are a group of marine coelenterates which are mostly colonial but some are solitary and occur only in polyp stages. Most (ture corals) of them belong to the order Madreporaria but some others belong to sub-class Octocorallia.

Besides a group of Hydrozoa, the *Millipora* and its allies also are called corals due to their skeletal structure but they have no real similarity with ture corals. Since the true corals are the builders of coral reefs and island so only the Madreporarian corals are considered here.

**Definition of Coral Reef.**

***Vaughan*** (1917) has defined a coral reef, "A coral reef is a ridge or mound of limestone, the upper surface of which is near the surface of the sea and which is formed of calcium carbonate by the action of organisms chiefly corals". *Bayer* and *Owre* (1969) mention, "The uppermost layer is a living stratum composed of growing corals, alcyonarians and millepores, together with a virtually endless array of other organisms that live on or in this frame work, some acting to break down the coral skeletons, others serving to cement the

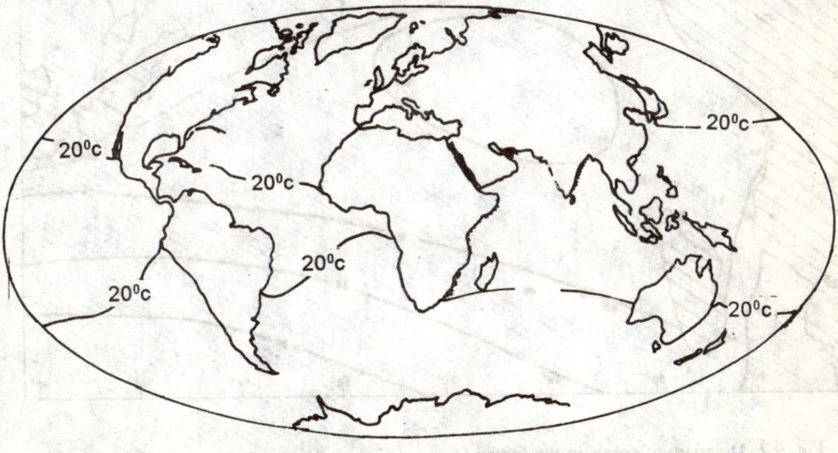


Fig. 9.1 Distribution of coral reefs today (heavy shading)

resulting debris together into a conglomerated mass that can act as a foundation for further coral growth. Boring sponges, molluscs, worms and brancles permeate the coral substance and open it to erosional forces that weaken it until it crumbles. Algae, sponges, hydroids, tunicates and other organisms, as well as chemical process, consolidate the fragments and provide a platform for new growth."

**Requirements and Distribution**

The reef-building corals require warm, clear, shallow water. Therefore, they are confined to continental and island shores in tropical regions (latitude 28° N - 28° S). They flourish best at temperature between 22°C- 28°C. As temperature falls with an increase in the depth of sea water, the reef building corals occur up to a depth of 30 metres, or at the most 50 metres. They fail to grow in shaded areas, which may otherwise be suitable. They inhabit water subject to strong wave action, because they cannot remove large amounts of sediment likely to accumulate on them in quiet waters. Excessive rains and fresh water are harmful for corals.

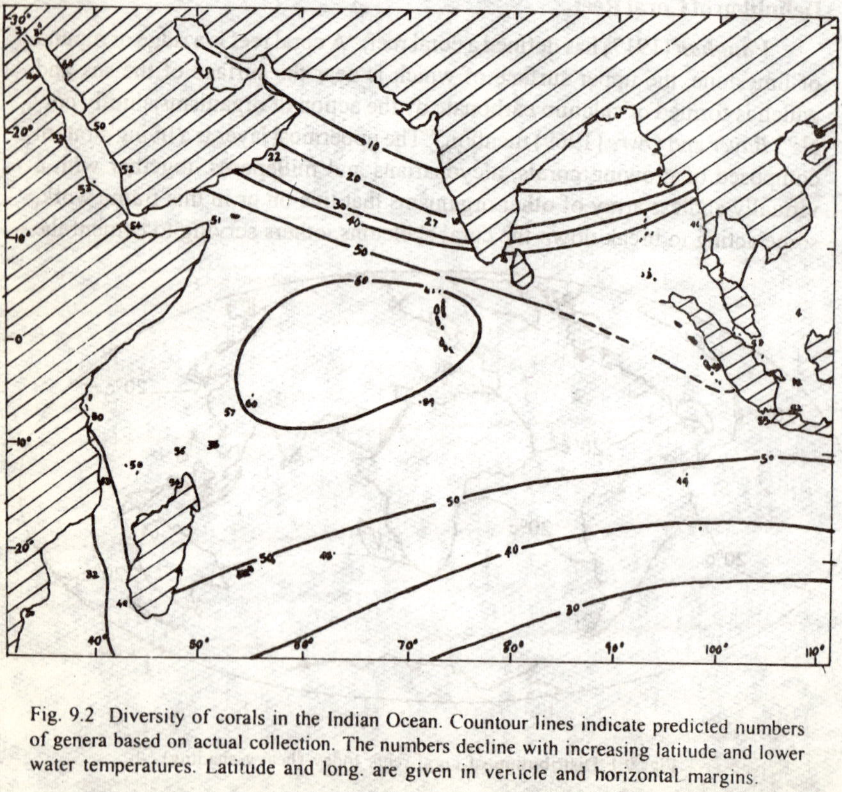


Fig. 9.2 Diversity of corals in the Indian Ocean. Countour lines indicate predicted numbers of genera based on actual collection. The numbers decline with increasing latitude and lower water temperatures. Latitude and long. are given in venicle and horizontal margins.

The reef building corals are found in two general regions : (i) the *Caribbean waters* including Florida, Bermuda, the Bhamas, and the West Indies; and (ii) the Indo-Pacific water from the east coast of Africa through the Indian Ocean and the Western Pacific as for as Hawaii. The second region specially abounds in coral reefs, and in fact the Pacific north-east of Australia is known as the coral sea.

**Classification of Corals**

The Madriporarian corals are divided into three groups :

*1.* ***Imperforate or Aporous Corals.*** They have a complete theca, compact sclerosepta and partitioned into loculae e.g., Astraeid corals, *Favia, Flebellum, Meandra* (=*Meandrina,* brain corals).

*2.* ***Perforate corals.*** In these corals the Corallum is extremely porous everywhere and is of loose construction e.g. *Porites, Acropora* (= *Madrepora*), *Montipora.*

*3.* ***Fungid Corals.*** These may be either perforate or imperforate e.g., *Fungia*

Corals can also be classified as follows

**(a)** ***Hydrozoan corals.*** Some of the animals such as *Millepora Stylaster etc.* belonging to the order Hydrocorallina are colonial and are surrounded by calcarcous exoskeleton. The skeleton is secreted by ectoderm. The individual has two types of polyps namely *gastrozooids* and branched *dactylozooids* lodged within the exoskeleton. The dactylozooids are arranged around the central gastrozooid. They help in the formation of coral-reefs.

**(b)** ***Octocorallian corals***. Include soft corals. The coral is formed of a colony of polyps with endoskeleton of separate calcareous spicules embedded in the massive mesogloea. In the colonial coral, *Tubiproa* or organ-pipe coral, the skeleton is made of calcareous spicules consisting of vertical tubes connected together by lateral platforms. The vertical tubes are also partitioned by smaller cross plates. The tubes contain polyps. In *Haliopora* or blue coral, the calcareous spicules form a massive skeleton of corallium. In *Gorgonia* or sea fan, the colony branches in one plane and the axial skeleton is made up of horny material intermixed with calcareous spicules arranged around the polyps.

**(c)** ***Hexacorallian corals.*** These constitute the *stony corals* or *true corals,* They may be solitary or colonial and assume a great variety of forms. They are the main constituents of coral reefs.

**Solitary corals.** *Fungia, Flabellum, Caryophylla,* etc., are the solitary corals or cup corals. The corallite is disc-like, cup-like or mushroom shaped in form and measures 5 mm. to 25 cm. across. It is often without a theca.

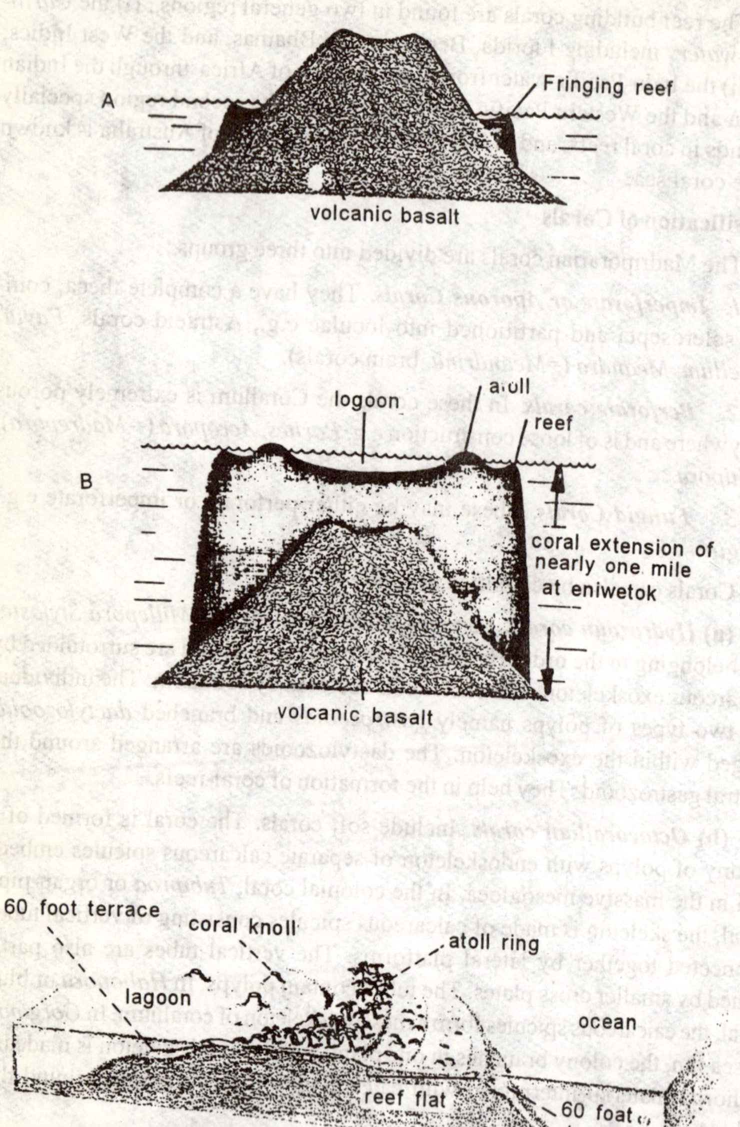


Fig. 9.3 Formation of an atoll. A, Fringing reef around an emergent volcano. B Cont. deposition of coral as volcanic cone subsides leads to the formation of a great coralline cap; emergent part of cap is atoll. C; Section through part of atoll.

**Colonial corals.** Majority of the stony corals is colonial with plate-like, spherical cup-like or vase-shaped skeleton (corallium) The typical examples of colonial corals are *Acropora, Oculina, Favia, Madrepora, Meandrina,* etc. The colony is produced by asexual methods from a single sexually produced polyp. The polyps live at the surface of the calcareos skeleton. Depending on the various methods of asexual budding, varying forms of colonies are produced. Some of the colonies are branched. In *Acropora*, there is always a primary polyp at the top of the colony with lateral branches on either side. In some corals, like *Oculina,* the polyps remain widely separeted, each occupying a separate theca. In others, like *Favia* and *Astraea,* the thecae are so close together as to have commmon walls. In the brain-coral, *Meandrina,* the polyps as well as the thecae become confluent, occupying valleys separated by ridges, on the surface of the corallium.

**Abode of Animals.** Coral reefs are the abode of several other animals as there are numerous interstices, cervices and other hiding places together with sheltering branches of the corals themselves. As such sponges, anemones, sea-urchins, starfishes, crabs, tubicolous-annelids, holothurians, snails and bivalves all live in coral reefs.

**Magnificant colouration.** *Hyman* (1940) writes that as coral beds consist of multitude of organisms of varied shapes and colours, viewed through the deep blue water of a lagoon, constitute one of the most beautiful sights in the world, excel the most splendid flower gardens. The coral polyps are yellow, green or brown due to zooxanthellae, but as they are sometimes permeated with red and green algae whose colours they take.

**Coral polyps.** The coral organism is a small anthozoan polyp measuring about 1 cm. in length. It lacks a pedal disc and the oral disc bears tentacles in cycle of six. The pharynx is devoid of siphonogloyphs. The mesenteries follow the hexamerous plan and are restricted to the upper part of the polyp. The muscles are poorly developed.

**Structure of Coral Polyp**

In structure the coral polyp is much like an Anemone except the skeletal partitions. The structure of the soft part is like sea anemone but pedal disc is absent. Basal region or pedal disc is occupied by skeletal cup. There is no oral cone. Mouth is situated in the middle of the oral disc. A typical oral disc with tentacles are simple and of moderate length ending in a terminal knob of nematocysts. A circular mouth surrounded by a flat peristomium leads into short stomodaeum or pharynx devoid of siphonoglyph. There are complete and incomplete septa or *mesenteries* arranged in cycles on the typical hexamerous plan. Coral polyp being small does not possess more than three cycles of mesenteries. Usually two pairs of directives are present and there are ostia on the septa. The polyps of colonial corals are all interconnected but the attachment is lateral rather than aboral, as in hydroids. The column wall folds outward above the skeletal cup and connects with similar folds of adjacent polyps. Thus, all the members of the colony are connected by a horizontal sheet of tissue which represents folds of body-wall and so it contains an extension of the gastrovascular cavity as well as an upper and lower layer of endoderm and ectoderm. The lower ectodermal layer secretes the part of the skeleton that is located between the cups in which the polyps lie. The living coral colony thus lies entirely above the skeleton and completely covers it.

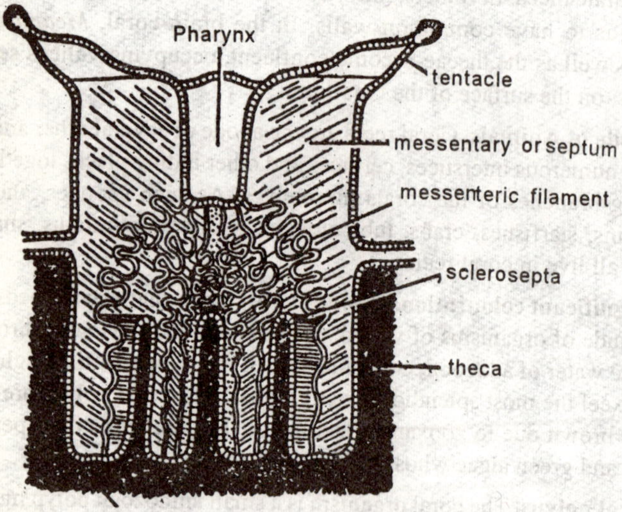


Fig. 9.4 V.S. through a coral in its theca (3 sclerosepta are shown).

## Structure of Coral Skeleton

The skeleton of a colony is termed as *corallum* and that of coral polyp a *corallite.*

***Structure of Corallite :*** Coral polyps remain fixed in a cup like exoskeleton, the *theta*. The theca and other parts of the skeleton wholly to the outside of the body of the polyp but remain in contact with its ectoderm throughout. According is *Voucoch* the skeleton of coral is made up of calcium carbonate and is formed by the precipitation of calcareous crystals in a colloidal matrix secreted by the ectodermal cells outside the body wall for the protection of the polyp.

A typical polyp coral has the following parts :

**(1)** ***Basal Plate :*** A basal plate lies between the polyp and the substratum. It is the bottom of the cup.

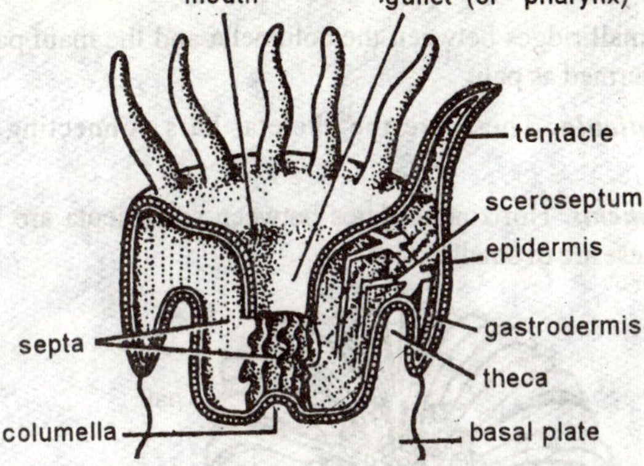


Fig. 9.5 One half of a simple coral polyp showing relationship of soft parts of theca (corallite)

**(2)** ***Theca.*** It is a cup like structure from which the polyp projects outside and into which it can be retracted.

**(3)** ***Sclerosepta.*** These are calcareous ridges or partitions arranged vertically or projecting radially inwards and are connected at the base by basal plate and at the side with the cup like theca. They look like mesenteries but in fact lie between the mesenteries and thus lie in definite relation to them, Like the msenteris, the sclero-septa typically, occur in hexamerous cycle, 12 tertiaries, 24 quaternaries etc; but other numbers are also met with. The sclero septa are usually endocoelic i.e. each skeletal ridge pushes up between the two septa of a pair. The original second cycle is usually, however, exocoelic. The formation of the sclerosepta usually preceedes that of the corresponding septa (mesenteries) and thus there are often move sclerosepta than mesentry pairs. The first cycle of sclerosepta reach the columella, while the later one fall short of it and may fuse with adjacent larger sclerosepta. It may be emphasized that the corallite lies entirely outside the body of the polyp and the polyp base is pushed into ridges over the sclerosepta and goes down into blind pockets between scelerosepta and mesentry.

The sclerosepta are commonly spiny or thorny with toothed upper edges.

**4.** ***Columella.*** It is a pillar like irregular central skeletal mass which may be either an independent outgrowth from the basal plate or one formed by the union of the central ends of the sclero-septa, then called *pseudocolumella.* The columella may be solid or trabeculae.

**5.** ***Epitheca.*** It is distinct calcareous layer which surrounds the base of the theca in a ring-like manner.

6. ***Costae.*** The space between the theca and epitheca is crossed by continuation of the sclerosepta called costae.

7. ***Pali.*** Small ridges between the columella and the main parts of the sclerosepta are termed as pali.

8. ***Synapticula.*** These are the skeletal bars connecting adjacent sclerosepta.

9. ***Disepiments.*** Horizontal plates between sclerosepta are known as disepiments. These are of small extent.

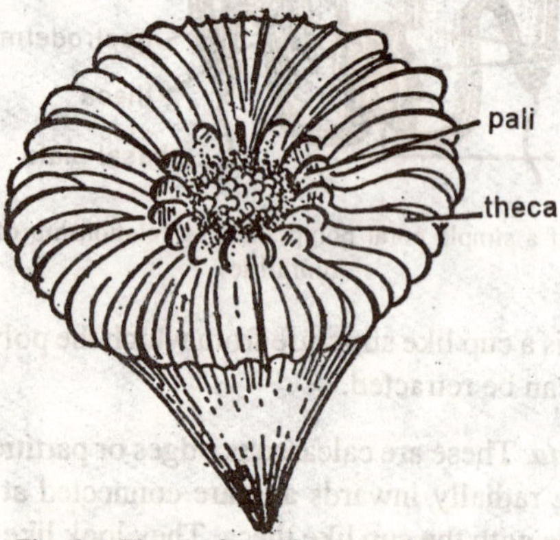


Fig. 9.6 Theca of a solitary coral showing pali.

10. ***Trabeculae.*** When the horizontal plates between sclerosepta are large and extend completely the corallite they are termed as trabeculae.

**Formation of coral.**

The coral polyp develops from a planula which settles down and begins to secrete a skeletal rudiment or *prototheca.* It is secreted by ectoderm first as a basal plate. Following it, the larva develops radial folds which secrete septa (sclerosepta) and at the same time a rim is built up as a thecal wall around the polyp, laying at the top. Meanwhile, further skeletal material is added into the gaps between the septa. The septa of the skeleton usually alternate with the mesenteries of a living coelenterate.

In living condition, the polyp fills the whole of the interior of the corallite and projects beyond its edge. The proximal portion of its body wall is in contact with the theca which is a product of the epidermis. The free part of the body-wall of polyp is folded over the edge of the theca so as to cover its distal portion. Each skeletal septum is covered by an inturned portion of the body-wall. Thus the septa are actually external and are in contact with the epidermis throughout. The space between the theca of the coral colonies is occupied in life by an extension of the polyp walls, *coenenchyme* continuous with the later above the upper edge of the theca and containing a gastrovascular space continuous with gastrovascular cavity of the polyps. The lower surface of the coenenchyme secretes the part of the corallum between the theca, and this part is called *coenosteum*. In addition, in many corals, the polyps may be connected by canals coming from the bases and passing through openings in the loosely connected constructed thecae. Corals in which the corallite is perforated like this with many openings are termed as perforate corals e.g., *Madrepora* whereas those corals in which the corallite are of solid texture and the polyps are connected by coenenchyme only over the upper edge of the theca are called imperforate or a porous corals e.g. *Flabellum, Astraea* etc.

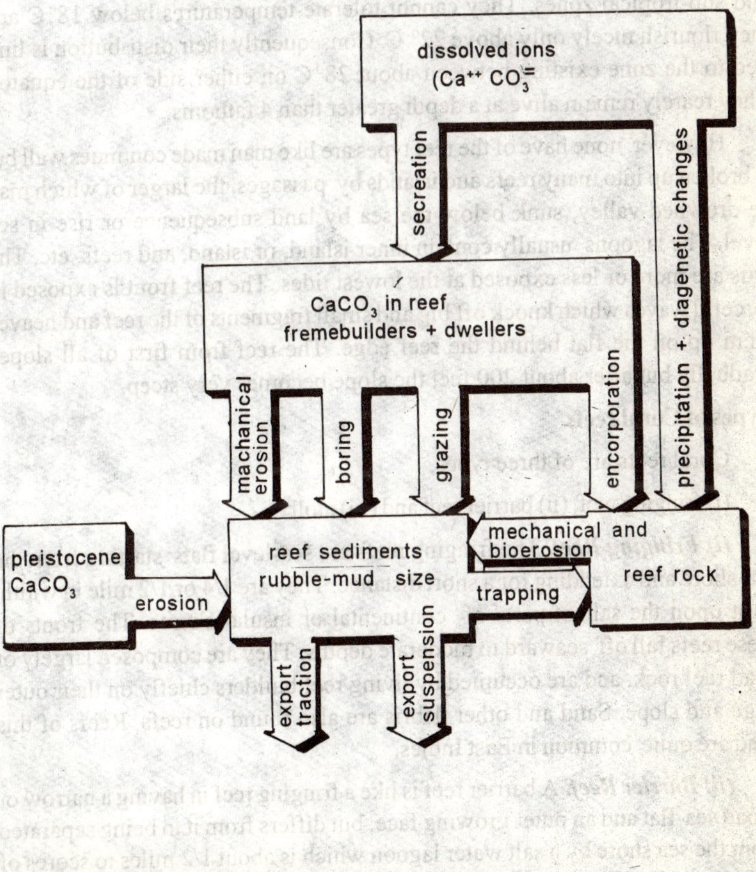


Fig. 9.7 Fate of calcium carbonate on a coral reef.

**Coral Reefs**

A coral-reef is a ridge or mound of lime stone the upper surface of which is near the surface of the sea and which is formed of calcium carbonate by the action of organisms chiefly corals (*Vaughan,* 1917). Though the reefs are built by stony corals but other organisms such as foraminifera, millipora, tubipores, heliopores, the molluscs, echinoderms, some coralline, algae and sponges contequeute together in the formation of compact structure, the coral reefs. The coral reefs are formed by incrusting their skeletal parts on the deposited lime. Coral reefs composed of multiple or organisms vary in shape and colour. The Zooxanthelae is responsible for the rich colouration of corals, which may be brown, yellow, or green. The reef building corals require warm, shallow waters, and consequently are limited to continental island shores in tropical and sub-tropical zones. They cannot tolerate temperatures below 18° C and they flourish nicely only above 22° C. Consequently their distribution is limited to the zone existing between about 28° C on either side of the equator. They rearely remain alive at a depth greater than 4 fathoms.

However, none have of the reef types are like man made continues wall but is broken up into many reefs and islands by passages, the larger of which may be drowned valley, sunk below the sea by land subsequence or rise in sea level. The lagoons usually contain inner island, or island, and reefs, etc. The flats are more or less exposed at the lowest tides. The reef front is exposed to forceful waves which knock off big and small fragments of the reef and heaves them up on the flat behind the reef edge. The reef from first of all slopes gradually but after about 200 feet the slope becomes very steep.

**Types of Coral reefs**

Coral reefs are of three types,

(i) fringing reef, (ii) barrier reef and (iii) atoll

***(i) Fringing Reef.*** The fringing reefs are sea-level flats starting from the sea-shore and extending for a short distance. They are 1/4 or 1/2 mile in width, built upon the salient parts of continental or insular shores. The fronts of these reefs fall off seaward to moderate depths. They are composed largely of dead reef rock, and are occupied by living reef builders chiefly on their outer edge and slope. Sand and other debris are also found on reefs. Reefs of this kind are quite common in East Indies.

***(ii) Barrier Reef.*** A barrier reef is like a fringing reef in having a narrow or broad sea-flat and an outer growing face; but differs from it in being separated from the sea shore by a salt water lagoon which is about 1/2 miles to scores of miles in width. The lagoon may be 20 to 40 fathoms or even more in depth. The inner shore is generally occupied by a fringing reef, the growth of which is less vigorous than that of the non-enclosed fringes. The outer or growing face of the barrier is continued in gentle slop 40 or 50 fathoms in depth follow a steep pitch to great depths.

Barrier reefs are frequently interrupted by *"passes"* or *passages* through which even the ships may enter. Barrier reefs may encircle whole islands. Sometimes small islands may make their appearance in the protected lagoons. These reefs are sometimes a great danger to shipping. The Great Barrier reef of north-eastern Australia is about 1350 miles in lenght, about 20-70 miles wide and encloses a channel from 10 to 25 fathoms deep. At places it is about 90 miles away from the shore. Much of the Reef is well below low tide. It is not a single reef but a long string of separate reefs whch are not always in a line. Inside the Great Barrier Reef are found various types of coral growths. There are hundreds of patches of oral or elongated coral islands built upto the surface. All have their own reefs.

***(iii) Atoll.*** It is a coral island and consists of a belt of coral reef having a central shallow lake communicating with the sea. It is a horse shore-shaped or circular reef enclosing a *lagoon*. These lagoons are 40 or 50 miles across. Several of such atolls occur in the South Pacific. The *Atoll of Biknii* is very famous, which has a land of 287 square miles. Another atoll which is very famous due to prominence in World War II, is *Atoll of Tarawa*.

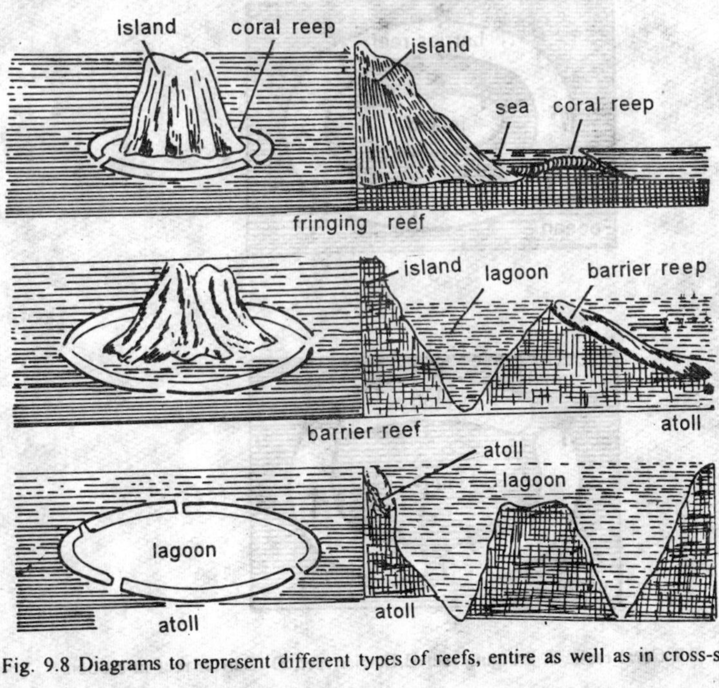


Fig. 9.8 Diagrams to represent different types of reefs, entire as well as in cross-section.

**Growth of the coral to form reefs.** Rate of coral growth is variable and of great importance. Slow growing massive types grow 5 mm. per year and fast-growing ones from 10 to 20 cm. per year. *Vaughan* estimated that a reef of 50 mm. deep is formed in 1,000 to 7,000 years and the age of the present reefs is 10,000 to 30,000 years.

**Formation of Reefs**

The reefs are largely built by tiny polyps of the stony corals (order Madreporaria). The polyps secrete around them limestone cups, which coalesce to form large masses. The later, with the passage of time, take the form of

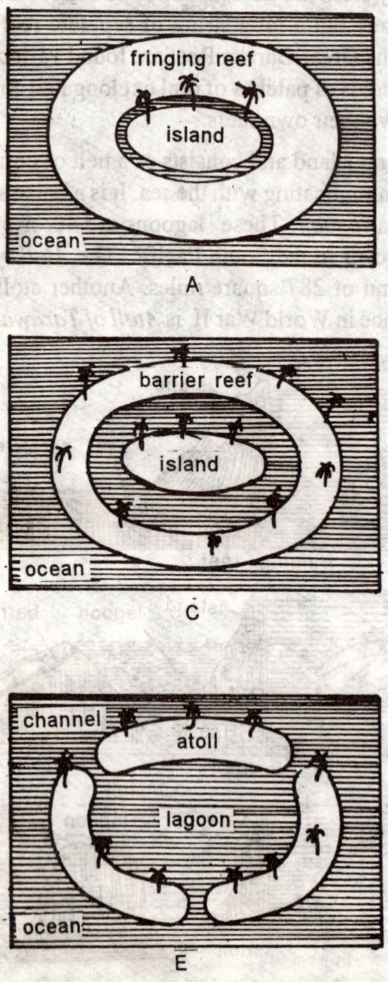


Fig. 9.9 Coral reefs. A-Fringing reef B-Barrier reef. D- Barrıer reef in section.

huge rocks. Some other animals also take part in the formation of reefs. The hydrozoan *Millepora*, calcified alcyonarians *Tubipora* and *Heliorpora*, and gorgonians add their skeletons to the reefs. The clams and tubeworms contribute their calcareous shell and tubes to the reefs. Lime secreting algae repair the windward edge of the reef by cementing loose pieces of coral rock into nearly solid remparts. Sinking shells of foraminiferans fill the tiniest pores in the reefs. Whereas some organisms build up the reefs, others, like burrowing animals, break them down. This turns the reefs into labyrinths of ledges, crannies grottoes crevices and tunnels.

The coral reefs grow very slowly. Most of them expand at the rate of 10-200 mm. per year. The existing reefs seem to have been formed in 15,000 to 30,000 years.

**Theories to Explain Great Vertical Thickness of Coral Reef.**

Since the present reef building corals do not grow below 150 feet at the outside, and becuase geological evidance indicates that corals of the past ages were also littoral, in their habits, it becomes essential to explain the great vertical thickness often attained by the coral reefs. Many theories have been propounded, which the following four are the main ones.

*1.* ***Darwin's Subsidence Theory.*** Charles Darwin's theory explains the formation of reef. According to him the reefs begin as fringe around slowly sinking shores, which continue to grow upward and outward as the land sinks. Thus the fringing reef turns into a barrier reef when coast subsides. The lagoon separating it from the reef becomes wider and wider. Islands surrounded by barrier reefs finally sink beneath the lagoon. Thus the encircling reef is left on which islands of wave tossed. Atolls are formed as accumulation of loose fragments of rocks.

*2.* ***Semper-Murray Solution Theory.*** *Sir John Murray* was the chief biologist on the 'Challenger', the British ship that sailed into the oceans from 1874 to 1876 to explore extensively the conditions and life in the sea. *Murrary* proposed that (i) Corals grow on high summits of the ocean bottom when these have been built up to the right level, (ii) the high summits are built by deposition of sediments and (iii) barrier reefs edge and through solution of the inner coral rock. This theory is now not accepted at all.

*3.* ***Submerged Bank Theory.*** This theory has been supported by many recent studies. According to this theory the coral formation grow on flat, pre-existing surface during or after the submergence of surfaces.

*4.* ***Daly Glacial-Control Theory.*** The main points of this theory are as follows : (i) During the last glacial period much water of the ocean turned into ice forming glaciers due to a very low temperature and thus the level of the ocean was lowered by 60 to 70 metres below the present level. (ii) Various terraces were then cut or islands levelled by wave action. (iii) Later, with the rising, temperatures, corals began to grow upon these platforms and kept pace with the rising sea-level as the ice melted.

This theory explains nicely about the uniform depth of coral lagoons, whose bottoms, below the debris since deposited, would consist of the platforms cut when the ocean was at its low level.

Theories 3 and 4 supplement each other and at present are most favoured by the students of the problems, although Darwin's idea (Darwin-Dana subsidence theory) continues to find much support. The submergence theory agrees with Darwin's subsidence theory in that both consider the reef foundations to be now at great depths than they were when the coral growth started. But the submergence theory does not admit any relationship between that various kinds of reefs and postulates that the barrier reef and atoll have grown upon pre-existing flat platforms. The atolls are considered to have been shaped by winds, waves and currents.

Boring have been made to find out the truth of these theories.

(i) Boring of Funafuti atoll in the South Pacific north of Fiji was made in 1904 by an expedition of the Royal Society of London. The boring was 3 inches to 5 inches in diameter and went up to 1114 feet without reaching the reef base. Twenty eight genera of reef building corals were discovered, 2 of which are now living on the reef in that locality above 60 metres. The material obtained from the boring did not contain nay of the deep water corals lives in that locality at the depth to which the boring went. This finding supports subsidence theory.

(ii) *Cary* (1931) made 3 boring at different distances from the shore into a reef in Samoa (Pacific island) and concluded that the reef rested on a level platform cut by the action of waves. This supports the glacial control theory.

(iii) The Great Reef Committee made two boring, one in 1928 and the other in 1938 on the Great Barrier Reef. Both borning gave the same result that the coral material extended out only to 400 to 450 feet and below this there was nothing but sand containing shells of various animals. There was no evidence of any underlying platform. Therefore this finding also supports Darwin's subsidence theory. Thus Darwin's subsidence theory applies of many reefs but some reefs may have been laid down on pre-existing platforms.

***Some important facts about reefs.*** Tropical storms can modify the reefs very much. Quarrying can also damage reefs as in India, where in 1971 *Pillai* has estimated that nearly 250 cubic meters of reefs material are removed per day for use in the production of cement, calcium carbide, calcium carbonate etc. Recently much concern has been expressed for the safety of some oceanic islands and their reef food chains because some reefs are being destroyed by ever increasing population of the Crown-of-thorns starfish.

*Acanthaster planei* starfishes feed upon the living coral and have caused much reef destruction in some parts.

**Economic importance of coral reefs**

The coral reefs are of much importance to oil industry. They form highly favourable sites for the accumulation of petroleum deposits.

The coral reefs are of importance for *curio trade*. Many plant and animals like sponges, molluscs, fishes, echinoderms etc. grow on these reefs. Even some humans inhabit them.

Some corals form highly priced decorative pieces.

*Corellium rubrem* is considered a precious stone. The red coral and organ pipe coral are used in medicine in South India. Skeleton of a few corals are used as building material.

Coral skeletons are also used in the formation of lime, mortar and cement. The skeletons are also used in making ridges which act as natural barriers against sea erosion and cyclonic storms.

The coral reefs serve as good nursery grounds for commercially important fishes. They form more colourful and beautiful fishes.

♦♦♦

# 10

# HELMINTHES

*Helminthes* constitute a large assemblage of *"worms"* of comparatively simple and varied organization. Helminthes is derived from the Greek word *helmins* means *"worms"*.The term worms is not correct because it is applied to elongated invertebrates without appendages and with bilateral symmetary .In 18th Century all the worm-like animals were included in an old Phylum-Vermes (*Linnaneus* ,1735).But this was a very heterogenous group which included a highly diverse assemblage of forms.In the modern classification this group is divided in to several independent phyla such as Platyhelminthes, Aschelminthes, Annelida etc.

The term Helminthes is restricted to a few phyla of invertebrate animals, all of which are superficially worm like but they differ markedly in their morphology, life-history and bionomics.The helminthes are restricted to two phyla of animal kingdom. These are :

1. Platyhelminthes or flat -worms .
2. Aschelminthes or round -worms.

## PHYLUM -PLATHELMINTHES

Term Platyhelminthes was proposed by *Gegenbaur* in 1859,means flatworms which refers to their characteristic contour of flattened body . The term derived from two Greek words -*platys* = flat ; *helminthes* = worms . The animals show low organization as they are without annus ,skeletal ,respiratory and circulatory systems. The body is filled with mesenchymal cells which are mesodermal in origin.

## GENERAL CHARACTERS

1. Bilateral symmetrical body ,with dorso -ventral flattening.
2. Body generally shows anterior cephalization that is deferentiation of anterior region into head.
3. Body generally worm like either long flat ribbon shapped or leaf-like. The

outline of body bizaree shape in trematodes and segmented tape-like body in cestodes.

4. Most of the flat worms are of small to moderate dimensions to size varies from microscopic to the extreme elongation as much 10 to 15 meters.
5. General colour creamy white but on account of the presence of food the body may acquire colour. Some of the free-living flat worms are brilliantly coloured ,often brown, black, gray.
6. Body with various types of attchment in the form of adhesive secretions, suckers, an hooks.
7. Body consists of three embryonic layers viz. ectoderm, mesoderm, endoderm.
8. In practice forms, the body is covered with a thick cuticle but free-living turbellarians are clothed with cellular or syncytial epidermis.
9. The exo-and endoskeleton are completely absent therefore the body is generally soft with hooks ,spines ,spicules and thorns.
10. Body space is filled with a mesenchyma that is perenchy-matous tissue, thererfore, flat worms are *acoelomate* that is without coelom.
11. The digestive system is generally absent in tapeworms and acoela. But in other consists of mouth, pharynx and variously modified intestime.
12. Respiratory and circulatory systems are absent but some of the trematodes have system of tubes the so called *lymphatic system* of uncertain function.
13. The nervous system is of primitive type with missing of nervous tissue as *brain* in anaterior-most region with several longitudinal ganglionated cord. Numerous transverse connections occur between the longitudinal cords. So the nervous system is generally ladder type.
14. Parasitic forms are without sensory organs but free-living forms are provided with sensory organs in the form of chemo-and tango-receptors, ciliated pits, statocysts and ocelli or eyes.
15. Well developed excretory system with *proto-nephridial tubules* having flame bulbs or cells. But accoela are without protonephridia.
16. Hermaphrodite rarely with separate sex. The eggs are generally devoid of yolk but with yolk cells. Varieties of copulatory organs are present. Cross-fertilization and self-fertilization both are common.
17. Development direct i.e. without larval stage or indirect with larval stages. Life cycles complicated with succession of larval stages involving one to three intermediate hosts.

18. Asexual reproduction common among fresh water forms.
19. Parthenogenesis and polyembryony common among trematodes and cestodes.
20. Endogenous and exogenous budding common in tapeworms.

Phylum Platyhelminthes is divided into three classes. The classification followed here is from *Hyman* (1951).

I Class - Turbellaria

II Class - Trematoda

III Class - Cestoda

## I. CLASS-TURBELLARIA

Turbellaria is derived from Latin word *'turbella'* means a stirring. The animal produce turbulance in water by body cilia. They posses following characters:

1. The members of this class are broad, flat and leaf - like in shape.
2. They lead a free living life in fresh or salt water, in moist soil, some members may be commensal or parasite.
3. Body unsegmented, dorso-ventrally flattened and covered with cilia.
4. The anterior end is differentiated into head.
5. Epidermis cellular or *syncytial* contains rod-like structures called *Rhabidites*.The rhabidites are curved structure capable of discharging a chemical secretion.
6. Epidermis and mesenchyma are provided with unicellular glands of two types (a) *cyanophilous* which secrete adhesive secretion and (b) *eosinophilous*.
7. Turbellarians are abundantly provided with adhesive organs which are of two types(a) *glandulo-epidermal adhesive organs* and (b) *glandulo-muscular adhesive organs*.
8. Alimentary canal incomplete i.e. with single opening, the mouth. Mouth is mid-ventral, sometimes near the centre of body. Intestine is blind or absent.
9. Excretory system with protonephridia.
10. Sense organs consist of tangoreceptors and chaemoreceptors.
11. Mostly bisexual or hermaphrodite or unisexual with very few exceptions.
12. Reproduction by asexual, sexual methods and by regeneration.
13. Cleavage spiral. Development direct without larval stage.

## Classification of Turbellaria

The members of this class are found in almost all places. It includes more than 1500 species. Class Turbellalria is divided into 5 orders.

1. Acoela
2. Rhabdocoela
3. Allocoela
4. Tricladida
5. Polycladida

***Order I- Acoela (Gr.a = without, koιlos = cavity)***

1. They are minute, delicate turbellarians with a ventral mouth.
2. Mouth and pharynx are present. Pharynx simple when present but usually absent and so the intestine is also absent.
3. Parenchyma syncytial not differentiated into mesoderm and endoderm.
4. Protonephridia absent.
5. Food is taken in solid form and digested by endodermal cells.
6. Gonoducts, gonads and yolk glands are absent.
7. Exclusively marine forms living under stones, algae etc. at the sea bottom. some live in the intestine of holothurians.

   Examples: *Convoluta, Hoplodiscus.*

***Order 2- Rhabdocoela (Gr.rhabdos = rod; koils = cavity)***

1. Small tubellarian usually less than 3mm in length.
2. Mostly free-living, marine or fresh water. Some of them inhabit in the respiratory cavities of various fresh water crustaceans, viz. *Temnocephala*.
3. Body is elongated, cylindrical or compressed with simple pharynx and simple unbranched sac-like intestine.
4. Yolk-glands may or may not be present.
5. Male with two to many testes, whereas the female with 1 to 2 ovaries.
6. Nervous system consists of two longitudinal nerve cords. Eyes are present in most of them. Order Rhabdocoela is divided into four sub-orders:

***Sub-order (i) Notandropora***

1. Exclusively fresh warter forms.

2. Pharynx simple.
3. Single median protonephridium.
4. Single compact testis mass, penis unarmed.
5. Yolk-gland is absent.
6. Asexual fission with chains of zooids.

   Examples: *Catenula, Rhycoscolex, Stenostomum.*

***Sub-order (ii) Opisthandropora***

1. Fresh water or marine forms.
2. Pharynx is simple.
3. Excretory system consists of paired protonephridia.
4. Compact testes, penis armed with a stylet.
5. Yolk glands absent.
6. Asexual reproduction with chains of zooids.

   Examples: *Microstomum, Macrostomum.*

***Sub-order (iii) Lecithopora***

1. Fresh water, marine or terrestrial, commensel or parasitic.
2. Pharynx bulbose.
3. Protonephridia paired.
4. Separate ovaries and yolk glands.
5. Exclusively sexual reproduction.

   Examples: *Graffilla, Anoplodium, Mesostoma, Gyratrix.*

***Sub-order (iv) Temnocephalida***

1. Freshwater, ectocommensal forms.
2. Anterior end having 2-12 tentacles.
3. Posterior end with 1 or 2 adhesive discs.
4. Pharynx doliform.
5. Gonopore simple.

   Examples: *Temnocephala, Monodiscus.*

***Order 3- Alloecoela (Gr.alloios = different; koilos = cavity)***

1. Mostly marine, few freshwater and brakish water forms. Some are ectoparisite or ectocommensal in habits.

2. The pharynx is bulb-like or plicate.
3. Intestine is straight or branched.
4. Excretory system consists paired protonephridia with two or three branches and nephridiopores.
5. Nervous system with 3 or 4 pairs of longitudinal nerve cords provided with transverse commissures.
6. Reproductive system consists of numerous testes and a pair of ovaries. Penis papilla mostly present. Order Alloecoela is divided into four Sub-orders.

***Sub-order (i) Archophora***

1. Marine animals.
2. Pharynx plicate.
3. Female reproductive system primitive.

   Example: *Proporoplana*

***Suborder (ii) Lecithoepitheliata***

1. Marine,freshwater or terrestriala forms.
2. Pharynx simple or bulbose.
3. Penis with a cuticular stylet.
4. Female duct simple or none. Typical yolk glands absent, nutritive cells may surround the ova.

   Examples: *Hofstenia, Prorhynchus.*

***Suborder (iii) Cumulata***

1. Freshwater or marine forms.
2. Pharynx plicate or bulbose.
3. Intestine mostly without diverticulae.
4. Penis unarmed.
5. Separate ovaries and yolk glands or germovitellaria.

   Examples: *Hydrolimax, Plagiostomum.*

***Suborder (iv) Seriata***

1. Mostly marine and a few are freshwater forms.
2. Pharynx plicate.
3. Intestine with lateral diverticula.

4. Statocyst mainly present.
5. Ovaries and yolk glands are separate.

   Examples: *Monocelis, Otoplana.*

**Order 4 Tricladida (Gr.treis= three; klados = branch)**

1. Marine, freshwater or terrestrial forms.
2. Large turbellarians, measuring 2-60 cm in length.
3. Pharynx plicate usually backwardly directed.
4. Intestine three - branched sacs.
5. Eyespots usually present.
6. Two ovaries, 2 or numerous testes.
7. Female reproductive system provided with one or two copulatory bursa.
8. Penis with penis papilla, common genital aperture.

   Order Tricladida is divided into three suborders.

**Suborder (i) Maricola**

1. Exclusively marine forms.
2. A pair of eyes and auricular grooves.
3. Typical penis papilla sometimes armed with stylet.
4. Rounded copulaltory bursa is generally present.
5. Only sexual reproduction.

   Examples: *Bdelluora, Ectoplana, Gunda.*

**Suborder (ii) Paludicola**

1. Mostly freshwater rarely brackish water forms.
2. Two to many eyes or eyeless.
3. Bursa with bursala canal, anterior to penis.
4. Reproduction mostly asexual.

   Examples: *Dugesia, Kenkia, Dendrocoelum.*

**Suborder (iii) Terricola**

1. Terrestrial tropical and subtropical forms.
2. Two or many eyes.
3. Bursa, if present, lies behind the penis.
4. Reproduction may be asexual.

   Examples: *Geoplana, Dolichoplana.*

***Order 5 - Polycladida (Gr.Poly = many; klados = branch)***

1. Size modrate, from 2-20 mm. Dorsoventrally flat leaf-like body; they are exclusively marine.
2. Intestine with many branches.
3. Eyes and nerve cords are present.
4. There are numerous ovaries and testes.
5. Genital apertures separate.

It includes two suborders:

***Suborder (i) Acotylea***

1. Pharynx usually vertical curtain like.
2. No suckers are present behind the female pore.
3. Tentacles nuchal type.
4. Eyes concentrated at anterior margin, not in pairs.

Examples: *Stylochus, Notoplana.*

***Suborder (ii) Cotylea***

1. Sucker present behind the female pore.
2. Pharynx tubular.
3. A pair of marginal tentacles bearing eyes or with a cluster of eyes in place of tentacles.

Examples: *Thysasmozoon, Yungia*

## CLASS II TREMATODA

1. They are ecto-or endoparasites and commonly called *'flukes*
2. Body is usually unsegmented, dorso-ventrally flattened and leaf-like.
3. Size varies from 0.5 mm to 1 or several cm. long.
4. Body is covered by thick cuticle and no cilia.
5. Suckers and sometimes hooks area present. Oral sucker present at anterior end for feeding, ventral sucker for attachment.
6. Epidermis absent. *Rhabdites* are absent.
7. Digestive system consists of mouth, muscular pharynx and intestine usually with two main branches. Anus is absent.
8. In some cases there is a system of parenchymal vessels showing the presence of primitive circulatory system.
9. Excretory organs consists of protonephridia.

10. There are three pairs of longitudinal nerve cords. Sense organs are poorly developed.
11. Bisexual or hermaphrodite except blood fluke (*Schistosoma*).
12. Single ovary and many testes. Cross fertilization takes place. Development direct or indirect.
13. Life-cycle simple or complicated with more than one hosts.

There are about 6000 species in class Trematoda. This class is divided into three orders:

Order 1 - Monogenia

Order 2 - Aspidobothria

Order 3 - Digenia

***Order 1 - Monogenia***

1. They are ectoplarasite mostly in cold-blooded animals (fishes).
2. The life-cycle is simple and passed in one host only, *monogenetic*.
3. Oral sucker absent, when present is poorly developed.
4. A pair of adhesive structures, the *opisthaptor*, is found on the posterior end.
5. A large disc-shaaped structure, the opisthaptor, is found on the posterior end. It contains hooks and is lined with chitin.
6. Paired excrecory pores are present in the anterior part of the body.
7. Male and female genital pores are separate.
8. One or two vaginae are present. The uterus is small and contains few eggs.
9. Development is indirect through a larvala stage called *onchomiracidium*. Order Monogenea is divided in two suborders:

***Suborder (i) Monopisthocotylea***

1. Opisthaptors are present without suckers but with hooks.
2. Genito-intestinal tract is absent.
3. Posterior adhesive organ is single or divided by septa.
4. Eye-spots usually present.

Examples: *Gyrodactylus, Monocotyle, Heterocotyle, Amhibdella.*

***Suborder (ii) Polyopisthocotylea***

1. Opisthaptor contains many suckers and also hooks.

2. Oral suckers two, opening into oral cavity.
3. Genito-intestinal tract present.
4. Eye-spots are absent.

   Examples: *Polystomum, Diplozoon, Microcotyle.*

***Order 2- Aspidobothria***

1. Endoparasite in the digestive tract of fishes and reptiles.
2. Oral sucker is absent.
3. The ventral sucker is divided by septa into many small parts. The hooks are absent it it.
4. Single excretory pore.
5. Only one testis is present.
6. Development direct, without alternation of hosts.

   Order Aspidobothria is divided into two families:

***Family (a) Aspidogastridae***

1. Sucker enormous circular, oval or elongated, divided by septa into one, three or four longitudinal rows of depressions termed *alveoli.*
2. Protonelphridia have separate bladders and separate or a common nephridiopore situated near the posterior end.
3. Testis single.
4. *Laurer's canal* may be present.
5. Endoparasite in the mantle, pericardial and renal cavities of clams and snail and in digestive system of fish and turtle.

   Examples: *Aspidogaster, Cotylaspis, Cotylogaster.*

***Family (b) Stichocotylidae***

1. It consists of only one genus which lives in bile passage or spiral valve of fishes.
2. The life cycle is complex one, involving at least two hosts.

   Example: *Stichocotyle.*

***Order 3 - Digenia***

1. They are endoparasite in invertebrates and verterbrates.
2. Two to four hosts in the life-cycle.
3. Two suckers, without hooks, mouth within anterior sucker.

4. There is only one excretory pore present in the posterior region of the body.
5. There is a common gonopore for male and female reproductive systems. Vagina is absent.
6. Uterus is long branched with many eggs.
7. Their larval forms reproduce asexually before metamorphosis.

Order Digenea is divided into many families; some important ones are as :

***Family (a) Bucephalidae***

1. They are commonly called *gastrostomes.* The mouth is located in the middle of the body.
2. Anterior end bears a simple or lobate oral sucker or has a generally lobed hood-like extension.
3. Intestine sacciform .
4. Acetabulum absent.
5. Genital pore posterior
6. The cercaria larvae known as oxhead, have a pair of long tails.
7. The primary host is fish and secondary host is clam.

Examples: *Bucephalus, Bucephalopsis.*

***Family (b) Paramphistomidae***

1. Body thick and fleshy with a powerful posterior sucker.
2. Cirrus pouch is present.
3. Oral sucker is absent.
4. Gonopore, situated in the anterior body half and provided with in some genera with a genital sucker.
5. Nephridiopore dorsal infront of acetabulum.
6. Uterus long having ascending limb. Vitelline glands follicular.
7. These are the endoparasites of gut and bile ducts.

Examples: *Paramphistomum, Gastrothylax.*

***Family (c) Opisthorchidae***

1. These are distomes of small to large size.
2. They live in the bile passage of fish-eating reptiles, birds and mammals.
3. Body flat and semitransparent.

4. Excretory bladder 'V' shaped.
5. Testes posterior. Genital pore infront of acetabulum.

   Examples: *Opisthorchis, Amphimerus, Metorchis.*

***Family (d) Fasciolidae***

1. Commonly known as *liver flukes*.
2. Body leaf-like and acetabulum near the oral sucker.
3. Gonopore just infront of the acetabulum.
4. Testes, ovary, vitelline glands all are branched.

   Examples: *Fasciola hepatica, Fasciplopsis buski.*

***Family (e) Echinostomatidae***

1. Body slender, elongated with spiny or scaly cuticle.
2. Acetabulum large behind the oral sucker.
3. Collar of spines is present in cercarial and metacercarial stages.
4. It is commonly found in the intestine of birds.

   Examples: *Echinostoma, Echinoparyphium, Parochis, Himasthla.*

***Family (f) Schistosomatidae***

1. These are *blood flukes* that inhabit the hepatic portal system and pelvic veins of birds and mammals.
2. Saxes are separate.
3. Female lodges in a *gynaecophoric canal* of male.
4. Testes numerous and ovary tubular.

   Example: *Schistosoma.*

   Besides above families the others are:

| | |
|---|---|
| Family - Zoognidae | Examples: *Zoogonus*, *Nassa* |
| Family - Allocredidae | Examples: *Bunodera*, *Plagioporus.* |
| Family - Azygiidae. | Examples: *Azygia*, *Otodistomum.* |
| Family - Hemiuridae. | Examples: *Haliplegus*, *Derogenes.* |
| Family - Didymozoonidae. | Example: *Dadymocystis.* |
| Family - Gorgoderidae. | Examples: *Gorgoderina*, *Gorgodera.* |
| Family - Plagiorchidae. | Examples: *Plagiorchis*, *Stylphlodora.* |
| Family - Dicrocoelidae. | Examples: *Dicrocoelium*, *Eurytrema.* |
| Family - Heterophyidae. | Examples: *Stictodera*, *Heterophyes.* |
| Family - Microphallidae. | Examples: *Spelotrema*, *Maritrema.* |

Family - Troglotrematidae. Examples: *Paragonimus*, *Troglotrema*.
Family - Cyclocoelidae. Examples: *Cyclocoelus*, *Typhlocoelum*.
Family - Clinostomatidae. Example: *Clinostomum*.
Family - Notocotylidae. Examples: *Notocotylus*, *Catatropis*.
Family - Strigeidae. Examples: *Ophiososma*, *Cyanthocotyle*.
Family - Brachylaimidae. Example: *Brachylaima*.
Family - Sanguinicolidae. Example: *Sanguinicola*.
Family - Spirorchidae. Examples: *Spirorchis*, *Hapalorhynchus*.
Family - Heronimidae. Example: *Heronimus*.

## CLASS III CESTODA

1. They are found as endoparasites in the intestine of vertebrates and are called *tapeworms*
2. Body is flat, elongate and ribbon-like, measures from 1 mm to 10 metres in length
3. Body is without epidermis, rhabidites, cilia but a cuticle is present.
4. Body is usually divided into a scolex or head, neck and few to many proglottids.
5. Scolex is provided with hooks or suckers or both.
6. Scolex is followed by neck which forms new '*proglottids*' by transverse budding called *strobilization*.
7. Mouth and digestive tract are absent.
8. Excretory system consists of protonephridia with flame-cells.
9. Nervous system consists of ganglionated nerve ring in scolex and two pairs of lateral nerve cords in progiottids.
10. Each proglottid contains one or two sets of male and female reproductive organs. Thus they are hermaphrodite.
11. Several testes in the form of follicles and single-lobed ovary.
12. Uterus presents a great variety of structures.
13. Fertilization internal and self.
14. Life-cycle is complicated with hooked embryo, and is passed in more than one host.

Class Cestoda has about 3400 species. The class is divided into two subclasses:

Sub-class(A) Cestodaria or Monozoa

Sub-class(B) Eucestoda or Merozoa

## *SUB-CLASS (A) CESTODARIA*

1. Endoparasites in the intestine and coelomic cavities of fishes and reptiles.
2. Body is formed by one segment and without a scolex.
3. No alimentary canal.
4. Single set of reproductive organs.
5. Larva with ten hooks, known as *pycophore.*

Sub-class Cestodaria is divided into two orders

### *Order 1 - Amphilinidea*

1. They are endoparasites in the body cavity of ganoid fishes.
2. Frontal gland and protrusible pharynx are present in the anterior part of the body.
3. Suckers are absent.
4. A proboscis is present.
5. Male and female pores present in the posterior part of body.
6. Uterus is coiled and opens near the anterior end.

Example: *Amphilina*

### *Order 2 - Gyrocotylidea*

1. They are endoparasites of chimaeroid fishes.
2. Anterior end has cup-like sucker.
3. Eversible proboscis at anterior end.
4. Male and female pores are situated at the anterior half of the body.
5. Uterus not coiled.

Example: *Gyrocotylus.*

## *SUB-CLASS (B) EUCESTODA*

1. Elongated, flat body is divided into many proglottids.
2. Body is divisible into scolex, neck and strobila.
3. Scolex with suckers, spines and hooks for attachment to host.
4. Each proglottid with male and female reproductive organs.
5. Larva with six hooks.

Sub-class Eucestoda is divided into following orders.

***Order 1 - Tetraphyllidea***

1. Endoparasite, exclusively in the intestine of elasmobranch fishes.
2. Scolex provided with four bothria (sessile suckers) often with hooks.
3. Testes lie infront of ovaries.
4. Cirrus armed with hairs, spines or hooks.
5. Common genital atrium marginal.

   Examples: *Phyllobothrium, Acanthobothrium.*

***Order 2 - Lecanicephaloidea***

1. Intestinal parasites of elasmobranch fishes.
2. Scolex divided by a transverse groove in two parts upper with disc and lower with four suckers.
3. Vitellaria in two lateral bands.

   Examples: *Lecanicephalum, Polypocephalus.*

***Order 3 - Proteocephalidea***

1. These are endoparasites of freshwater fishes, amphibians and reptiles.
2. Scolex mobile with four lateral suckers, an apical sucker in some.
3. Ovary bilobed. Uterus with many branches and pores.
4. Vitellaria scattered.
5. Common genital atrium marginal.

   Examples: *Proteocephalus, Ophiotaenia.*

***Order 4 - Diphyllidea***

1. Parasitic in the intestine of elasmobranch fishes.
2. Proglottids only 20 or less.
3. Scolex with two bothria and a spiny head stalk.
4. Vitellaria scattered.

   Example: *Echinobothrium.*

***Order 5 - Trypanorhyncha***

1. Parasitic in the spiral valve of digestive tract of elasmobanch fishes.
2. Strobila is segmented and is of moderate size.
3. Scolex with two or four sessile bothria and four protrusible spiny proboscis.

4. Testes extend beyond ovary posteriorly.

   Examples: *Haplobothrium, Tetrarhynchus.*

***Order 6 - Pseudophyllidea***

1. Endoparasites in the intestine of fishes, birds and mammals.
2. The larva lives in crustaceans.
3. Generally six shallow groove-like suckers and hookless bothria are found on the scolex.
4. Proglottids may be absent or present.
5. Neck is not developed.
6. Many yolk glands are present.

   Examples: *Amphicotyle, Dibothriocephalus.*

***Order 7 - Nippotaenioidea***

1. Endoparasite in the intestine of Japanese freshwater fishes.
2. Scolex with single terminal sucker.
3. Proglottids few, each with one set of genitalia.
4. Vitellaria compact.

   Examples: *Nippotaenia, Amurotaenia.*

***Order 8 - Taenioidea***

1. Endoparasites in the intestine of birds and mammals.
2. Elongated body, with proglottids.
3. Scolex bears four suckers often with an apical rostellum armed with hooks.
4. Excretory system with four longitudinal vessels and flame cells.
5. Complete sets of reproductive organs in each proglottids. Single compact yolk gland.

   Examples: *Taenia, Echinococcus.*

***Order 9 - Aporidea***

1. It is endoparasite in swan (birds).
2. Scolex with four suckers, rostellum armed.
3. No segmentatioin, ootype, yolk glands and sex-ducts or pores.

   Examples: *Nematoparataenia, Gastrotaenia.*

## POLYCHOERUS

| | | |
|---|---|---|
| Phylum | - | Platyhelminthes |
| Class | - | Turbellaria |
| Order | - | Acoela |
| Genus | - | *Polychoerus* |

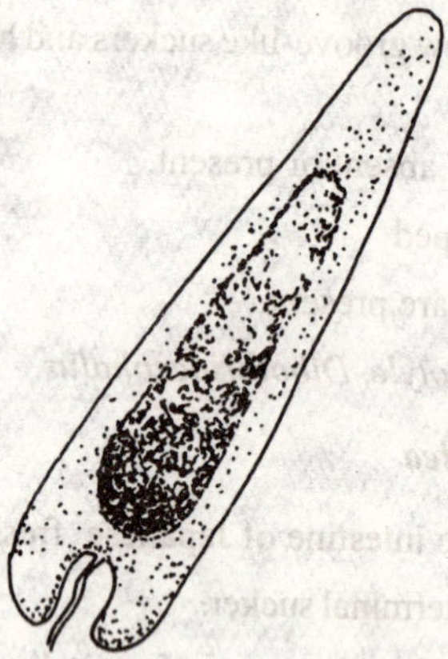

Fig. 10.1 *Polychoerus.*

1. It is found in littoral zone of ocean in colder winter.
2. It is bright orange or orange red in colour.
3. Body flat, oval, changeable shape. The anterior end is round while posterior end is not ched bearing tail filament.
4. Entire body is covered with cilia.
5. Statocyst is present in median line.
6. Digestive tract is absent.
7. Definite gonads are wanting. Yolk glands are absent. Definite oviduct absent.
8. The seminal bursa opens ventrally by female gonopore.
9. Penis is present.

## CONVOLUTA

| | | |
|---|---|---|
| Phylum | - | Platyhelminthes |
| Class | - | Turbellaria |
| Order | - | Acoela |
| Genus | - | *Convoluta* |

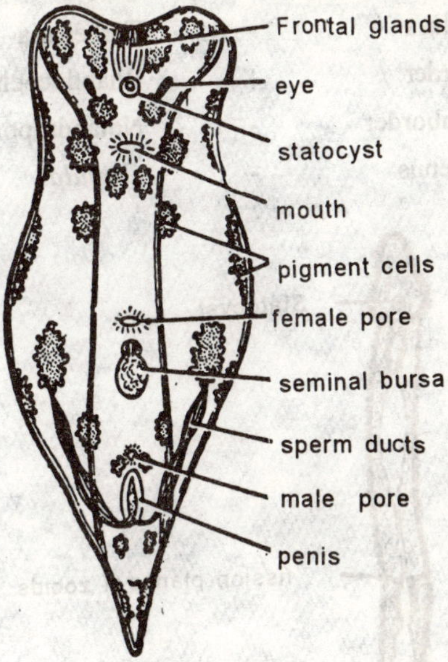


Fig. 10.2 *Convoluta*

1. *Convoluta* is found in sea water.
2. Living under stones, among algae amd most commonly on the bottom mud.
3. Small worm with sides of the body are held curved ventrally without any projections.
4. The body has pigment cells.
5. The anterior or cephalic end has two eyes, a statocyst and cluster of frontal glands.
6. The mouth is located near the anterior half of the ventral surface.
7. A digestive cavity is lacking.
8. The excretory system is completly absent.
9. Hermaphrodite but protandry is common.
10. The eggs are enclosed in a thin gelatinous capsule and are struck to some objects.
11. The eggs are developed directly into young worms that escaple from the jelly.

## CENTULA

| | | |
|---|---|---|
| Phylum | - | Platyhelminthes |
| Class | - | Turbellaria |
| Order | - | Rhabdocoela |
| Suborder | - | Notandropora |
| Genus | - | *Centula* |

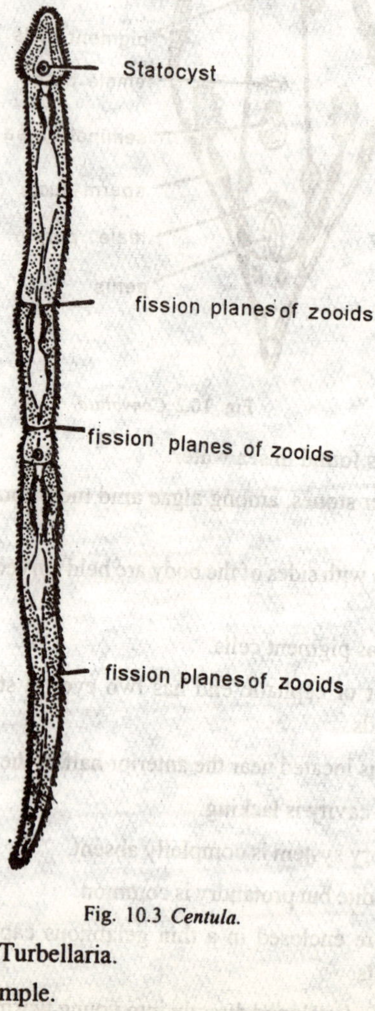


Fig. 10.3 *Centula.*

1. It is a freshwater Turbellaria.
2. The pharynx is simple.
3. Single median protonephridium.

4. Nervous system consists of four pairs of longitudinal nerve cords and a statocyst.
5. Asexual reproduction by fission with chains of zooids.
6. No sexual reproduction.

## MICROSTOMUM

| | | |
|---|---|---|
| Phylum | - | Platyhelminthes |
| Class | - | Turbellaria |
| Order | - | Rhabdocoela |
| Suborder | - | Opisthandropora |
| Genus | - | *Microstomum* |

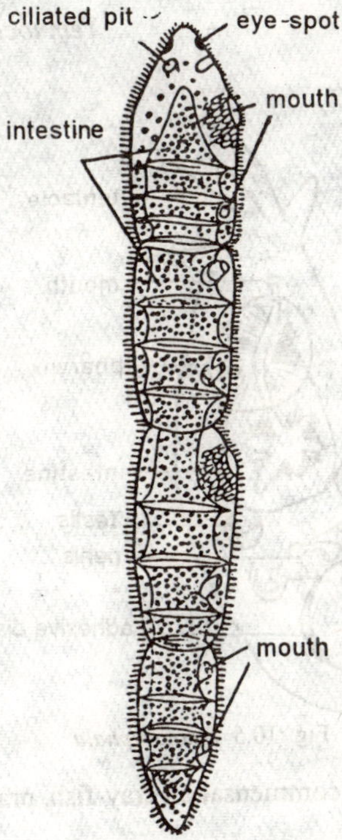


Fig. 10.4 *Microstomum.*

1. It is a microscopic, freshwater form.
2. It swims by means of cilia present on the body surface.
3. Its epidermis contains nematocysts derived from hydroid food.

4. It is found that *Microstomum* attacks *Hydra* only when it needs nematocysts.
5. The digestive cavity is simple, unbranched and straight.
6. It reproduce asexually by repeated transverse fission forming a linear chain of 4, 8 or 16 zooids. Each zooid separates to become an individual.

## TEMNOCEPHALA

| | | |
|---|---|---|
| Phylum | - | Platyhelminthes |
| Class | - | Turbellaria |
| Order | - | Rhabdocoela |
| Suborder | - | Temnocephalida |
| Genus | - | *Temnocephala* |

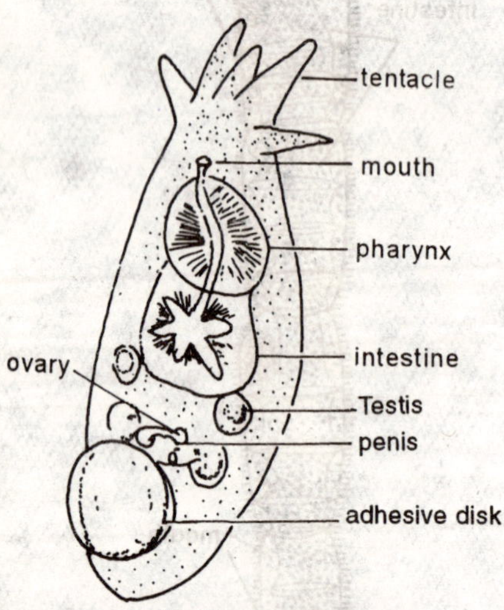


Fig. 10.5 *Temnocephala*

1. It is found as an ecto-commensal on cray-fish, prawns and other crustaceans.
2. It contains 12 finger-like projections or tentacles at anterior end and an adhesive disc at the posterior end.
3. The border of syncytial epidermis is apparently cuticle devoid of cilia.
4. Mouth anterio-ventral. Pharynx thick walled and simple lobulated intestine present.

5. It feeds upon diatoms, rotifers and other small animals.
6. A pair of nephridiopores is present on the dorsal surface of the anterior part of the body.
7. Cerebral ganglion bears two dorsal eyes.
8. There are four testes and single ovary. Their gonoducts leading into a common genitala atrium.

## PLAGIOSTOMUM

| | | |
|---|---|---|
| Phylum | - | Platyhelminthes |
| Class | - | Turbellaria |
| Order | - | Alloecoela |
| Suborder | - | Cumulata |
| Genus | - | *Plagiostomum* |

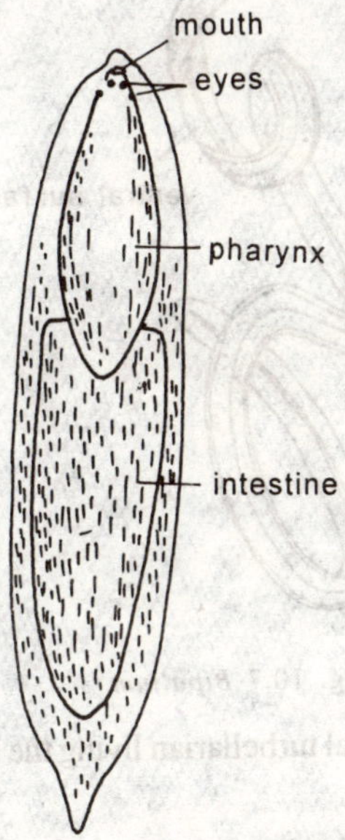


Fig. 10.6 *Plagiostomum*

1. It is present in both fresh and marine waters.
2. Mouth is anterior.
3. Pharynx is bulbose. Intestine with short diverticulae.
4. Gonopore posterior.
5. Sperm is characteristic in having wing-like lateral expansions.

## BIPALIUM

| | | |
|---|---|---|
| Phylum | - | Platyhelminthes |
| Class | - | Turbellaria |
| Order | - | Tricladida |
| Suborder | - | Terricola |
| Genus | - | *Bipalium* |

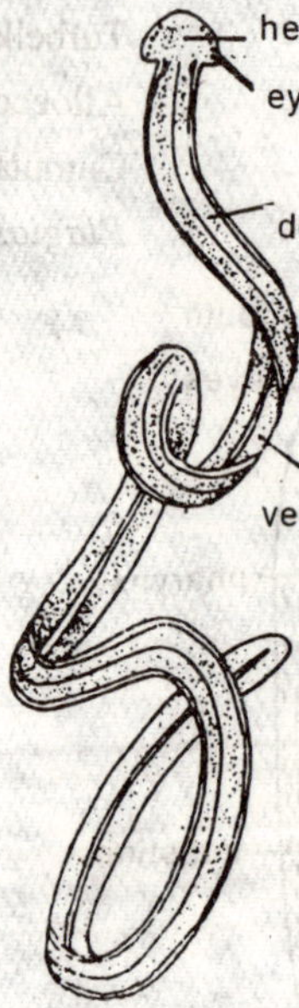


Fig. 10.7 *Bipalium*

1. *Bipalium* is a long terrestrial turbellarian living the humid soil on the floor of tropical forests.
2. It is cosmopolitan in distribution.
3. The animal measures 20-50 cm in length.
4. It consists of an expanded lunate head and cylindrical long body.
5. Numerous eyes are present on the margin of the head and sides of body.
6. There are stripes on the dorsal surface and a creeping sole on the ventral surface.
7. The stripes are purple, black, yellow, olive and grey in colour.

8. Reproduction generally asexual. It never becomes sexual in temperate climate and it propogates by fragmentation.

9. *Bipalium adventium* breeds sexually.

**GUNDA**

| | | |
|---|---|---|
| Phylum | - | Platyhelminthes |
| Class | - | Turbellaria |
| Order | - | Tricladida |
| Suborder | - | Maricola |
| Genus | - | *Gunda* (*Procerodes*) |

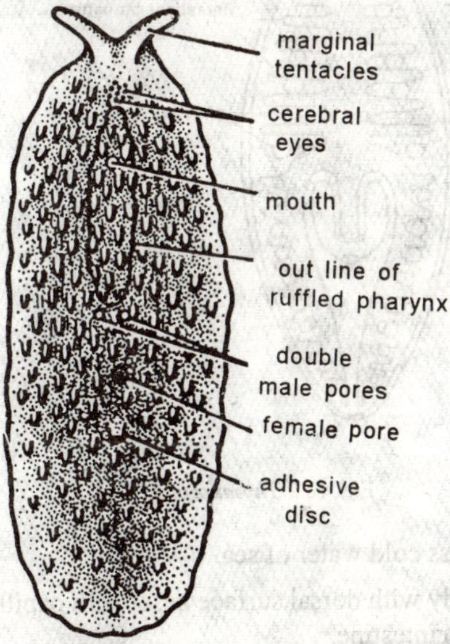


Fig. 10.8 *Gunda segmentina.*

1. It is a marine animal.
2. The auricle and a pair of eyes or ocelli are present in the anterior end.
3. Intestine has regularly arranged intestinal diverticulae. The gonads alternate with the diverticulae giving the appearance of segmentation.
4. Incapable of sexual reproduction.

5. *Lang* used the feature of intestine as a classical example in support the gonocoel theory of coelom formation by *Haeckel*.

## THYSANOZOON

| | | |
|---|---|---|
| Phylum | - | Platyhelminthes |
| Class | - | Turbellaria |
| Order | - | Polycladida |
| Suborder | - | Cotylea |
| Genus | - | *Thysanozoon* |

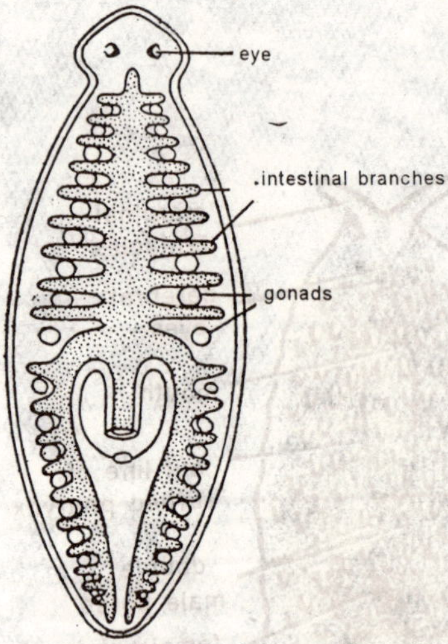


Fig. 10.9 *Thysanozoon*

1. It is found in a less cold water of sea.
2. It has an oval body with dorsal surface beset with papillae, each receiving a branch from intestine.
3. Anterior end bears a pair of marginal tentacles and numerous cerebral eyes.
4. Pharynx is tubular.
5. Glandulo-muscular adhesive organ is present on the dorsal surface behind the female gonopores.
6. A single female genital pore lies behind the male genital pores.
7. Seminal bursa is absent.

## ASPIDOGASTER

| | | |
|---|---|---|
| Phylum | - | Platyhelminthes |
| Class | - | Trematoda |
| Order | - | Aspidobothria |
| Genus | - | *Aspidogaster* |

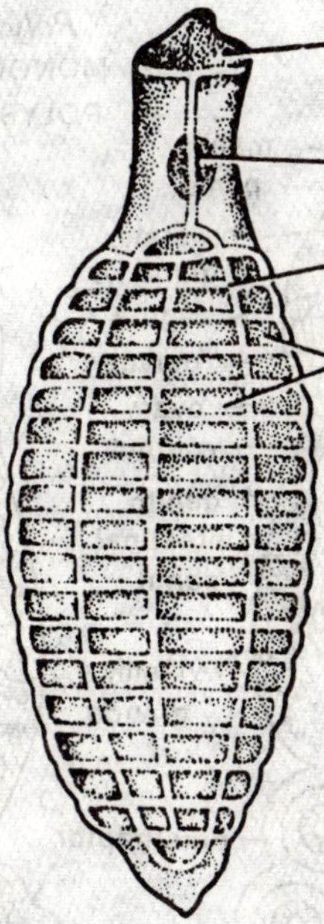


Fig. 10.10 *Aspidogaster*

1. It is an endoparasite in the pericardial and renal cavities of freshwater mussel and in the gut of fishes and turtles.
2. Body is elongated and dorsoventrally flattened with anterior narrow end.
3. The narrow anterior end has a subterminal mouth which is devoid of oral sucker.
4. A large sucker appears which occupies great space in the posterior region of body. The sucker is sub-divided into four longitudinal rows of sucking cups or alveoli.
5. Gut is simple and straight.
6. Excretory system consists of protonephridia with excretory bladder.
7. Bisexual life-cycle without change of host.

# POLYSTOMUM

| | | |
|---|---|---|
| Phylum | - | Platyhelminthes |
| Class | - | Trematoda |
| Order | - | Monogenia |
| Suborder | - | Polyopisthocotylea |
| Genus | - | *Polystomum* |
| ORDER | - | MONOGENEA |
| GENUS | - | POLYSTOMUM |

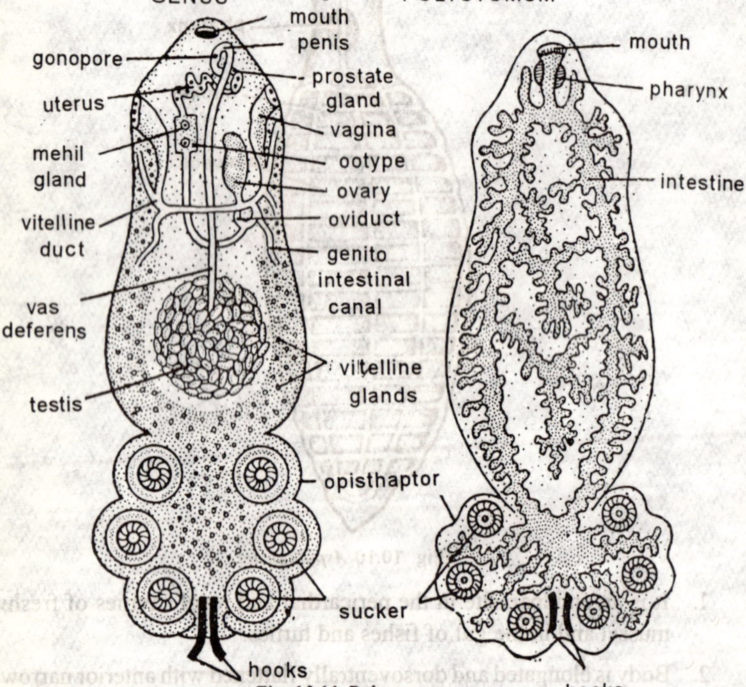


Fig. 10.11 *Polystomum*

1. It is an endoparasite in the urinary bladdar of frogs, and turtles.
2. Body is leaf-like and dorso-ventally flattened.
3. The posterior region has a disc-shaped opisthaptor with six cup-like suckers and 2 to 4 large chitinous hooks.
4. Mouth is surrounded by an oral sucker.
5. Intestine is bifurcated.
6. Hermaphrodite with single ovary and numerous testes.
7. Breeding season starts in spring. Eggs are produced in the urinary blad-

der of frog and discharged into the water alongwith the urine of the host.

8. Hatching takes place after 4 to 5 weeks and larva attaches to the gills of tadpole and later develop into adult in the alimentary canal of metamorphosed tadpole.

## DIPLOZOON

| | | |
|---|---|---|
| Phylum | - | Platyhelminthes |
| Class | - | Trematoda |
| Order | - | Monogenia |
| Suborder | - | Polyopisthocotylea |
| Genus | - | *Diplozoon* |

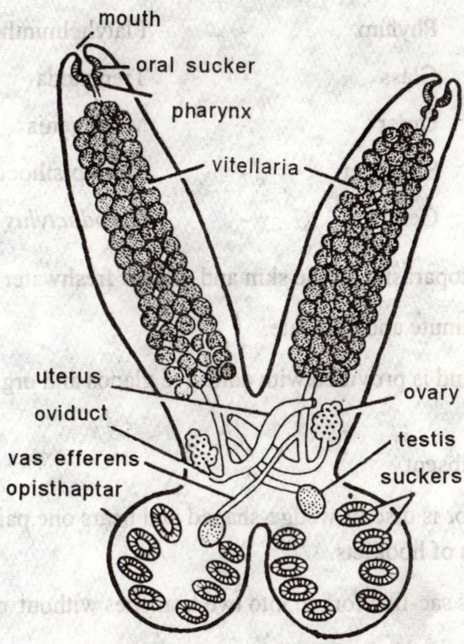


Fig. 10.12 *Diplozoon*. two individuals.

1. *Diplozoon* is an ectoparasite, found attached to the gills of freshwater fishes, feeding on their blood.
2. Two individuals are always found attached permanently together in the form of 'X'.
3. Opisthaptor bears eight suckers arranged in two rows of four each.

4. Digestive system consists mouth surrounded by oral sucker, pharynx and intestine. Intestine is not bifurcated.

5. Reproductive system consists of a testis and ovary along with their usual ducts.

6. The sperm duct, uterus and vagina of one individual cross with those of other and thus ensures cross fertilization. Uterus surrounded by *Mehlis' glands* and contains one egg.

7. Egg differenciates into diporpa larva, two of which unite and metamorphosed into adult individual. Larva has two eyes digestive system and a pair of suckers.

## GYRODACTYLUS

| | | |
|---|---|---|
| Phylum | - | Platyhelminthes |
| Class | - | Trematoda |
| Order | - | Monogenia |
| Suborder | - | Monopisthocotylea |
| Genus | - | *Gyrodactylus* |

1. It is an ectoparasite on the skin and gills of freshwater fishes.
2. Body is minute and elongate.
3. Anterior end is provided with adhesive glands and organs. Oral sucker is absent.
4. Eyes are absent.
5. Opisthaptor is disc or wedge-shaped and bears one pair of anchors and eight pairs of hooklets.
6. Intestine is sac-like forked into two branches without diverticula.
7. Genito-intestinal canal is absent. Genital pore is median.
8. Development is intra-uterine i.e. viviparous.
9. A remarkable feature is that a second embryo is produced within the first one, a third within the second and a fourth within the third. The first embryo, on completing its development passes out, still enclosing the other embyros, and attaches directly to a host fish.

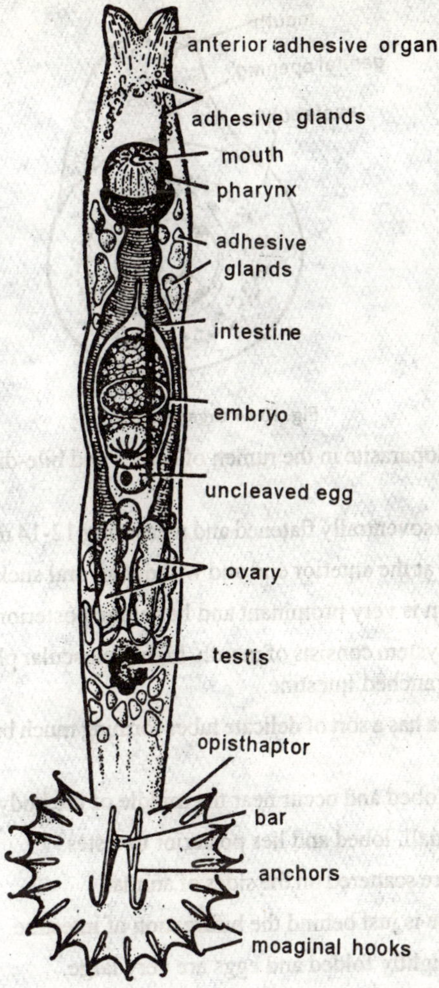


Fig. 10.13 *Gyrodactylus*

## GIGANTOCOTYLE (PARAMPHISTOMUM)

| | | |
|---|---|---|
| Phylum | - | Platyhelminthes |
| Class | - | Trematoda |
| Order | - | Digenia |
| Family | - | Paramphistomidae |
| Genus | - | *Gigantocotyle* |

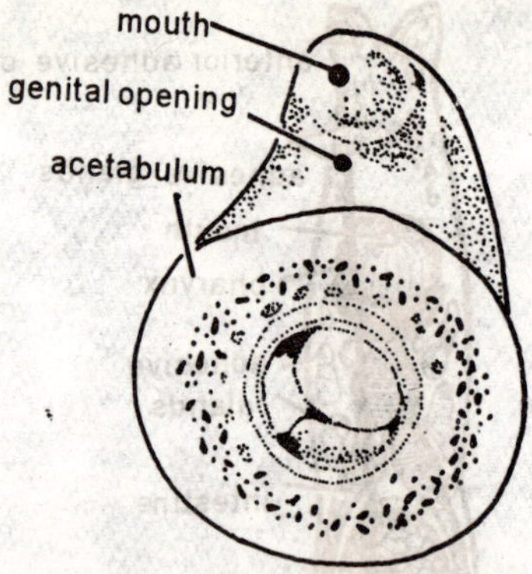


Fig. 10.14 *Gigantocotyle*

1. It is an endoparasite in the rumen of cattles and bile-duct of sheep, goat etc.
2. Body is dorsoventrally flatened and measuring 12-14 mm in length.
3. Mouth lies at the anterior end and without an oral sucker.
4. Acetabulum is very prominant and lies at the posterior end.
5. Digestive system consists of mouth, highly muscular pharynx and bifurcated, unbranched intestine.
6. Parenchyma has a sort of delicate tubes forming much branched lymphtic system.
7. Testes are lobed and occur near the middle of the body.
8. Ovary is small, lobed and lies posterior to testes.
9. Vitellaria are scattered on the sides of animal.
10. Genital pore is just behind the bifurcation of intestine.
11. Uterus is slightly folded and eggs are very large.

## GASTRODISCOIDES HOMINIS

| | | |
|---|---|---|
| Phylum | - | Platyhelminthes |
| Class | - | Trematoda |
| Order | - | Digenia |
| Family | - | Paramphistomidae |
| Genus | - | *Gastrodiscoides* |
| Species | - | *hominis* |

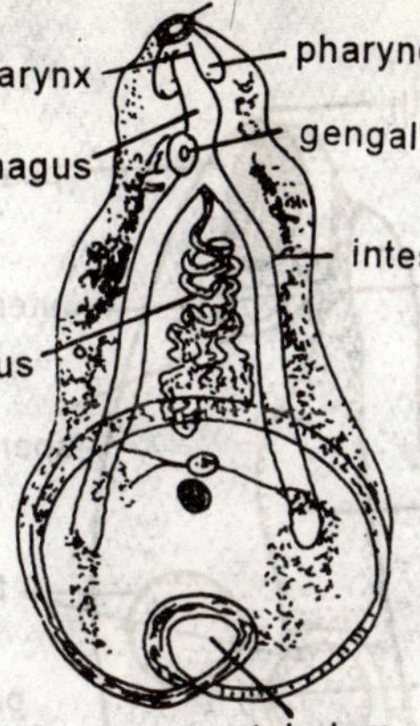


Fig. 10.15 Gastrodiscoides

1. It is an endoparasite in the colon and caecum of man.
2. It is found in India and Indo-China.
3. It measures 5-7mm in length.
4. Body is differentiated into anterior and posterior part.
5. Terminal mouth is present an anterior part.
6. Oral sucker is absent.
7. Posterior part bears a prominant acetabulum on ventral side.
8. Gonopore is in the anterior half of the body on ventral side.
9. Excretory pore lies on the dorsal side infront of acetabulum.
10. The secondary host is snail.

## GASTROTHYLAX

| | | |
|---|---|---|
| Phylum | - | Platyhelminthes |
| Class | - | Trematoda |
| Order | - | Digenea |
| Family | - | Paramphistomidae |
| Genus | - | *Gastrothylax* |

1. It is an endoparasite in the rumen of cattles.
2. It is elongated animal possessing a deep ventral pouch which extends upon the ventral sucker and it is supposed to be elongated genital chamber.

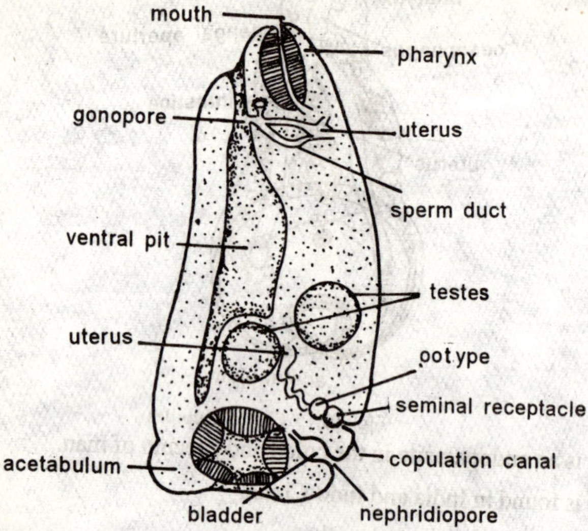


Fig. 10.16 Gastrothylax

3. Digestive system includes mouth, pharynx, oesophegus and simple non-diverticulated intestine.
4. Two rounded testes near the base of the pouch.
5. Female reproductive organs consists of single ovary, uterus, ootype, vagina and seminal recepticle.
6. Single excretory pore is present.
7. Posterior end is divided with large acetabulum or adhesive sucker.

## FASCIOLOPSIS BUSKI

| | | |
|---|---|---|
| Phylum | - | Platyhelminthes |
| Class | - | Trematoda |
| Order | - | Digenea |
| Family | - | Fasciolidae |
| Genus | - | *Fasciolopsis* |
| Species | - | *buski* |

1. It is a large intestinal fluke of man, dogs, pigs etc.
2. It is found in India and China.
3. It measures about 70 mm in length and 20 mm in width.
4. Oral sucker and acetabulum both are well developed.

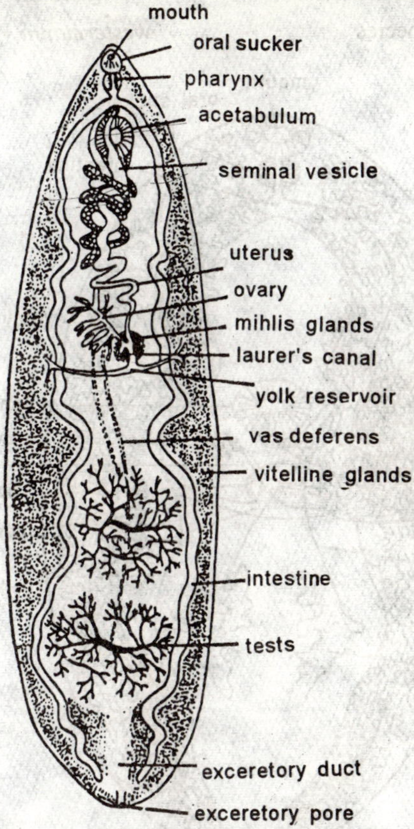


Fig. 10.17 *Fasciolopsis buski.*

5. Mouth surrounded by oral sucker leads into small pharynx. Oesophegus is short and leads into non-diverticulated caeca.
6. Gonads are present in posterior region of body.
7. Intermediate host is snail.
8. Cercaria larvae encyst on tubers of water caltrop or singhara. The singhara is eaten raw and thus infection occurs.
9. It causes inflammation and haemorrhage of the intestine.

## PARAGONIMUS WESTERMANNI

| | | |
|---|---|---|
| Phylum | - | Platyhelminthes |
| Class | - | Trematoda |
| Order | - | Digenea |

| | | |
|---|---|---|
| Genus | - | *Paragonimus* |
| Species | - | *westermanni* |

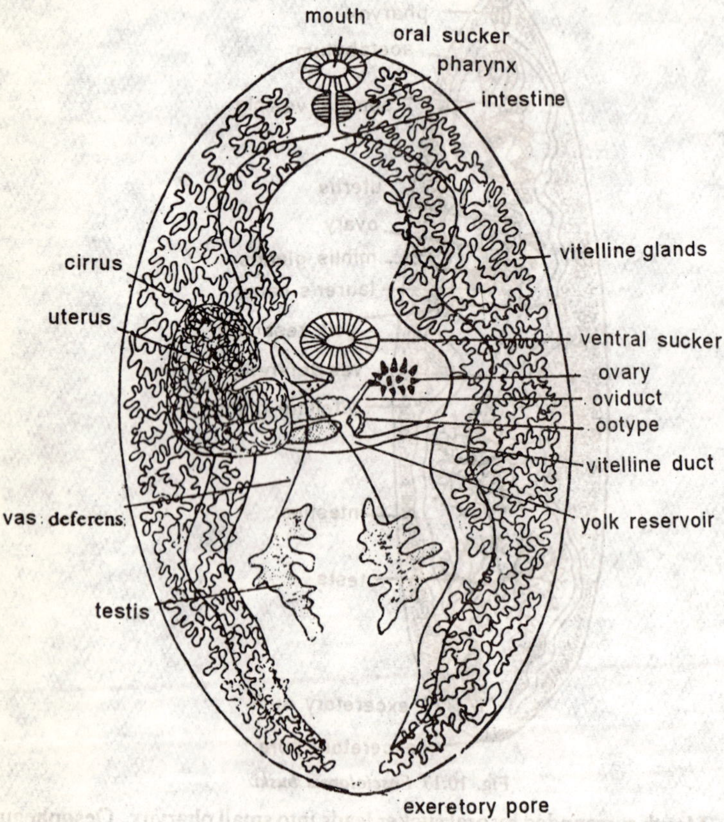


Fig. 10.18 *Paragonius*

1. *Paragonimus westermanni or lung fluke* is parasite in the lungs of mammal including man.
2. The body of the fluke is oval with well developed oral sucker surrounding the mouth and a ventral sucker in the centre of the body.
3. The internal organs are very much like that of *Fusciola hepatica.*
4. The cyst like pockets are formed around the parasite which rupture and liberate the eggs into bronchial tubes to be executed with sputum.
5. The miracidia live only few hours after hatching.
6. The cercariae are long, have small knob-like tail and spiny cuticle.

7. They do not swim but creep in leech-like manner.
8. They pierce the cuticle of cray-fish where they become encysted inside the tissues and gradually develop into mature infective metacercariae.
9. The infection of the new host usually takes place by eating infected crab or cray-fish.
10. They cause or help in tuberculosis, cough and blood stained sputum.

## AMPHILINA

| | | |
|---|---|---|
| Phylum | - | Platyhelminthes |
| Class | - | Cestoda |
| Subclass | - | Cestodaria |
| Order | - | Amphilinida |
| Genus | - | *Amphilina* |

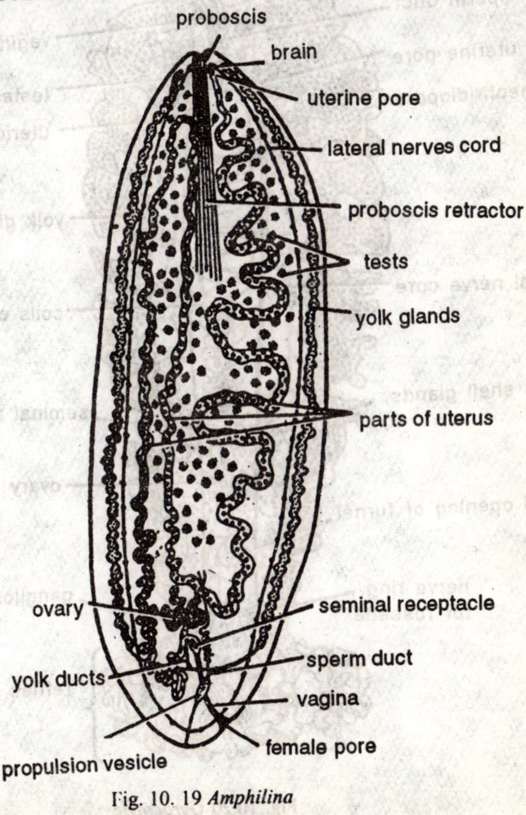


Fig. 10. 19 *Amphilina*

1. It is an endoparasite found in the coelom of fish, *Acipenser*.
2. Body flat and leaf-like. Scolex is absent.
3. Anterior end is provided with weakly developed protrusible proboscis.
4. Testes are scattered and cirrus is armed.
5. Vagina lies behind the ovary and opens on the left side. Uterus is coiled.
6. The intermediate host is a crustacean *Gammarus*, which is eaten by *Acipenser*.

## GYROCOTYLE

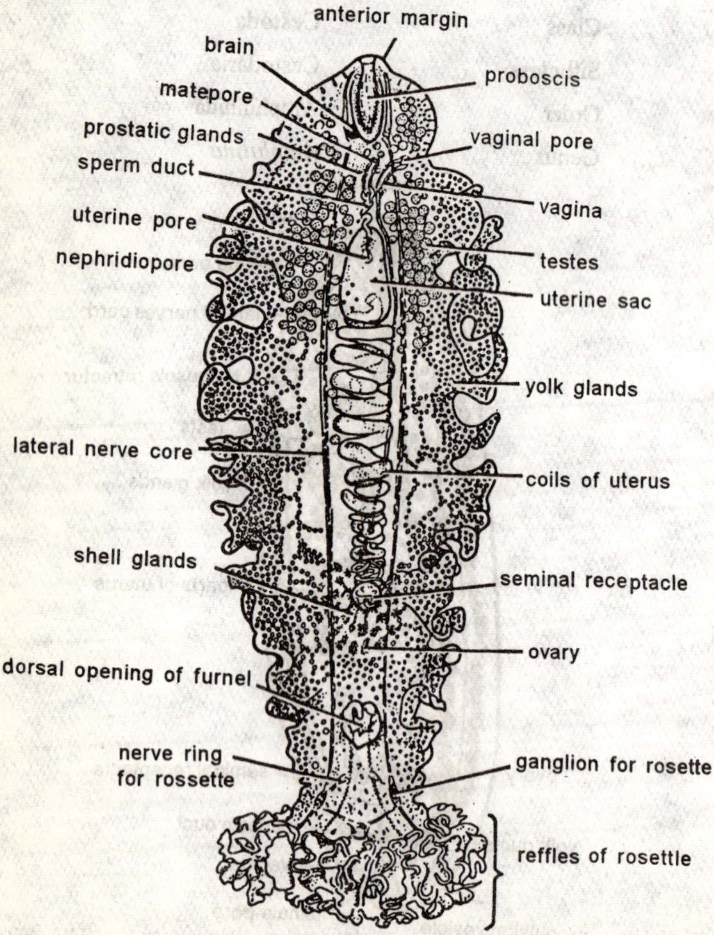


Fig. 10.20 *Gyrocotyle*

| | | |
|---|---|---|
| Phylum | - | Platyhelminthes |
| Class | - | Cestoda · |
| Subclass | - | Cestodaria |
| Order | - | Gyrocotylidea |
| Genus | - | *Gyrocotyle* |

1. These are endoparasite in the intestine of chimaeroid fishes.
2. Body is elongated and dorso-ventrally flatened.
3. Anterior end bears a large opening that leads into a highly muscular protrusible mass, the proboscis.
4. Posterior end bears a rosette-shaped adhesive organ, surrounding a funnel shaped depression.
5. The margins of body are generally ruffled.
6. Bisexual. Male reproductive system comprises scattered testes, sperm duct and penis papilla. Female reproductive system comprises ovary, oviduct and uterus.
7. Follicles of yolk gland scattered throughout.

## TETRARHYNCHUS

| | | |
|---|---|---|
| Phylum | - | Platyhelminthes |
| Class | - | Cestoda |
| Subclass | - | Eucestoda |
| Order | - | Trypanorhyncha |
| Genus | - | *Tetrarhynchus* |

1. It is an endoparasite in the intestine of elasmobranch fishes.
2. Body is long divisible into scolex and proglottids.
3. Scolex is long and differentiated into a long proximal part containing the proboscis apparatus and a distal part bearing the bothria or sucker.
4. Scolex bears four suckers each with a eversible proboscis armed with spines. The proboscis are retractile in muscular sheaths which end in muscular bulbs.
5. Each proglottid contains a complete set of reproductive organs.

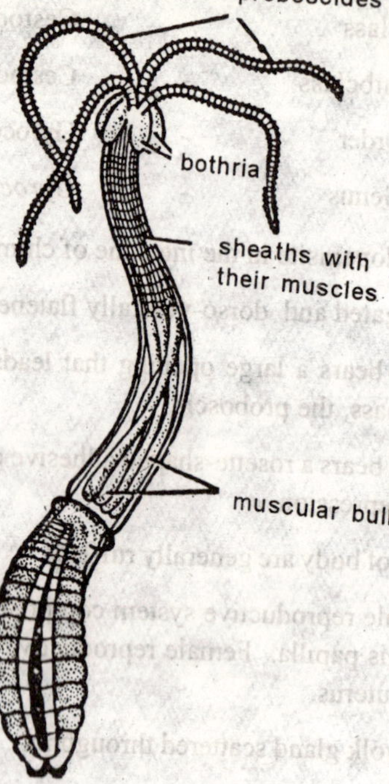


Fig. 10.21 *Tetrarchynchus.*

## DIPHYLLOBOTHRIUM (DIBOTHRIOCEPHALUS)

| | | |
|---|---|---|
| Phylum | - | Platyhelminthes |
| Class | - | Cestoda |
| Subclass | - | Eucestoda |
| Order | - | Pseudophyllidea |
| Genus | - | *Diphyllobothrium* |
| Species | - | *Latum* |

1. It is found in the intestine of man. It is world-wide in distribution.
2. It is the largest and most pathogenic cestod of human beings, measuring 20 metres in length.
3. Body is differentiated into scolex, neck and proglottids. The number of proglottids is 3000-4000.
4. Scolex is fusiform and bears two slit-like suckers or bothria.
5. Mature proglottid broader than long.

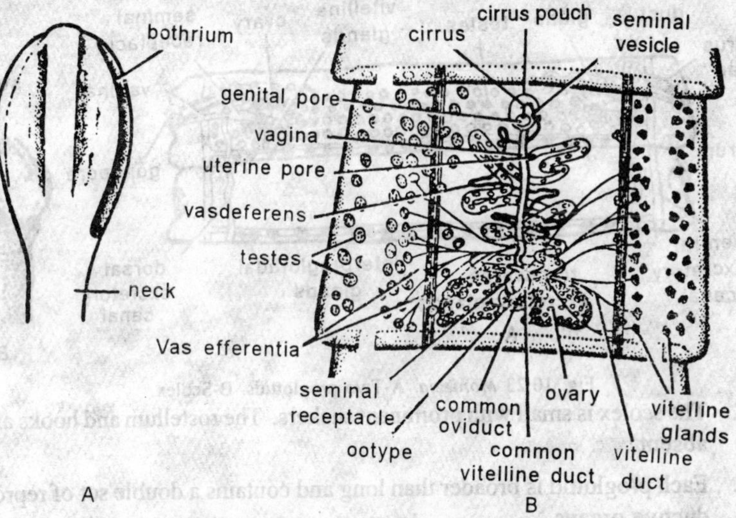


Fig. 10.22 *Diphyllobothrium* A- Scolex; B-A mature proglottid.

6. Each proglottid is hermaphrodite and contains complete sets of male and female reproductive organs.
7. Testes are numerous. Vasa efferentia unite to form coiled sperm duct.
8. Ovary is trilobed. Vagina runs posteriorly forming seminal vesicle. Uterus very much twisted.
9. Gonopores ventral.
10. Life-cycle involves two intermediate hosts, one is *Cyclops* (a copipod) and other a fish.
11. It causes bothriocephalus anaemia, erythropenia and haemorrhage.

## MONIEZIA

| | | |
|---|---|---|
| Phylum | - | Platyhelminthes |
| Class | - | Cestoda |
| Subclass | - | Eucestoda |
| Order | - | Taenioidea |
| Genus | - | *Moniezia* |
| Species | - | *expansa* |

1. It is found in sheep, cattles and other ruminants.
2. Body is divisible into scolex, neck and proglottids.

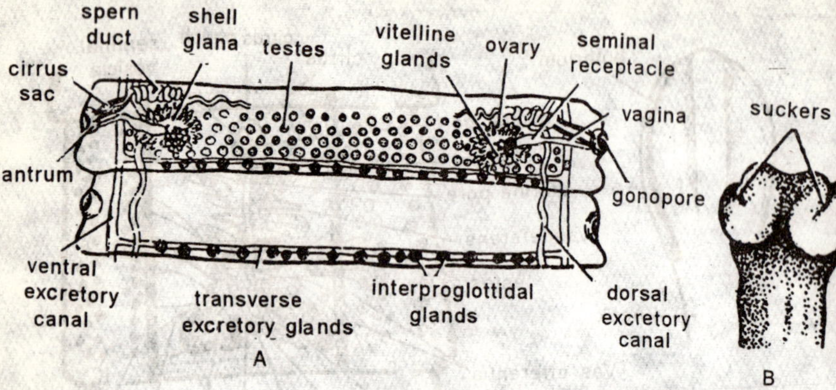


Fig. 10.23 *Moniezia.* A-Two proglottids. B-Scolex.

3. The scolex is small with prominant suckers. The rostellum and hooks are absent.
4. Each proglottid is broader than long and contains a double set of reproductive organs.
5. Genital ducts are dorsal to osmoretgulatory canals.
6. Testes are numerous scattered throughout the centre of proglottid. Cirrus in cirrus pouch.
7. Ovaries are in the form of an open fan. Uteri form networks.
8. The uterus develops as a network of canales, but latter breaks up into a large number of uterine capsules, each enclosing 3-30 embryos.
9. Intermediate host is dog-fleas, *Ctenocephalus canis* and *Pulex irritans*.

## TAENIA SAGINATA

| | | |
|---|---|---|
| Phylum | - | Platyhelminthes |
| Class | - | Cestoda |
| Subclass | - | Eucestoda |
| Order | - | Taenoidea |
| Genus | - | *Taenia* |
| Species | - | *saginata* |

1. It is commonly called the *beef tapeworm*. It lives in the intestine of man where beef is eaten.
2. The body is dorso-ventrally flattened and much larger than *Taenia solium*. It measures 15-20 feet in length.
3. Body is divisible into scolex, neck and proglottids.

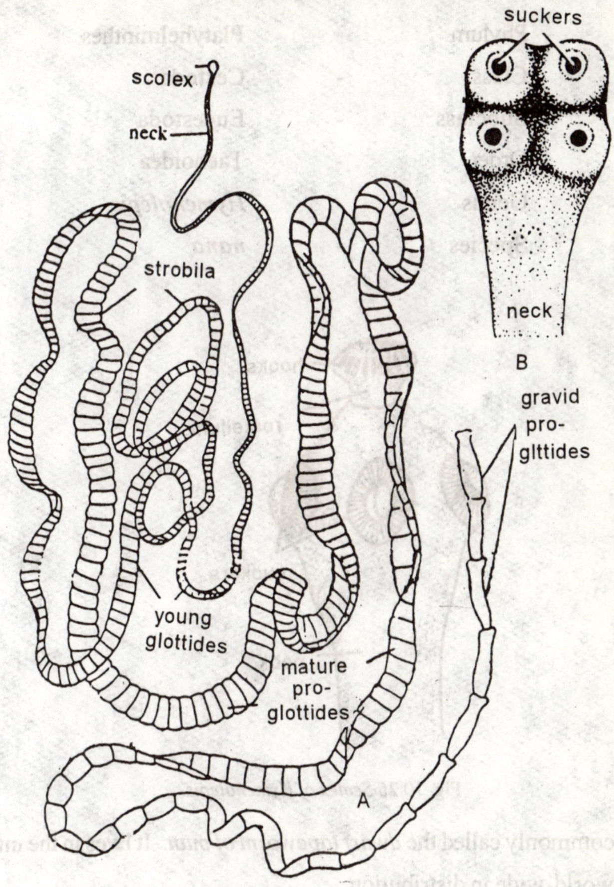


Fig. 10.24 *Taneia saginata*

4. Scolex bears four large suckers for adhesion and devoid of rostellar hooks.
5. About 1000 proglottids are present.
6. Testes are about 300-400. Ovary in the form of two lobes. Ootype is present surrounded by *Mehlis gland.*
7. Life-history is similar to that of *T. solium.*
8. Primary hosts are cattles.
9. The intermediate host is mite (*Galumna*).
10. It is not so pathogenic but sometimes may cause *diarrhoea or anaemia.*

## HYMENOLEPIS NANA

| | | |
|---|---|---|
| Phylum | - | Platyhelminthes |
| Class | - | Cestoda |
| Subclass | - | Eucestoda |
| Order | - | Taenoidea |
| Genus | - | *Hymenolepis* |
| Species | - | *nana* |

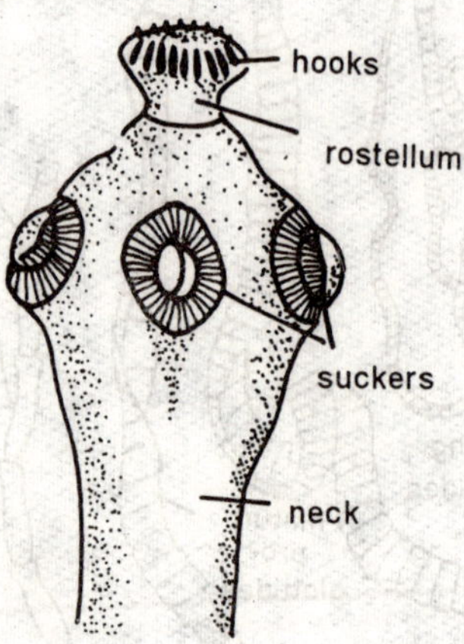


Fig. 10.25 *Scolex of Hymenolepis.*

1. It is commonly called the *dwarf tapeworm of man*. It lives in the intestine.
2. It is world-wide in distribution.
3. It measures 25-40 mm in lengths and has about 100-200 proglottids.
4. The rostellum is well-developed and retractile. It bears a single circlets of 20-30 hooks.
5. Neck is long and slender.
6. There are three testes with varying arrangement. Cirrus pouch is large with external and internal seminal vesicle and with an accessory sac.
7. Life-history is simple without any intermediate host.
8. The inner membrane of the hexacanth bears long, wavy filaments.
9. Larvae develop in the intestinal villi while the adult lives in the intestine.
10. It causes *abdominal pain* and *diarrhoea*.

## DIPYLIDIUM CANINUM

| | | |
|---|---|---|
| Phylum | - | Platyhelminths |
| Class | - | Cestoda |
| Subclass | - | Eucestoda |
| Order | - | Taenioidea |
| Genus | - | *Dipylidium* |
| Species | - | *caninum* |

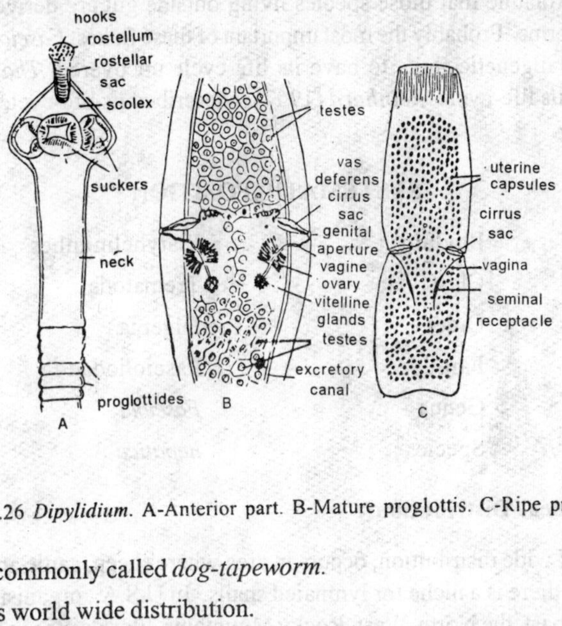


Fig. 10.26 *Dipylidium*. A-Anterior part. B-Mature proglottis. C-Ripe proglotis.

1. It is commonly called *dog-tapeworm.*
2. It has world wide distribution.
3. It lives in the intestine.
4. It measurers 15-40 cm in length and has under 200 proglottids.
5. The scolex is somewhat triangular and bears four deeply cupped suckers.
6. Rostellum is retrctile into a rostellar sac in scolex. It bears about 60 hooks in 3-7 rings.
7. Mature proglottids are long and barrel shaped having a double set of male and female sex organs. The genital pores at the middle of both lateral margins.
8. Numerous testes are present before or behind the female genitalia.

▲▲▲

11

# FASCIOLA HEPATICA

*Fasciola* belongs to class Trematoda. Most of the trematods living in other animals inhabit the digestive tract or its accessory tubes or cavities, and it seems probable that those species living outside gut are derived from gut dwelling forms. Probably the most important of these flukes, *Fasciola hepatica*, is the first digenetic fluke to have its life-cycle uncovered. *Thomas* (1883) described its life-cycle. *Reinhard* (1957) redescribed its life-cycle in an interesting way.

**SYSTEMATIC POSITION**

| | | |
|---|---|---|
| Phylum | - | Platyhelminthes |
| Class | - | Trematoda |
| Order | - | Digenia |
| Family | - | Fascioliodae |
| Genus | - | *Fasciola* |
| Species | - | *hepatica* |

### Geographical Distribution

World wide distribution, occurs in area where sheep, cattle and goats are raised and there is a niche for lymnaied snails. In U.S.A., organisms found in the Gulf Coast, the North-West, Rocky Mountains, Florida etc. In U.S.A., it is endemic.

### Habits and Habitat

*Fasciola hepatica* is derived from Greek and Latin words - (L.,*fasciola* = small bandage; Gr.,*hepar* = liver) is a *digenetic* parasite, i.e., its life history is completed in two hosts - a primary vertebrate host and a secondary or intermediate invertebrate host. Adult *F. hepatica* occurs in the liver and bile passages of the primary host which is usually sheep, but, at times, it may occur in some other vertebrates, like goat, horse, dog, ass, ox, deer, antelope, rabbit, elephant, monkey, etc. Humans are occasional host for *F. hepatica,* and infections are most often obtained through eating watergrass (*Nasturtium officinale*) in salad. A single host may harbour as many as about 200 of these flukes in its liver, which may consequently cease to function normally. This effect is known as *liver-ı ɔt.F.hepatica* spends a part of its life history in an intermediate host , a freshwater snail which is either *Limnea truncatula* or some specific species of *Planorbis, Fossariamodicella, Pseudosuccinea* etc.

**Shape, Size and Colouration.**

The body is large, soft and fleshy. It is dorsoventrally flattened and leaf-like having somewhat oval appearance. The size varies from 1.8-5.1 cm in length and about 0.4-1.5 cm. in breadth near the mid body. The colour of body is generally pinkish. The margins are blackish or brownish due to the presence of vitelline glands. The alimentary canal is brownish in colour due to the presence of ingested bile of the host.

**Structure.**

The anterior end of *F. hepatica* is triangular while posterior end is tapering. The anterior end terminates into a projection, which is called *apic¬l or cephalic cone or head lobe*. At the apex of the head lobe is situated a small

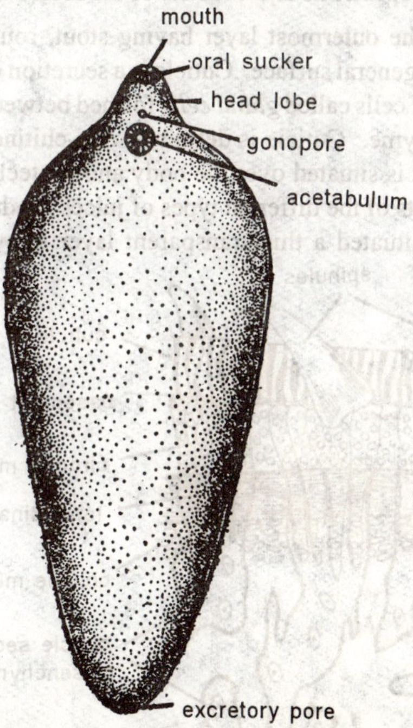


Fig. 11.1 *Fasciola hepatica*. External features.

opening, the mouth, which is surrounded by an anterior muscular sucker, called *oral or anterior sucker*. At a short distance behind the head lobe is another saucer-shapted sucker known as *ventral sucker or acetabulum*, situated on the ventral side. The ventral sucker is superficial and has no connection with the interior of the body.

*Fasciola hepatica* was formerly known as *Distomum hepaticum* and the name of the group to which it belongs was known as *Distomidae*. These names were adopted as the mouth and anterior sucker were regarded as two suckers or two mouths (di,two;stoma,mouth). The word *stoma* was used in trematodes for suckers, which are characteristic structures. But now it is certain that the mouth cannot be claimed as a sucker.

On the dorsal surface there is a single aperture in the middle line at about one-third of the length of the animal from the anterior end. This aperture is the opening of *Laurer's canal*. This aperture appears during breeding season.

**Body Wall**

Body wall in *Fasciola* is characteristic and adapted to parasitic mode of life. It lacks a cellular epidermis and consists of the following layers:-

***1. Cuticle.*** It is the outermost layer having stout, rough, blunt scales which project over the general surface. Cuticle is a secretion of mesenchyme, either of special type of cells called *gland cells* formed between muscle-fibres or of general mesenchyme. Cuticle is definitely non-chitinous and is made up of scleroproteins. It is situated over the body and protects the liver-fluke from the harmful effects of the different types of juices produced in the host. Below the cuticle is situated a thin transparent layer known as *basement membrane*.

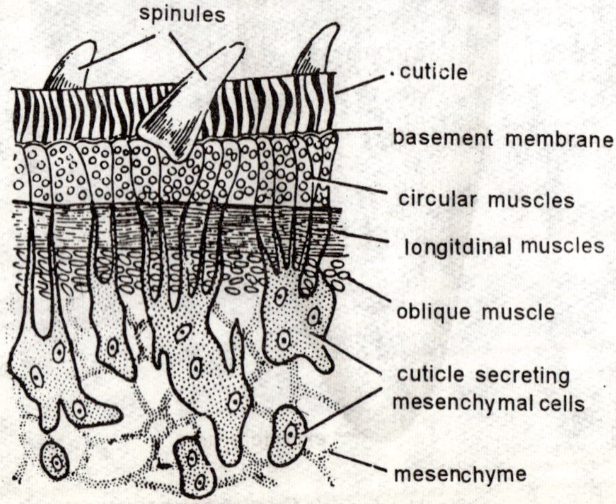


Fig. 11.2 *Fasciola*. V.L.S. body wall seen under light microscope.

***2. Musculature.*** Immediately beneath the cuticle are three layers of muscles-the outermost *circular* layer, the median *longitudinal* layer and the inner *digonal* layer. In addition to these muscles others make up the suckers and pharynx. Still others, the parenchymal muscles, are dorsoventrally oriented along the lateral regions of the body.

***3. Parenchyma or mesenchyma.*** Immediately benath the musculature are numerous cells, known as *beta cells or mesenchymal cells*. Some of the beta cells possess cytoplasmic tubules that penetrate through the muscular layers and reach the cuticle. These specialized beta cells are thought to secrete cuticular materials that reinforce the existing cuticle.

Electron microscopic studies by *Threadfgold* (1968) have revealed that the so called cuticle of *Fasciola* is actively a thick layer of multinucleated protoplasm containing mitochondria, canals of endoplasmic reticulum, vacuoles and vesicles. It is termed as *tegument*. It is continuous with tegument secreting cells lying embedded in the mesenchyme. The outer surface of tegument is thrown into numerous folds. These increase the surface area considerably for the absorption of fluids of the host. Further, tegument is perforated by fine pore canals. Through these canals fluids are absorbed into the mesenchyme.

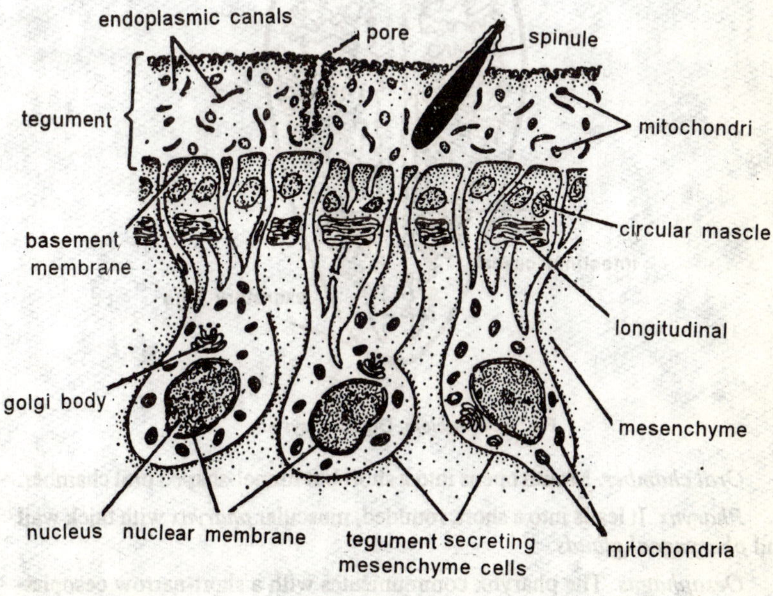


Fig. 11.3 *Fasciola*. V.L.S. body wall seen under electron microscope.

**Function.**

The body wall of *Fasciola* is more than an inert protective layer,it is a dynamic structure of great importance in the metabolism of the organism. Although the animals derive a great deal of their nutrients through ingestion, certain chemicals such as certain amino acids, which are necessary as energy sources and as raw materials for the synthesis of complex body chemicals, are absorbed through the body wall. It is also possible that certain metabolic wasters are discharged through cuticle.

## Digestive System

The digestive system is complex and highly branched with mouth but no anus. It consists of the following parts:

*Mouth.* It is a small opening situated ventrally at the apex of head lobe and surrounded by oral sucker.

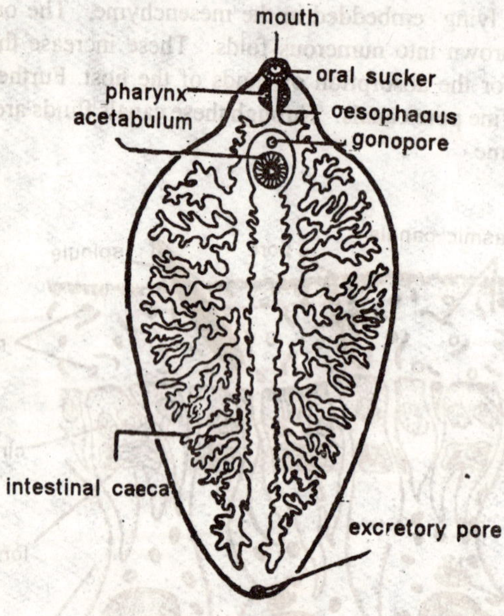


Fig. 11.4 *Fasciola.* Digestive system.

*Oral chamber.* Mouth opens into a short but funnel-shaped oral chamber.

*Pharynx.* It leads into a short, rounded, muscular *pharynx* with thick wall and *pharyngeal glands*.

*Oesophagus.* The pharynx communicates with a short-narrow oesophagus joining the intestine.

*Intestine.* The intestine immediately bifurcates into two lateral branches, running one on either lateral side of the body upto the posterior end where these end blindly. A large number of caeca or diverticula are given out along either side of its whole length. The diverticula of the outer side are large and highly branched,while those of the inner side are short and unbranched.

***Food and feeding.***

The food of *Fasciola* is bile and blood as it lives in the bile passage of sheep and cattles. It takes about 0.18-0.2 ml of blood per day. Small microscopic debris is also taken as food. The food is taken by the sucking effect of the pharynx and is pumped into the intestine.

***Digestion.*** The exact manner of food digestion is not well understood. Digestion is supposed to be extracellular, carried in the intestinal ceaca. Proteolytic enzymes are reported in the animal. The digested food is diffuses into the surrounding mesenchyme. The mesenchyme distributes digested food to every part of body. As earlier mentioned the tegument also helps in the absorption of certain amino acids as well as monosaccharides (glucose and fructose). The reserve food material is in the form of glycogen and fat which are stored in mesenchymal cells and muscles.

***Egestion.*** As there is no anus, the undigested food materials are egested out through mouth.

***Respiration.*** In the *Fasciola*, the respiratory organs are absent so that the respiration takes place through general body surface by the process of diffusion. Since the animal is an endoparasite, the rate of respiration is very low and is anaerobic. In this process, there is no necessity of oxygen. Glycogen undergoes anaerobic glycolysis to form pyruvic acid. Pyruvic acid is decarboxylated to form carbon dioxide and an acetyl group. The latter combines with coenzymes to form acetyl coenzymes A. Acetyl coenzyme A is finally condensed and reduced to form fatty acids. During the process energy is released.

**Excretory System**

The excretory system consists of flame cells or protonephridia connected with an intricate system of excretory ducts.

1. *Excretory ducts.* A single and median longitudinal excretory canal which runs behind the ventral sucker and terminates into the *excretory pore*. It is ramified throughout the body. In the anterior region it receives two ventral and two dorso-lateral excretory canals. The dorsal ducts before opening into the main longitudinal excretory canals are united so as to form a common duct. Main excretory canal as well as dorsal and ventral-lateral excretory canals divide into small capillaries or tubules, each of which ends internally into a *flame cell.*

2. *Flame cells.* The flame cells are characteristic, each has a thin elastic

wall with a nucleus and a cavity containing many long cilia arising from basal granulels or blepharoplasts. The cilia in a living animal vibrate like the flickering of the flame, hence named as flame cells. The cavity of flame cell is continued with the cavity of capillary. The shape of these cells is irregular due to pseudopodial processes into the mesenchyma.

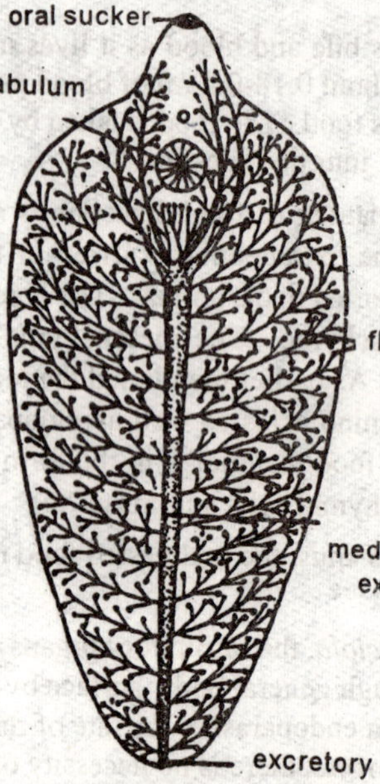


Fig. 11.5 *Fasciola.* Excretory system.

***Physiology of excretion***

The excretory products are carbon dioxide, fatty acids and ammonia. These are diffused into the flame cells from the surrounding mesenchyme. The fluid moves through the tubule by the action of cilia into the excretory ducts and finally passes out of the body through the excretory pore. All the ducts except the median longitudinal duct are lined with cilia. The course of the flow of excretory products is as follows:

Excretory products mesenchyme flame cells

Main longitudinal canal vessels twigs capillaries

Excretory pore outside.

**Nervous System**

The nervous system is well developed in spite of the fact that due to sluggish and parasitic habitats such a large correlation is not need. The nervous system includes:

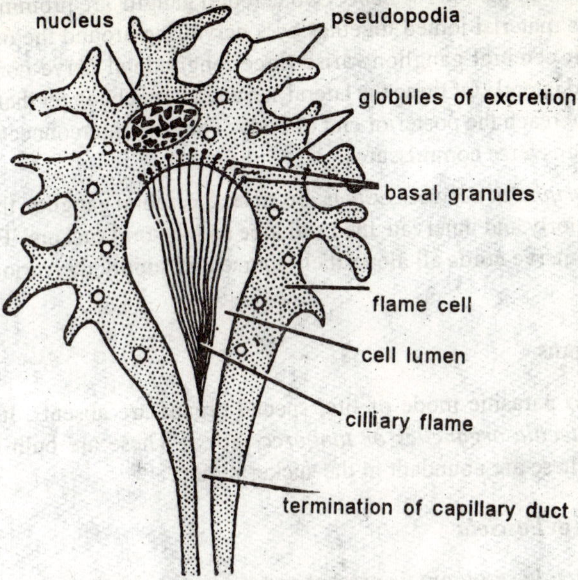


Fig. 11.6 *Fasciola.* A flame cell.

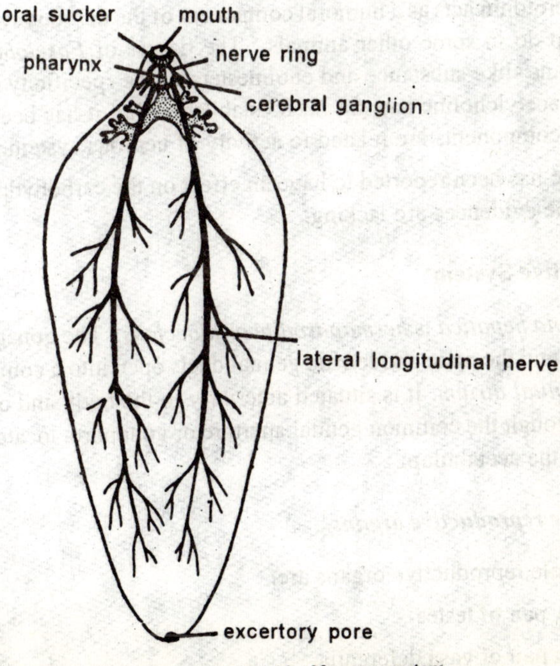


Fig. 11.7 *Fasciola.* Nervous system.

*(a) Central nervcus system.* Two cerebral ganglia are prominant masses of nervous material joined together by a nerve ring around the oesophagus. From each cerebral ganglion arise three longitudinal nerve cords, dorsal, ventral and lateral. Of these the lateral nerves are much longer than the other. They alone reach the posterior end of body. They are interconnected at intervals by transverse commissures

*(b) Peripheral nervous system.* From each cerebral ganglion three nerves arise anteriorly and innervate the head lobe and the oral suckers. Fine nerves arise from nerve cords all along their course and supply the various parts of the body.

## Sense Organs

Due to parasitic mode of life, special organs are absent. It however, possesses *tectile organelles or tangoreceptors.* These are bulb-like nerve endings. These are abundant in the sucker region.

## Hormones in *Fasciola*

In *Fasciola* serotonin is present and it seems to play a role in the tissue comparable to that played by adrenalin in mammalian tissue. It is not known whether serotonin acts as a humoral component of the nervous system, as it is reported to do in some other animals. The tissues of *Fasciola* contain an acetylcholine -like substance and cholinesterase; the specificity of cholinesterase for acetylcholine has not been established, but it has been suggested that these components are related to activity of nervous system.

Insulin has been reported to have an effect on the carbohydrate metabolism but the evidences are lacking.

## Reproductive System

*Fasciola hepatica* is *hermaphrodite or dioecious.* The gonands are well developed and the male and female genital ducts open into a common chamber, the *genital atrium.* It is situated anteriorly in the body and opens to the exterior through the common genital aperture or gonopore, located ventrally in front of the acetabulum.

### *1. The male reproductive organs*

The male reproductive organs are:

(i) A pair of testes

(ii) A pair of vasa deferentia

(iii) Seminal vesicle

(iv) An ejaculatory duct

(v) Cirrus and cirrus sac.

*(i) Testes.* These are greatly ramified and tubular structures situated in the middle part of the body, one behind the other (tandem arrangement).

(ii) *Vasa deferentia.* A narrow delicate duct, the vas deferens, emerges out from each testis and runs forward to meet the fellow of opposite side

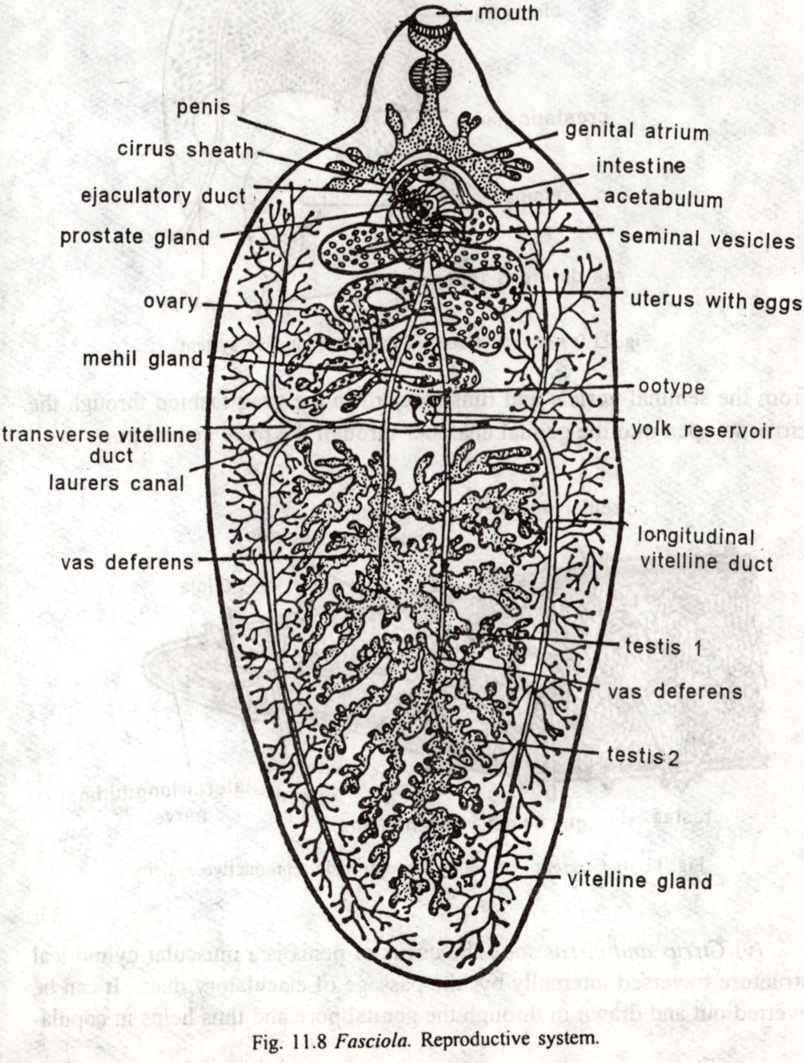


Fig. 11.8 *Fasciola.* Reproductive system.

forming the common sperm duct just below the ventral sucker.

(iii) *Seminal vesicle.* The common sperm duct immediately leads into a large muscular pear-shaped sac, the seminal vesicle, which lies in front of ventral sucker.

*(iv) Ejaculatory duct.* The ejaculatory duct arises as fine convoluted tube

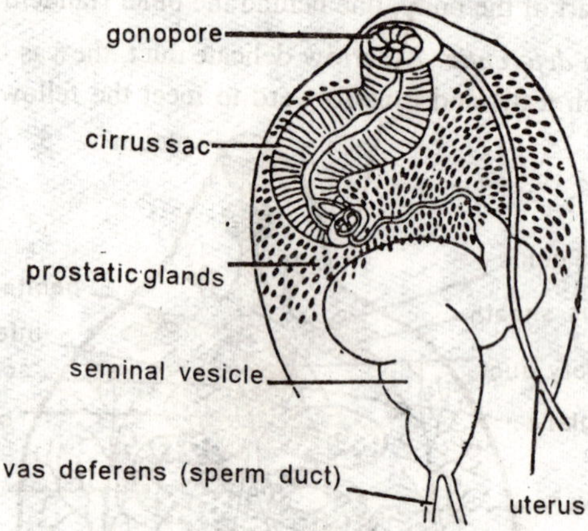


Fig. 11.9 *Fasciola.* Detail of male reproductive system

from the seminal vesicle and runs forward in a zigzag fashion through the cirrus to open into the genital chamber through the male genital pore.

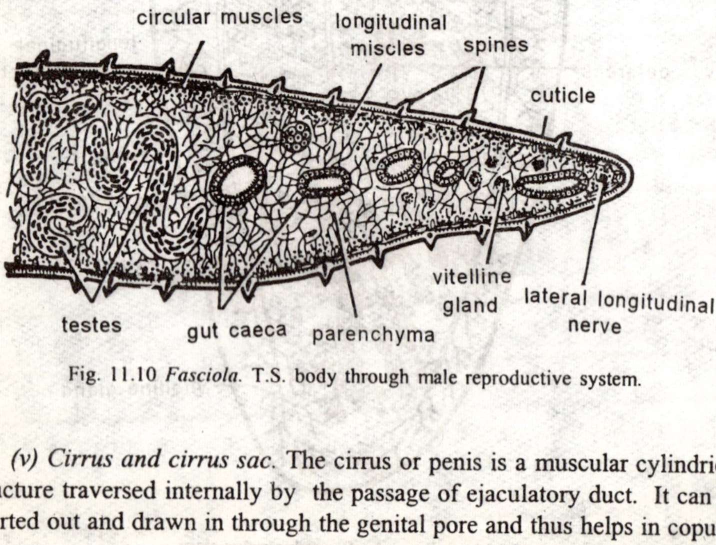


Fig. 11.10 *Fasciola.* T.S. body through male reproductive system.

*(v) Cirrus and cirrus sac.* The cirrus or penis is a muscular cylindrical structure traversed internally by the passage of ejaculatory duct. It can be everted out and drawn in through the genital pore and thus helps in copula-

tion. The cirrus and the seminal vesicle both are enclosed in a bag-like cirrus sac.

*(vi) Prostate glands.* Numerous unicellular prostate glands are found around the ejaculatory duct.

### *(B) Female Reproductive Organs*

The female reproductive organs are:

(i) Ovary

(ii) An oviduct

(iii) Uterus

(iv) Laurer's Canal

(v) Mehlis' gland

(vi) Vitelline glands and vitelline ducts

(i) *Ovary*. The ovary is a large, highly branched tubular structure which is situated on the right side in front of testes in the mesenchyme. It occupies the anterior one-third of the body.

(ii) *Oviduct*. From the inner side of ovary arises a short, narrow and convoluted duct, the oviduct. It runs down ward to join the uterus.

(iii) *Uterus*. It is a long wide and highly convoluted tube that extends upto the genital atrium, opening into it through the female genital aperture, close to male genital pore on the left-side. It contains a large number of capsules containing fertilized eggs.

(iv) *Laurer's canal.* The Laurer's canal arises from the oviduct and acts as a sperm duct. It develops a temporary opening on the dorsal body surface during breeding season. It may serves as vagina

(v) *Mehlis' gland.* At the junction of oviduct, vitelline duct and uterus a mass of unicellular gland lies known as *Mehlis' glands*. The function of these glands remains unclear. It was believed that these glands were merely responsible for the formation of shell but *Stephenson* (1947), *Rennison* (1953) and *Johri* and *Smyth* (1955) have shown definitely that this cannot be the case, because the shall of eggs is rich in various proteins, phenols and phenolases that are not present in these glands. Several views have been proposed to explain the function of *Mehlis' glands:*

(a) These glands secrete a fluid enchances the hardening or tanning process of newly formed eggs. (b) The secretion of these glands cause release of the shell globules from the vitelline glands. (c) The secretion forms a thin membrane around the cells forming eggs, and the shall globules than built up from within this membrane (d) The secretion lubricates the uterus facilitating

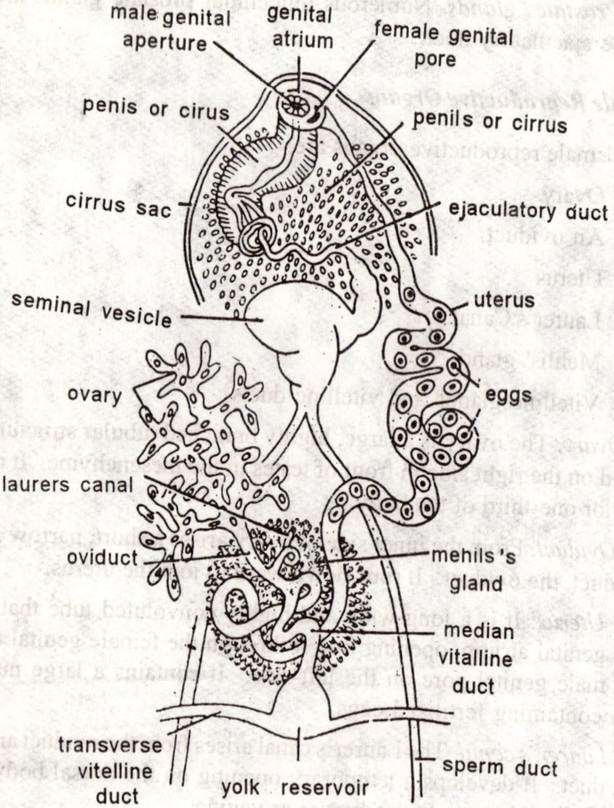


Fig. 11.11 *Fasciola*. Details of male and female reproductive system.

passage of the eggs. (e) The secretion activates sperms which are passed down. All or some of these hypothesis may be valid.

*(vi) Vitelline glands and vitelline ducts*. The vitelline glands or the yolk glands are in the form of numerous minute vesicles or follicles arranged on either side of the body from one end to the other. A fine duct arises from each vesicle and after uniting with similar ducts from other vesicles opens into the lateral longitudinal ducts. The two longitudinal ducts are connected in the middle line by a transverse vitelline duct. A median duct arises medially from the transverse vitelline duct and joins the oviduct.

The functions of vitelline glands is now know. In addition to contributing yolk material for incorporation within the egg, the vitelline glands also secrete large globules known as shell globules-which envelop the developing egg and eventually coalesce and become hardened to form the shall. The hardening process involves the tanning of protein present within the coa-

Phenolase
Phenol ------------------> quinone ----------> Tanned protein
Protein ------------------> protein ——→

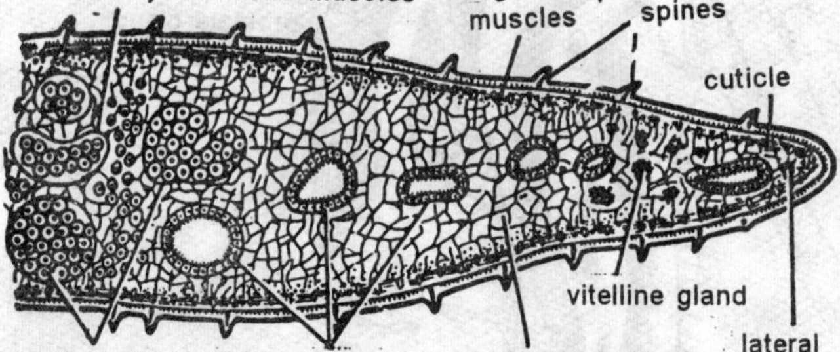


Fig. 11.12 *Fasciola*. T.S. body through ovary and uterus.

lesced globules by quinone. Thus vitelline glands contribute both yolk and shell material.

*Ootype*. In some cases the oviduct after receiving *Laurer's canal* and a common vitelline duct swelles to form ootype. In *F. hepatica* according to *Stephenson* (1947) the ootype is absent.

## Life History

### *Copulation and fertilization.*

Cross-fertilization takes place by insertion of cirrus of one fluke into the *Laurer's canal* of another fluke. Transference of biflagellate sperms directly into the oviduct is possible. Secretion of prostate glands and as mentioned above, the secretion of *Mehlis' gland* keep the sperms active for fertilization. Laurer's canal become invisible after copulation. Fertilization takes place in oviduct. Self-fertilization is also seen here the sperms enter the uterus of same fluke through female genital pore and pass down to reach oviduct where fertilization takes place.

Fertilized egg gets about 30 or so yolk-cells and shell globules from vitelline glands by the way of various vitelline ducts. The shell is formed with a lid or operculum on one side. *Rowen* (1956,57) demonstrated that there is a viscous granular cushion at the opercular end of the egg, situated immediately beneath the operculum. The zygote and a mass of yolk - cells enveloped by the shell constitute the capsule. When complete the capsule

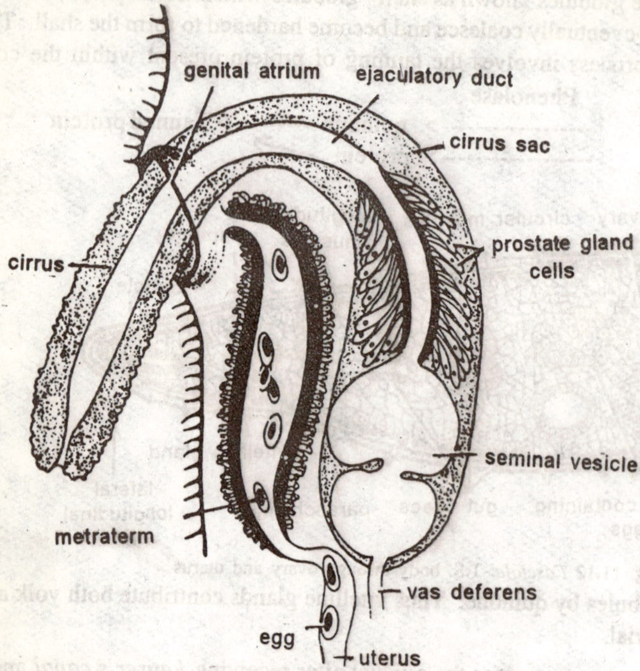


Fig. 11.13 *Fasciola*. Shows cirrus coming out of gonopore.

passes into the uterus.

Capsule formation occurs throughout the year and one fluke may produce about 50,000 capsules. A single fluke may produce a total of 2,00,000 capsules in about 11 years.

The capsules differ in shape and size in different animals. Hence, with the size and shape of capsules the parent flukes are recognized.

***Development.***

Cleavage starts immediately after fertilization. The fertilized egg divids into a small granular propagative cell and a large ectodermal somatic cell. *Somatic cell* divides and gives rise to the body-wall of the larva: *Propagative cell* later on divides into a cell similar to somatic cell, which takes part in the development of mesoderm, endoderm, and a cell called *germ cell* afterwards reaches the posterior part of the body and forms germ-ball consisting of 14 to 20 germ cells.

Capsules then escape through the gonopore and are liberated into the bile capillaries from where they reach the intestine of the host and are passed out with the faeces.

The development remains arrested while the embryos remain in the faeces, they may survive in wet faecal matter for several months. If washed free, the development of the embryo proceeds. The optimum temperature for development ranges from 10- 30°C. At 30°C the encapsulated embryo differentiates into *miracidium larva* within eight days. The availability of oxygen

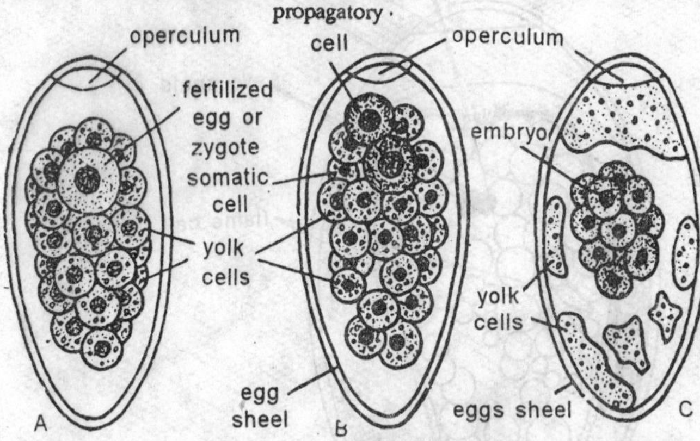


Fig. 11.14 *Fasciola*. Early stages of development. A= Fertilized egg, B= Two cell-stage, C= Many cell stage.

is also a factor. The fully formed miracidium is a more or less conical little animal covered with ciliated epithelium. It does not look at all alike an adult fluke but does show resemblance to some free-living flat worm.

*Miracidium larva*. As the miracidium is formed the hatching enzyme is secreted in response to exposure to light. The enzyme dissolves the cementing material by which the operculum is attached, thus releasing the operculum. Expension of the granular cushion ,accompanied by exosmosis of salts and other materials from within the egg pushes off the operculum. The hatching occurs in water and free-swimming miracidium penetrates the intermediate host.

*Structure of miracidium*. It is an elongated, about 0.07 mm long creature with a broad anterior end and a narrow posterior end. Anteriorly there is a *conical or apical papilla or terebratorium*. It is ciliated except in the region of apical papilla. However, anterior cilia are slightly longer.

The ectoderm is formed by 21 hexagonal plate like cells which are arranged in 5 rows. The I row consists of 6 small plates. The II row contains again 6 plates, III row has 3 plate cells, 4 plates in IV row and 2 plates in V row. The arrangement of the ciliated plates is quite constant and is of considerable value in determining phylogenetic relationship. In *F. hepatica* the cells of first row arranged are two dorsal, two ventral and two lateral in position. The cells of second row are three dorsal and three ventral. In the third row the cells are one dorsal and two venterolateral. The fourth row has two right and two

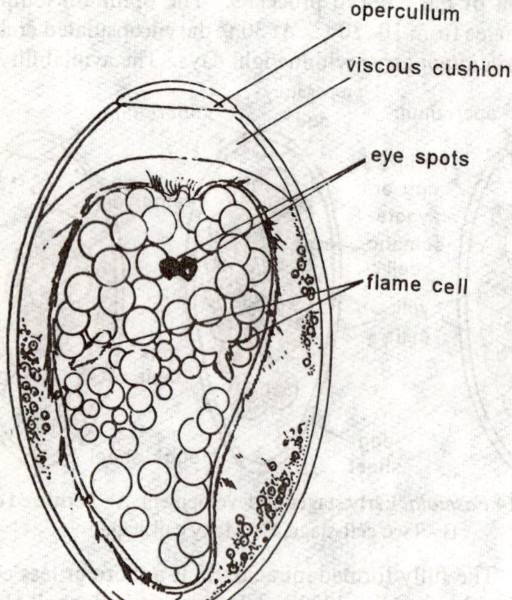


Fig. 11.15 *Fasciola*. Miracidium just before hatching.

left cells and fifth row has one left and one right cells.

The ectoderm is followed by musculature consisting of outer circular and inner longitudinal muscle fibres. The musculature is followed by a layer of cells forming the sub-epithelium. The ectoderm, musculature and sub-epithelium forms the body wall of larva.

At the anterior tip of the miracidium is found a mobile apical papilla. A pair of cephalic glands or penetration glands, located in the mesenchyme, empty to the outside near the apical papilla. The secretion of these glands aids in dissolving the host's tissue during penetration process. Besides a sac-like multinucleated apical gland is located between the two cephelic glands. Previously it was supposed to be the gut of larva.

The sensory organs of miracidium includes two eye-spots. Eyes contain pigments and are situated at the surface of the second row of epidermal plate over the brain. The eyes may be primary receptors involved in the process of hatching. When kept in dark few miracidia hatch, but illumination rapidly induces hatching. The brain of the miracidium is a large cephalaic ganglion

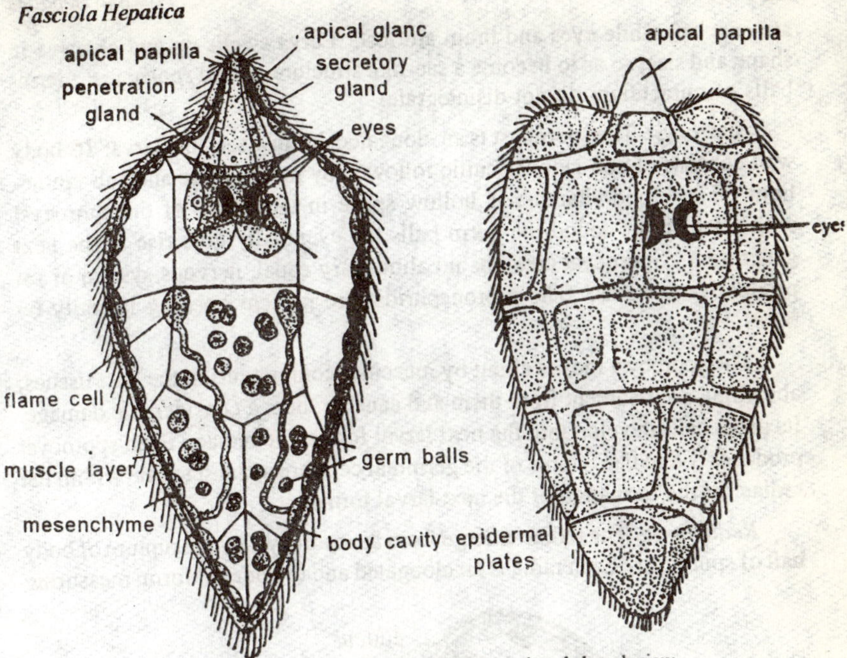


Fig. 11.16 *Fasciola*. Miracidium in ventral and dorsal view.

lying in the parenchyma behind the apical region. It is X-shaped. From this nerve centre, fibres are oriented in all directions to innervate the various body tissue.

There is a pair of flame-cells or protonephridia, which are excretory organs and their ducts open to the outside between IV and V rows of epidermal plate cells.

During the differentiation of the miracidium within the egg shall, certain germinal cells become trapped in the parenchyma. These develop into germ balls by increasing in number and size and becoming enveloped, eventually giving rise to the next larval generation. Such germ balls are usually located in the posterior region of the body.

Miracidium swims about in water or moisture film, where cattle usually pass out their faeces. It dies in 30 hours unless it is able to reach intermediate host, which is always an amphibious snail of the genus *Limnea*. Intermediate host in case of *Fasciola indica* is *Limnea acuminatisata* while in the case of *Fasciola hepatica* intermediate host in *Limnea truncatula*. When larva enters the tissue of the intermediate host with the help of apical papilla, it reaches soft parts, such as mantle,pulmonary sac and hepato-pancreas. Penetration into the intermediate host is helped by the secretion of the penetration glands or other glands. On reaching the intermediate host, ciliated epidermal cells are cast off, while eyes and brain are lost. Larva swells up and changes in shape and size, so as to become a sac-like structure, called *sporocyst*. Germ-balls of miracidium do not disintegrate.

*Sporocyst.* The sporocyst is an elongated, hollow,sac-like larva. Its body wall consists of non-cellular cutile followed by musculature and sub-epithelium. The brood chamber is hollow space in the centre of the sporocyst within which are found the germ balls the eventually give rise to the next generation. Sporocyst contains no alimentary canal, nervous system or reproductive systems. The protonephridia are present opening laterally by pores.

The sporocyst moves about by muscular contractions in the host tissues, absorbing nourishment from them and causing them a considerable damage. Its germ balls develop into the next larval forms, the rediae. Each sporocyst produces 5-8 rediae. Some of the germinal cells are again set aside within the rediae for the formation of the next larval forms.

*Redia larva.* The rediae emerge from the sporocyst by the rupture of body ball of sporocyst. Each radia is an elongated and cylindrical form, measuring

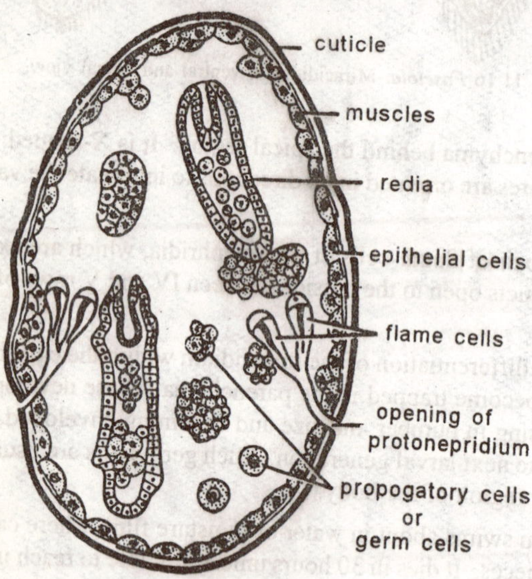


Fig. 11.17 *Fasciola.* Sporocyst

1.3-1.6 mm in length. It has a circular ridge, called *collar* near the anterior end and a pair of projections termed *lappets* near the posterior end. The body-wall consists of the usual cuticle, musculature and subepithelium. Unlike the sporocyst, the redia possess a mouth located at the anterior terminal of the elongated body. The mouth leads into a muscular pharynx,which in term leads a blind-sac, the gut . Numerous unicellular pharyngeal glands open into the pharynx.

Protonephridia divide further and form a much branched system. All the flame cells of one side open out through a common excretory duct. A birth pore is present, lateral to and in proximity of the mouth. Within the brood chamber are found germ balls.

Redia is a very active little creature. It moves about by the contractions of the body and lappets. It nourishes itself by sucking fluid and cells of the host tissues with its muscular pharynx. It soon migrates into the digestive glands of the snail. During summer months when the food is sufficiently

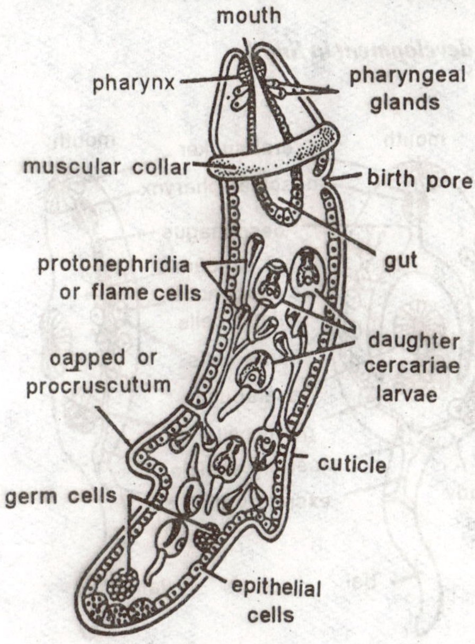


Fig. 11.18 *Fasciola.* Redia.

available, mother rediae mature in 12-18 days, the germ balls give rise to a second generation of daughter rediae. Germ balls of daughter rediae in the winter develop into next larval stage, the cercariae.

*Cercaria larva.* Cercariae leave the mother or rediae through their birth pores. From each redia 14-20 cercaria are produced. Cercaria is about 0.25-0.35 mm long and is an oval larva with a tail. The body wall consists of cutile, musculature and subepithelium. On the anterior side, there is a mouth with a sucker, a suctorial pharynx, long oesophagus and a bifid intestine. A ventral sucker or acetabulum is present on the ventral side. A large number of dark, brown cells, called *cystogenous glands* are found beneath the cuticle. There is a pair of excretory ducts, each with several flame cells. These open near the hind end of the larva into a bladder that leads out through a pair of nephridiopores. A nervous system similar to that of adult has also appeared. Rudiments of reproductive organs develop from the germ cells near the hind-end.

From the digestive glands of the snail, the cercariae pass into the pulmonary sac and then escape into the surrounding water. Cercaria swims about 5 minutes to an hour in summer. Afterwards, it settles down on some green leaves of water plants. Tail is cast off and the body becomes rounded and forms around itself a thick brownish cyst wall from the cystogenous glands. The encysted cercaria is now called metacercaria.

***Factors affecting development in Snail***

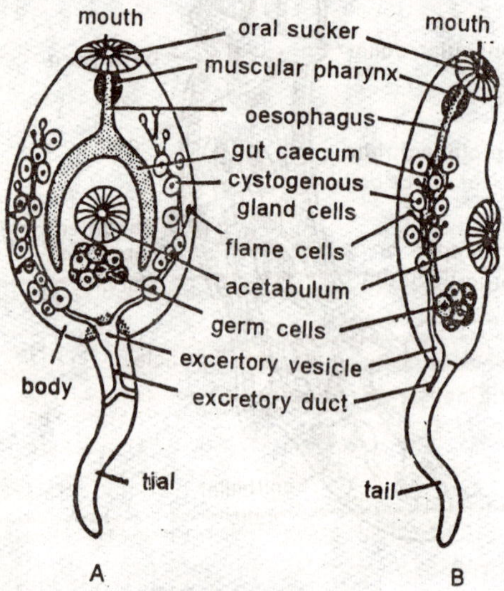


Fig. 11.19 *Fasciola.* A=Cercaria in ventral view. B=Cercaria in lateral view.

The rate of development in the snail host is a function of the environmental temperature. *Kendell* reported a curious effect of temperature on the development of *Fasciola*. In snails maintained at laboratory temperature, daughter rediae were not produced, the first generation of rediae giving rise to cercariae. On the other hand snails exposed daily to a period of low temperature i.e. 4-5°C, daughter rediae were produced from rediae.

The state of snail nutrition also affects development. The effect of food availability on development in the snail host is probably the explanation for the observation that fewer larvae develop in small snails than in large ones.

*Metacercaria.* It is a juvenile fluke having a rounded form with a diameter of about 0.2 mm. It resembles cercaria larva but differs from later in the absence of tail, cystogenous glands and presence of thick cyst wall. Excretory bladder opens directly through a single pore.

The metacercaria, encysted on vegetation, may live for as long as a year at low temperature, or for as short time as two or three weeks at 25°C.

***Inspection of final host***

The final excystment in the host requires an elevated temperature i.e. 38°C, low redox potential, a high carbon dioxide tension, and the presence of bile. The factors necessary for the excystment of *Fasciola* matacercariae seem to serve as stimuli to activate the worms, but the mechanism is not clearly

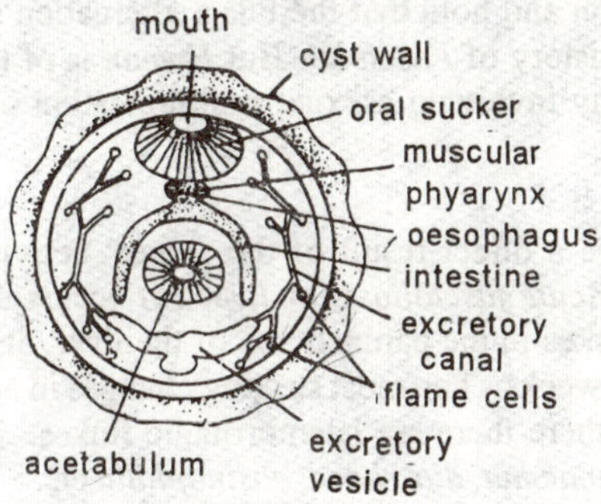


Fig. 11.20 *Fasciola.* Metacercaria.

understood. When metacercaria reaches the intestine of host , the cyst is digested and young fluke is liberated. The young worm penetrate the wall of gut and passes into the peritoneal cavity. In the body cavity the young worms seem to move about rather aimlessly until they encounter the capsule of liver. On contact with the liver, the flukes burrow into it, devouring liver parenchyma as they move. It grows in size within 5-6 weeks and finally settles in the bile duct. It attains sexual maturity and begins to lay eggs 11-13 weeks after ingestion by the mammalian host.

***Nature of life-history***

The life-history of *Fasciola hepatica* is much complicated due to the occurrence of several larval forms in a series. Formarly, it was held that the rediae and cercariae develop by parthenogenesis from the propagatory or germ cells, which were believed to be the eggs (*Grobben*, 1882). This kind of asexual parthenogenetic reproduction by larval forms is known as *heterogamy*. This view was widely accepted for sometime and under its way, three terms were coined. These are *parthenitae* for sporocyst and rediae larvae, *adolescariae* for cercaria larvae, and *maritae* for adult flukes. This view, however, is now considered erroneous and has been given.

It has been found that the germ cells present in the larval forms are not eggs. Instead they are diploid cells derived by the mitotic divisions of propagatory cell which is separated from the zygote at its first cleavage. Thus it can be said that they are parts of zygote. It, thus follows that one zygote produces several larvae. Thus, this process of reproduction in sporocysts and rediae has been looked upon as *polyembryony* by *Ishii* (1934), *Chen* (1937), *Rees* (1940) and *Cart* (1944). Since all the larvae resulting from a zygote are not formed simultaneously but a intervals, the phenomenon should be called delayed *polyembryony*.

*Steenstrup* (1942) and some other workers interpret polyembryony as a sexual multiplication and hold that there is a alternation sexual and asexual generation in life-history of *Fasciola*. But *Hyman* is of the view that it is a continuous ontogeny involving asexual multilplication in the larval stages.

**Pathogenesis**

Pathogenesis as a direct result of the flukes' activities may be either chronic or acute. *Acute fascioliasis or liver rot* occurs during the preadult migration of the flukes in the parenchyma of the liver, and sometimes other organs, for about 8 weeks. The most striking changes in acute fasciolosis are seen in the liver, where there are haemorrhagic flukes. The liver becomes swollen and cause *anaemia, diarrhoea, eosinophilia* etc.

*Chronic fascioliasis* occurs beyond 12 weeks, when the flukes have reached the bile ducts and are maturing sexually. Althogh milk and wool production looses are the usual results of chronic infections, deaths also occur.

The liver performs a vast array of essential functions in vertebrates, and it is axiomatic that liver damage will have a variety of detrimental effects. The adult fluke causes anemia, haemorrhage and suppression in production of new red blood corpuscles. The result of liver damage is failure to gain weight in young animals, weight losses in older animals, reduced milk production and lowered fleece weights in sheep. The osmotic pressure of the blood is lowered and fluid is retained in the tissue spaces.

*Treatment.* Carbon tetrachloride has been used in sheep at 500 mg/kg. Hexachlorethane was developed specially for use in cattles. Other drugs used are *emetine hydrochloride, filcin, tetrachloroethane*, etc.

***Control.*** Control of *liver flukes* is effected in three major ways:

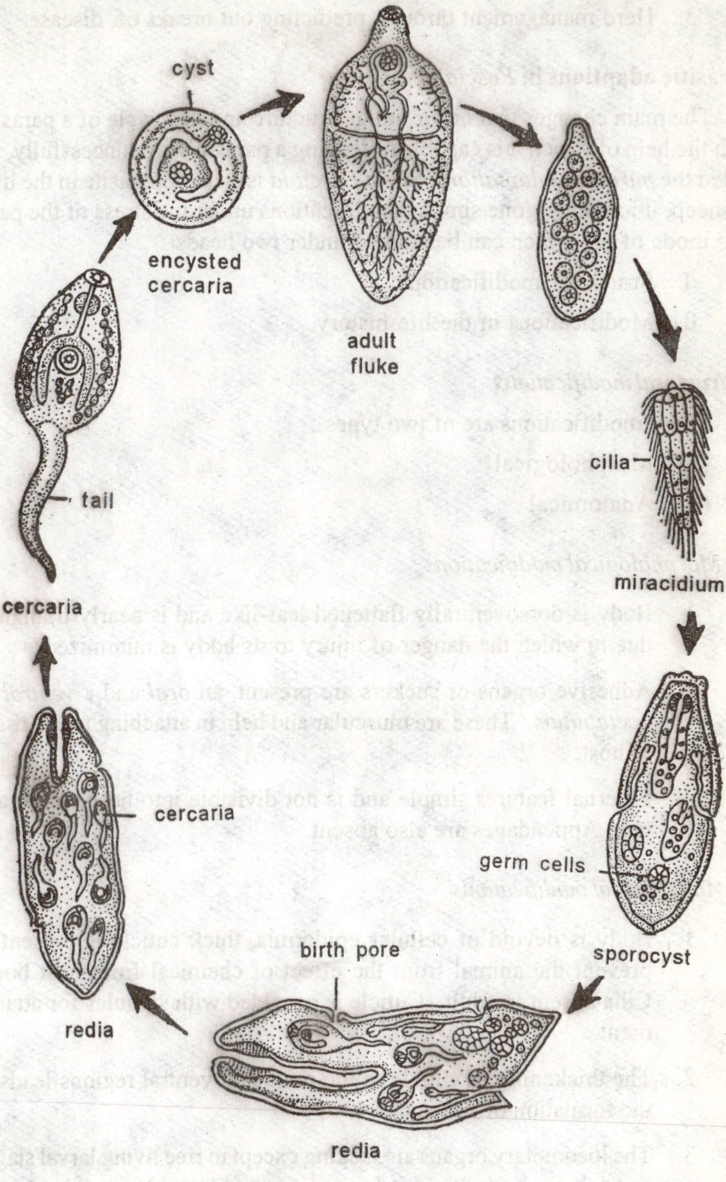


Fig. 11.21 *Fasciola*. Life - cycle.

1. **Snail control.**
2. **Treatment of infected sheeps.**

3. Herd management through predicting out breaks of disease.

## Parasitic adaptions in *Fasciola*

The main changes that occur in the structure and life-cycle of a parasite, with the help of which it is capable of leading a parasitic life successfully, are called the *parasitic adaptations*. Since *Fasciola* is an endoparasite in the liver of sheep, it has undergone similar modifications under the stress of the parasitic mode of life which can be studied under two heads:

I. Structural modifications.

II. Modifications in the life-history.

### *I. Structural modifications*

These modifications are of two types.

(A) Morphological

(B) Anatomical

#### *(A) Morphological modifications*

1. Body is dorsoventrally flattened leaf-like and is nearly triangular due to which the danger of injury to its body is minimized.

2. Adhesive organs or suckers are present, an *oral* and a *ventral or acetabulum.* These are muscular and help in attaching to the tissue of host.

3. External features simple and is not divisible into head, trunk and tail. Appendages are also absent.

#### *(B) Aanatomical modifications*

1. Body is devoid of cellular epidermis, thick cuticle is present to prevent the animal from the effect of chemical from host body. Cilia absent in adult. Cuticle is provided with spinules for attachment.

2. The thickening of cuticle in the orala and ventral regions leads to the formation of two suckers.

3. The locomotary organs are lacking except in free living larval stages as adult *Fasciola* is an endoparasitic and it need not make excursions for food and safety.

4. The digestive system is incomplete. The mouth is anterior and terminal and surrounded by oral sucker. The pharynx is also muscular and sectorial. Both are adapted to suck the nutrient.

5. Intestine is highly branched so that digested food can be distributed to different parts of body. Anus is absent.

6. Circulatory and respiratory systems are absent. Anaerobic respiration takes place.

7. Excretory system is well developed. It consists of extremely branched excretory canals which terminate into flame cells.

8. Nervous system is poorly developed. Sense organs are absent.

9. Reproductive organs are extensively developed. The highly branched ovary and tests are capable of producing large number of gametes. Bisexuality offers full changes of fertilization.

***II. Modifications in life-history***

1. The production of eggs in a very large number, about 10 million, due to which the threat of survival of parasite is removed.

2. Eggs are shelled and each is provided with an operculum.

3. A secondary host is present for the dispersal of species.

4. Multiplication takes place by *paedogenesis* during larval stage.

5. Miracidia are free swimming larval stages because they have to search new hosts. The cercariae are encysted when out of the body of secondary host which ensures protection from external effects.

❄❄❄

# 12

# PHYLUM - ASCHELMINTHES

There is variety of opinions among zoologists about the systematic position of nematodes. Some authors (*Gagenbour* etc.) have designated a separate Phylum Nemathelminthes, which includes or does not include Acanthocephala and Nematomorpha. Recently, nematodes have been grouped in a class Nematoda in the Phylum Aschelminthes. The term Aschelminthes is coined by *Grobben*.

**CHARACTERS**

1. Round worms have elongated, cylindrical and vermiform body with tapering ends.
2. Body is unsegmented but may be wrinkled. It is not distinguished into regions.
3. Body unpigmented being either white or with a yellowish tinge.
4. Caudal end of the body generally straight in female but coiled in males, and the males are shorter than females.
5. Anterior cephalization is not prominent, therefore, the body is without any definite regions.
6. Mouth terminal surrounded by lips. In strongyloides the lips are modified into teeth known as leaf-crown.
7. Amphids and papillae are the main sensory organs and are of great taxonomic value in case in free-living forms.
8. Marine nematodes are annulated due to the presence of transverse striations or longitudinal striations which are of common occurence.
9. Body is covered by rough resistant cuticle, having bristles, spines, warts and papillae etc. Sometimes cuticle modified as cephalic, cervical and lateral cuticle and in the form of vesicle.
10. Caudal end with a pouch of cuticular nature known as phasmids, very common in parasitic form.
11. Cuticle in the strongyloides is modified into an umbrella-like form known as *bursa* which is generally supported by muscular rays.

12. The layers of body wall generally made up of cuticle, sub-cuticle or epidermis and muscle layer. The cuticle is generally of keratin which allows only glucose and urea etc.
13. Epidermis is syncytial where they are usually divided in four sections in the four longitudinal chords. One dorsal, one ventral and 2 lateral in position.
14. Muscles consist of longitudinal muscle fibres with variable arrangements which is generally *holomyarian, meromyarian* and *polymyariana* etc. The muscles are of great taxonomic value.
15. The body cavity is pseudocoel.
16. Digestive tract well developed generally made up of mouth, buccal cavity, pharynx or oesophagus are of various type, intestine and anus.
17. Nervous system consists of a nerve-ring encircling the oesophagus. From it, nerves are given out anteriorly and posteriorly.
18. Protonephridia absent but the excretory system is made up of canals or gland-like organs.
19. Sexes are separate. Testes and ovaries are tubular and coiled. Usually there is a single testis. The duct from the testis opens into the cloaca and the cloaca is associated with accessory structures such as circular spicules. Ovaries, oviduct and uteri are double.
20. Aschelminthes are ovo-viviparous, oviparous or viviparous.
21. Life-cycle complicated may be with or without intermediate host.
22. Larval stage four. Third stage of larva is infective.

**CLASSIFICATION**

Phylum Aschelminthes is divisible into one class-Nematoda.

**CLASS-NEMATODA**

1. Body is slender, cylindrical, tapering towards both the ends.
2. Body is covered with cuticle.
3. Intestine is well formed.
4. Body cavity is not lined with epithelium.
5. Cloaca is absent in female.
6. Male and female reproductive organs are well developed.

The Classification of Nematoda is not stable as they have been studied mainly independently. The scheme given below will be found simple and convenient.

Class Nematoda is divided into two sub-classes:

Sub-class I-Aphasmidea

Sub-class II-Phasmidea

**Sub-Class I-Aphasmidea**

1. Phasmids or caudal sensory organs are absent.
2. Amphids or anterior sensory organs many.
3. Excretory organs either reduced or absent.
4. Coelomocytes and mesenterial tissue are well developed.
5. Caudal papillae are absent in females.

The Sub-class is divided into following orders:

***Order (1) Chromadoridea***

1. Oesophagus is not elongated but differentiated into three regions.
2. Cuticle smooth or ringed. Ornamented and sometimes with bristles.
3. Amphids spiral.
4. Mainly marine and free-living.

Examples : *Druconematids*, *Walichoanolaimus*.

***Order (2) Enoploidea***

1. Mainly marine and free-living.
2. Six sensory papillae on the lip.
3. Ten to twelve sensory bristles arranged in one or two circlets.
4. Cyathiform (goblet-shaped) amphids.
5. A pair of cephalic slits.
6. Size small to moderate.

Examples: *Pelagonema*, *Enoplus*, *Metonchidiamus*.

***Order (3) Dorylaimoidea***

1. Found in fresh-water and soil.
2. Cuticle smooth without bristles.
3. Head with an inner circlet of six and an outer circlet of ten sensory papillae.
4. Odontostylet, the piercing apparatus derived from a modified tooth in the oral region.
5. Amphids cyathiform.
6. Posterior part enlarged.

Examples: *Dorylaimus*, *Xiphinema*.

***Order (4) Mermithoidea***

1. Anterior end with 16 labial papillae.
2. Sensory bristle absent.
3. Cuticle is smooth.
4. Amphids either rudimentary or cyathiform.
5. Oesophagus opening in blind intestine.
6. Larva parasites in invertebrates whereas adults are free- living

   Examples: *Mermis*, *Paramermis*.

***Order (5) Trichuroidea***

1. Intestinal muscles, caudal gland or suckers are absent.
2. Mouth without lips.
3. Pharynx very long and narrow.
4. Females with one ovary.
5. Males with one or no copulatory spicule.

   Examples: *Trichinella*, *Trichuris*.

***Order (6) Desmocolecoidea***

1. Small sized with heavily ringed cuticle.
2. The head is armoured. Anterior end with four sensory bristles.
3. Amphids are crescent shaped or pump-shaped.
4. They are marine and free-living.

   Examples: *Tricoma*, *Desmoscolex*.

***Order (7) Monohysteroidea***

1. Small forms with smooth or ringed cuticle having bristles.
2. Amphid circular.
3. Anterior end with 4, 6, 8, or many sensory bristles.
4. Free living forms, few are marine, some are fresh water and others terrestrial.

   Examples: *Monohystera*, *Siphonolaimus*.

**Sub-class II- Phasmida**

1. With phasmids and pore like amphids.
2. Excretory organs with paired lateral canals.
3. Coelomocytes 4-6.
4. Caudal or hypodermal glands absent.

5. Reproductive system well developed.

The sub-class has been divided into following orders:

***Order (1) Rhabditoidea***

1. Mouth without lips and with 6, 12, or 18 papillae.
2. Cuticle is ringed or smooth.
3. Pharynx, long with posterior bulb.
4. Single ovary in females and males with copulatory spicules.
5. Mostly parasite on birds and mammals.

Examples: *Rhabditis*, *Diplogaster*.

***Order (2) Oxyuroidea***

1. Pin-shaped small or medium sized.
2. Cervical papillae absent.
3. Pharynx with posterior bulb.
4. Female with slender tail.
5. Males with one or two copulatory vesicles and sometimes with papillae forming a bursa.
6. Parasites in invertebrates and vertebrates.

Examples: *Enterobius*, *Oxyuris*.

***Order (3) Ascaroidea***

1. Large parasitic.
2. Mouth surrounded by three rerely six lips.
3. Cervical papillae present.
4. Pharynx with or without valve.
5. Female short tail.
6. Male with ventrally coiled tail and with two equal copulatory spicules and usually without alae.

Example: *Ascaris*.

***Order (4) Strongyloidea***

1. Mouth without distinct lips but has leaf crowns.
2. Pharynx is without bulb.
3. Males are provided with a distinct caudal bursa which is supported with rays.
4. Females with a ovejactor.

Examples: *Ancylostoma*, *Necator*, *Strongylus*.

***Order (5) Dracunculoidea***

1. There are no definite lips or buccal capsule.
2. Mouth is surrounded by ring of papillae.
3. Pharyngeal bulb is absent. Pharynx is muscular anterior but glandular posteriorly.
4. Copulatory bursa is absent in males.
5. They are parasite of vertebrates.

Examples: *Dracunculus*, *Micsopleura*.

***Order (6) Spiruroidea***

1. Mouth with two lateral lips, sometimes 4 or 6.
2. Pharynx with bulb with anterior muscular and posterior glandular part.
3. Males without bursa but have spirally coiled tail.
4. Parasitic forms.

Examples: *Gnathostoma*, *Spiroxys*.

***Order (7) Rhabdiasoidea***

1. Cuticle smooth.
2. Definite pharyngeal bulb is absent.
3. Parasitic stage is hermaphrodite or parthenogenetic free- living stage may develop into males and females.

Example: *Rhabdias*.

***Order (8) Filariodea***

1. Slender filiform worms.
2. Mouth simple, lips are absent.
3. Females larger having an anterior vulva and a long tubular vagina.
4. Males with two unequal spicules. Alae present or absent.
5. Found in blood or lymphatic system. Transmitted by blood sucking insects.

Examples: *Wuchereria, Microfilaria, Loa.*

## TRICHURIS

| | | |
|---|---|---|
| Phylum | - | Aschelminthes |
| Class | - | Nematoda |
| Sub-class | - | Aphasmidea |

Order - Trichuroidea

Genus - *Trichuris*

1. It is commonly called as *whip-worm*. It is an endoparasite found in the caecum and large intestine of man.
2. It has world wide distribution but common in tropical countries.
3. Female is slightly larger than males.
4. The anterior part of the body having pharynx is usually more slender and this portion is buried in the mucosa of host's alimentary canal.
5. Mouth is simple without lips.
6. It feeds on blood and epithelial cells.
7. Life-cycle is simple with no intermediate host.

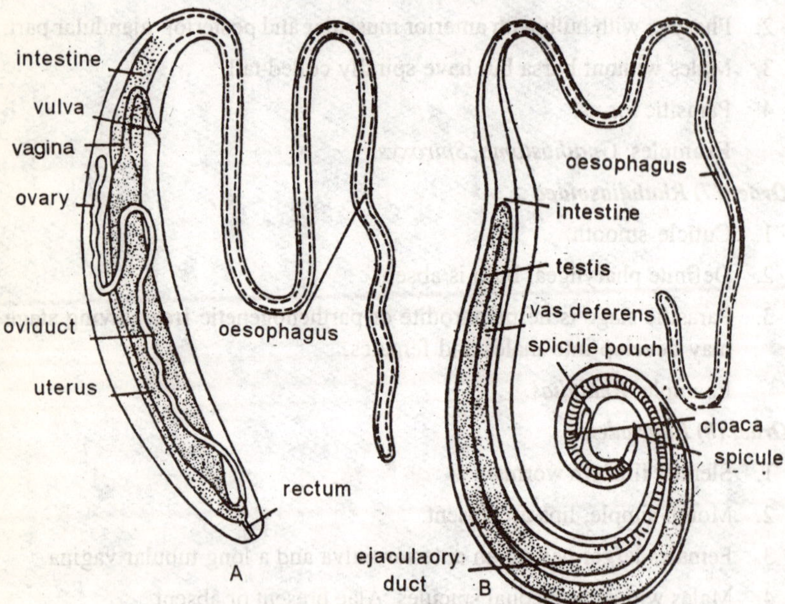


Fig. 12.1 *Trichuris*

8. Development of eggs take place in moist soil. Embryos develop in about three weeks.
9. Contaminated food and water responsible for infection in man by swallowing eggs with juveniles.
10. Liberated juvenil grow into adults within a month.
11. Excretory system is absent.

12. Heavy infections is responsible for diarrhoea and anaemia.

## TRICHINELLA

The systematic position is the same as that of *Trichuris*.

1. It is a parasite of man, pig, rat and other vertebrates. It is commonly known as *trichinia worm*.
2. It is commonly found in Europe. United States and Arctic regions.
3. Sexes are separate, males are about 1.5 mm. in length and the female are 3 mm. in length.
4. In both the sexes, the mouth is without lips.
5. Pharynx is long, slender and non muscular and is embedded in stichosome.
6. The oesophagus is tubular, well developed and occupies 1/3 of the total length of body.

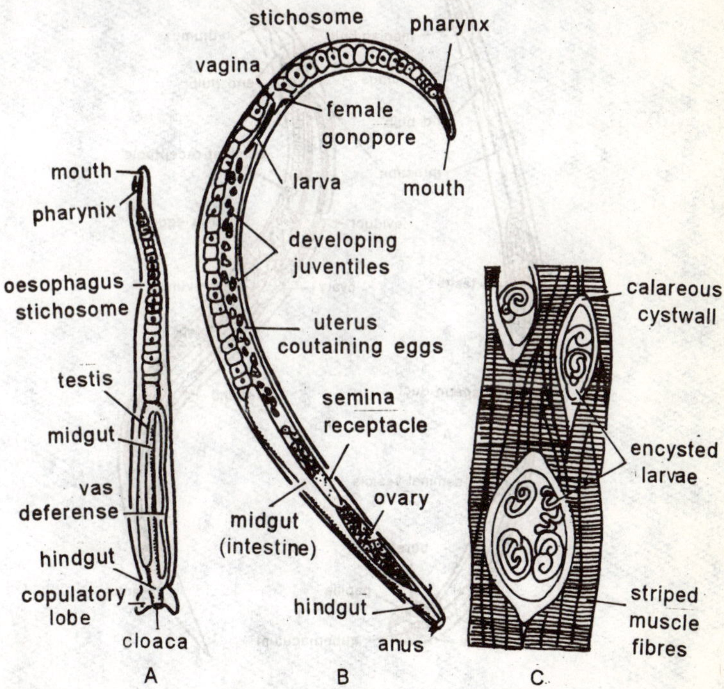


Fig. 12.2 *Trichinella*. A-Adult male. B-Adult female, C-Encysted larvae in striped (Voluntry) muscle of pig.

7. The female worms which are viviparous enter the lymphatic and reach the muscular system where they secrete a cyst around them. This is called encysted sexless stage.
8. The encyated sexless trichinae are ingested by man along with pork-flesh, where the cyst wall is dissolved by the action of digestive juice and the reproductive organs start developing.
9. The developing worm begins to migrate into the muscles and produces the symptoms of a dreaded disease call*Trichinosis*.
10. Patients feel gastro-intestinal pain, and weakness with or without fever.

## RHABDITIS

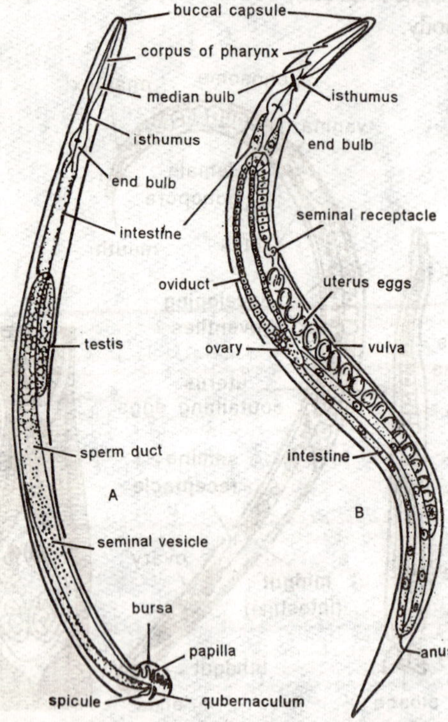


Fig. 12.3 *Rhabditis*
A= male B= female

| | | |
|---|---|---|
| Phylum | - | Aschelminthes |
| Class | - | Nematoda |
| Sub-class | - | Phasmidea |
| Order | - | Rhabditoidea |
| Genus | - | *Rhabditis* |

1. Genus *Rhabiditis* includes numerous free-living, semiparasitic forms. They are commonly found in organic matter, faeces of man. They feed on the fluid formed by bacterial decying of organic matters.
2. The pharynx has two bulbs with an isthmus joining them.
3. The excretory system is H-shaped.
4. The reproductive system is J-shaped.
5. The large female has a pointed tail and didelphic sex- organs i.e. having paired ovaries, oviducts, seminal recapticles and uteri.
6. The male has a flattened copulaltory bursa, two equal spicules and monorchic sex organs i.e. single testis.
7. Life-cycle is simple and direct. The juveniles adhere to the bodies of dung beetles for transport to fresh dung.

## ENTEROBIUS VERMICULARIS

| | | |
|---|---|---|
| Phylum | - | Aschelminthes |
| Class | - | Nematoda |
| Sub-class | - | Phasmidea |
| Order | - | Oxyuroidea |
| Genus | - | *Enterobius* |
| Species | - | *vermicularis* |

1. It is commonly called *'Pinworm'* or *'seatworm'*.
2. It has world-wide distribution, but specially found in Europe, India, Africa, Canada, Chile and U.S.A.
3. Adult worms live in the caecum, appendix and adjacent parts of intestine of man.
4. Sexes are separate. The male measures 2-5 mm. in length and the female measures 8-13 mm. in length.
5. Anterior end is provided with three lips and a pair of cephalic expansion.

6. Male has ventrally curved tail with a single spicule and a pair of cuticular expansions known as *caudal alae*. It is minorchic.
7. Female has a straight tail which becomes long and pointed. It is didelphic. Vulva is anterior.
8. Life-cycle is simple there is no intermediate host. Gravid females migrate to anus causing itching sensation. They crawl and deposit eggs on the perineal skin. Due to scratching, gravid females rupture and thus small eggs are liberated. The re-infection in children takes by putting fingers in mouth. Eggs hatch. Juveniles inhabit small intestine. They

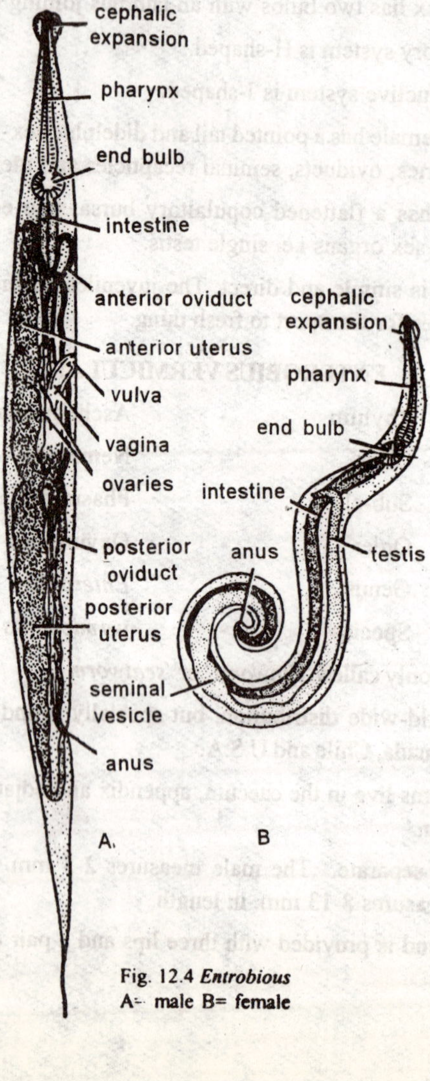


Fig. 12.4 *Entrobious*
A= male B= female

moult twice and turns into adults which reach their normal habitat. The infection is called *Enterobiasis*. Symptoms are inflammation of caecal mucosa, abdominal pain, insomnia, hysteria, restlessness, loss of appetite and sometimes appendicitis.

## LOA LOA

| | | |
|---|---|---|
| Phylum | - | Aschelminthes |
| Class | - | Nematoda |
| Sub-class | - | Phasmidea |
| Order | - | Filarioidea |
| Genus | - | *Loa* |
| Species | - | *loa* |

1. It is commonly known as *African eyeworm*, chiefly found in Africa.
2. They live in the subdermal connective tissue of man in West Africa.
3. The male and female measure 20-35 mm. and 20-70 mm. in length respectively.

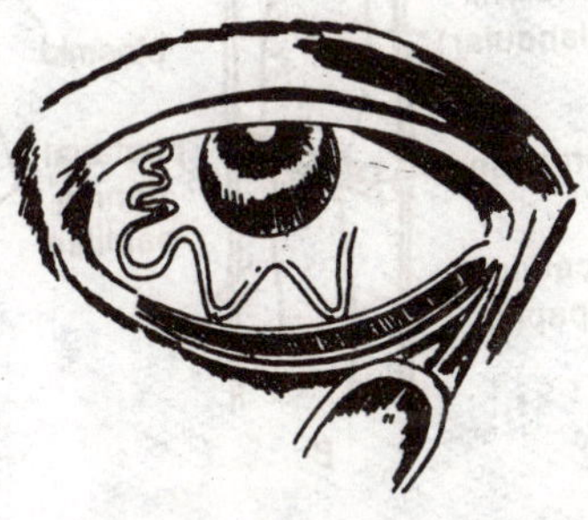

Fig. 12.5 *Loa loa* in cornea of man.

4. Body is covered by numerous warts.
5. The intermediate hosts are mango flies, *Chrysops*. The larva moult and develop up to infective stage in fly. These are transmitted in blood of man through proboscis.
6. The adult worm moves about in the body in the subdermal tissue and sometimes passes across the eye ball causing intense itching and swelling known as '*Calabar Swellings*'.

7. These are injurious and fatal when they penetrate brain and spinal cord.

## DRACUNCULUS

| | | |
|---|---|---|
| Phylum | - | Aschelminthes |
| Class | - | Nematoda |
| Sub-class | - | Phasmidea |
| Order | - | Dracunculoidea |
| Genus | - | *Dracunculus* |
| Species | - | *medinensis* |

1. It is commonly called '*Guinea Worm*'. It is found in tropical countries like India, Africa, South America and West Indies.
2. It lives in the deeper subcutaneous tissues of man.
3. The anterior end is blunt while the posterior end is pointed.

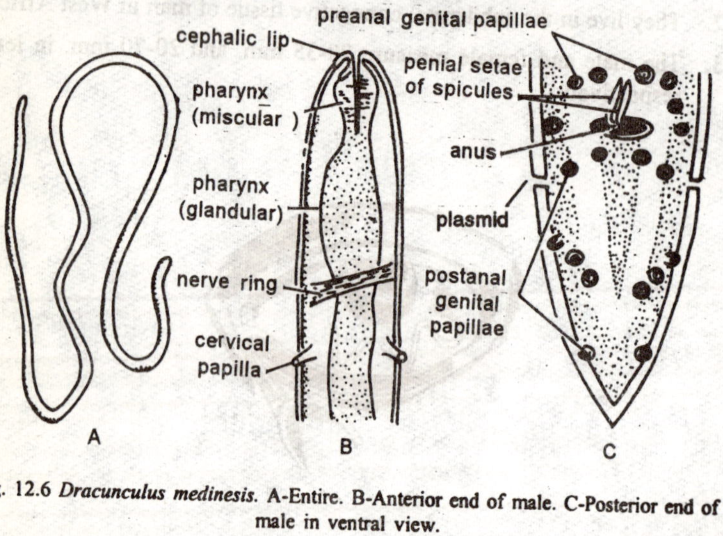


Fig. 12.6 *Dracunculus medinesis.* A-Entire. B-Anterior end of male. C-Posterior end of male in ventral view.

4. The adult male measures 4 cm. in length. Its hind end is curved and bears 10 pairs of genital papillae and a pair of pineal spicules.
5. Adult female measures 70-120 cm. in length. Its hind end is straight and bears a spine-like process.
6. The oesophagus is distinguished into an anterior narrow muscular region and a posterior broader glandular region.
7. The female reproductive system is didelphic with opposed uteri.

8. Females are viviparous. The alimentary canal and vulva are atrophied in mature females.

9. Mature female migrates to the superficial layers of the skin in those parts of the body which are likely to come in contact with cold water. Toxic substances are produced under the skin which results in the formation of a blister, leading to an ulcer, which breaks on coming in contact with water. On coming in contact with water, the worm ruptures, releasing larvae in water. The larvae swims about but do not develop further unless swallowed by female *Cyclops*, which serves as intermediate host. Here the female larvae moult twice and become infective. Infection occurs when *Cyclops* are swallowed by a new host alongwith drinking water.

10. It causes guinea worm disease characterised by *asthma, nausea, vomiting, diarrhoea* etc.

♦♦♦

# 13

# ASCARIS LUMBRICOIDES

Much is known about the morphology and life-cycle of Nematoda but out of them *Ascaris* is most known. *A.lumbricoides* is the most common nematode parasite found in the human intestine. Being of large size and found abundantly, *Ascaris* is studied as the main example of the nematodes.

### SYSTEMATIC POSITION

| | | |
|---|---|---|
| Phylum | - | Aschelminthes |
| Class | - | Nematoda |
| Sub-class | - | Phasmidea |
| Order | - | Ascaroidea |
| Genus | - | *Ascaris* |
| Species | - | *lumbricoides* |

**Habits and Habitat**

*A.lumbricoides*, a large nematode parasite of man, is world wide in distribution. It occurs more commonly in the small intestine of children than in adults because of the children's custom of playing in contaminated, egg-laden soil adjacent to dwellings. As many as 1000-1500 adult worms may be present in a single host. Because of their strong, pointed and cuticulate bodies and their habit of wandering, they may perforate the gut wall and cause peritonitis or they may enter the bile or pancreatic ducts or the vermiform appendix and interfere with normal digestive functions.

**External Features**

***Shape and Size.*** Both the sexes are elongated , cylindrical and pointed at both the ends. Males are 15-31 cm. long and 2-4 mm.in diameters, while the females are 20-35 cm. long and 4-6 mm. in width. The males are further distinguished by a ventrally curved posterior end, from the slit-like opening of which often protrude two copulatory spicules.

***Colouration.*** In the living condition, the colour of the animal is milky-white or reddish-yellow. There is a characteristic lustre on the body.

***Longitudinal lines.*** There are four longitudinal lines or streaks, running throughout the length. One is mid-dorsal,one mid-vertral and two lateral. The dorsal and ventral lines are white in colour, the lateral streaks are thicker and broader than dorsal and ventral streaks and are brown in colour.

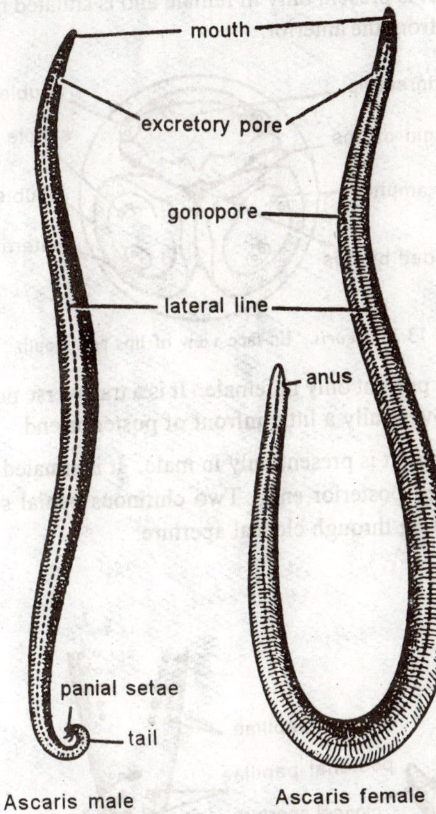


Fig. 13.1 *Ascaris lumbricoides.*

***Mouth.*** A triangular mouth is present at the anterior end of the body. It is surrounded by three denticulated lips. One of these is mid-dorsal and other two are latero-ventral in position. The mid-dosal lip is almost elliptical and has a fork fleshy core in the centre. A circle of four sensory papillae is found on it. The remaining two lips also have a forked core and two rows of *sensory papillae*. Each sensory structure is called *amphids*. The sensory papillae of nematodes are very useful in the identification of the species and assigning their position in the classfication. The sensory papillae are *tangoreceptors*. The amphids are *chaemoreceptors*.

The posterior end of the male bears ventrally 50 pairs of *pre-anal papillae* in front of the cloacal aperture and five pairs of *post-anal papillae* behind it. The first two pairs of post-anal papillae are double. These papillae of the male help in copulation.

***Excretory pore.*** It is situated mid-ventrally, a little behind the mouth.

***Genital pore.*** It is present only in female and is situated mid-ventrally at about 1/3 distance from the anterior.

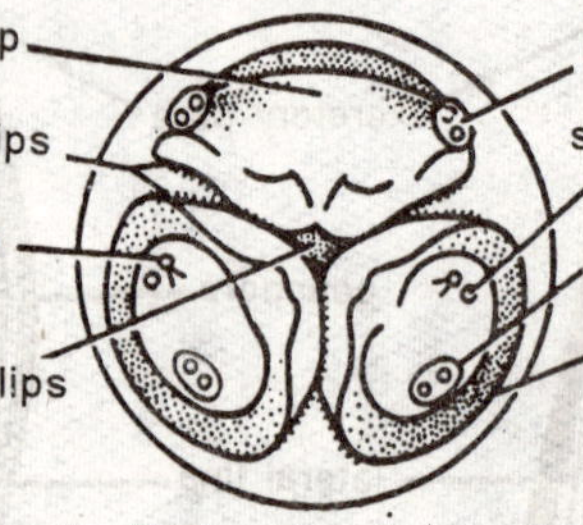


Fig. 13.2 *Ascaris.* En-face view of lips and mouth.

***Anus.*** It is also present only in female. It is a transverse pore guarded by thick lips, lies mid-ventrally a little infront of posterior end.

***Cloacal aperture.*** It is present only in male. It is situated mid-ventrally infront of the curved posterior end. Two chitinous penial setae or penial spicules are protruding through cloacal aperture.

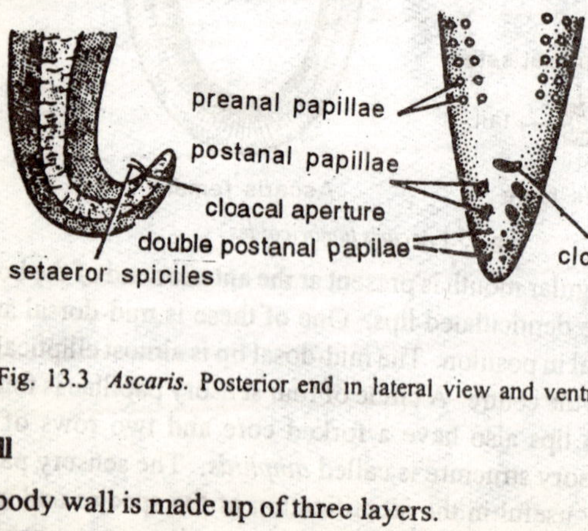


Fig. 13.3 *Ascaris.* Posterior end ın lateral view and ventral view.

**Body Wall**

The body wall is made up of three layers.

(1) Cuticle

(2) Epidermis

(3) Muscle layer

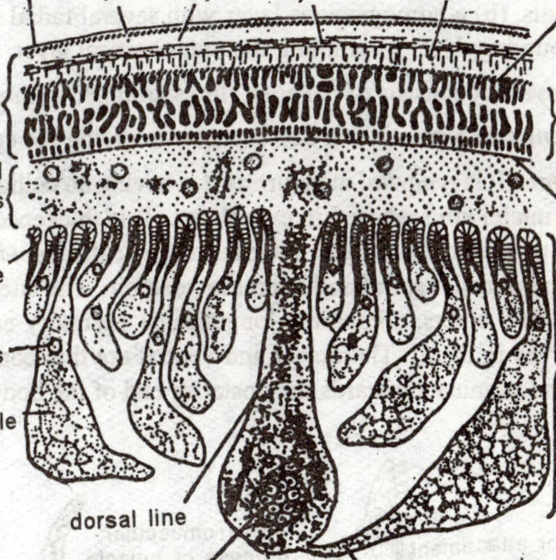


Fig. 13.4 Body wall in section.

1. ***Cuticle.*** The cuticle forms a thin transparent, elastic but resistant covering around the body. It protects the body against the action of digestive juices. It is secreted by underlying epidermis. Histologically, the cuticle is formed of several layers and can be distinguished into (i) Cortical layer, (ii) Matrix, (iii) Fibrous layer and (iv) Basement membrane.

   (i) *Cortical layer.* It is made up of dense keratin. It protects the animal from the digestive juices of host.

   (ii) *Matrix.* It is an elastic layer containing sulphur rich matricin.

   (iii) *Fibrous layer.* It is made up of collagen fibres that cross each other obliquely.

   (iv) *Basement membrane.* It forms the inner limit of cuticle.

Under electron-microscope the cuticle is made up of following layers:

(1) *Lipoid layer.* It is in the form of a thin osmophilic membrane, about 1000 A° thick.

(2) *Cortical or cortex layer.* It consists of (a) an outer cortical layer which lies as discontinuous strips or rings around the body, and (b) an inner cortical layer.

(3) *Matrix layer.* It consists of (a) an outer fibrillar layer tranversed by branching canals, (b) a homogeneous layer with several radial striations, and (c) a boundary layer resembling the outer cortical layer.

(4) *Fibre layer.* It consists of collagen fibres arranged in three strata.

(5) *Basement membrane.* It is a thin layer surrounding the epidermis.

2. ***Epidermis.*** It is *syncytial* in nature as it is having no cellular boundaries i.e. the nuclei are scattered in cytoplasm. It is thickened and projects into the pseudocoel as four longitudinal ridges called, *epidermal chords*. These are dorsal, ventral and lateral in position. The lateral chords are thicker than others, containing longitudinal excretory canals and lateral nerves in them. The dorsal and ventral are thin containing nerve cords and terminate towards the posterior end of the body.

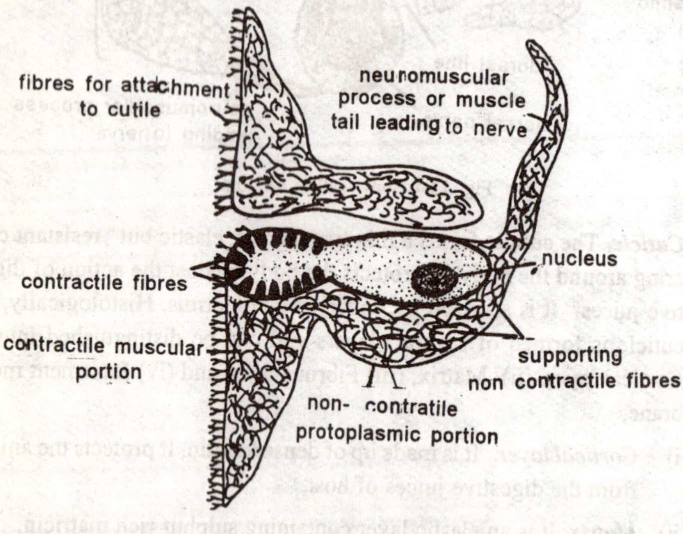


Fig. 13.5 Body wall in section.

3. ***Muscle layer.*** Beneath the epidermis is a single layer of longitudinal muscles which is distinguished into four quadrants. Of them two are dorso-lateral and two ventro-lateral. Each containing about 150 muscle cells. The muscle cells are highly specialised spindle-shaped cells,

distinguished into two parts lying below the ectoderm.

(a) *Fibrillar region.* Formed of contractile and non-contractile fibres arranged alternatively, it is found wrapped around the protoplasmic part.

(b) *Protoplasmic region.* It is the inner part of the muscle cell situated towards the pseudocoel. It is almost bladder-like in appearance and consists of a single nucleus.

There is a complete absence of circular muscles in nematodes.

**Pseudocoel**

There is a large fluid-filled space between the body wall and the viscera. The fluid contains fibrous tissue in the form of fenestrated membranes and fixed cells. The epithelial lining of this cavity is absent hence cannot be a true coelom. The fenestrated membranes extend over the viscera and muscle layer from a single giant *pseudocoelocyte* placed on the dorsal side where pharynx connects with the intestine. Four more star shaped giant cells lie in relation to epidermal cords. Thus in all there are five pseudocoelocytes. The

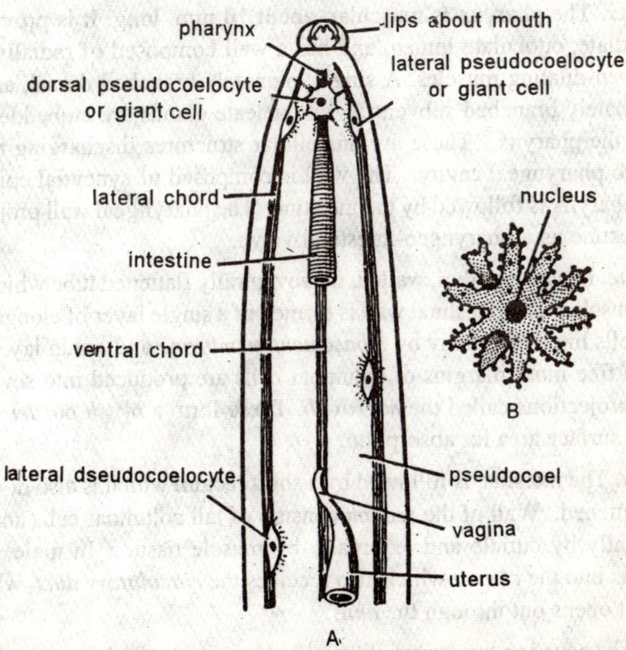


Fig. 13.6 *Ascaris.* A-portion of gaint cells in psedocoel. B-single gaint cell.

pseudocoelocytes and fenestrated membranes may be regarded as a kind of mesenchyme.

In fact, the spaces are said to be large vacuoles in a few, enormous cells. A protein-rich fluid of characteristic, unpleasent odour fills the cavity. This fluid may produce some of the toxic and inflammatory symptoms of ascarid infections.

**Locomation**

Locomotion is by means of longitudinal muscles which produce undulating movements of the body in the dorsoventral waves. The circular muscles are absent to counteract against longitudinal muscles, the hydrostatic pressure of body fluid against the cuticle serves the purpose.

**Digestive System**

The alimentary canal of *Ascaris* is complete. It comprises mouth, pharynx, intestine, rectum and anus.

***Mouth.*** It lies at the anterior end of the body. It is bounded by three lips-one mid dorsal and two latero-ventral in position. It leads into the pharynx.

***Pharynx.*** The pharynx is muscular, about 10 mm. long. It is provided with a triradiate, cuticulate lumen, and has a wall composed of radially arranged, lumen-dilating muscles. A single, pinnately branched dorsal, and a pair of palmately branched subventral, uninucleate glands are embedded in the wall of the pharynx. These are unicellular structures discharging their contents into pharyngeal cavity. The wall is composed of syncytial epithelium. The pharynx is followed by the intestine. The pharyngeal wall projects into the intestine as a pharyngeo-intestinal valve.

***Intestine.*** It is a long, thin, walled, dorsoventrally flattened tube which is devoid of muscles. The intestinal wall is formed of a single layer of elongated columnar cells lined externally by a *basement membrane* and a thin layer of *cuticle*. The free inner margins of columnar cells are produced into several finger-like projections called the *microvilli*. These form a *brush border* and increase the surface area for absorption.

***Rectum.*** The intestine is followed by a short rectum which is also dorsoventrally flattened. Wall of the *rectum* consists of tall columnar cells and is lined internally by cuticle and externally by muscle tissue. In male, the rectum opens into the *cloaca* which also receives the *ejaculatory duct*, while in female,. it opens out through the *anus*.

***Anus.*** The anus is a transverse slit guarded by anterior and posterior lips and is provided with a special muscle, the *depressor ani*.

***Digestion.*** *Ascaris* gets semi-digested or digested food from its host. So it does not require to digest food. However, it can digest semi-digest food by means of enzymes secreted by gland cells of pharynx and cells of anterior part of intestine. These enzymes can digest carbohydrates, proteins and fats.

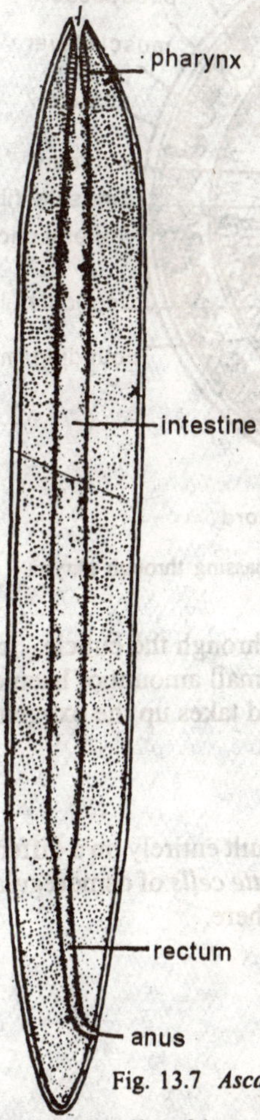


Fig. 13.7 *Ascaris.* Alimentry canal.

***Absorption.*** The cells of intestine generally absorb digested food. Excess fat and glycogen are stored in different tissue.

***Egestion.*** By the contraction of dilator muscles attached to rectum undigested food is expelled out.

## RESPIRATION

There being no special organs of respiration, but the parasite carry on anaerobic respiration by breaking down glucose and forming carbon dioxide

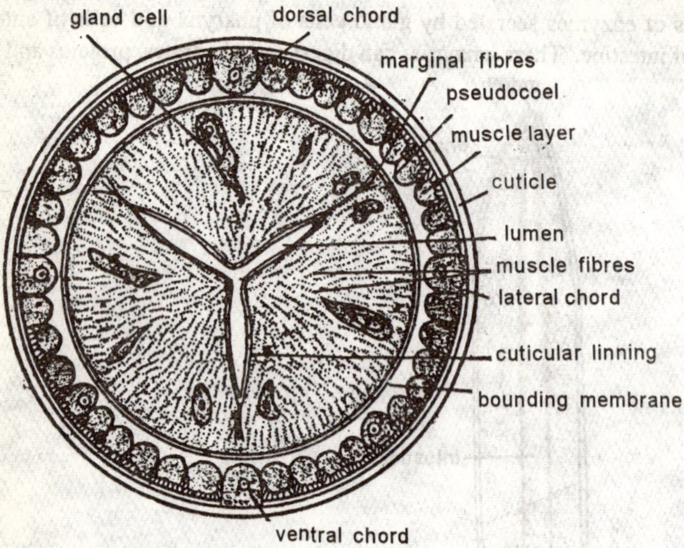


Fig. 13.8 *Ascaris.* T.S. body passing through phrynx.

and fatty acid which are eliminated out through the cuticle. In the presence of oxygen, it respires aerobically. The small amount of haemoglobin in the body wall (muscles) and pseudocoel fluid takes up the oxygen even when it is present in low tension.

## EXCRETORY SYSTEM

The excretory system of *Ascaris* is built entirely on a different plan. The flame cells or nephridia are absent. *Renette cells* of osmoregulatory function found in other nematodes are not found here.

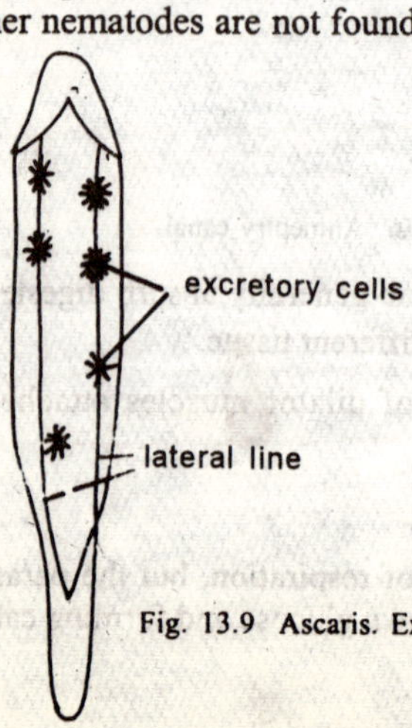


Fig. 13.9 Ascaris. Excretory system.

In *Ascaris*, the excretory system is made up of a syncytium containing four or more nuclei. It consists of:

1. ***Excretory duct.*** There is no excretory duct having two or more nuclei embedded into its walls. The excretory duct opens out through excretory pore.
2. ***Excretory cell.*** There is an excretory cell comprising two lateral excretory filaments running along each side of the body adjacent to the lateral chord.
3. ***Commissure.*** An excretory bridge is formed where two lateral filaments join. Each excretory filament contains an intracellular excretory canal. The intracellular excretory canals divide up into a network of canaliculi

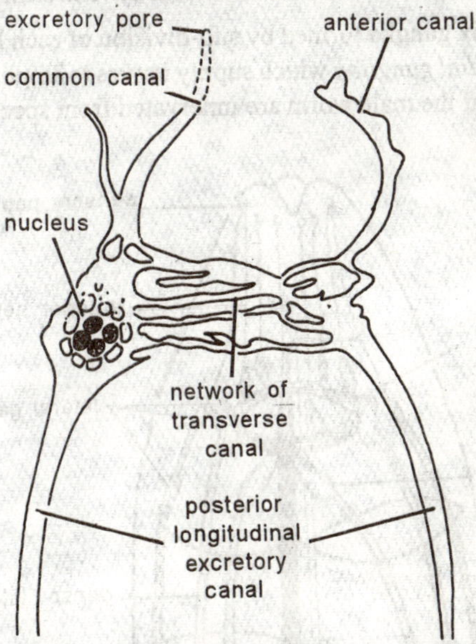


Fig. 13.10 *Ascaris.* H-shaped excretory system in anterior region of body.

in the commissural region. Excretory products are believed to diffuse into the canals which secrete *urea*, which is eliminated out through the excretory pore.

**Nervous System**

The nervous system consists of ganglia and nerves, all in close association with the ectoderm. The nervous system of *Ascaris* has been thoroughly studied by *Goldschmidt*. The number of cells taking part in the formation of entire system is apparently fixed and it is 162. There is a distinct *circumpharyngeal nerve ring* around the pharynx which connects a pair of *lateral ganglia*, a *dorsal ganglion*, a pair of *subdorsal ganglia* (each sub-divided into six smaller ganglia) and a pair of large *ventral ganglia.* From these ganglia of the nerve-ring there proceed backwards eight nerves: a *mid-dorsal nerve* that passes through the dorsal epidermal chord; a *mid-ventral nerve* that passes through the ventral epidermal chord; a pair of *dorsolateral nerves* that run just above the lateral chords; a pair of *lateral nerves* that extend through the lateral epidermis chords and a pair of *ventro-lateral nerves* that run just below the lateral epiaermal chords. Of these, the ventral nerve is most prominant and is ganglionated through out its length, therefore, it is regarded as nerve cord. Posteriorly the nerve cord joins a *pre-anal ganglion.* The dorsal and ventral nerves are connected at intervals by asymmetrical *dorso-ventral connectives* throughout their length. Similarly, the ventral and lateral nerves are also connected at interval by *ventro-lateral commissures.*

One of the six ganglia formed by sub-division of each lateral ganglion is known as *amphidial ganglion* which supply nerves to amphid of its side. The genital papillae of the male worm are innervated from special *bursal nerves.*

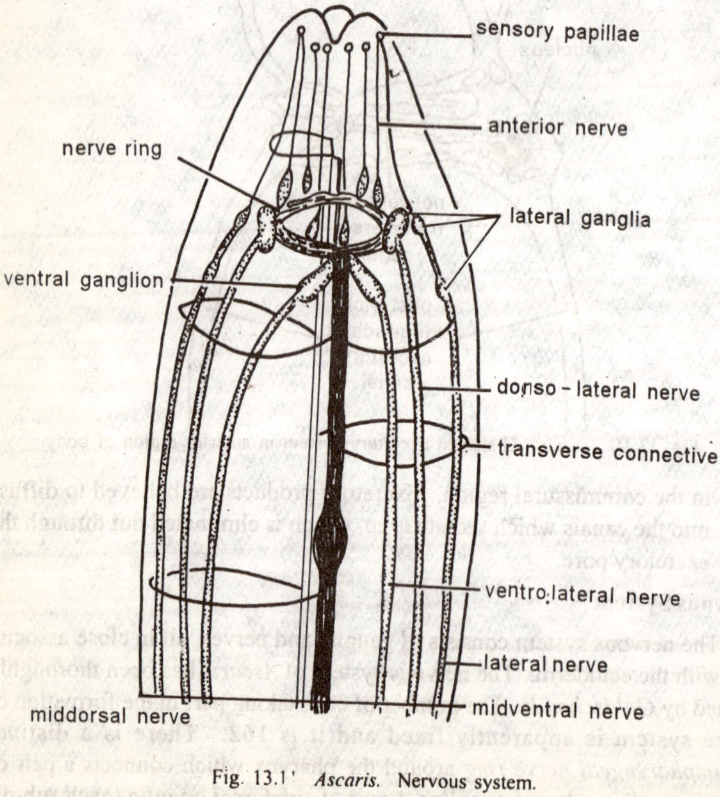


Fig. 13.1' *Ascaris.* Nervous system.

## Sense Organs

The sense organs include papillae amphids and phasmids.

***Papillae.*** Various types of papillae such as *labial, cervical* and *genital*

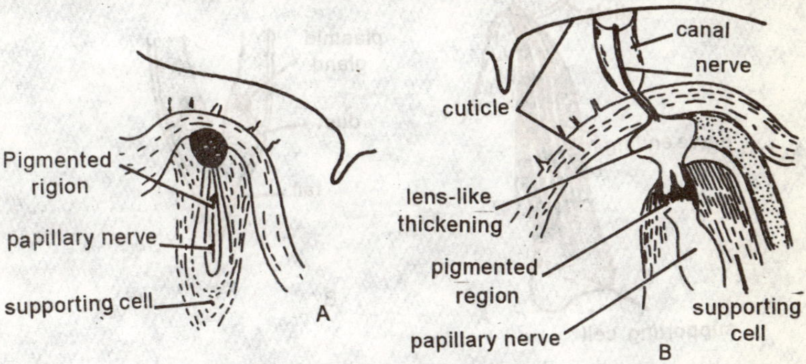


Fig. 13.12 *Ascaris.* Caplatic and labial papillae

*papillae* are present in *Ascaris*. A labial papilla consists of a nerve axis placed inside a cuticular elevation surrounded by supporting cells. The dorsal lip

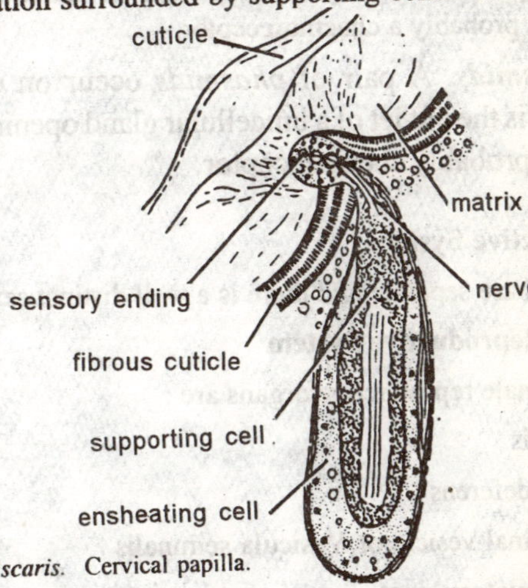


Fig. 13.13 *Ascaris.* Cervical papilla.

has two double papillae, each consisting of a labial papilla on the medial side and a cephalic papilla on outerside. Each ventro-lateral lip has a labial papilla near its upper margin and a double papilla near its lower margin. The double papilla, a labial papilla towards the medial side and a cephalic papilla towards the outerside. Thus the lips bear six labial and four cephalic papillae in all. *Genital and caudal papillae* have a similar structure.

***Amphids.*** These are two in number, one on each *ventro- lateral lip* a little

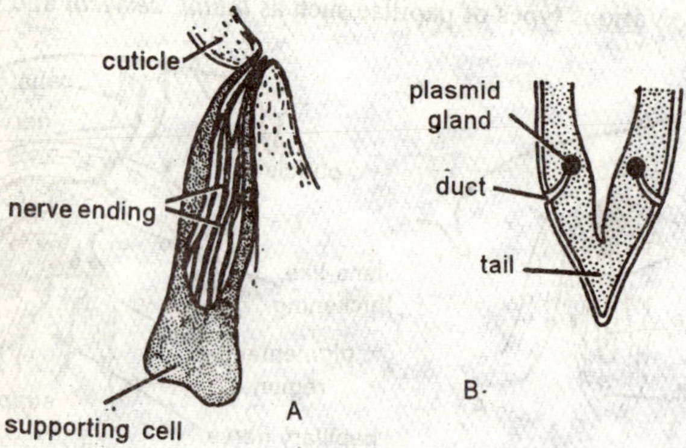


Fig. 13.14 *Ascaris.* A- amphid. B-phasmid.

above the labial papilla. An amphid consists of supporting cell opening out side by a pore and having a bundle of nerve-fibres that unite into one fibre. These are probably a chaemoreceptor.

***Phasmids.*** A pair of *phasmids* occur on each side of the tail. Phasmid is the out let of a unicellular gland opening over a small papilla which is probably chemoreceptor.

**Reproductive System**

Sexes are separate and there is a well distinct sexual dimorphism.

**1. Male Reproductive System**

The male reproductive organs are :

1. Testis
2. Vas deferens
3. Seminal vesicle or Vesicula seminalis
4. Ejaculatory duct
5. Penial sac with penial setae.

1. ***Testis***. The testis is a long thin, coiled and thread-like structure in the anterior part of pseudocoel. It represents *monorchic condition* (presence of single testis) and *telogonic* mode of origin of germ cells (i.e. germ cells developing only from the proximal part of gonad). The testis can be distinguished into three parts:-

(a) anterior solid part consisting of sex cell, (b) middle part and (c) posterior part. The central axis of the tube is in the form of a solid *cytoplasmic rachis*, around which the clusters of amoeboid sperms in various stages of development.

Testis is continued posteriorly into the vas deferens.

2. ***Vas deferens.*** The vas deferens is as wide as the testis and the two cannot be marked externally. It is very short and leads into the seminal vesicle.

3. ***Vesicula seminalis.*** The vesicula seminalis is a wide tube present in the posterior 1/3rd part of the body. It joins the ejaculatory duct.

4. ***Ejaculatory duct.*** It is a short, narrow but muscular tube opening into the cloaca, which leads outside by cloacal opening.

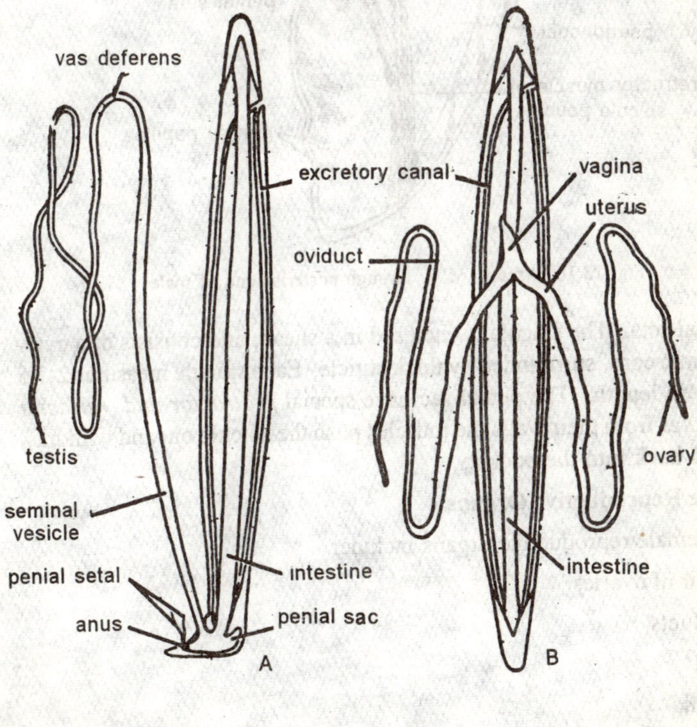


Fig. 13.15 Ascaris. A-male reproductive system. B-female reproductive system.

5. ***Penial sac with penial setae.*** On the dorsal side of the cloaca are a pair of sac-like structures called *penial sac*. These are formed by evaginations

of the cloaca. Each pouch secretes and lodges a club-shaped spicule or

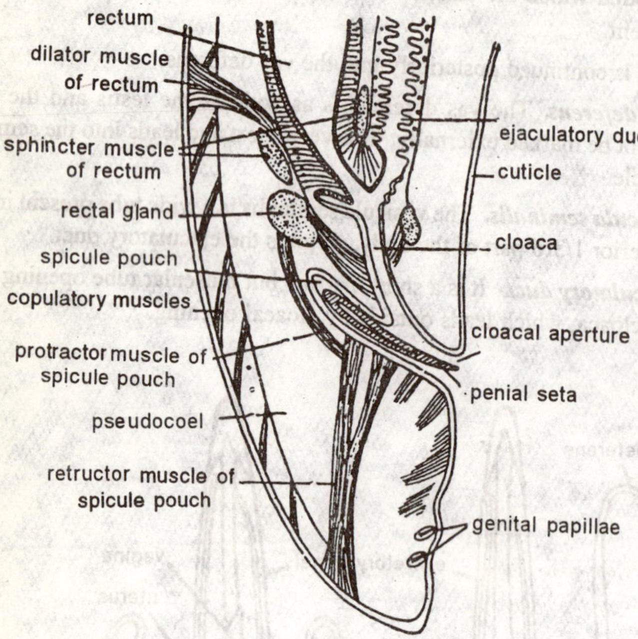


Fig. 13.16 Ascaris. L.S. through posterior end of male.

penial seta. The spicula is enclosed in a sheath and consists of a cytoplasmic core surrounded by thick cuticle. Each spicule measures 2-3.5 mm. in length. The penial sac have special *protractor and retractor muscles* from their walls, the muscles push the spicule out and withdraw them back into the pocket.

## II. Female Reproductive Organs

The female reproductive organs include:

1. A pair of ovaries
2. Oviducts
3. Uteri
4. Vagina

1. ***Ovaries.*** The ovaries represent *didelphic* condition i.e. there are two ovaries. Each ovary is a long, thread-like, much twisted and blind tubule occupying posterior 1/3rd of the body. Its walls consists of a single layer of cuboidal epithelial cells lined externally by basement

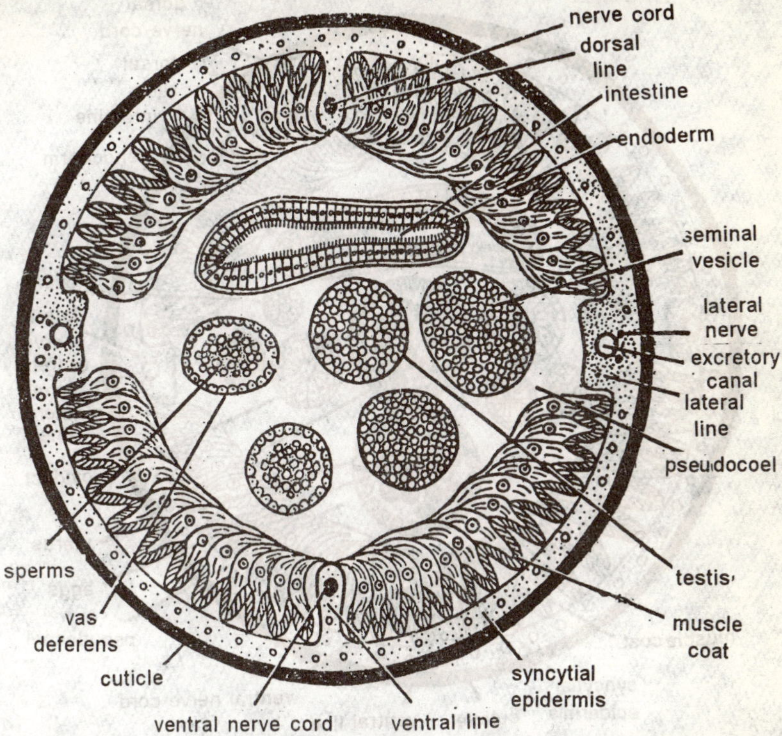


Fig. 13.17 *Ascaris.* Male T.S.

membrane. There is no lumen, the central part is occupied by the *cytoplasmic rachis,* around which young eggs are clustered.

2. ***Oviducts.*** The oviducts are two thin ducts very similar to the ovaries. These open into the uteri.

3. ***Uteri.*** Each oviduct continues into a much wider, thicker and almost untwisted tube, the uterus. The wall of the uterus is composed of tufted cells, surrounded by a muscle layer with inner *circular* and outer *oblique fibres*.

4. ***Vagina.*** Two uteri join to form one vagina which opens to the outside by female genital pore or *vulva,* lying on mid-ventral side about 1/3 of the length of the body posterior to anterior end.

**Life-Cycle**

*Ascaris is monogenetic* i.e., it requires only one host to complete its life-cycle.

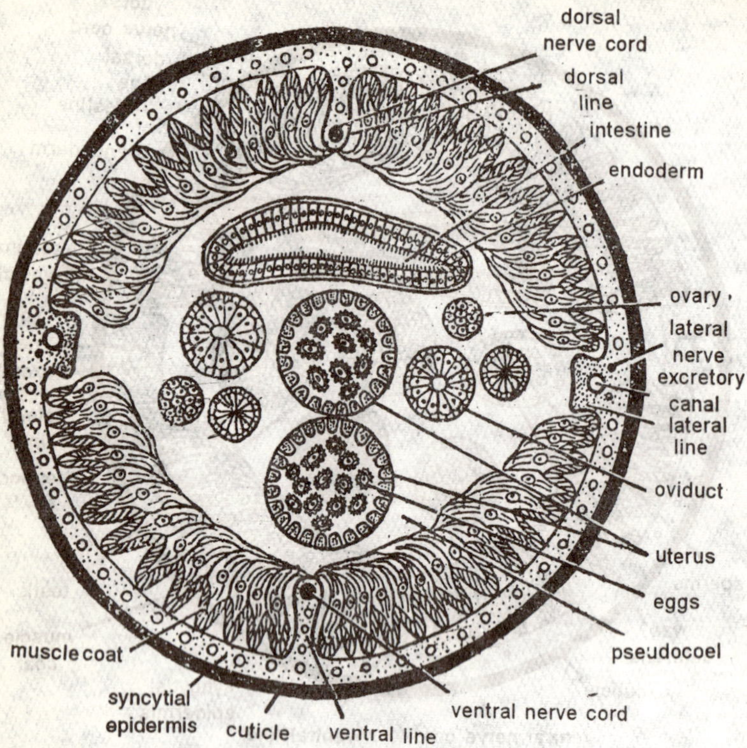


Fig. 13.18 Ascaris. Female T.S.

1. ***Copulation and Fertilization.*** *Ascaris* copulate in the intestine of man. During the process sperms are ejected by male into vagina of the female. These move in the uterus where fertilization occurs. While moving further the eggs become covered by a chitinous cells and an outer irregular *albuminous coat.*

The eggs are round or oval (60 × 45 ), light brown in colour and float in saturated solution of common salt. The unfertilized eggs laid by the mother, may mix up with the fertilized eggs but the unfertilized eggs are larger (90 × 45 ) and sink in saturated salt solution.

2. ***Egg-laying.*** The uteri of a single female may contain as many as 2,70,000,00 eggs (*Cram*, 1927), with an average daily production of about 2,00,000 eggs. The fertilized eggs are released in the intestine of host through the vulva. From the intestine of host they are eliminated along with faecal matters.

3. ***Development.*** The development takes place at 85°F i.e. outside the

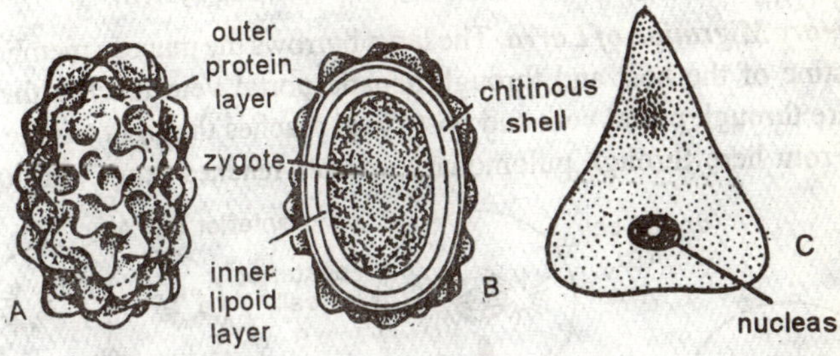


Fig. 13.19 *Ascaris.* A-egg, B-insection, C-a sperm.

body of man. It is stopped at 60° F and eggs gradually degenerated at 100° F. They can remain alive for years in moist soil.

***Cleavage.*** The cleavage is spiral and determinate. The first division is transverse producing an upper dorsal cell *(AB)* and a lower ventral cell *(P)*. The dorsal cell *(AB)* divides vertically into an anterior cell *(A)* and a posterior cell *(B)*. Simultaneously the ventral cell divides horizontally into upper cell *(EMST)* and lower cell *($P_2$)*. Thus the four-celled embryo has T-shaped apperance. The *P2* cell shifts towards right and comes to lie in front of the upper cell *(EMST)*. The A and B cells divide together into right and left cells and by further divisions form the ectoderm of the embryo. The *EMST* and *P2* cells divide into *E, MST, P3* and *C* cells. The offsprings of cell *E* form the endoderm, while the descendants of *EMST* constitute the mesoderm and a part of ectoderm. The cells *P3* divides into *P4* and *D*. The cells *C* and *D* divide and contribute to the ectoderm and mesoderm. The cell *P4* divides into *G1* and *G2* and forms the primordial germ cells. Thus blastula stage is attained. It undergoes gastrulation by invagination and finally develops into an active juvenile, representing the rhabditoid stage. Under favourable conditions temperature, moisture and oxygen supply, it requires 10-14 days to reach the *Ist stage larva*. After a week or so it moult inside the egg shell and changes into the *second stage larva*. The hatching and further development of the larva is possible only if the egg enters the final host.

***Infection and Emergence of Larva.*** The eggs containing ***rhabditiform*** *larva* contaminate food and water. If someone takes this contaminated food or drink contaminated water the eggs reach the healthy man. This type of *transference* is called *contaminative transference or infection*. The shell of the egg is weakened in oesophagus and is disssolved in the intestine of the new host. The larva is about 200-300 in length and 14 in breath. It has fully formed alimentary canal, excretory and nervous system.

***Primary Migration of Larva.*** The larva burrows the mucous membrane of the intestine of the host and through hepatic portal vein reaches the liver. From here through portal vein and post-caval reaches the right auricle of the heart. From here through pulomonary arteries reaches lungs. In lungs it

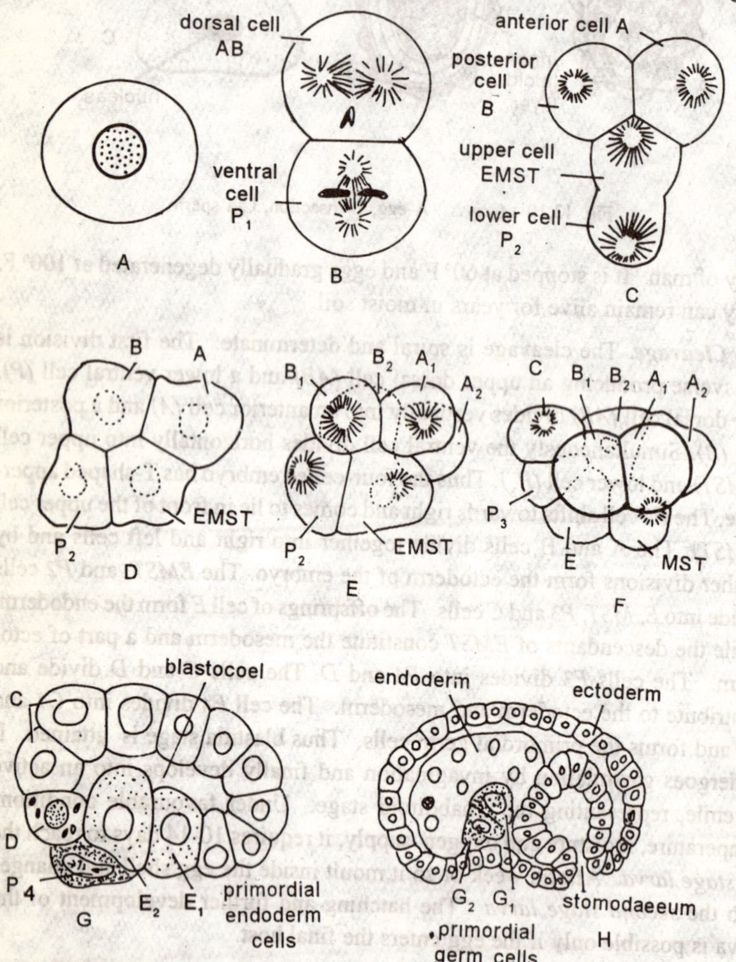


Fig. 13.20 *Ascaris.* Stages showing early development. A-Zygote, B--cell stage, C-4 cell stage(T-shaped), D-4 cell stage(rhomboidal), E-6-cell stage, F-8-cell stage, G-median sagittal section through blastula, H-Median sagittal section through the embryo after invegination of stomodaeum and the primordial germ cells.

moults twice. The first moulting takes place after 5-6 days (*third stage larva*) and second moulting after 10 days forming *fourth stage larva.* This larva now increases in size and is about 2 mm. in length.

***Secondary Migration.*** From lung the larva reaches bronchii, trachea, larynx and finally to pharynx where it irritates causing coughing by which it re-enters the alimentary canal when the cough is swallowed. After reaching the intestine through stomach it moults fourth or last time. The young *Ascaris* becomes adult sexually mature within 6-10 weeks and begins its life-cycle again, the whole migration is completed within 10 days.

**Pathogenesis**

*Ascaris* is pathogenetic to man. When larvae migrate through the lungs, they cause lung-inflammation and fatal pneumonia. Sometimes, *fever, anaemia, leucocytosis* (presence of excess of white corpuscles in the blood) and *eosinophilia* result. In mature form, *Ascaris* may or may not disturb life-activities. Some abdominal discomfort and acute colic pains with *vomitting, diarrhoea* and *slight temperature* occur. Often masses of *Ascaris* block the intestine, in which condition they can only be removed by surgical operations. Several hundred thousands of worms are responsible for blockage and they are fatal. *Appendicitis* is also caused by the blocking of appendix by *Ascaris*. Toxic substances produced by *Ascaris* cause *convulsions*, (an involuntary contraction of the involuntary muscles of the body), *delirium*, (insanity or disorder in mind), *coma* and *general nervousness*. *Sang* (1922) found that toxins produced by *Ascaris* cause interference with the digestion of proteins.

**Therapy**

Several drugs like *Santonin, Antepar, Chenopodium* and *Hexylresorcinol* are used effectively under physician's directions. A mixture of oil of *chenopdium and tetrachlorothyene* is usually successfully used. *Lamson* discovered that *hyxylresorcinol* is also very effective particularly when given as crystoids (in gelatin capsules) with fasting for 12 hours before treatment and 4 hours afterwards followed by sodium sulphate for removal of deal worms.

*Hetrazan syrup* is also very effective and nontoxic especially for children. *Piperazine hydrate, Alcopar and Banocida* are also recommended by several workers as antihelminthes.

**Prevention**

Adequate sanitation plus education should prevent *ascariasis*. Lack of adequate fertilizer in many parts of the undeveloped world means that night soil must be used to fertilize crops. Due to the resistance of *Ascaris* eggs to standard disinfectants, the situation would appear to be hopeless. There is a realistic solution. *Thitasut* (1961) reported that aqueous solution containing 120 ppm free iodine killed *Ascaris lumbricoides* eggs within 5 min or less and did not effect the flavour of leafy vegetables.

One of the problems in control is preventing children from ingesting eggs with contaminated soil. Given the habits of children throughout the world, including dirt eating, playing in contaminated soil and then placing their fingers in their mouths, *ascariasis* will unfortunately be with us until adequate sanitation is accomplished.

♦♦♦

# 14

# PARASITISM AND PARASITIC ADAPTATIONS

Parasitism is such an association between two animals of the same or different species in which one lives on or inside (parasite) the body of the other (host) receiving nourishment and shelter without any compensation for the host. Therefore, in this association the parasite is benefited and the host is harmed.

**Origin and evolution of parasitism**

Parasitism is a secondary mode of life and has arisen from pre-living condition. Parasites living inside the bodies of their hosts (*endoparasites*) are the most successful ones. Between the endoparasite life and free-living life, intermediate stages are not uncommon, proving that parasitism did not evolve all of a sudden but followed a series of steps in succession.

Living together or *symbiosis* started when smaller animals occasionally taking shelter was repeatedly used, the association of two organisms become more intimate and the small oraganism gets not only the shelter but also free transport and nourishment in the form of bits from the host, without interfering with the activities of the host or injuring it. Such a simple symbiosis is known as *commensalism* and the benefactor is called a *commensal.*

Still later, the commensal started living within the body of the host without doing any harm to the host. This a commensal may be firstly an ectocommensal and later an endocommensal. Symbiosis sometimes proves beneficial to the host. Thus both the organisms benefitting each other mutually, and the host becomes so dependent upon the commensal that it cannot get along without it. This kind of association is called *mutualism.*

Commensalism usually evolves not in the direction of mutualism but towards a deeper association, the *parasitism.* A commensal that at first takes shelter, then bits of food and finaly starts to feed on body tissues of the host so that the latter has to suffer with some harm.

**Different types of parasitism**

The parasitism can be classified into following categories :

1. *Accidental.* When a free living animal accidently reaches inside the

body of the host then it leads a parasitic life for some time.

2. *Faculative.* When the animal can live both as parasitic or free-living, they are known as fuculative parasite.
3. *Obligatory.* These animals can not lead a free life, therefore, they need a suitable host for them.The obligatory parasitism is of different types :
   (a) *Ectoparasitism.* The animal lives on the external surface of the body of the host.
   (b) *Endoparasitism.* The animal lives inside the body of the host in various oragns and system.

**Effect of parasitism upon the parasite.**

Changes from a free-living condition to a parasitic life involves changes in the body of the animal. Due to parasitic mode of life certain new structures are developed due to new environmental requirements similarly certain characters, which are of no use for new environment, and lost. All such changes include with effect of parasitism on the parasite. The helminths are modified morphologically as well as physiologically to live in their particular environment.

I. Morphological Adaptations.

II. Physiologically Adaptations.

***I. Morphological Adaptations.***

These are of two types (i) Structure which are partially or wholly lost. (ii) Strutures which have developed a fresh.

*Degenerations.*

1. *Organs of locomotion.* Locomotion is actually an effort for procuring food. But parasites habitually inhabit such places in the host's body, Where sufficient food is available without effort. Thus, the organs of locomotion, such as the cilia of free-living turbellarians, are absent in the parasitic forms. It is interesting to note that the locomotory organs are dully present in free-living larvae of parasitic forms; the miracidium possesses cilia and the cercaria possesses a tail for locomotion.
2. *Organs of nutrition.* Food of the parasite comprises the readily available digested or semidigested food of the host. Therefore, there is a reduction in trophic organs. In trematodes, there is an incomplete gut and in most of them, suctorial pharynx is present. In cestodes, the mouth and alimentary canal are lacking because digested food is readily available in the gut of the host.
3. *Nervous System.* Living safely in a more or less stable environments,

parasites do not need a complex nervous system.

4. *Sense Organs.* Progressive parasitism is accoplished by the loss of sense organs. The sense organs, necessary for quick and effcient response to the stimuli, are associated with the free active life. In the host the environment is more or less uniform and so the sense organs are not essential. Hence they are reduced. The endoparasites are generally provided with tangoreceptors.

*Neoformation.*

1. *Protective covering.* There are no epidermal cells in the adult, instead of which the body is covered by a several-layered thick cuticle which protects the parasite from the juice of the host.

   The outer surface of the cuticle is formed of a fibrous protein called *keratin*. The cuticle may be provided with scales.

2. *Organs of attachment.* For a firm grip on or in the host's body, some special organs of adhesion are necessary. The flatworms, for this purpose, are variously armed with suckers, hooks and spines. The suckers themselves may be with or without hooks and spines.

3. *Organs for penetration.* The parasites must penetrate the host's body for entering it. So the parasites develop certain structures for penetration. Miracidium larva has a conical process at the anterior end called apical papilla. These are a pair of penetration glands inside the body near the anterior end.

4. *Suctorical Pharynx.* All the endoparasites except cestodes have a suctorial pharynx by means of which liquid food is sucked into the intestine. Intestine is always branched.

5. *Reproductive system.* It is higly developed and the production of eggs is prolific. The parasitic flatworms are hermaphordite. In cestodesthe reproductive organs are much more elaborate and are repated in each proglottid. Each mature proglottid possesses one (*Taenia solium*) or two sets (*Diphylidium*) of male and female reproductive organs. In each gravid proglottid all other oragns of the system degenarate to make room for the uterus which is greatly enlarged and branched to accomodate a large number of eggs.

***II. Physiological adaptations.***

1. *Protective mechanism.* In the alimentary canal of the host the parasite protects itself from the action of digestive juice of the host.

The tapeworms accomplish this (a) by stimulating the walls of the gut to secrete mucus, which then forms a protective clothing around the parasite, (b) by secreting antienzymes to neutralize the digestive enzymes of the host, and

(c) by probably continually renewing their protective body covering i.e. tegument. Lime cells in the body. Blood parasites are known to withstand the effect of antibodies and phagocytes of blood by some mechanism.

2. *Osmoregulation.* The osmotic pressure of body-fluid of endoparasites, especially in case of trematodes, is almost the same as that of the host so that there is no necessity of osmoregulation. But in the intestinal tape-worms, the osmotic pressure absorption of nourishment from their general body surface from the surrounding medium.

3. *Anaerobiosis.* The pH tolerance of the parasite is high 4-11. The intestinal parasite lives in an environment completely devoid of oxygen. The respiration is of the anaerobic type consisting of extracting oxygen from the food stuffs. In the absence of oxygen, energy is obtained by the fermentation of glycogen which is broken into lactic acid.

4. *Periodic appearence.* There are some parasites which appear at definite period. The larvae of *Wuchereria bancrofti* circulate into the peripheral blood circulation in the night between 10 P.M. to 4 A.M. for further development they require *Culex* mosquitoes. The culex bites a man at night i.e. nocturnal in habit. Their migration is correlated with the nocturnal habit of *Culex.*

5. *Neoteny.* It is a special phenomenon in which the larvae are capable of reproduction and they produce young ones. *Lingula* (Cestoda) exhibits neoteny and this can be considered an adaptation for parasitic mode of life.

6. *High fertility.* Parasites develop enormous fertility producing millions of fertile eggs. This is correlated with the passive transference of the infective stages of the parasites from one host into another. While passing through the complex life- cycle, these potential offsprings face several hazards as a result of which a very small percentage of the total eggs produced reaches adulthood. For example in *Fasciola*, if the eggs fall in water, and the first larva (miracidium) finds a suitable snail within a limited period, the cercaria encysts, on vegetation within the reach of final host and the metacercaria happens to be ingested by the proper final host, only then the life-cycle is completed. Not only this, the larval stages themselves multiply say for example for sporocyst produces a number of rediae. A single redia produces a number of cercaria larvae. Similarly a single hexacanth may multiply by generating buds which produce daughter and grand daughter scolices (Hydratids of *Echinococcus*).

7. *Transferance of eggs or infective stages.* This is another major problem for which the parasites have to develop adaptations. Transference from

one host to another is either :

(i) **Active transfer.** This is a less common method. The infective stages are actively transferred to final host. The larval stages of most parasites bore directly through the skin of their host such as the larvae of hookworms and schistosomes.

(ii) **Passive transfer.** This is the most common method and usually results from the careless habits of the host and is achived in several ways such as contamination, contagion and inoculation. Oral infection occurs when the host takes uncooked food or contaminated water. The intermediate host, such as mosquitoes, inoculate parasites when they suck the blood of hosts.

**Effect of parasitism on the host.**

Although a given host may immune to a parasite by virtue of being constitutionally unstable as a habitat, it is important to recognize that the establishment of a parasite in a host also involve the capacity of the parasite to resist reactions of the host that might be deleterious. These host reactions would include inflammatory tissue response and immune responses of a humoral character. Both types of host reactions occur in most parasitisms involving characters. The immune system in higer vertebrates is composed of several tissue components. These include fixed cells of spleen, liver, lymph nodes and thymus. There is also a system of mobile phagocytic cells, originating mainly in the body, such as undifferentiated mesenchyme cells. Some of these cells has the ability to elaborate antibodies, which may may be released from cells into the body fluid. These globulins may combine with parasitic organisms or their products and produce an inhibitory effect. A peculiar organisms or their products and produce an inhibitory effect. A peculiar type of immunity develops in man due to the presence of tapeworms; after the first infection no further infection is possible as long as the primary infection lasts, this is termed *premunition*.

The possible role of interferons in the immunity of hosts of animals parasites should be also mentioned. These substances are produced in response to the presence of a foreign agent. Unlike antibodies, they seem to render host cells until as habitations for intracellular parasites.

There are other types of host response that cannot be clearly recognized as immune in characters. For example, the changes in the fatty acids found in the alimentary canal of animals infected with the tapeworm, *Hymenolepis,* or the marked modification of the amino acid composition of bile in sheeps harboring *Fasciola hepatica* are obviously host response to the presence of parasites concerned. It would be of great interest to determine whether such responses are beneficial to the parasites, to the host or to neither.

15

# CHARACTERS & CLASSIFICATION OF ANNELIDA

All the invertebrate animals, who possessed a soft-body, usually bearing a general resemblance to the common earthworm were included in the old phylum Vermes and were called *worms*. Thus the old phylum Vermes represented a heterogenous assemblage of no exact classificatory value. *Cuvier* was first who pointed out the fundamental differences between the lower unsegmented worms and the higher segmented worms in 1798. Now-a-days lower unsegmental worms divided into three phyla Platyhelminthes (includes flat worms), Nemathelminthes (round worms) and Trochelminthes (wheel worms). *Lamarck* gave the name Annelida (ring worms) to the higher segmented worms.

Term Annelida is derived from two words one is Latin (*annelus* = a ring) and a Greek (*eidos* = form).

Annelids are typical animals that burrow or crawl upon the bottom in the ocean or in fresh water. Some are free swimming, at least during the breeding season, and others are adapted to life in moist soil. Many marine species build permanent tube which they inhabit; in other species the tube is temporary, like a burrow, and may be abandoned for another. Some annelids are actively predaceous, seizing their prey in well developed jaws, others notably permanently tube dwelling polychaetes, feed upon microscopic particulate matter brought in by ciliated tentacles or entrapped in sheets of mucus.

## CHARACTERS

1. Annelids are bilaterally symmetrical with elongate vermiform body.
2. The body is metamerically segmented and ringed appearance with the serially repeated segments give to the body. The body is composed of segments which are usually of similar structure internally as well as externally.
3. Presence of cuticles secreted by the underlying epidermis containing unicellular gland cells and sensory cells.
4. Appendages unjointed and are minute, rod-like, chitinous setae, few to

many per somite generally absent in *Hirudinea*.

5. Body wall consists of layers of circular and longitudinal muscles.
6. Coelom well developed, extensive, separated into segmental compartments by transverse partitions,or septa, but coelom is reduced in *Hirudinaria*.
7. The digestive tract complete, tube like , extending entire length of body from mouth to anus.
8. A closed circulatory system of longitudinal blood vessels with lateral branches in each somite, blood plasma contains dissolved haemoglobin and free amoebocytes.
9. Respiration by epidermis, by gills in some tube dwellers.
10. Excretion by segmental paired nephridia, each removing wastes from the coelom and blood stream directly to the exterior.
11. Nervous system with ventral nerve cord composed of paired ganglia at segmental intervals connected anteriorly with the dorsal "brain" by a pair of circumpharyngeal connectives.
12. Sensory organs in the form of sensory cells, organs for touch, taste and light perception are present.
13. In few the sexes are separate, while in others sexes are united (hermaphrodite).
14. Gonads may be simple structure as in *Nereis*, where they develop only during breeding season. In some forms gonads are well developed.
15. Cleavage is spiral and determinate.
16. Development direct or indirect through a larval stage, the *trochophore*.
17. Process of regeneration is common.

**CLASSIFICATION**

There are about 8,700 known species of segmented worms which are divided into four main classes. The classification is primarily based on the basis of presence or absence of setae, parapodia and metamerism. The 4 classes are:

Class- I. Polychaeta

Class- II. Oligochaeta

Class-III. Hirudinea

Class- IV. Archiannelida.

**Class. I. Polychaeta**

(Gr. *Poly* = many; *chaete* = bristles)

1. These are marine and carnivorous annelids.
2. Body usually elongated, cylindrical and distinctly segmented into numerous similar metameres.
3. Head consists of prostomium and peristomium and bears eyes, tentacles, cirri and palps.
4. Setae are numerous and are borne up on leteral prominences of the body wall known as *parapodia.*
5. Clitellum is absent.
6. Parapodia highly vascular and respiratory in function.
7. Coelom is spacious usually divided by intersegmental septa.
8. The blood vascular system is well-developed and does not communicate with coelom.
9. Alimentary canal is provided with an eversible buccal region and protrusible pharynx.
10. Sexes are separate. Gonads are simple and developed by active proliferation of coelomic epithelium.
11. Fertilization external, Free-swimming trochophore larva is present.
12. Asexual reproduction by lateral budding.

Class Polychaeta is divided in two orders:

***Order (i) Errantia***

1. Free swimming, often pelagic while some are tubicolous (live in tubes).
2. All segments are similar except head and anal regions.
3. Parapodia provided with cirri, are equally developed throughout. These are organs of locomotion.
4. The head is usually definite which bears eyes, tentacles and cirri.
5. Pharynx is usually protrusible and armed with chitinous jaws and teeth.
6. Clitellum absent.
7. The branchiae or gills are present but not restricted to anterior end of body.

Examples: *Aphrodite, Polynoe, Nereis, Glycera, Eunice, Syllis.*

***Order (ii) Sedentaria***

1. Sedentary in habit. Burrowing and tube dwelling.

2. Body is distinguished into two or three regions due to differences in shape of segments, parapodia and setae.
3. Head is small or much modified, without eyes and tentacles, prostomium indistinct.
4. Gills are situated only in the anterior part of the body, when present.
5. Pharynx is non-protrusible devoid of jaws or teeth. Feeding on plankton or organic detritus.
6. The parapodia are peculiar which are confined in the posterior region of body and without cirri.

   Examples: *Chaetopterus, Sabella, Arenicola, Terebella, Amphitrite, Serpula, Spirorbis.*

**Class. II. Oligochaeta**

(L; *oligos* = few; *chaete* = bristles)

1. Mostly terrestrial, some live in fresh-water.
2. Body with conspicuous external and internal segmentation.
3. Head not distinct, prostomium is usually small, without eyes and tentacles.
4. Parapodia and cirri are absent. Setae are usually arranged segmentally.
5. Clitellum usually present.
6. Pharynx is not protrusible and without jaws.
7. Hermaphrodite i.e. bisexual. Gonads are complicated. Ovaries and testes are compact structures.
8. Development direct without larval stage.

Class Oligochaeta is divided in two orders:

***Order (i) Archioligochaeta***

1. Mostly fresh water forms.
2. Body has few segments. Setae present.
3. Gizzard is either absent or poorly developed.
4. Simple clitellum, consists of single layer of cells situated for forwards.
5. Eye spots are usually present.
6. Reproduction is both asexual and sexual.
7. Male reproductive openings lie in front of female reproductive openings.

   Examples: *Tubifex, Aelosoma.*

***Order (ii) Neooligochaeta***

1. Mostly terrestrial forms found in moist soil.
2. Body is large with many segments.
3. Setae are arranged in lumbricine manner.
4. Well developed gizzard.
5. Clitellum after 12th segment and is composed of two or more layers of cells.
6. Female genital apertures before male genital pore and usually lie in 14th segment.
7. Eye spots are developed.
8. Reproduction sexual only.
9. No free larval stage.

Examples: *Pheretima, Lumbricus, Megascolex, Eutypheus.*

**Class III. Hirudinea**

1. They are marine, aquatic, terrestrial or free living. Some are ectoparasitic and blood-sucking annelids.
2. Body is segmented and is marked externally by ring-like annuli.
3. Presence of anterior and posterior cup-shaped suckers which are the organs of adhesion and locomotion.
4. The ventral mouth lies in the cup-shaped hollow cavity of anterior sucker. The anus is dorsal and situated just above the posterior sucker.
5. The coelom is greatly reduced due to the presence of connective tissue and is represented by haemocoelomic sinuses.
6. The true blood vessels with definite muscular walls present.
7. The nervous system includes a brain united by small oesophageal connectives to a double ganglionated ventral nerve cord.
8. Nephridia are segmentally arranged and open independently to exterior by nephridiopore, which are the organs of excretion.
9. Hermaphrodite, the testes are many and segmentally arranged, whereas ovary is single paired. Eggs are laid in cocoons.
10. Fertilization internal, development direct without larval stage.

Class Hirudinea is divided into following orders:

***Order (i) Acanthobdellida***

1. Mostly ectoparasite on the fin of salmon fishes.
2. Anterior sucker wanting, posterior sucker well formed.
3. Proboscis is small and not well developed.
4. Nephridia without internal openings.
5. Body generally composed of twenty or twenty-one segments.
6. It forms a connecting link between Oligochaeta and Hirudinea.

   Examples: *Acanthobdellida.*

***Order (ii) Rhynchobdella***

1. Aquatic leeches, parasite on snails, frogs and fishes.
2. The anterior part of body may be protruded and retracted so as to form a proboscis or introvert. Jaws are absent.
3. Each typical body segment consist of 3, 6 or 12 rings.
4. The mouth is minute slit-like opening placed inside the anterior sucker.
5. Blood is colourless. Coelom is reduced to sinuses without botryoidal tissues.

   Examples: *Branchellion, Glossiphonia, Pontobdella, Piscicola.*

***Order (iii) Gnathobdellida***

1. Fresh water and terrestrial forms. Ectoparasites and blood sucking.
2. Each typical segment consists of five rings or annuli.
3. Anterior sucker has three jaws, one median dorsal and two ventro-lateral.
4. Proboscis non-protrusible or absent.
5. Blood is red in colour.
6. Coelom is reduced. Botryoidal tissue present.

   Example: *Hirudo, Hirudinaria, Haemopis.*

***(Order (iv) Pharyngobdellida***

1. Freshwater amphibious carnivourous leeches.
2. Each body segment is divided into five rings or annuli.
3. Pharynx non-protrusible. Jaws are absent.

   Examples: *Dina, Erpobdella, Trocheta.*

**Class. IV Archiannelida**

1. This group represents the primitive members of Annelida
2. Mostly marine worms.
3. Body is elongated, cylindrical with internal segmentation.
4. The head is distinct bearing the tentacles, eyes and ciliated pit.
5. Parapodia are wanting.
6. Hermaphrodite or unisexual; gonads are simple and develop temporarily only during breeding season by active proliferation of coelomic epithelium.

About 45 species, one dozen genera are included in this class. The class is divided into five families.

(i) Polygordidae *e.g. Polygordius, Protannelis.*

(ii) Protodrilidae *e.g. Protodrilus.*

(iii) Saccocirridae *e.g. Saccocirus.*

(iv) Dinophilida *e.g. Dinophilus, Diurodrilus, Trilcbodrilus.*

(v) Nerillidae *e.g. Nerilla, Nerillidium, Troglochaetus.*

**POLYNOE**

| | | |
|---|---|---|
| Phylum | - | Annelida |
| Class | - | Polychaeta |
| Order | - | Errantia |
| Genus | - | *Polynoe* |

1. It is a bioluminescent polychaet found under stones in sea-water near tide mark.
2. It is found in U.S.A. (South Carolina to Massachusetts) and Europe.
3. Body is short, broad and oval with few segments. Body shows considerable degree of dorso-ventral compression.
4. It has a covering of elytra which are modified dorsal cirri. Elytra do not contain setae and have rich nerve supply.
5. Elytra are bioluminescent and the illuminating elytra are cast off when distrubed or elytra are shed-off to deceive the enemy.
6. Head bears three tentacles and two pairs of peristomial cirri.

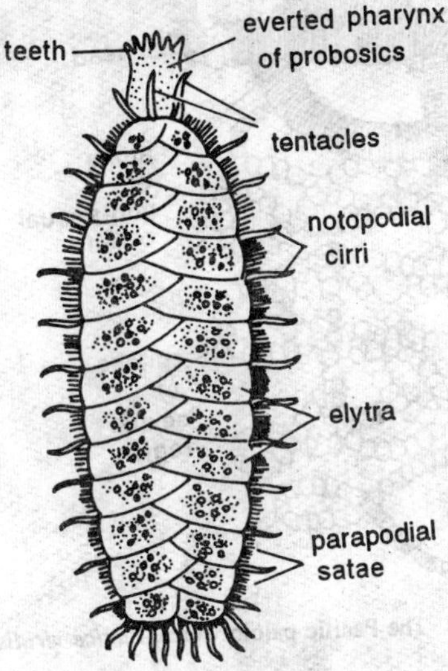


Fig.- 15.1 *Polynoe.*

7. Pharynx is protrusible.
8. It is carnivorous and may even eat each other (cannibalism).
9. Fertilized eggs and resulting embryos adhere in masses to the dorsal surface under the shelter of elytra.

## EUNICE

The classification is same as that of *Polynoe.*

1. It is tube-dwelling marine polychaete found among the coral reefs at sea-bottom.
2. It is found in Atlantic and Pacific oceans, West Indies and Fiji Islands.
3. It is commonly known as 'Palolo-worm'.
4. Body elongated with numerous segments. Prostomium consists of two segments with one pair of cirri, five tentacles and a large palp.
5. Gills begin usually on 6th segment.
6. Body yellowish or brownish and differentiated into anterior sexless *atoke* and posterior sexual *epitoke* regions. Epitoke projects from the tube and gets separated from atoke, and come to the surface, swim vigorously and finally discharge the gonads. The empty epitoke sinks to the bottom after swarming.
7. Fertilized eggs develop into a larval form.

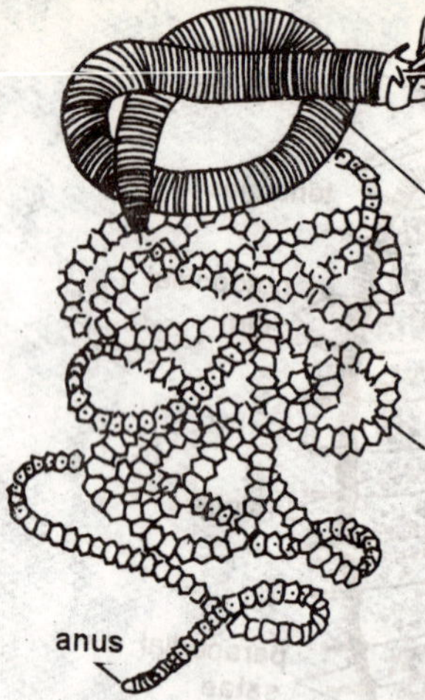


Fig.- 15.2 The Pacific palolo worm, *Eunice viridis*.

## SYLLIS

The classification is the same as that of *Polynoe*.

1. *Syllis* is a pelagic, marine polycheat.
2. *Syllis* is common in U. S. A., Woods Hole regions, Long Island, Sound to Bay of Fundy.
3. Small elongated worms, less than an inch in length.
4. Head small with prostomium and peristomium.
5. Prostomium with eyes, three tentacles and two fused palps.
6. Peristomium carries two pairs of cirri, tentacles and cirri are used.
7. Each segment of the body carries a pair of parapodia.
8. Parapodia without notopodium, represented by long jointed notopodial cirrus.
9. Neuropodium with setae and cirrus.
10. The pharynx has a single large tooth provided with a poison gland opening near its apex.
11. A muscular proventriculus generally present.
12. Asexual reproduction by lateral branching,fission and gemmation is common.
13. Sexual reproduction with alternation of generations.
14. The sexual forms develop from ovum and asexual forms develop by

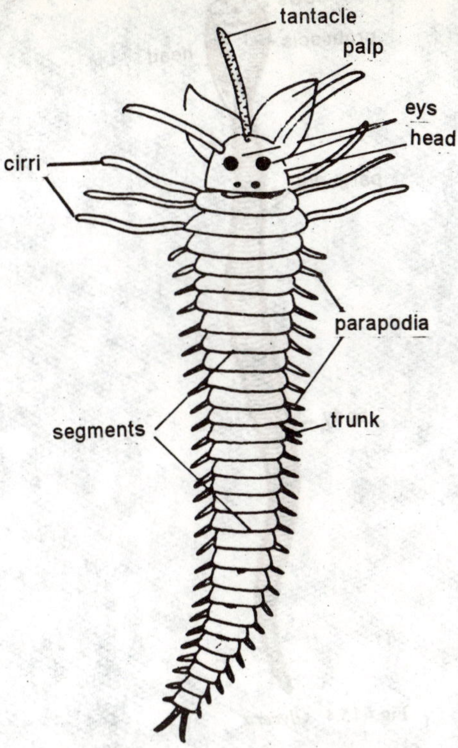


Fig.- 15.3 *Syllis.*

process of posterior constriction and proliferation.

15. The sexual forms detached from the colony and help in dispersal of the species.

## GLYCERA

The classification is the same as that of *Polynoe.*

1. It is a marine polychaet living in a cylindrical burrow which is made by the proboscis in the sand, mud in shallow water.
2. It is found in U.S.A., North Carolina to Bay of Fundy and South Carolina to Cape Cod.
3. The body of *Glycera* is elongated, cylindrical and meta merically segmented.
4. The head is distinct bearing a small conical prostomium with minute tentacles and rudiment palps.

Fig.- 15.4 *Glycera*

5. The parapodia are small out-growths of integument which contain special gills.
6. Presence of a long, tubular proboscis bearing teeth, which give the peculiar shape to the worm.
7. There are no blood vessels; each gill contains a diverticulum of coelom. However the periviseral fluid contains numerous red corpuscles which assume the function of blood.
8. It is generally used as a fish bait.

## APHRODITE

The classification is same as that of *Polynoe.*

1. *Aphrodite* or sea-mouse commonly found in deep water 10 to 100 fathoms in muddy bottom of almost all seas.
2. It generally burrows in mud and creeps on the sea bottom.
3. Fairly common in coral line.
4. It is commonly know as "sea mouse".

5. Body short almost oval and dorso-ventrally compressed.
6. Head small generally hidden, situated at the anterior end.

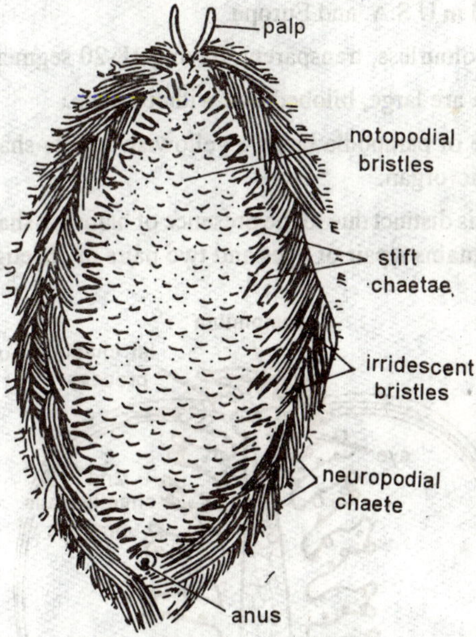


Fig.- 15.5 *Aphrodite*

7. The head carries a single median tentacle and two large lateral palps.
8. Body is covered by modified dorsal cirri the elytra, which are plate like.
9. The elytra is a respiratory organ.
10. Many of the notopodial chaetae are iridescent forming a blanket of flet-like covering of the whole body of animal.
11. Apart from these brilliantly coloured thread-like setae three are spine-like brown stiff bristles.
12. The anus is situated at the extreme posterior end on the dorsal surface.
13. When distributed the stiff iridescent bristles can be erected just like the spine of porcupine to frighten away the enemies.
14. The alimentary canal has caeca for the digestion of food.
15. When animal moves in water it shows beautiful iridescent colour.

## TOMOPTERIS

The classification is same as that of *Polynoe.*

1. It is a marine pelagic polychaete.
2. It is found in U.S.A. and Europe.
3. Body is colourless, transparent, having 18-20 segments.
4. Parapodia are large, bilobed and without setae.
5. Each lobe of parapodia bears a yellowish rosette-shaped or spherical photogenic organ.
6. The head is distinct due to the presence of hammer-shaped prostomium which contains a pair of eyes and two pairs of chaetigerous processes

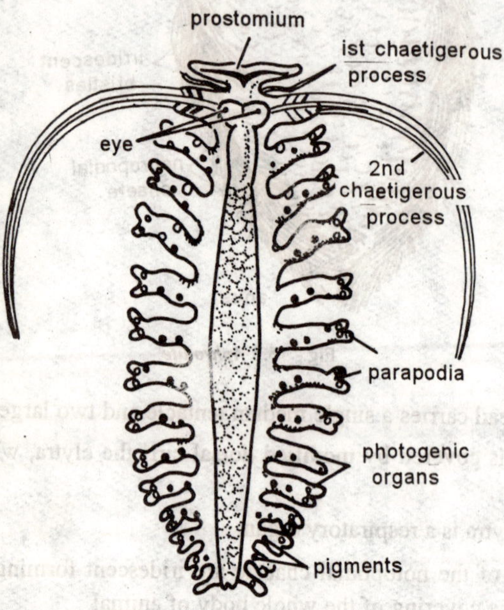


Fig.- 15. 6 *Tomopteris.*

or tentacular cirri. The anterior are very short while posterior ones are very much elongated.

7. The ventral mouth is without jaws.
8. Circulatory system is absent.

## MYZOSTOMA

The classification is the same as that of *Polynoe*.

1. It is an unsegmented, flattened polychaete, generally ectoparasites on the starfishes, sea-lilies or feather stars.
2. They are found in West Indies, Havana, Europe.
3. These are disc-shaped situated on convex ventral side of the host.
4. Five pairs of parapodia situated on convex ventral side.
5. It has four pairs of small suckers, probably sense organs, on ventral side.
6. Ten pairs of minute protuberances of short cirri extend from the edge of the body.

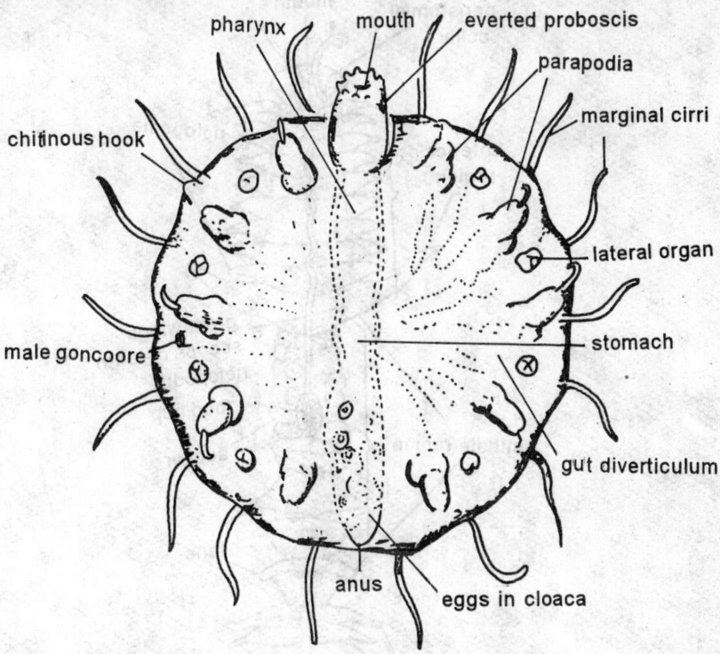


Fig.- 15. 7 *Myzostoma*

7. Mouth is ventral in position that leads into a protrusible pharynx.
8. A distinct head is absent.
9. Coelom or body cavity is absent.
10. Circulatory system and respiratory organs are absent.
11. Hermaphrodite, protandrous. Later on testes degenerate leaving behind female reproductive organs.

12. Development indirect through trochophore larva.

## CHAETOPTERUS

| | | |
|---|---|---|
| Phylum | - | Annelida |
| Class | - | Polychaeta |
| Order | - | Sedentaria |
| Genus | - | *Chaetopterus* |

1. It is a tubicolous worm which generally lives permanently in a 'U' tube made up of sand and mucous with excurrent and incurrent openings.

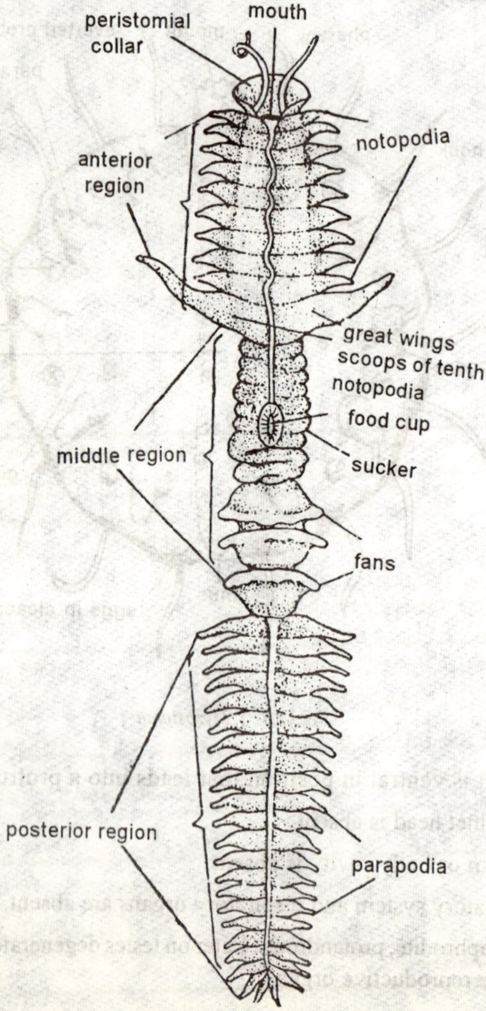


Fig.- 15. 8 *Chaetopterus. Entire worm*

2. It is found in Europe, U.S.A., North Carolina to Cape Cod.
3. It is commonly known as 'paddle-worm'.
4. Long worms, about 0.15 to 0.45 metre in length.
5. Body divided into three regions:
   (a) Anterior flattened region with a funnel-shaped peristomial collar having two tentacles called peristomial cirri. In the 4th segment setae are enlarged, while 10th segment has a pair of wing-like aliform notopodia.
   (b) Middle region possesses modified parapodia forming three pairs of fans. It consists of 5 segments.
   (c) Hinder or posterior region has smaller segments bearing parapodia. These segments are about 30 in number.
6. Feeding is very characteristic. Three paired fans of middle region and cilia of ciliated groove create a water-current, which flows from one end of the tube to the other. The water-current serves for nutrition and respiration. Notopodia secrete mucus, which forms mucous begs with food particles. It is rolled into a ball or bolus which is passed on to the mouth through dorsal groove.
7. *Chaetopterus* is highly bioluminescent emits blue-green light.
8. A great power of regeneration, the whole body can be regenerated from a single segment.

## ARENICOLA

The classification is the same as that of *Chaetopterus*.

1. It is found on coasts between tide marks. It lives in a 'U' shaped burrow made up of sand and mucous.
2. It is found in Europe and U.S.A. (from Florida to Cape Cod).
3. It is popularly known as 'lug worm'.
4. The body is elongated, worm-like, broadest at anterior end.
5. The length of the worm is 5 to 8 inches.
6. The body is divided into three regions.
7. The anterior or prebranchial region of the peristomium, an achaetous segment and six chaetiferous segments. Eyes, tentacles absent. Peristomium contains 9 pairs of lithocytes,

8. The middle region not so thick of 12 or 13 chaetiferous and branchiferous segments.
9. The posterior region much thinner without setae and branchiae, with segments less clearly marked and variable in number.

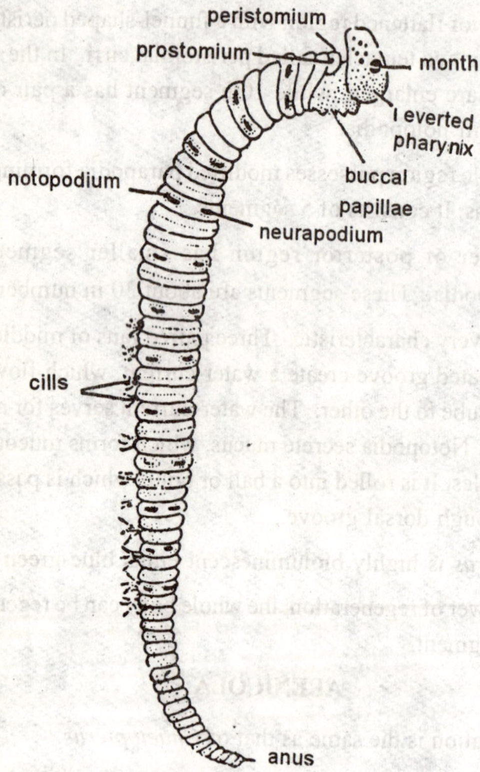


Fig.- 15. 9 *Aernicola.*

10. Mouth lies ventral to the prostomium. The buccal region and pharynx protrusible as proboscis which is covered by chitinized papillae.
11. Nephridia six pairs.
12. It is generally used as bait for tishes.

**AMPHITRITE**

The classification is the same as the of *Chaetopterus*.

1. It is a tubicolous marine worm, living in tubes or burrows lined with sand and mucous.
2. It is found in Europe and U.S.A. (from North Carolina to Cape Cod).

3. Body elongated and cylindrical, pinkish in colour measuring about 20-30 cm. in length
4. Body is differentiated into three distinct regions: anterior region (head), middle region (trunk) and posterior region (abdomen).
5. Head small; prostomium and peristomimum are fused to form a horse-shoe shaped preoral lobe.
6. The prostomium forms the upper lip and peristomium lower lip of the mouth. Eyes and palps are absent.
7. A tuft of long, yellowish and contractile tentacular filaments arises from a transverse ridge just behind the pre-oral lobe.
8. Tentacles are feeding organs bearing ciliated groove on its oıal surface.
9. The first three segments of thorax bear red and feather-like branched

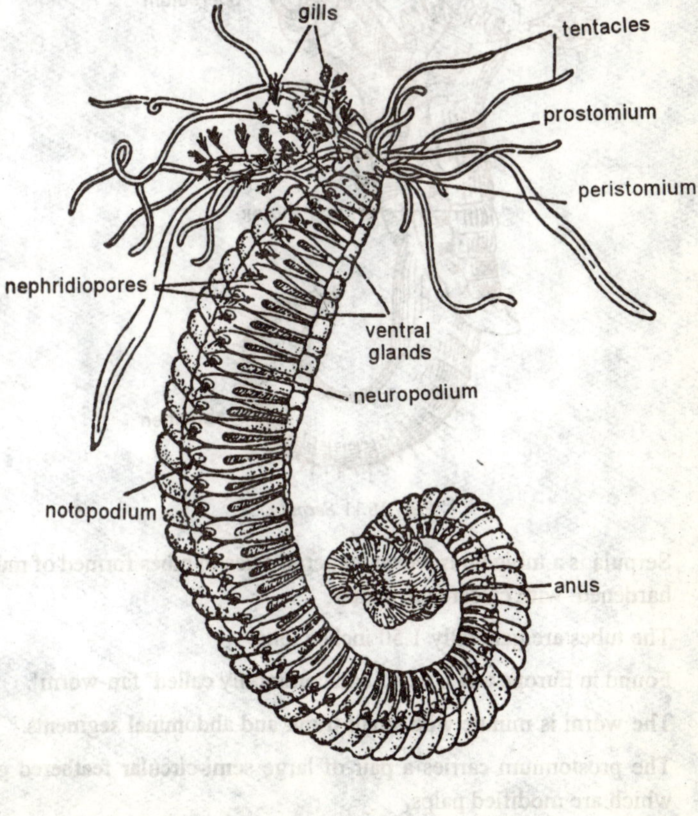


Fig.- 15. 10 *Amphirite.*

gills. Rest of the segments bear notopodia with setae. Some of the thoracic segments have shield glands for mucous secretion.

10. Abdomen is long and many segmented. Each segment contains notopodia and neuropodia.

11. The caudal segments lack notopodia and anus is found terminally in the last segment.

## SERPULA

The classification is the same as that of *Chaetopterus*.

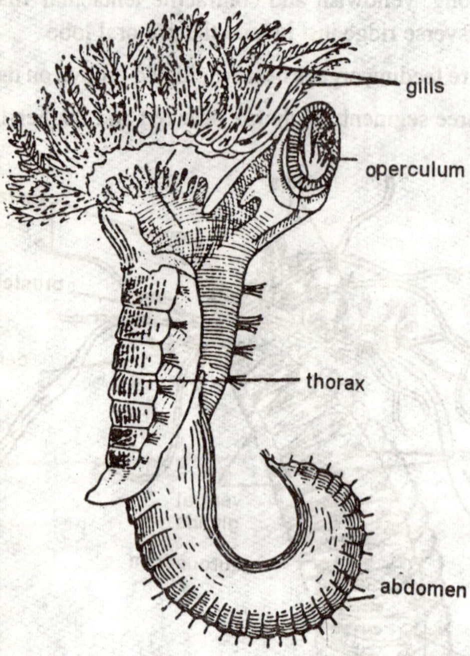


Fig.- 15.11 *Serpula.*

1. Serpula is a tubicolous worm, generally lies in tubes formed of mucin hardened with calcareous matter.
2. The tubes are generally 1.50 inches long.
3. Found in Europe and U.S.A. and commonly called' fan-worm'.
4. The worm is minute with small trunk and abdominal segments.
5. The prostomium carries a pair of large semi-circular feathered gills which are modified palps.

6. The gills are respiratory and also serve as food collecting organs.
7. One of the tips of gill filament is modified into stopper or operculum which lock the aperture of tube when *Serpula* withdraws inside the tube.
8. The dorsal and ventral cirri of thoracic segments are fused to form a thoracic membrane on each side which is used to smoothen the interior of the tube.
9. The nephridia of posterior segments act as genital ducts.

## SABELLA

The classification is the same as that of *Chaetopterus*.

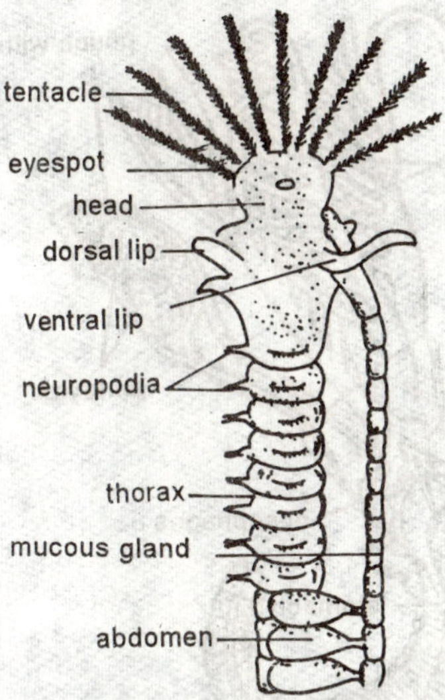


Fig.- 15.12 *Sabella.*

1. Tubicolous polychaetae, live in tubes of sea-mud or sand which they rarely or never leave.
2. It is distributed throughout the world but especially U.S.A. (North carolina to Cape Cod).
3. It is commonly called as 'peacock-worm'.
4. Body is elongated bearing equal number of segments.
5. Body is divisible into a small head, thorax and abdomen.

6. Head consists of prostomium with several large gill-filaments or feathery pinnules having mucus secreting cells.
7. Thorax formed of 8 segments and provided with gland-shields and parapodia. The gland secrete mucous. The first segment is called peristomium which is collor-like and has notopodium and setae but no neuropodium.
8. Abdomen is formed of 300 segments, and is marked with a faecal groove which serves to carry faeces out of the tube.
9. Buccal mass is not eversible.
10. Eyes are present on branchial filaments.
11. *Sabella* is a filter-feeder, feeding by the movements of tentacles.

## SPIRORBIS

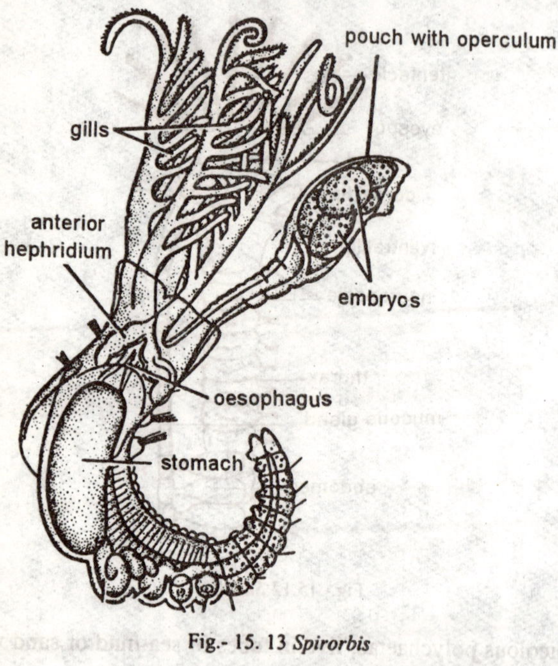


Fig.- 15. 13 *Spirorbis*

The classification is the same as that of *Chaetopterus*.

1. It is a minute tubicolous, sedentary polychaete, living in a small calcareous tube, coiled in a flat spiral manner.
2. It is commonly found in U.S.A. , Long Island to Bay of Fundy and northwards.

3. Body measure 3mm in length.
4. The body is cylindrical, elongated and is divisible into upper thorax and lower abdomen.
5. Each gill consists of an elongated stem on which are borne the short filaments; one of the gill filaments is enlarged to form an operculum which closes the mouth of tube.
6. The animal is hermaphrodite: anterior part of body gives rise to the ova while posterior part or abdomen produces the sperms.
7. Eggs are developed inside the operculum.
8. The prostomium and peristomium are compactly fused together to form a collar which secretes the tube.
9. Presence of paired anterior nephridia which are excretory in function.

## STERNASPIS

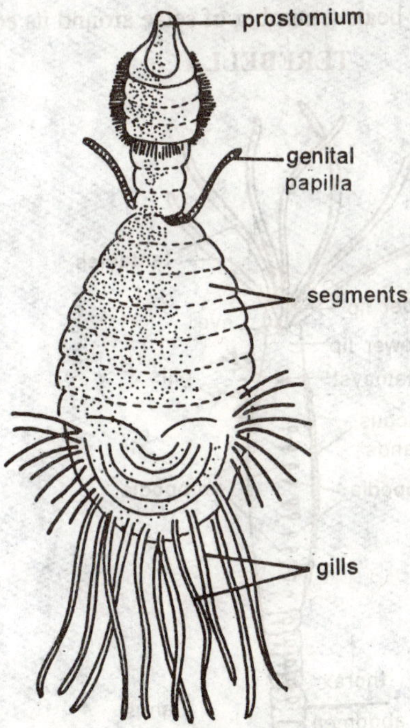


Fig.- 15.14 *Sternaspis*

The classification is the same as that of *Chaetopterus*.

1. It is a marine, tubicolous polychaete.

2. It is found in U.S.A. (Florida).
3. The body is short, broad and made up of about 30 segments.
4. Head is formed of prostomium and peristomium.
5. Prostomium is elongated and without any appendages.
6. Peristomium is knob-like and contains mouth.
7. First three segments possess imcomplete bristles, segments 8th-15th possess setae while segments 5th, 6th, 7th and 16th-24th are without setae.
8. Intestine is coiled.
9. Anus is terminal and surrounded by gills.
10. Presence of a single pair of nephridia bearing small ciliated funnels, which do not open to exterior.
11. A ventral shield formed by a thickening of the cuticle in thc posterior region of the body bears a number of setae around its edges.

**TEREBELLA**

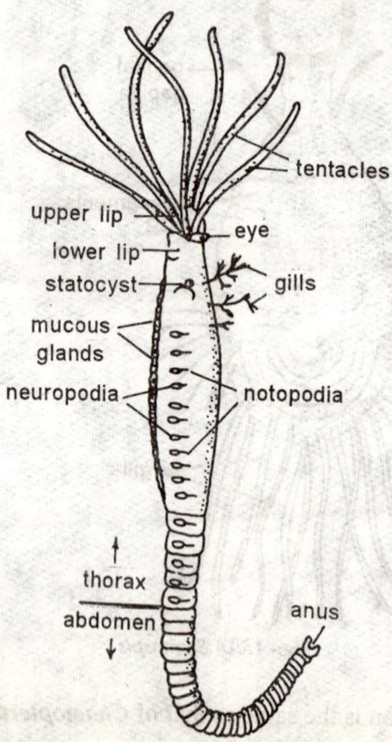


Fig.- 15. 15 *Terebella.*

The classification is the same as that of *Chaetopterus*.

1. It is a marine, burrowing, sedentary polychaete.
2. It is found in Europe, U.S.A., Woods-Hole region and Vineyard Sound to Bay of Fundy.
3. Body is long and vermiform differentiated into head, thorax and abdomen.
4. Head is horse-shoe-shaped made up of prostomium and peristomium.
5. Prostomium flattened into a mobile upper lip which carries a transverse series of numerous filiform tentacles and two pairs of eyespots. Palps are absent.
6. Peristomium forms lower lip and lacks cirri.
7. The gills or branchiae are found on the dorsal surface of few anterior segments and they form four wide comb-like branches on a single peduncle.
8. Thorax with gland shields.
9. Parapodia are feebly developed.
10. Nephridia are large and small and serve as both excretory and genital ducts.

## MYXICOLA

The classification is the same as that of *Chaetopterus*.

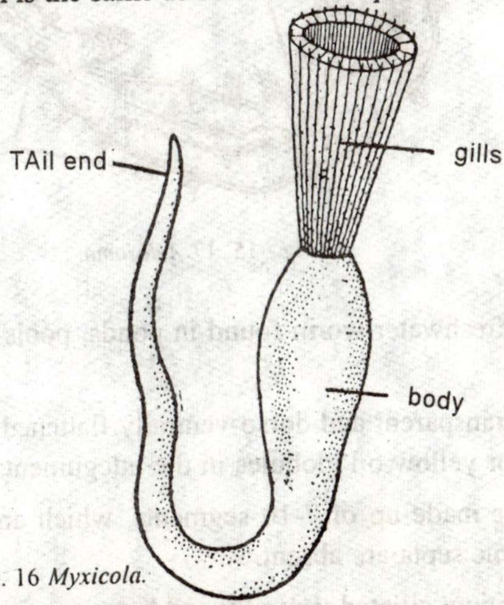


Fig.- 15. 16 *Myxicola*.

1. *Myxicola* is a sedentary tubicolous animal lives in a tube in the sea sand. Occasionally the worm leaves the tube and swims freely.

2. Body is elongated, dark green in colour and measures about 10 cm in length.
3. Eyes, mucous and shields are absent.
4. Gills are present which are well developed and protrudes out.

**AEOLOSOMA**

| | | |
|---|---|---|
| Phylum | - | Annelida |
| Class | - | Oligochaeta |
| Order | - | Archioligochaeta |
| Genus | - | *Aeolosoma* |

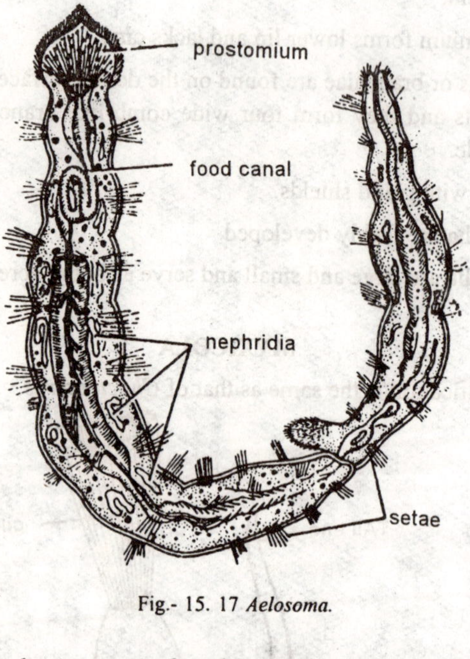


Fig.- 15. 17 *Aelosoma.*

1. It is a freshwater worm found in ponds, pools and living among the algae.
2. Body transparent and dorso-ventrally flattened and spotted with red, green or yellow oil globules in the integument.
3. Body is made up of 7-10 segments, which are not clearly marked. Coelomic septa are absent.
4. Prostomium ciliated ventrally and bears a pair of ciliated pit.
5. Each segment bears four bundles of hair-like setae.

6. Clitellum present only on the ventral side of the 5th-7th segments.
7. Nephridia are paired and metamerically repeated.
8. Nervous system is hypodermic, the ventral nerve cord is absent.
9. These are hermaphrodite worms but commonly reproducing asexually by transverse fission .
10. Nephridial ducts act as gonoduct. The genital products pass out through nephridiopores.

## TUBIFEX

The classification is the same as that *Aelosoma*.

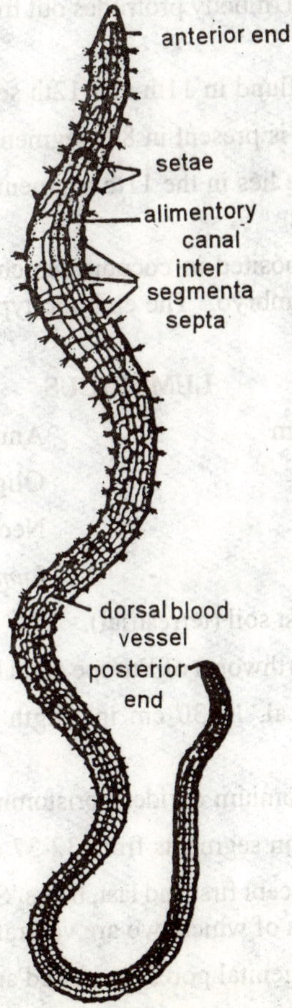


Fig.- 15. 18 *Tubifex*.

1. It is a tubicolous fresh water animal found on the bottom of deep lakes. It lives into tubes, made up of mud, minerals and mucus.
2. It is found in U.S.A. (Long Island Sound to Marine).
3. They occur in large numbers in the form of patches.
4. The body is elongated, somewhat cylindrical reaching upto the length of 4 cm. approximately.
5. It is reddish in colour and is divisible into a number of ring-like metameres.
6. Each segment of body except the few posterior segments bears the tufts of minute setae on its either side.
7. The posterior end of body protrudes out from the tube and waves back and forth.
8. The clitellum is flund in 11th and 12th segment.
9. Contractile heart is present in 8th segment.
10. Male genital pore lies in the 11th segment and female genital pore in the 12th segment.
11. The eggs are deposited in cocoon, which contain food yolk for the nourishment of embryo. The epibolic type of process occurs during development.

**LUMBRICUS**

| | | |
|---|---|---|
| Phylum | - | Annelida |
| Class | - | Oligochaeta |
| Order | - | Neooligochaeta |
| Genus | - | *Lumbricus* |

1. It is found in moist soil (terrestrial).
2. It is a common earthworm of Europe and U.S.A.
3. Body is cylindrical, 15-30 cm in length and made up of 100-150 segments.
4. The lobular prostomium divides peristomium into two parts.
5. Clitellum present in segments from 32-37 and is swollen.
6. Each metamere except first and last, bears 'S' shaped chitinous chaetae arranged in 4 pairs of which two are ventral and two ventro-lateral.
7. Female and male genital pores are paired and are situated in 14th and 15th segments.
8. Gizzard is absent.

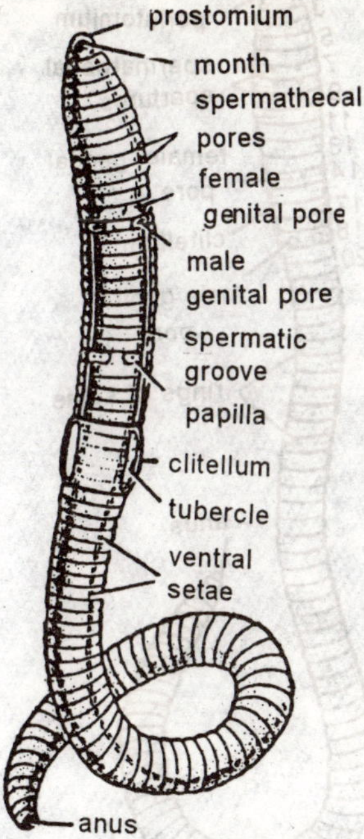


Fig.- 15.19 *Lumbricus.*

9. Prostate glands are absent.
10. Genital papillae absent.
11. There are two pairs of spermathecae in 9th and 10th segments. Three pairs of seminal vesicles in 9th, 11th and 12th segments. Testis sacs are two , one in 10th and other in 11th segment.
12. Each segment has a pair of nephridiopores, situated ventro-laterally.

**MEGASCOLEX**

The classification is the same as that of *Lumbricus*.

1. It is a terrestrial worm lives in moist soil.
2. It is found in South India, Sri Lanka, Australia, Tasmania and New Zealand.
3. It measures 8-12 cm. in length and bears 80-100 segments.
4. The dorsal body surface is dark purplish brown and the ventral surface is somewhat pale in colour.

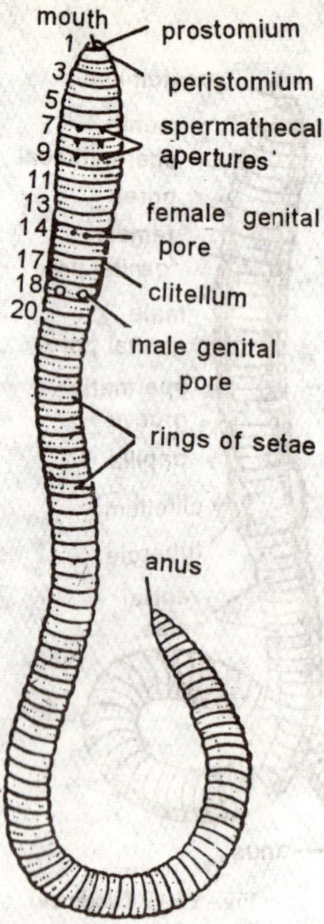


Fig.- 15.20 *Megascolex.*

5. Clitellum present in segments from 14-17.

434

6. Gizzard lies in 6th segment.
7. Longitudinal blood vessels are two, dorsal and ventral and they are connected by eight pairs of hearts in segments 6-13.
8. One pair of metanephridia is present in each segment behind the 19th segment.
9. Two pairs of seminal vesicles, one pair lies in 9th segment and other in 12th segment.
10. One pair of prostate glands lie in segments 18 and 19.
11. 3 pairs of spermathecae lie in segments, 7, 8 and 9. Each with two diverticula.
12. Two female genital pores on segment 14th.
13. Two pairs of penial setae help in copulation.

## ACANTHOBDELLA

| | | |
|---|---|---|
| Phylum | - | Annelida |
| Class | - | Hirudinea |
| Order | - | Acanthobdellida |
| Genus | - | *Acanthobdella* |

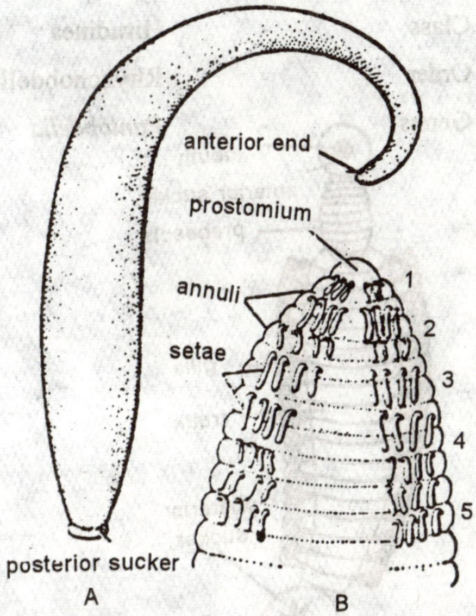


Fig.- 15.21 Acanthobdella. A= Entire leech, B= Anterior end in ventral view showing setae.

1. It is an ectoparasite on the caudal and anal fins of salmon fish.
2. It is found in West Siberia and North-East Europe.
3. Body is elongated dorso-ventrally flattened and both the ends pointed.
4. It contains only 20 segments.
5. Anterior sucker is absent, posterior sucker made up of four segments.
6. First five segments bear double rows of setae ventro-laterally.
7. Body cavity or coelom spaceous and is incompletely divided by septa.
8. Nephridia with external and internal openings.
9. Circulatory system made up of a dorsal and a ventral blood vessel.

10. Nervous system consists of 20 ventral ganglia.
11. Male genital pore in 7th segment and female genital pore in 8th segment.
12. Acanthobdella is a connecting link between Oligochaeta and Hirudinea.

**PONTOBDELLA**

| | | |
|---|---|---|
| Phylum | - | Annelida |
| Class | - | Hirudinea |
| Order | - | Rhynchobdellidr |
| Genus | - | *Pantobdella* |

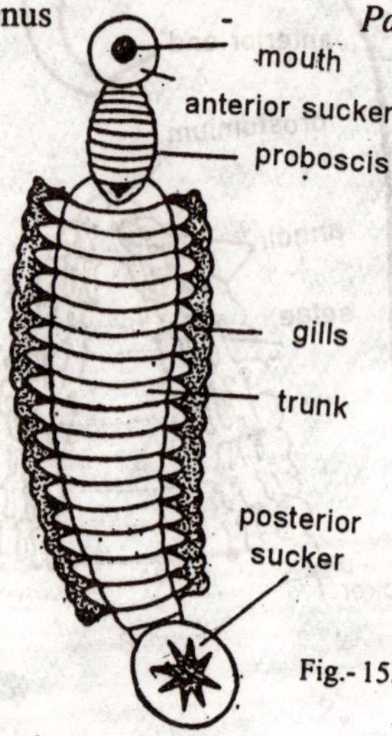


Fig.- 15.22 *Pontobdella.*

1. It is an ectoparasite on fishes chiefly skates and rays. Sometimes it is called 'skate sucker'.
2. It is found in Europe and U.S.A.
3. The body is cylindrical, elongated with anterior narrow and posterior blunt portion.
4. The anterior and posterior suckers are cup-shaped and well marked.
5. Anterior sucker is applied to the skin of fish and then makes a wound.
6. The body surface is rough which is covered by small rounded green tubercles.
7. Nephridia form a complex network on the ventral surface of body.
8. The anterior end of body protrudes out to form a retractile proboscis or introvert.

9. Setae and jaws wanting.

10. Hermaphrodite, sexual reproduction is common.

11. Eggs are laid in molluscan shells. Parental care is shown as the eggs are guarded by the parent till they hatch.

**BRANCHELLION**

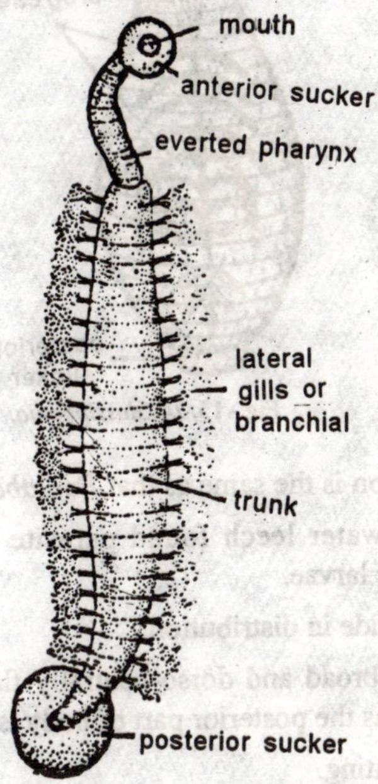


Fig.- 15. 23 ***Branchellion***

The classification is the same as that *Pontobdella*.

1. *Branchellion* is an ectoparasite on marine fishes specially *Torpedo*.
2. It is world wide in distribution but not reported in India.
3. Small, elongated body of about 10 cm. long.
4. The anterior end is produced into a proboscis and a wide abdominal region.
5. The abdomen bears 11 pairs of series of foliaceous non-digitate gills (branchiae) laterally in each segment.
6. The anterior and posterior suckers well developed and pedunculate.
7. Nephridia are absent but excretion by complex network of tubes.
8. Hermaphrodite, sexual reproduction is common.

## GLOSSIPHONIA

Fig.- 15. 24 *Glossiphonia.*

The classification is the same as that *Pontobdella.*

1. It is a freshwater leech found parasite of mussels and feeds on *Chironomous* larvae.
2. It is world wide in distribution.
3. The body is broad and dorso-ventrally flattened. The anterior end tapers whereas the posterior part of body is oval.
4. Setae are wanting.
5. Three pairs of eyes persist.
6. The posterior sucker is cup-shaped and well formed.
7. Presence of elongated crop chambers arranged in the middle of body segments.
8. In this leech copulation has never been observed, but one individual deposits spermatophores on any part of the body of another individual.
9. The male and female ducts open by a common pore located between eleventh and twelth segments.
10. Cleavage unequal. No. cocoon formation takes place.

## OZOBRANCHUS

The classification is the same as that of *Pontobdella.*

1. It is an ectoparasite on the river-turtle and fishes.
2. Body is sub-cylindrical that often bears lateral gills.

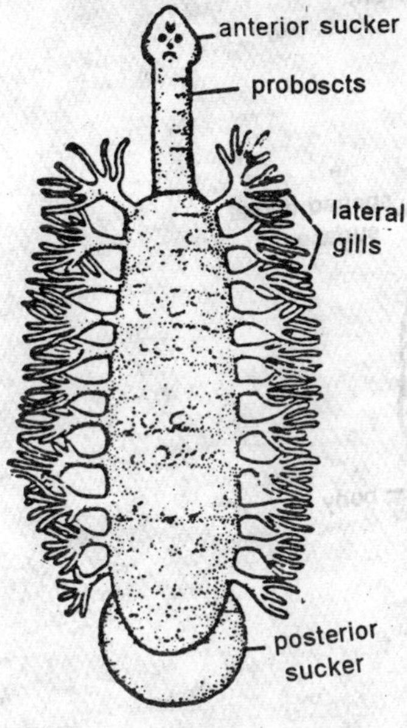


Fig.- 15. 25 *Ozobranchus.*

3. Usually more than three annuli per segment.
4. Anterior and posterior suckers are well developed.
5. Anterior region is produced into a long proboscis.

## HERPOBDELLA (NEPHELES)

| | | |
|---|---|---|
| Phylum | - | Annelida |
| Class | - | Hirudinea |
| Order | - | Gnathobdellidida |
| Genus | - | *Nepheles* |

1. It is found in fresh water ponds and streams under stones and in vegetation.
2. It has world wide distribution.
3. Body is slender about 8 cm. in length and brownish black in colour.
4. Anterior end is rounded and devoid of anterior sucker. Head bears 2 pairs of eyes.
5. Each body segment has 5 annuli. The annuli and true segments are not easily demarkated and consists of 19 segments.
6. Posterior end is flattened and bears a distinct small posterior sucker.

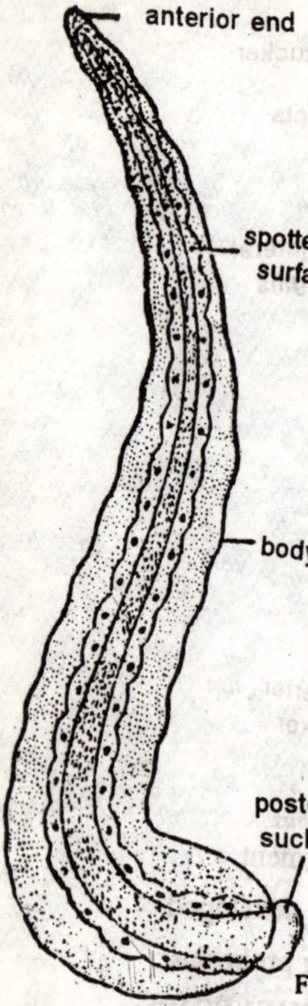


Fig.- 15. 26 *Nepheles.*

7. Setae, parapodia and jaws are absent.
8. It is carnivorous, feeds on aquatic larvae, worms and snails.
9. Sexual openings are separated from one another by two rings.
10. Development takes place in cocoon.

## HAEMOPIS

The classification is the same as that *Nepheles*.

1. It is found in ponds, streams and swampyu meadows.
2. It is cosmopolitan in distribution.
3. It is commonly called 'hungry horse leech'.
4. Body segmented, segments indicated externally by ring-like constrictions, which are much more numerous than the true segments.
5. Head is not distinct, jaws are armed with blunt teeth in two rows. A pair of eyes present.

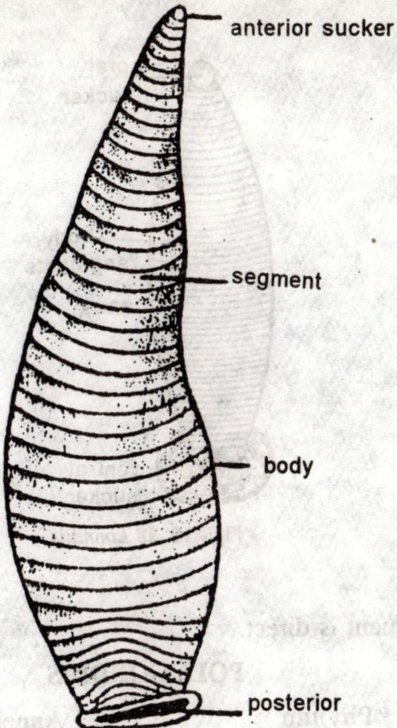


Fig.- 15.27 *Haemopis.*

6. Presence of one pair of caeca in crop.
7. It is carnivorous leech, feeding on earthworms, aquatic larvae of insects, tadpoles and small fishes.
8. Copulatory glands are absent.
9. Development takes place in cocoons.

## LIMNATIS

The classification is the same as that of *Nepheles*.

1. It is a fresh water ectoparasitic leech of domestic animal, horses, cattles etc.
2. Commonly found in Syria, Egypt and Lebanon.
3. Body is elongated never more than 4 mm. in length.
4. Anterior and posterior suckers are well developed.
5. Proboscis non-protrusible.
6. Oral sucker armed with three jaws.
7. Each segment with 5 annuli.

Fig.- 15. 28 *Limnatis.*

8. Development is direct.

## POLYGORDIUS

| | | |
|---|---|---|
| Phylum | - | Annelida |
| Class | - | Archiannelida |
| Genus | - | *Polygordius* |

1. It is found living in the sand of European Seas.
2. *Polygordius* is a marine worm with narrow, elongated cylindrical body.
3. The prostomium is small, the peristomium large and forms the head, which bears paired prostomial tentacles and ciliated pits.
4. Presence of a tooth-like process round the anus, and in front a circlet of adhesive papillae.
5. Segmentation is not distinct in front, but well marked behind and is without parapodia or chaetae.
6. Beneath the skin longitudinal muscle fibres in four quadrants. Circular muscles absent.
7. Presence of paired segmentally arranged nephridia leading into nephrostomes.
8. Vascular system is reduced and single nerve cord lies in the epidermis.
9. Sexes are separate.
10. Gametes released by the rupture of the body wall.

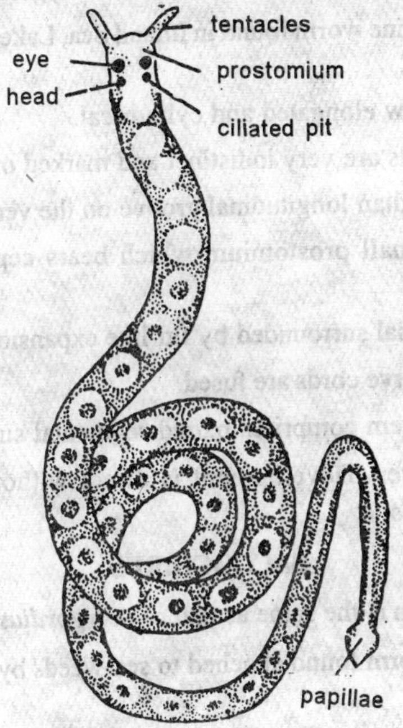


Fig.- 15.29 *Polygordius.*

11. Fertilization external. Development is indirect with Loven's larva, which resembles a typical trochophore larva.

## PROTODRILUS

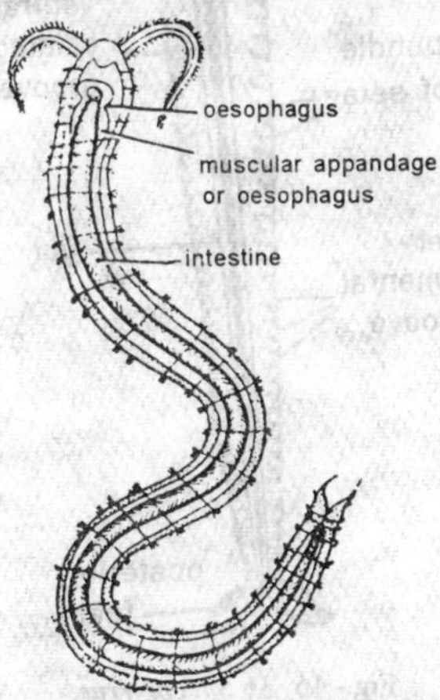


Fig.- 15.30 *Protodrilus.*

The classification is the same as that of *Polygordius*.

1. It is small marine worm found in Inland Sea, Lake at Faro near Messina (U.S.A.).
2. Body is narrow elongated and cylindrical.
3. Body segments are very indistinct and marked off by ciliated rings.
4. There is a median longitudinal groove on the ventral surface.
5. Body with small prostomium which bears cephalic tentacles and ciliated pits.
6. Anus is terminal surrounded by fin-like expansions.
7. The ventral nerve cords are fused.
8. Excretory system comprises paired segmental simple nephridia.
9. Hermaphrodite. Development is direct without larval stage and metamorphosis.

## SACCOCIRRUS

The classification is the same as that of *polygordius*.

1. It is marine worm found attached to sea-weeds by adhesive pad.

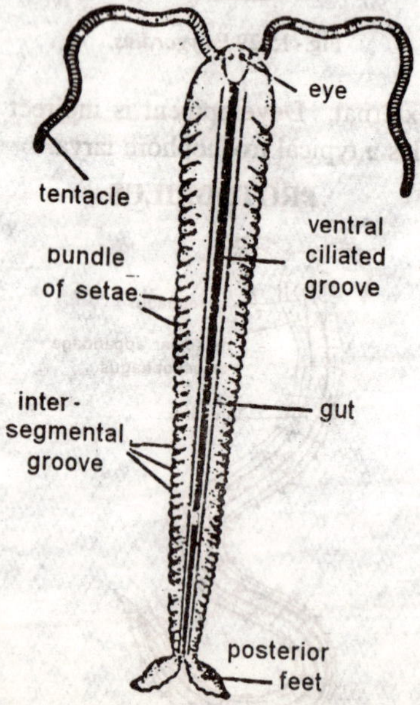


Fig.- 15. 31 *Saccocirrus*.

2. It is found in European countries and Madras.
3. Body long, cylindrical, 10-15 mm in length.
4. Anterior region bears a pair of cephalic tentacles and eyes.
5. Body segments marked by prominant inter-segmental grooves.
6. Each segment bears lateral setae.
7. Longitudinal ciliated groove is situated in the middle of ventral surface.
8. Anus terminal with adhesive pads on.
9. Nervous system comprises of two separate ventral nerve cords lying just beneath the epidermis.
10. Sexes separate, female with spermathecae and males with a pair of protrusible penis in each segment.
11. Development is direct.

## NERILLA

The classification is the same as that of *Polygordius*.

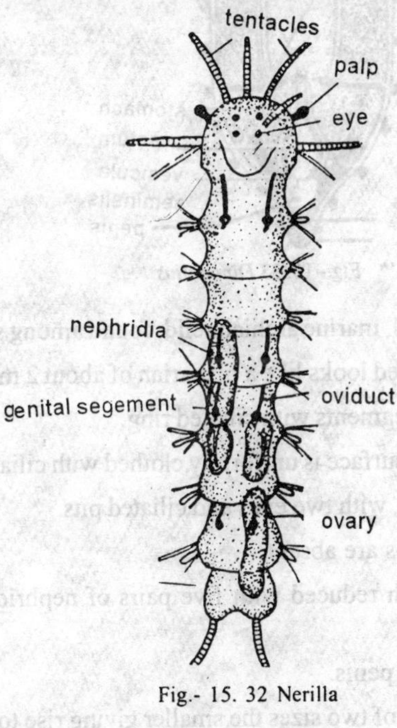


Fig.- 15. 32 Nerilla

1. It is a marine worm found in sea-water.
2. Body elongated worm-like measuring 1 mm in length.

3. Head with three prostomial tentacles, two palps and two pairs of setae.
4. Saxes are separate.
5. Three genital segments in male and one in female.
6. In male three pairs of sperm ducts uniting at a common median genital aperture.
7. Development is direct.

## DINOPHILUS

The classification is the same as that of *Polygordius*.

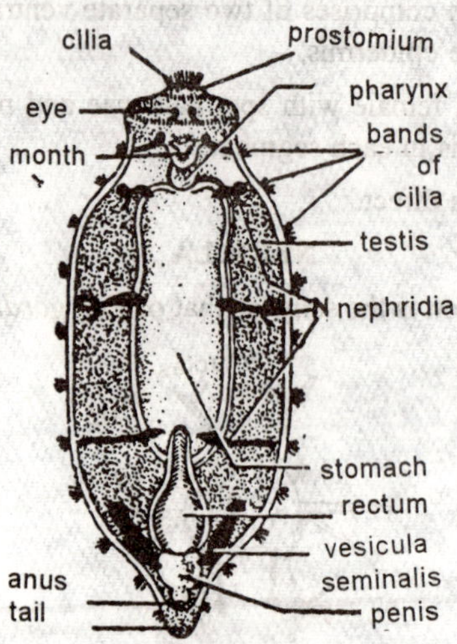


Fig.- 15. 33 *Dinophilus*.

1. Dinophilus is small, marine archiannelid, found among sea weeds.
2. Body short, flattened looks like a planarian of about 2 mm long.
3. Body with 5 to 6 segments with ciliated ring.
4. The entire ventral surface is uniformly clothed with cilia.
5. Prostomium broad, with two eyes and ciliated pits.
6. Tentacles and palps are absent.
7. Coelom very much reduced with five pairs of nephridia. Sexes are separate.
8. Male with median penis.
9. Females with eggs of two sizes the smaller giving rise to males and the larger to females.

♦♦♦

# 16

# NEANTHES-NEREIS

*Neanthes* (old name *Nereis*) is commonly known as *'sandworm'* because it is found buried in sand, or *'clamworm'* as it is found along with clams but has no ecological relationship with them. *Neanthes* is one of the largest and most common typical marine annelid with most of the characteristics of the phylum. It has a great number of species which are found in different parts of the world but resemble greatly with each other. Few important and common species are : *N.dumerilli, N.diversicolar, N.irrorata, N.cultrifera, N.virens, N.pelagica,* and *N. caudata.*

**SYSTEMATIC POSITION**

| | | |
|---|---|---|
| Phylum | - | Annelida |
| Class | - | Polychaeta |
| Order | - | Errantia |
| Genus | - | *Neanthes* (*Nereis*) |

**Habits and Habitat**

*Neanthes* has world wide distribution i.e. it is found throughout the world, in Europe, America, Alasks, North Pacific and other places. It inhabits the sea-shores between tide marks. It is also found among the sea-weeds and under stones, but normally it lives in burrows. The burrows are formed 60 cm. deep. The burrow is lined by mucous which helps in cementing the grains of sand, a mechanism by which the burrow cannot collapse. The worm lives within the burrow during day time. At night, it protrudes the anterior of its body and moves it all round in the search of food. Thus the worm is nocturnal. The worm is carnivorous in diet and feeds on small crustaceans, molluscs and annelids. Occasionally, the worm leaves its burrow and creeps about the stones. During breeding season, it leaves its burrow permanently and swims about actively in the surface layer of water.

**External morphology**

*Shape and size.* Body is elongated, cylindrical and vermiform. It is *bilat-*

**Fig. 16.1 Neanthes**

*erally symmetrical, metamerically segmented* and somewhat flattened dorsoventrally. It gradually tapers to the posterior end. The dorsal surface is convex but the ventral surface is flat. The size varies from few to 40 cms.

*Colouration.* Different species are differently coloured *N. virus* is steel-blue in colour; *N.limbata* is of brownish colour while jaws are light amber-coloured; *N.cultrifera* is greenish; *N.pelagica* is reddish-brown; *N.lamnicola* is also reddish-brown; and *N.vexilosa* is dark brown or blue-green with iridescence.

*Segmentation*-Body is divided into approximately 200 or more *segments* or *metameres.* In *N.cultrifera* there are 80 segments. These are marked externally by grooves and internally by septa. All the metameres are alike except the last which is somewhat rounded. It is known as *tail somite* or *pygidium* and carries a pair of *anal cirri.* Anus is terminal.

*Divisons of the body.* The whole body of *Neanthes* is divided into three well-defined regions.

1. Head
2. Trunk
3. Pygidium.

**1. Head.**

The anterior-most part of the body is specialized into a head, which

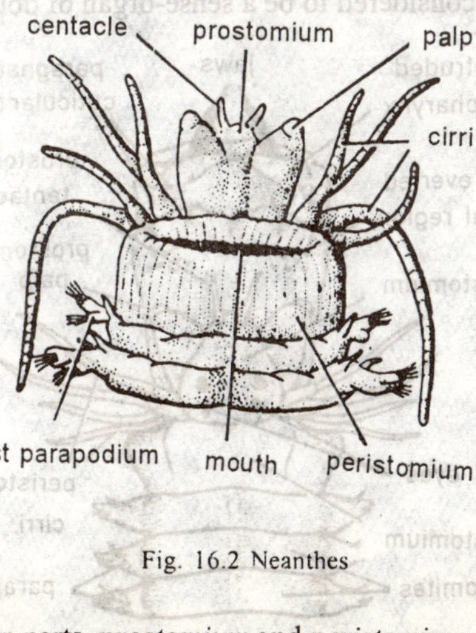


Fig. 16.2 Neanthes

consists of two main parts, *prostomium* and *peristomium.*

(i) *Prostomium* (Gr., *Pro* = before; *stoma* = mouth). It is the lobe like structure present infront of mouth and is a roughly triangular in shape. It is situated mid dorsally. Protruding forward from the anterior edge of the prostomium are two *tentacles* and from the ventral side project two *palpi*. Each palp is two-jointed. Tentacles are regarded as organs of touch, and palpi are organs of taste and smell. There are two pairs of

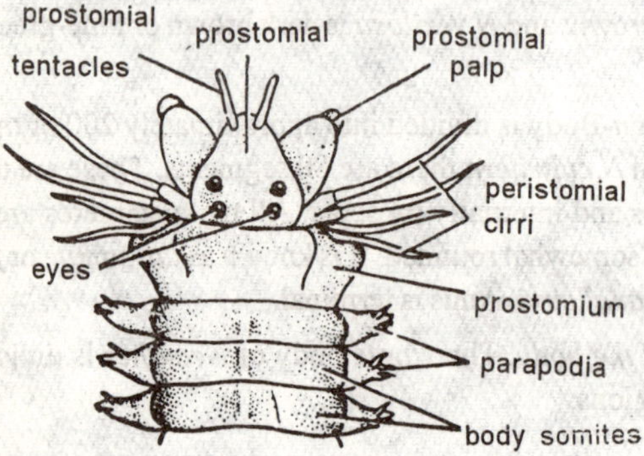


Fig. 16.3 Neanthes

simple eyes on the dorsal surface of the prostomium for the perception of light. On each side of prostomium is a ciliated pit, which is called *nuchal organ,* considered to be a sense-organ of doubtful importance.

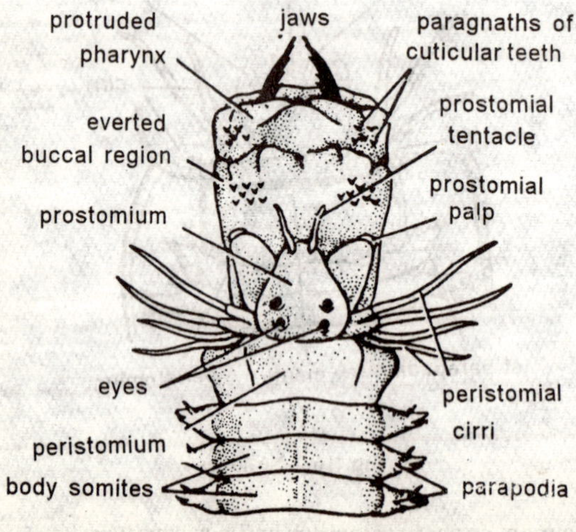


Fig. 16.4 Neanthes. Head in dorsal view with everted pharynx

(ii) *Peristomiuum* (Gr. *Peri* = around; *stoma* = mouth). It is a large ring-like round part present behind the prostomium. It bears on its ventral side a slit-like *mouth*. It differs from a trunk segment in being longer, in lacking *parapodia* and in the presence of four (two pairs) thread-like *peristomial cirri* on each side; of these one pair is dorso-lateral and the other ventro-lateral. These are homologous with the notopodial cirri and neuropodial cirri of the parapodia of trunk segments. Each cirrus is a long, slender, tactile structure, having a short proximal joint and a long distal joint. The dorsal cirri are longer than the ventral ones.

The study of the developmental stages of *Neanthes* shows that the prostomium is not a true segment of the body but develops as a dorso-anterior projection of the peristomium, whereas the peristomium is formed by the fusion of the first and second body segments.

**2. Trunk**

It comprises practically the entire body except the head and the last segment or pygidium. It comprises a number of segments which are all alike. Each trunk segment bears on either lateral sides a flat, fleshy, hollow and

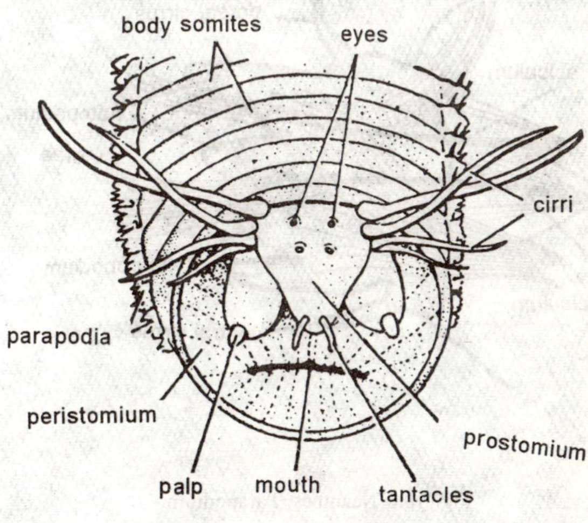


Fig. 16.5 Neanthes. Head in frontal view.

vertical flap-like structure, the *parapodia*. Term parapodia was coined by *Huxley* (Gr. *para* = besides; *podos* = foot)

Parapodium is a *biramous* structure and consists of an upper or dorsal part or *notopodium*, and a lower or ventral part or *neuropodium*. Notopodium and neuropodium are both bilobed structures i.e. each part is further subdivided into two leaf-like lobes or *ligulae,* a dorsal *superior ligula* and a ventral *inferior ligula* and from each appears a bundle of setae or chaetae. The setae lodged in an epidermal pit, the *setigerous sac.* Each seta develops from a single large formative cell, present at the base of setigerous sac. The setae can be protruded, rotated in all the directions and retracted by special muscles within the parapodia. Each set is two jointed structure with a proximal *shaft* and a distal *blade.* The setae are of two types (a) Typical type has a long stout shaft and a short stout blade, (b) Long blade type has a small shaft but a long, pointed blade with one edge serrated. In *Heteronereis phase* a third type is also seen called as oar shaped. Here the blade becomes oar-shaped. Besides long setae in each part, there is a stouter, dark-coloured seta known as *aciculum,* which projects only for a short distance but not beyond the outer edge of the parapodium. Setae help in locomotion, while *aciculum* has a skeletal function. The dorsal margin of notopodium is produced into a short, cylindrical tactile appendage, the *dorsal cirrus* and a similar structure is produced at the ventral margin of the neuropodium, the *ventral cirrus* of the two cirri, the dorsal one is much larger than the ventral.

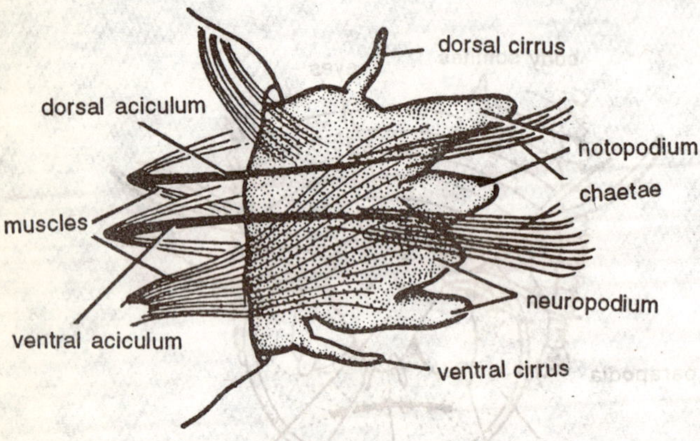


Fig. 16.6 Neanthes. Parapodium.

The parapodia are the largest in the mid-region of the body and decrease in size towards the anterior and posterior ends of the body. The first two parts of parapodia have no notopodial setae. The parapodia are highly muscular and primarily the organs of locomotion, used both in creeping and swimming. They are highly vascular and glandular and probably also serve as respiratory organs.

*Nephridiopores.* On the ventral side, near the base of parapodia, each segment bears a pair of minute excretory pores through which the nephridia communicate out.

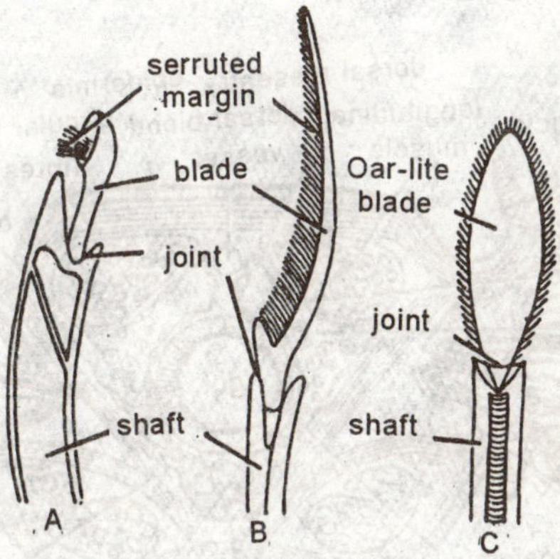


Fig. 16.7 Neanthes. Various types of setae. A=Typical B=Long blade, C=Oar shaped

**3. Pygidium or Tail.**

The last segment is tail which bears a terminal anus, a pair of long filamentous *anal cirri* and several minute sensory papillae. This segment does not bear parapodia.

**Body wall.**

The body wall consists of cuticle, epidermis, musculature and coelomic epithelium.

1. *Cuticle.* It is a thin, tough, chitinous and non-cellular layer covering the body externally. It is secreted by the epidermal cells underlying it. It is perforated by numerous pores through which the epidermal gland cells open out. The cuticle exhibits an irridescent lustrue due to the presence of two intersecting systems of fine striations in it.
2. *Epidermis.* It lies below the cuticle. It is made up of a single layer of columnar supporting cells resting on a basement membrane. Except the ventral side, the epidermis is thick at places. Many gland cells are found on the ventral surface near the base of parapodia. These gland cells secrete mucous which is used in forming the tubes in which the animal lives.

3. *Musculature.* Below the epidermis are present the muscles which are of three types :

(a) Circular, (b) Longitudinal and (c) Oblique muscles.

**(a) *Circular muscles.*** These form a continuous layer beneath the epidermis, which is slightly thicker on the ventral side. In the parapodia, the circular

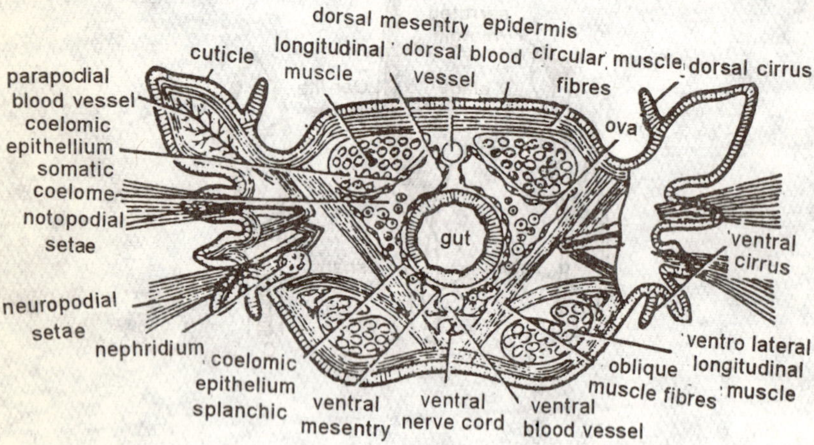


Fig. 16.8 Neanthes. T.S. of a segment with parapodia.

muscles are modified to form a complicated system of parapodial muscles comprising the *protractor* and*retractor* muscles. The protractors radiate from the bases of the setigerous sacs to the layer of circular muscles on all sides. The retractors extend from the outer part of the setigerous sacs to the dorso-lateral body wall. By the contraction of protractor muscles protrusion of acicula and central lobes of parapodia occur. The retractor muscles bring about withdrawl of the acicula-bearing lobes of the parapodia.

**(b) *Longitudinal muscles.*** The longitudinal muscles are present inner to the circular muscles and are much better developed. They do not form a complete layer but occur as four bundles, two dorso-lateral and two ventro-lateral. Contraction of this muscle makes the body short and stout.

**(c) *Oblique muscles.*** Two pairs of oblique muscles are found in each segment. They arise from the ventral body wall on the lateral sides of the never cord extending forwards to attach with the bases of the parapodia. The oblique muscles are divided into two parts, one part is attached to the dorsal part of the base of the parapodia and the other to the ventral part. Bending of parapodia in the vertical and horizontal planes is brought about by differential contraction of the various oblique muscle-bundles.

All the muscles are composed of smooth or unstriped muscles fibres.

4. *Coelomic epithelium.* It is the inner most layer of body wall and is composed of single layer of thin cells.

*Functions of body wall.* The body wall performs a number of functions which are as follows :

(i) The cuticle protects the body from mechanical injuries and dessication.

(ii) The epidermal gland cells secrete mucous which lines the burrows to prevent the collapse.

(iii) The epidermis of dorsal side is highly vascular which helps in respiration.

(iv) The sensory cells make the animal sensitive to external stimuli.

(v) Muscles help in movements and locomotion.

(vi) The coelomic epithelium secretes coelomic fluid.

**Coelom**

Between the body wall and gut intervenes a fairly spacious *coelom* or *perivisceral cavity.* This is not, however, a continuous cavity but, on the contrary, is broken into a series of coelomic chambers by the intervention of intersegmental septa, corresponding to the external grooves. However, the septa are incomplete, especially round the alimentary canal, so the continuous coelomic chambers communicate with each other. The coelom is further partitioned by the presence of a pair of oblique muscles running obliquely dorsoventrally from the ventral to the dorsolateral body wall. Ventrally, these muscles are inserted into the body wall below the ventral nerve cord; dorsally they are attached to the body wall near the bases of parapodia. The perivisceral cavity is lined by coelomic epithelium or peritoneum. Peritoneum that lines the inner surface of body wall is called *parietal peritoneum* and that covers the outer surface of viscera is called *visceral peritoneum.* A double fold of peritoneum called *mesentery* suspends the alimentary canal into the coelom from mid-dorsal body wall.

The coelom is filled with *coelomic fluid.* This fluid contains amoeboid corpuscles. The coelomic fluid more or less freely circulates throughout the body.

*Functions of coelom.* The coelom performs following functions :

1. Food and oxygen are supplied to different parts of the body with the help of coelomic fluid.

2. It protects the body from external shocks.

3. The coelomic fluid also helps in the protrusion of parapodia during locomotion.

**Locomotion**

Two structural elements are especially important in the locomotion of *Neanthes,* the longitudinal muscles and parapodia. The longitudinal muscles do not form a continuous layer. Instead they are broken up into two pairs of blocks, one pair dorsal and other ventral. Because of this the muscles of the two sides of a segment can be in opposite phases, one contracted and other relaxed so that the passage of waves of contractions along the body can be thrown it into lateral undulations. Assisted by the parapodia the lateral undulations displace the body forward.

The parapodium in *Neanthes* is clearly quite versatile in its mode of action. The animal remains dependent upon the hydrostatic skeleton to pro-

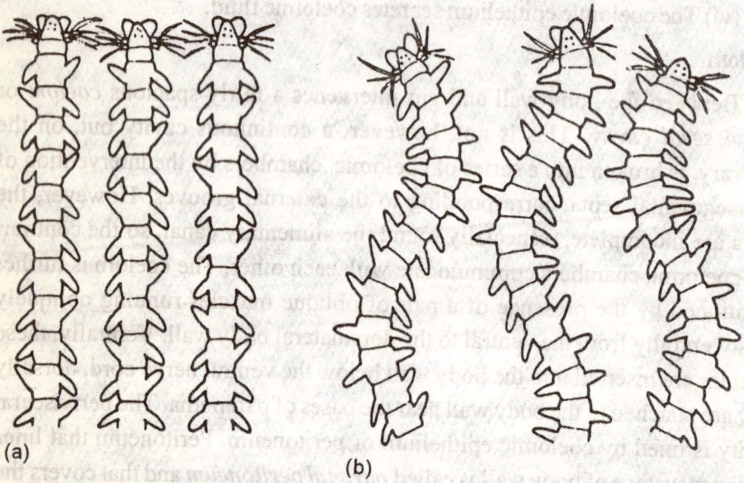


Fig. 16.9 Diagrammatic representation of locomotion in Neanthes. (A) Slow crawing crawling . (B) Rapid crawling.

vide firm basis for muscular action. The properties of musculature and parapodia interact to provide for several different types of movements. These movements are of three types :

1. Slow creeping
2. Fast creeping

3. Swimming

1. *Slow creeping.* Slow creeping depends almost entirely upon the use of parapodia as a series of levers. Initially a parapodium moves forwards, with its tips lifted from the ground and with the aciculum withdrawn, this is called the *recovery stroke* or *preparatory stroke.* At the end of this stroke the parapodium makes contact with the substratum, the aciculum is protruded, and the oblique muscles contract so that the body is pulled forwards.The parapodium is brought to backwards during this phase which is called *effective stroke* or *power stroke.* The movements of parapodia are so integrated that the two members of any one segmental pair alternate with each other in phase. Moreover, the movement of any one parapodium begins slightly after that of the one next behind it.

2. *Fast creeping.* Slow creeping readily passes into rapid creeping. This involves a similar rhythmic pattern of waves, but with the difference that longitudinal muscles are now of primary importance. They contact serially in parallel with the movement of parapodial. The oblique muscles are now probably of much less importance than in show creeping.

3. *Swimming.* Here the pattern of movement is essentially the same as that of fast creeping. Sinusoidal waves of the body interact with parapodial movement, but there is a marked increase in length of the wave and also in their amplitudes and frequency.

**Digestive System.**

It includes alimentary canal and associated digestive glands.

***Alimentary Canal.*** It is a complete straight tube of varying diameters extending through the entire body from the mouth in front to the anus behind. It can be divided into three regions :

(i) *Stomodaeum* or *foregut* : It is the anterior region of the alimentary canal, lined internally by ectoderm and cuticle and comprises the buccal cavity and pharynx.

(ii) *Mesenteron* or *midgut* : It is middle region of the alimentary canal and it is lined internally by endoderm and comprising the oesophagus and stomach-intestine..

(iii) *Proctodaeum* or *hindgut* : It is the posterior-most part of the alimentary canal, lined internally by ectoderm and cuticle and comprising only rectum.

(a) *Mouth.* It is a transverse slit on the ventral side of peristomium. It is over hung by the prostomium. It leads behind into buccal cavity.

(b) *Buccal cavity and pharynx.* The anterior part of the buccopharyngeal regions of *Neanthes* forms an eversible *proboscis.* It is sorrounded by several coats of muscles. The buccal cavity is situated in the peristomium and leads

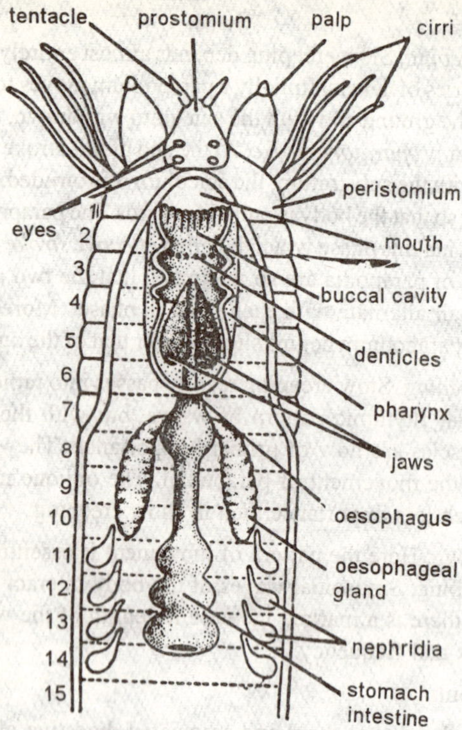


Fig. 16.10 Neanthes. Anterior end dissected dorsally to show alimentry canal.

behind into the pharynx which extends upto 4th segment. Both of them are lined internally by cuticle. The cuticle of buccal cavity is thckened at various points into dark brown *denticles* or *paragnaths*. The pharynx is thickwalled and muscular. In its posterior part are found a pair of large, powerful, chitinous and movable jaws. Each jaw has a hollow *base* and an incurved, pointed, somewhat notched *apex*. The inner margins of the jaws are serrated. The buccal cavity and pharynx are wrapped in a common muscular coat and together they extend up to the fourth or fifth trunk segment. They can be fully everted to form the *proboscis* or the *introvert*. This operation exposes the tips of the jaws for capturing the prey.

(c) *Oesophagus*. The pharynx narrows posteriorly to lead into the oesophagus. It is a narrow tube extending through five segments behind the pharynx. A pair of long, unbranched, sacculated glandular pouches, the *oesophageal glands* or *caeca*, open into it laterally at the anterior end. The oesophagus opens into stomach-intestine; the opening between the two is guarded by a sphincter muscle. A distinct stomach is absent.

(d) *Stomach-intestine.* It is a wide tube extending through the remaining length of the body. It is straight, thin-walled tube and segmentally constricted by septa. The stomach-intestine is the principal site of digestion and absorption. The stomach-intestine opens into a rectum which lies in the last segment.

(e) *Rectum.* It is the last part of alimentary canal, which opens posteriorly by the anus. It is very short chamber and occupies the last body segment or pygidium.

*Digestive glands.* As described above that into the oesophagus open a pair of laterally placed caecal or pouches called *oesophageal glands.* They secrete digestive juices probably proteolytic in nature. Besides micro-scopic gland cells are present in stomach-intestine. These glands secrete digestive juices that contain enzymes to digest food.

*Histology.* The gut wall consists of an outer most layer of serosa or visceral peritonium, followed by longitudinal muscles, circular muscles and an

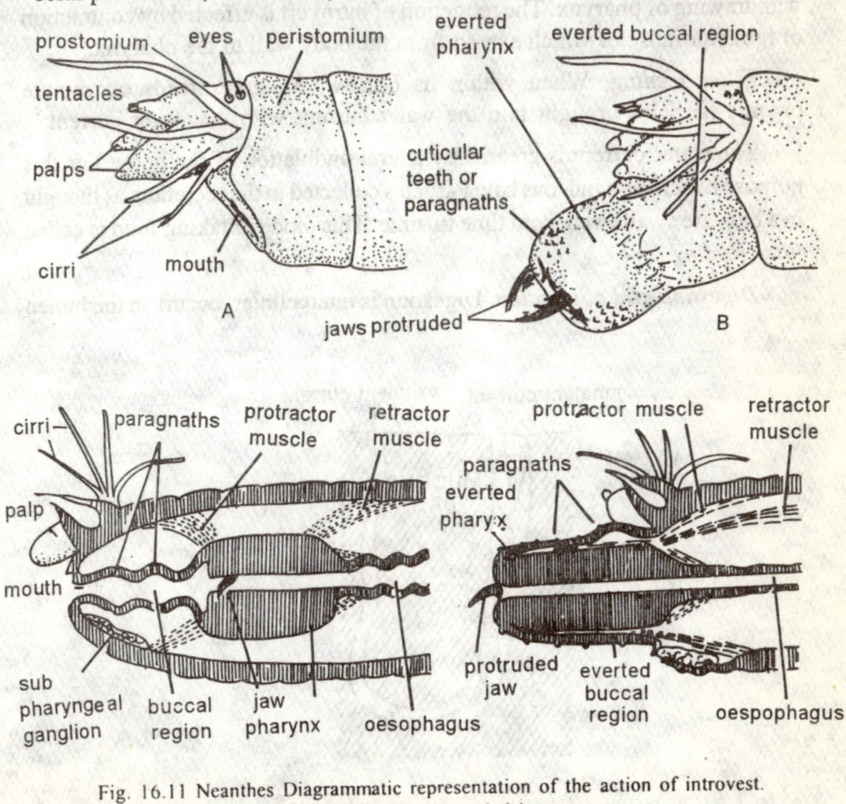


Fig. 16.11 Neanthes Diagrammatic representation of the action of introvest. (A) = at rest (B)= protruded introvest.

enteric epithelium which is endoderm in the case of mid-gut and ectoderm in the case of fore-gut and hind-gut.The ectoderm is lined by cuticle.

*Food. Neanthes* is a *carnivorous* and *voracious* animal and feeds on small animals, such as molluscs, crustaceans, sponges and others. It is also called clam-worm, because it is generally found, where clams occur. But it never eats the living clams as they are of bigger size. However, it feeds on small animals.

*Feeding.* To capture the prey, *Neanthes* everts its stomodaecum. The *eversion* of the pharynx is brought about by the contraction of *protractor muscles* extending from the introvert to the wall of peristomium and also by the pressure of coelomic fluid. The muscles contract and the coelomic fluid gets filled in the buccal region, so that the buccal cavity everts out like a bag and the pharynx protrudes out. The two pharyngeal jaws are thrust with their dentate margin facing outwards. The jaws come close together and grasp the prey. The food is now dragged inside the burrow and then swallowed by the withdrawing of pharynx. The retraction of introvert is effected by contraction of retractor muscles which extend from the body wall to the pharynx.

*Filter feeding.* When within its burrow, *Neanthes* feeds on minute organic particles brought into the water current with the water current.

The water current is created by lateral undulation of the body. For this purpose it secretes a mucous bag. Particles collected in the secretion, as thought in a bag, are swallowed from time to time. This mode of taking food is called *filter feeding.*

***Digestion and absorption.*** Digestion is intercellular, occurs in the lumen

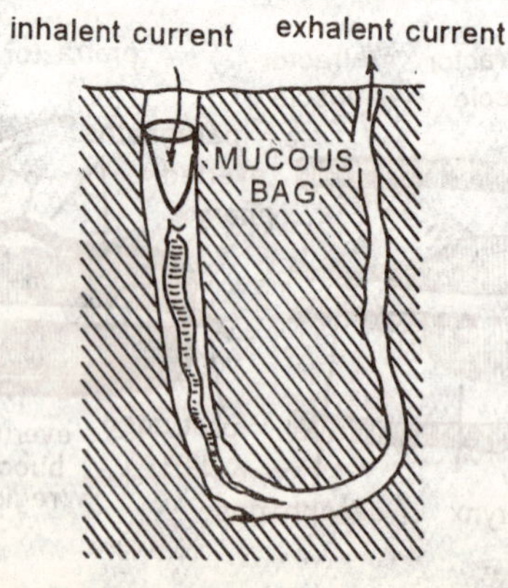


Fig. 16.12 Neanthes. Filter feeding in burrow

of mid-gut. It is brought by the enzymes secreted by oesophageal glands and gland cells of stomach-intestine. The digestive enzymes are *proteases, amylases and lipases.* Absorption mainly occurs in the stomach-intestine.

**Respiratory System.**

In *Neanthes*, respiratory organs are absent. Respiration occurs by diffusion through body surface. The parapodia with their rich-blood supply and body wall with its plexuses of blood vessels subserve the function of respiration. When the blood passes through these net-works, it obtains oxygen from the surrounding water and gives out carbon dioxide. Blood contains a respiratory pigment *heamoglobin*, which increase the absorptive capacity of blood for oxygen and carbon dioxide.

**Circulatory System.**

In *Neanthes* a closed type of circulatory system is present which helps in transportation of materials from one region to other region of body. The function of transporting materials is performed by the blood. The blood flows in a system of tubes, the blood vessels. The blood vessels are of two types distributing and collecting. Both the vessels are connected to each other by a system of extremely fine vessels, the *capillaries.* By the peristaltic contraction, the blood is maintained in constant circulation.

*Blood vessels.* There are three longitudinal vessels which are as follows:

(i) *Dorsal vessel.* It runs from one end of the body to the other situated in the dorsal mesentery above the alimentary canal. Its walls are highly contractile which drive blood from posterior to the anterior end. It acts as a collecting vessel, but in the 5th segment it bifurcates and both the branches from a plexus to supply the oesophageal wall. Thus dorsal vessel collects blood from the stomach-intestine region by means of two pairs of *dorso-intestinal* or *efferent intestinal vessels* in each segment. It also collects blood from the body wall, parapodia and nephridia through lateral vessels.

(ii) *Ventral vessel.* It is main distributing vessel. It is situated below the gut and in it the blood flows anteriorly backwards. In the last segment it communicates with the dorsal vessel by a simple *circum-rectal ring.*

In each segment, behind the oesophagus, the ventral vessel is connected with the dorsal vessel by a pair of loop-like *lateral commissural vessels.* Each of these vessels shortly divides into *afferent* branches, which carry blood to the body wall by *afferent cutaneous,* nephridium by *afferent nephridial* and parapodium by *afferent parapodial* of its side. In these organs, the afferent branches break up into a network of capillaries which reunite to form the corre-

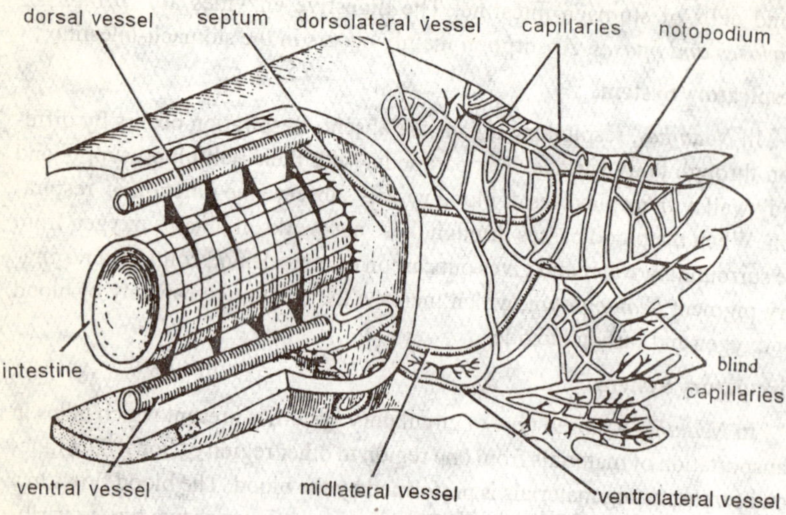


Fig. 16.13 Neanthes. Digrammatic representation of blood vessels in T.S. of a segment.

sponding efferent branches. The efferent branches finally unite and open the dorsal vessel.

In each segment, behind the oesophagus, the ventral vessel also supplies blood to the gut wall through a pair of *afferent intestinal* or *ventro-intestinal vessels*. These break up into a capillary network in the gut wall. As stated above that the blood from gut is collected by a pair of dorso-intestinal or efferent intestinal vessels and poured into dorsal vessel.

The ventral vessel is a distributing vessel except in the anterior region, where it is a collecting vessel and collects blood from oesophageal wall.

(iii) *Peri-neural vessel.* It surrounds the nerve cord. It collects blood from the ventral body wall and sends it to the ventral vessel.

*Blood.* The blood is red in colour due to the presence of dissolved respiratory pigment, the haemoglobin. In the blood are present blood corpuscles. The blood corpuscles are colourless, nucleated, amoeboid and resemble the leucocytes of vertebrate blood.

Blood performs a number of functions. It helps in transporting materials, removal of waste products, removing of microorganisms and respiration i.e. exchange of gases.

**Excretory system.**

Excretory system consists of metamerically or segmentally arranged pairs of coiled and ciliated tubes, called *segmental orgnas* or *nephridia*. Each segment, except the peristomium and the pygidium, possesses a pair of nephridia.

Each nephridium is made up of a syncytial mass of protoplasm containing many nuclei. Each nephridium is divisible into two parts :

1. Body of nephridium
2. Neck of nephridium

1. *Body of nephridium.* It is a roughly oval, compact gland shaped structure which is placed transversely in each segment and contains a coiled ciliated tube in it. The cilia are not found in its last part called *terminal duct.* The terminal duct opens to the outside by a minute aperture, the *nephridiopore.* The nephridiopore is situated on the ventral surface of parapodia near the ventral cirrus. Nephridiospore can be opened or closed.

2. *Neck of nephridium.* The distal part of nephridium is called the neck. In the neck the wall of the ciliated tube is expanded. This part crosses the septum of segment to open into the coelomic cavity of the other where it forms a ciliated funnel-like structure called *nephrostome.* The wide margin of

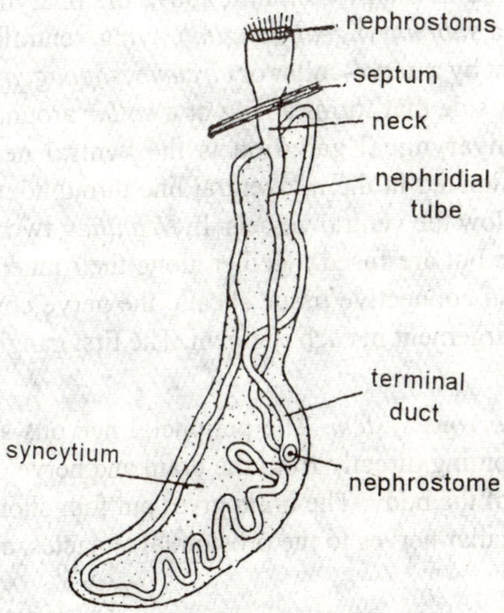


Fig. 16.14 Neanthes. A nephridium.

nephrostome is provided with long delicate cilia. Such an open type of nephridium with a ciliated nephrostome is called a *metanephridium.*

In each segment of the body close to the dorsal wall, there is found a pair of *ciliated organs*. These are considered to be the gonoducts of mesodermal origin, which, however, have lost their external openings and have ceased to function as such.

Excretion is possible by two methods. The outer surface of each nephridium is densely covered with blood capillaries. The cells forming the ciliated tube collect the waste materials from the blood and eleminate it through the nephridiopores. Secondly, ciliated funnel helps in removal of dead coelomic corpuscles, which have been perished by bacteria or some foreign bodies.

**Nervous System.**

*Neanthes* is an active animal, therefore, a well developed nervous system is present. It consists of :

(i) Central nervous system

(ii) Peripheral nervous system

(iii) Visceral nervous system

(i) *Central Nervous System.* Central nervous system consists of brain or *cerebral* or *supra-pharyngeal* ganglion, which is a large bilobed mass of nervous tissues situated in the prostomium above the pharynx in the head. This is connected to a *subpharyngeal ganglion,* lying ventrally to the pharynx in the 3rd segment by a nerve-collar or *circumoesophageal connective* or commissure on either side that forms a *ring* or a *collar* around the pharynx. Following the subphyaryngeal ganglion is the ventral nerve, which is ganglionated and is situated in the midventral line throughout the complete length of the body below the ventral vessels. In *Neanthes,* two ventral nerve-cords are not separate but are fused together along their inner sides and are enclosed in a common connective tissue sheath. the nerve cord is provided with a ganglionic enlargement in each segment. The first ganglion is situated in the 4th segment.

(ii) *Peripheral nervous system.* The peripheral nervous system mainly includes the nerves coming directly from the brain and nerve cord to supply the different regions of the body. The brain gives out four short optic nerves to the eyes, two tentacular nerves to the prostomial tentacles, and two palpal nerves to the palps.

The ganglia of the circumpharyngeal connectives supply two pairs of *tentacular nerves* to the two ventro-lateral pairs of peristomial tentacles. The ganglia of the accessory connective innervate the two dorso-lateral pairs of prostomial tentacles by a pair of tentacular nerves. From the subpharyngeal ganglion a pair of nerves goes to the first parapodia Three pairs of nerves arise from each ganglion of the nerve cord, the first goes to the anterioı segment and the remaining two pairs innervate the organs of the segment, parapodia and body wall.

(iii) *Visceral nervous system.* It includes ganglia situated above the pharynx and a few fine nerves. These ganglia and nerves are connected on

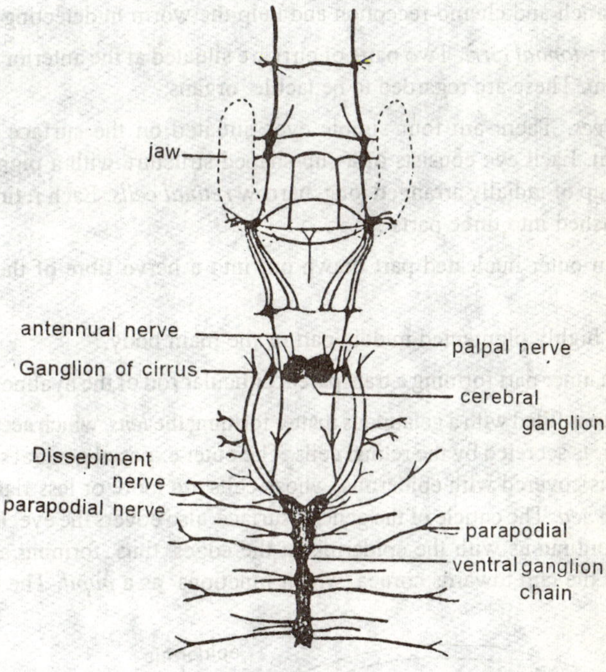


Fig. 16.15 Neanthes.Nervous System.

one side with the brain and on the other side with the ventral surface of the circumpharyngeal connectives.

**Sense organs.**

In *Neanthes* the sense organs are specialized and well developed. Following types of sense organs are present :

1. *Prostomial palps.* On either side of the prostomium from ventroleteral surface originate a pair of short, thick and muscular palps. Each palp is made up of two pieces; a basal piece, which is large and a terminal small piece. The terminal piece is capable of retraction into the basal piece. These are tactile in function.

2. *Prostomial tentacles.* These are a pair of small, cylindrical arise from the anterior border of the prostomium. The surface of tentacles bear numer-

ous sensory spiral organs, each consisting of about 100 photoreceptive cells, spirally arranged within a cuticular pit. The prostomial tentacles are probably tactile.

3. *Nuchal organs.* These are a pair of small pits in the posterior part of the prostomium on its dorsal side. They are lined with ciliated epithelium and gland cells. They are connected with the posterior part of brain. They serve as organs of smell and chemo-reception and help the worm in detecting prey.

4. *Peristomial cirri.* Two pairs of cirri are situated at the anterior end of peristomium. These are regarded to be tactile organs.

5. *Eyes.* There are four simple eyes situated on the surface of the prostomium. Each eye consists of a cup-shaped structure with a pigmented wall made up of radially arranged long, narrow *retinal cells*. Each retinal cell is distinguished into three parts :

(a) An outer nucleated part drawn out into a nerve fibre of the optic nerve.

(b) A highly pigmented middle part or the main body.

(c) An inner part forming a transparent cuticular rod of the hyaline layer.

The cup is filled with a gelatinous matter forming the *lens,* which accoridng to *Andrews,* is secreted by the retinal cells. The outer exposed general surface of the eye is covered with epidermis, whose cells are more or less flattened, forming *cornea.* The cuticle of the general surface also covers the eye. Retinal cells are continuous with the epidermis at the edges, thus, forming a small opening of the cup towards cornea, which functions as a *pupil.* The retinal

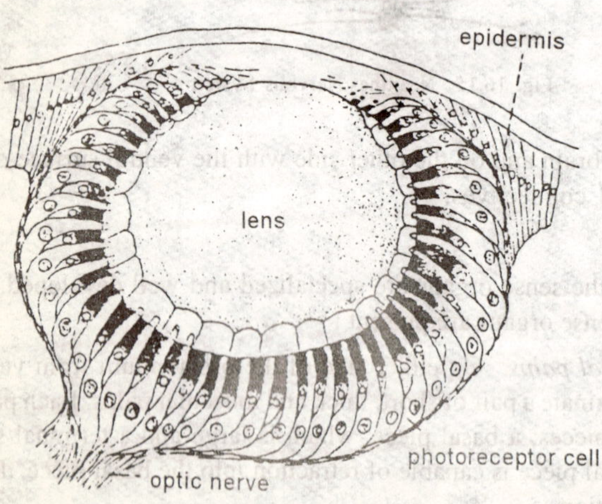


Fig 16.16 Neanthes. V.S. of eye.

cells join by means of processes with the brain.

**Reproducitve System.**

*Neanthes* is unisexual and the male and female may exhibit slight sexual dimorphism. The gonads i.e. ovaries and testes are temporary organs, appearing only during-breeding season. The breeding season is summer. These gonads disappear after the liberation of gametes. In male *N. dumerilli* there is only one pair of testes lying in each segments. But in *N. virens* the testes extend in many segments around blood vessels. The gonoducts are absent.

During breeding season the cells which detach from the coelomic epithelium in male constituting the spermatogonium becomes disseminated throughout the coelom. These cells undergo rapid divisions to form spermatids which undergo spermiogenesis to form spermatozoa. A mature spermatozoon consists of a minute rod-shaped *head* and a vibratile *tail.* Similarly the oogonia in female separate into the coelomic fluid and after divisions from the ova which are found floating in it.

The gametes are escaped through the nephridiopores or by the rupture of body wall. A pair of dorsal *ciliated organs* are found in each segment of the body. Each ciliated organ is a small, ciliated tract of folded funnel shaped structure which opens into coelom by means of a wide aperture. It is believed

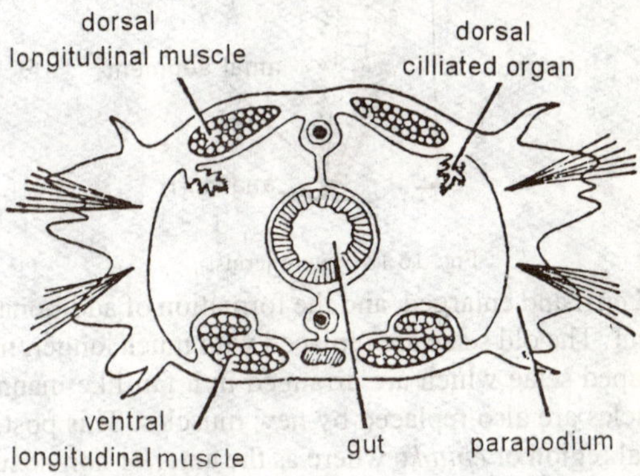


Fig. 16.17 Neanthes. T.S. of a segment showing ciliated organs.

that they are like coelomoducts and act as gonoducts. During breeding season, they open to the outside by temporary apertures.

**Heteronereis phase.**

*N. dumerilli,* undergo remarkable changes in external form and behaviour during breeding season. At sexual maturity, most of the posterior segments are filled with gametes. The parapodia are also modified. The notopodia,

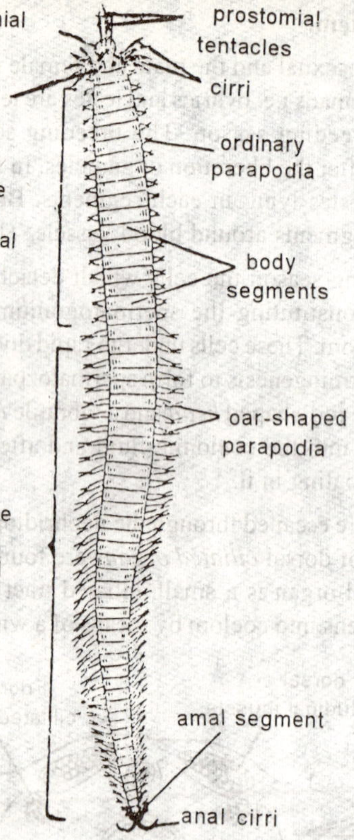


Fig. 16.18 Heteronereis

neuropodia and cirri and enlarged, and the formation of additional foliaceous outgrowths occur. The old setae are replaced with much longer, more numerous and oar-shaped setae which are arranged in a fan-like manner. The old parapodial muscles are also replaced by new muscles. This posterior region of body is sexual region or *epitoke* where as the anterior non-sexual region is referred to as *atoke.*

Following characteristics are noted in heteronereis phase :

1. Heteronereis does not crawl on sea bottom but it swims actively on water surface.
2. The eyes become enlarged and the animal properly precieves the light stimulus.
3. The peristomial cirri are comparatively larger than *Neanthes.*
4. The intestine becomes compressed and functionless due to much devel-

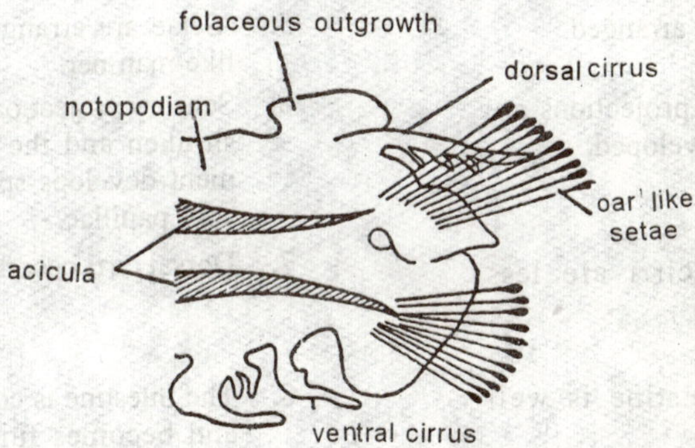


Fig. 16.19 Parapodium of Heteronereis.

opment of the gonads.

5. Special sensory papillae are formed on the anal segment.

**Swarming**

*Neanthes* swarms on moonless night during summer months after sunset. The swarms rapidly about the females. Females produce some substance called *fertilium* (the fertilizin produced by eggs) which attracts males. This fertilizin is responsible for stimulating and shedding of gametes (*Barnes*, 1968). Swarming continues for about an hour or so. Females die after liberation of eggs and probably males also.

**Table 16.1**

Differences between *Neanthes* and *Heteronereis*

| *Neanthes* | *Heteronereis* |
|---|---|
| 1. *Neanthes* is a non-sexual sluggish creature, creeping at the bottom of sea. | 1. Heteroneris is an active swimmer on the surface of water. |
| 2. Eyes normal and the worm is less sensitive to light. | 2. Eyes become enlarged and conspicuous so that the heteroneris is very sensitive to light. |
| 3. The parapodia are thin, flap-like and non-vascular. | 3. The parapodia of the posterior or the sexual part are large, highly vascular and develp |

| | |
|---|---|
| | leaf-like outgrowths. |
| 4. The setae on the parapodia are normal. | 4. The normal setae are replaced by large, flattened and oar-shaped setae. |
| 5. Setae are arranged. | 5. Setae are arranged in fan-like man ner. |
| 6. Sensory projections are well Developed. | 6. Sensory projections beocme shruken and the anal segment develops special sensory papillae. |
| 7. Dorsal cirri are less curved. | 7. Dorsal cirri become altered. |
| 8. The intestine is well formed. | 8. The intestine is compressed and becomes functionless due to development of gonads. |
| 9. Less developed for swimming. | 9. Better adopted for swimming. |

**Fertilization**

Fertilization is external and takes place in sea water except *Platynereis megalops* where internal fertilization takes place. Here the male inserts its anal segment in the mouth of female. The sperms are passed into the coelom as the intestine is eroded. In the coelomic fluid fertilization takes place.

**Development**

The development starts after fertilization of eggs. Each egg is covered in a thick transparent gelatinous envelope. Inside this membrane there are two membranes :

(i) Quter membrane is very thin and delicate.

(ii) Inner membrane is thicker and distinctly striated radially, therefore, it is known as *zone radiata.* The cytoplasm contains numerous yolk spherules and oil drops.

Fertilization produces several changes in the egg. The yolk spherules move at the lower side or vegetal pole, leaving a granular protoplasmic area at the upper or animal pole.

**Cleavage.** The first two cleavages of the zygote are the equal and vertical and result in four cells or *blastomeres* lying in the same plane. The cleavage is *determinate,* i.e., the fate of the blastomere is fixed and after the four-cell stage, each blastomere gives rise to only one quardrant (quarter) of the em-

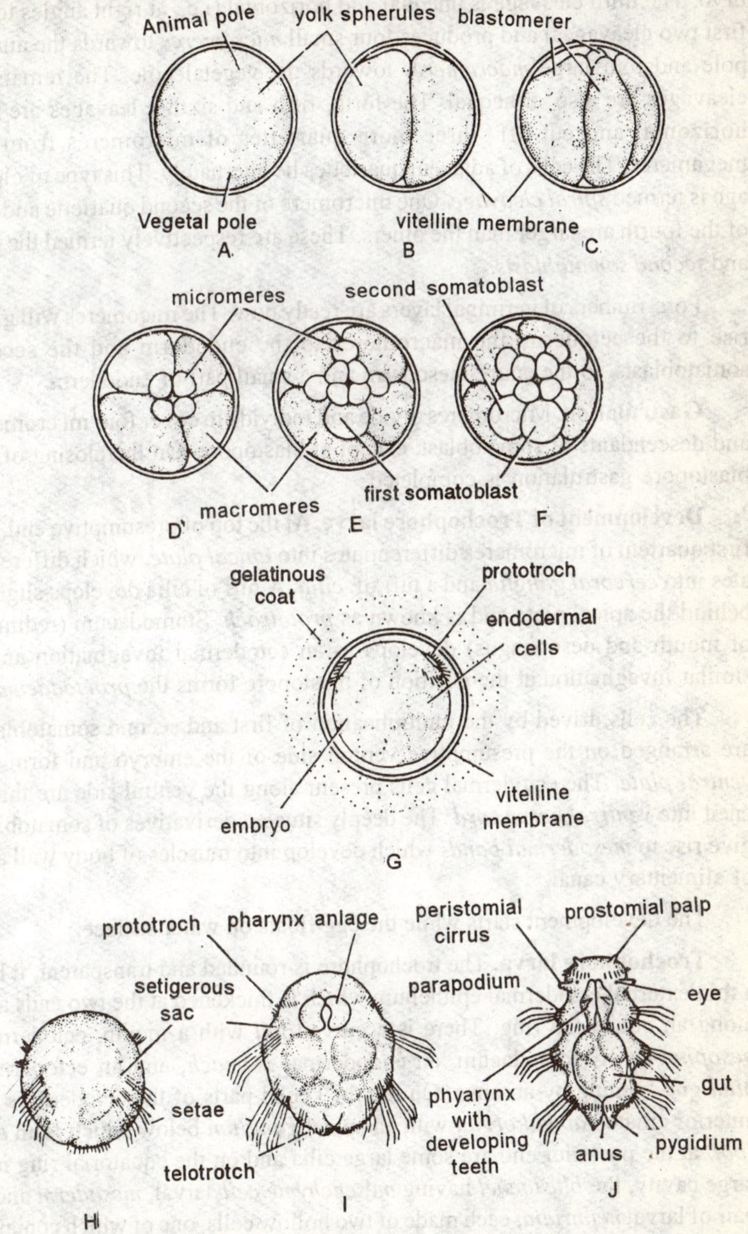


Fig. 16.20 Neanthes. Stages in development. A-Fertilized egg, B- Two cell stages C-two cell stage D-Eight cell stage, E-Ist somatoblast formed, F-2nd somatoblast formed G Young trocophore before hatching, H-Post trocophore larva with three setigerous segments, I-Older larva, J-Later larv, after three weeks.

bryo. The third cleavage is unequal and horizontal (i.e., at right angles to the first two cleavages) and produces four small *micromeres* towards the animal pole and four large *macromeres* towards the vegetal pole. The remaining cleavages are also u nequal. The forth, fifth and sixth c leavages are also horizontal and cut off three more quartettes of micromeres from the megameres. The cells of adjacent quartettes lie alternately. This type of cleavage is termed *spiral cleavage.* One micromere of the second quartette and one of the fourth are larger than the others. These are respectively termed the *first* and *second somatoblasts.*

Fore-runners of germinal layers are ready now. The micomeres will gives rise to the ectoderm, the macromeres to the endoderm and the second somatoblasts to the entire mesoderm and a small part of endoderm.

**Gastrulation.** Micromeres divide and redivide to cover four mi cromeres and descendants of somatoblast, except at blastopore. On the closing of the blastopore gastrulation is completed.

**Development of Trochophore larve.** At the top of presumptive end, the first quartent of micromeres differentiates into *apical plate,* which differentiates into *cerebral ganglia* and a tuff of *cilia*. A tuff of cilia develops slightly behind the apical plate and is known as *prototroch.* Stomodaeum (rediment of mouth and oesophagus) develops as an ectodermal invagination and a similar invagination at the position of blastopore forms the *proctodaeum.*

The cells drived by the multiplication of first and second somatoblasts are arranged on the presumptive ventral side of the embryo and form the *ventral plate.* The ectodermal cells present along the ventral side are thickened into *ventrral nerve cord.* The deeply situated derivatives of somatoblast give rise to *mesodermal bands* which develop into muscles of body wall and of alimentary canal.

The development starts while the egg floats on water surface.

**Trochophore larva.** The trochophore is rounded and transparent, it has a thin external ectodermal epithelium which is thickened at the two ends and along an equatorial ring. There is a curved *gut* with a mouth, ectodermal *oesophagus* or stomodaeum, an endodermal *stomach*, and an ectodermal *hind gut* opening by an anus. On the thickened parts of the ectoderm is an anterior ciliated *apical organ* with an apical *ganglion* below which is an *eye spot,* at the posterior end are some large cilia and on the equatorial ring is a large cavity, the *blastocoel* having *mesenchyme cells*larval, *mesoderm* and a pair of larval *nephridia,* each made of two hollow cells, one of which contains a flame of cilia. There lies an otocyst near the nephridia. The trochophore is a small, ciliated larva which is without segmentation and also lacks coelom.

An important point to be noted in the development of *Neanthes* is that the fate of every cell in the early cell in the early embryo is predetermined. Study of following the fate of cells is called *cell-lineage*.

In *Neanthes,* the trochophore stage is passed within the egg membrane and the ciliated *nectochaetes larva* comes out of the egg. It prossesses three segment and bristles.

**Metamorphosis.** The nectochaetes larva swims about freely for a few days, feeding on microorganisms and then undergoes metamorphosis to become a young worm. Metamorphosis begins with the elongation of the lower region becomes constricted into segments which soon develop bristles. The ciliated bands disappear and the upper part of the larva becomes the head. The part of larva behind prototroch elongates and develops on its sides two pair of invaginations called *setigerous sacs.* Long setae grow out from the setigerous sacs. A third pair of setigerous sac with projecting seate is also added. The mesenchymal cells multiply and form a pair of mesodermal bands one on either side of the gut. Superficial division of body into segement is accompanied by internal segmentation of mesodermal bands into three pairs of somites. Each somite develops a cavity, the *coelom.* Both the somites of each segment unite above and below the gut forming the dorsal and ventral

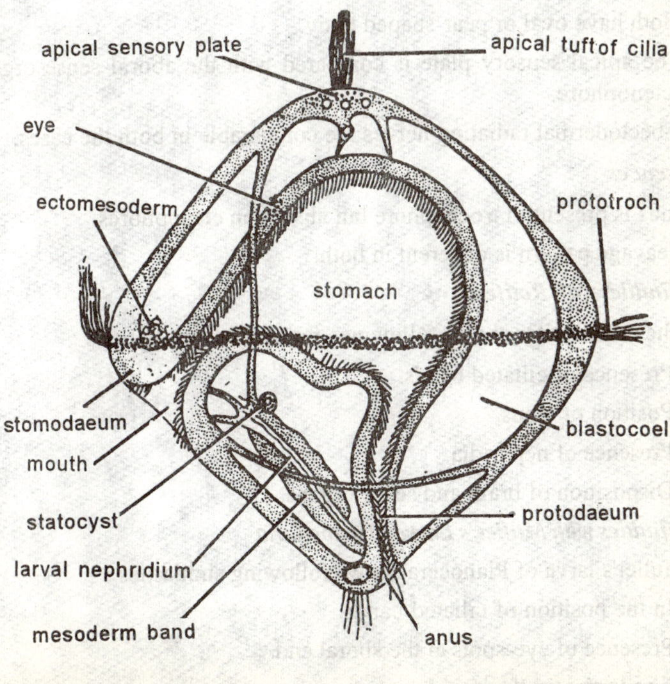


Fig. 16.21 Neanthes. A typical trochophore larva.

mesenteries. Later on the ventral mesentray is disappeared. The coelom obliterates the blastocoel. The part anterior to prototroch modified into the prostomium. The part behind prototroch gives rise to the peristomium. The part bearing the anus is continually pushed backwards as new segments are formed in front of it. Tentacles, palps, parapodia with their cirri and permanent setae are formed. Larval setae are dropped. Larval nephridia are replaced by permanent nephridia. The young worm settles down on the bottom and changes into the burrowing adult worm.

**Significance of trochophore**

The trochophore larva has much significance since it is found widely in different groups such as Turbellaria, Mollusca, Bryozoa, Nemertinea, Annelida etc. It has much resemblence with adult rotifers. The early development of the animals of these groups is almost identical and their trochophore stages are also similar in many respects. The differences in their development arise after the trochophore stage. The affinities of annelid trochophore can be discussed as follows :

***I. Affinities with adult ctenophore***

*Similarties*

1. Both have oval or pear-shaped body.
2. The apical sensory plate is compared with the aboral sense organ of ctenophore.
3. Subectodermal radiating nerves are comparable in both the cases.

*Differences*

1. Anus is present in trochophore but absent in ctenophores.
2. Cleavage pattern is different in both.

***II. Affinitiies with Rotifera***

The adult rotifer shows followings features of simillarities :

1. Presence of ciliated bands.
2. Position of anus.
3. Presence of nephridia.
4. Disposition of brain and sense organs.

***III. Affinities with Muller's Larva of trubellaria***

Muller's larva of Planocera shows following similarties :

1. In the position of ciliated bands.
2. Presence of eye-spots at the aboral end.
3. Similarties in the development.

♦♦♦

# 17

# HIRUDINARIA GRANULOSA

Class Hirudinea (Lat. *Hirudo* = a leech) form a relatively small but well-defined group of segmented worms which are generally adapted to an ectoparasitic mode of life. These are commonly called 'leeches'. About 300 species of leeches are known. They abound in tropical and sub-tropical regions. Most of them are freshwater, while some are marine or terrestrial. Majority are blood suckers and parasitic, others are predators, feeding on worms, snails, insect larvae etc.

Genus *Hirudinaria* includes four Indian species namely, H.*viridis*, *H.manillensis*, *H.javanica* and *H.granulosa*. *H.granulosa* is the more common species and commonly known as cattle leech.

**SYSTEMATIC POSITION**

| | | |
|---|---|---|
| Phylum | - | Annelida |
| Class | - | Hirudinea |
| Order | - | Gnathobdellida |
| Genus | - | *Hirudinaria* |
| Species | - | *Granulosa* |

**Habits and Habitat**

*Hirudinaria* is a common Indian leech found in India, Burma, Sri Lanka, Bangla Desh and Pakistan. It occurs in freshwater ponds, lakes, slow streams etc., where it either swims by vertical undulations or grips objects with its suckers and moves by looping. It is sanguivorous in diet i.e. feeds on the blood. A single full meal can last for more than a year.

Leech is bisexual but always cross-fertilization occurs. Cross fertilization involves pairing of two individuals in head-to-tail position. Eggs are deposited in cocoons. Development is direct i.e. without any free larval stage in the life history.

**External Morphology**

***Shape and size.*** Body is elongated dorsoventrally flattened and vermiform with flat ventral and convex dorsal surfaces. A fully stretched leech appears to be ribbon-shaped, whereas in contracted condition it is cylindri-

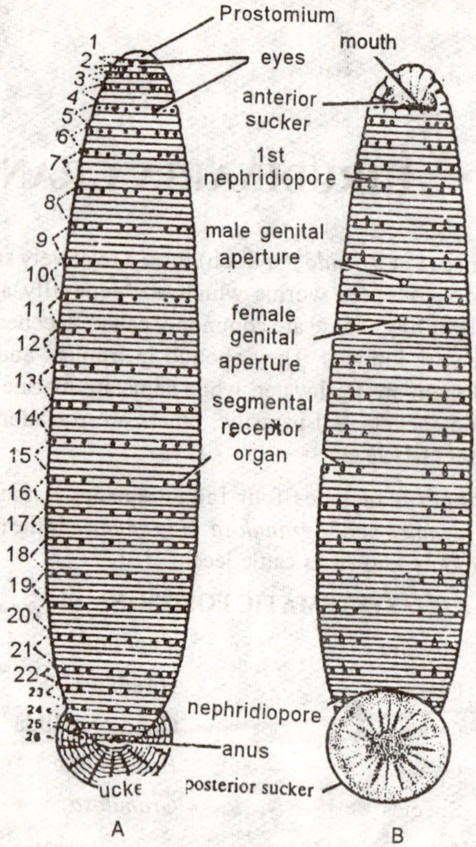


Fig.17.1 *Hirudinaria granulosa.* (A) Dorsal view (B) Ventral view

cal. Often it measures about 4-6" but might reach upto 12" or more.

***Colouration.*** The body is beautifully coloured with characteristic markings. The dorsal surface is generally olive-green and the ventral surface is orange-yellow or orange-red and the two sides bear distinct stripes of orange or yellow and black. On the dorsal side is a median longitudinal black strip.

***Division of Body.*** Body is divided into 33 segments but the external surface of each segment is further divided into annuli by transverse grooves. The number of annuli is different in different segments. The entire body can be distinguished into 6 regions.

1. *Cephalic regions.* It is composed of first five (Ist-5th) segments and a prostomium, of which Ist and 2nd are uniannulate, 3rd biannulate, and 4th and 5th are triannulate. Each segment possesses a pair of eyes. On the ventral

surface of cephalic region is present anterior sucker and mouth.

2. *Preclitellar region.* It comprises of 6th, 7th and 8th segments. The 6th segment is triannulate and 7th and 8th consists of five annuli. Ventrally, the last annualus of each segment possesses a pair of nephridiopores.

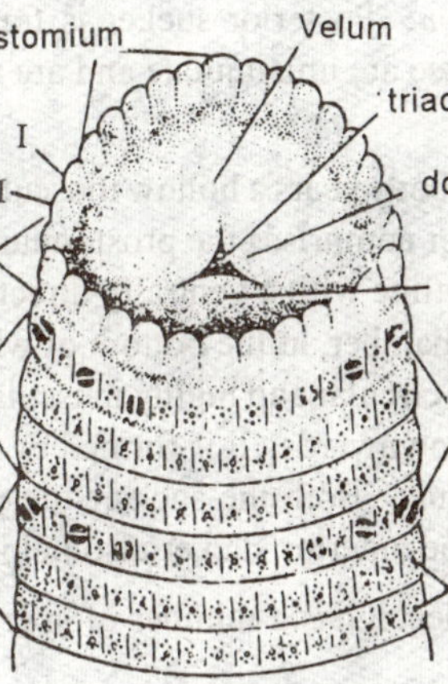


Fig. 17.2 *Hirudinaria.* Cephalic region in ventral view.

3. *Clitellar region.* It extends from 9th to 11th segments. Each segment consists of five annuli and carries a pair of nephridiopores. Clitellum is distinct only during reproductive season.

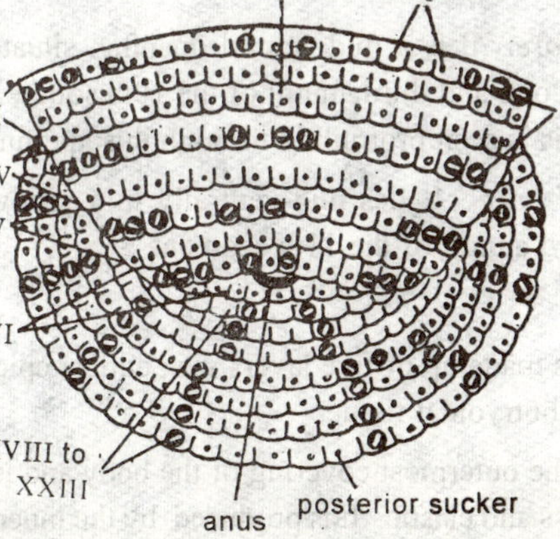


Fig. 17.3 *Hirudinaria* Caudal region in ventral view.

4. *Middle region*. It extends from 12th to 22nd segment. All the thirteen segments are quinquannualate (5 annulate) and possess nephridiopores.

5. *Caudal region*. The caudal region comprises of three segments from 23rd to 26th. These are without nephridiopores. Segment 23 is triannulate whereas the rest are biannulate. Anus is located middorsally in 26th segment.

6. *Posterior sucker region*. Posterior sucker is formed by seven segments from 27th to 33rd. These are uniannulate and are arranged in concentric rings.

***Suckers.*** Each end of the body bears a hollow muscular organ, the sucker. The anterior or cephalic sucker comprises the prostomium and three anterior body-somites. It is oval in outline with a ventrally directed, cup-like hollow structure called the pre-oral chamber, at the bottom of which lies the mouth. The posterior end also bears a circular and highly muscular disc, the posterior or anal sucker, also directed downwards. It is formed by the complete fusion of the last seven body segments. Both the suckers are primarily meant for adhesion and locomotion. The anterior sucker also helps in feeding.

***Apertures.*** The body of leech has a number of apertures. These are as follows:

(i) *Mouth*. It is a narrow triradiate aperture situated in the centre of the preoral chamber of anterior sucker.

(ii) *Anus*. It is a small aperture situated mid-dorsally on the 26th segment at the junction of the body and posterior sucker.

(iii) *Nephridiopores*. 17 pairs of minute nephridiopores present on the ventral side of the body, a pair on the last annulus of each of the segments 6-22.

(iv) *Male genital pore*. It is a mid-ventral opening, situated in a groove between the second and third annuli of the 10th somite. Sometimes, a thread like penis is seen protruding through this aperture.

(v) *Female genital pore*. It lies mid-ventrally in a groove between the second and third annuli of the 11th segment.

**Body Wall**

The body wall is made up of five layers viz., cuticle epidermis, dermis, muscular layers and botryoidal tissue.

***1. Cuticle.*** It is the outermost covering of the body and is thin, delicate, transparent, colourless and elastic. It is produced by the inner epidermis and is periodically shed especially, after a heavy meal. The cuticle is also shed off, if leech is kept in dirty water.

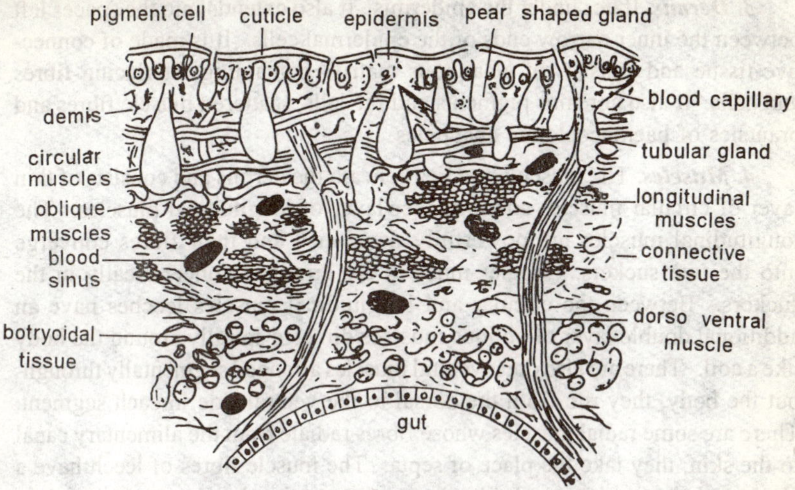


Fig. 17.4 *Hirudinaria* Section of body wall.

***2. Epidermis.*** Lying below the cuticle is the epidermis, which consists of a single layer of hammer-shaped cells. These cells are broad,flattened and pentagonal towards outer ends and narrower towards their inner ends. The outer ends fit closely while the nucleated inner ends have interspaces containing fibres, connective tissue, pigment cells and haemocoelomic capillaries, forming a vascular and respiratory membrane.

These are several glands found in the epidermis. These are modifications of epidermal cells.

(i) *Slime-glands.* These are distributed all over the body and produce a kind of slimy mucus, which covers whole of the body. The tubular body of the slime-gland reaches deeper into dermis and possesses a nucleus, a thin layer of cytoplasm and several secretory granules. Through the narrow neck, each gland opens to the outside by a minute aperture in the cuticle.

(ii) *Sucker glands.* These are also pear-shaped or rounded in form and occur in clusters in both the suckers. Their secretion serves to smoothen the surface for the attachment of suckers during locomotion.

(iii) *Prostomial glands.* These occur in groups in the prostomium. Their secretion plugs the ends of ootheca or egg-case.

(iv) *Clitellar glands.* These are found in the clitellar region (segments 9 to 11 ) and become active during the breeding season. They are of two special types, the *chitogenous glands* forming the wall of the cocoon, and the *albumen gland* secreting albumen that fills up the cocoon. The former lie between circular and oblique muscles and the latter in longitudinal muscles.

***3. Dermis.*** It lies under the epidermis. It also extends into the spaces left between the inner narrow ends of the epidermal cells. It is made of connective tissue and consists of gelatinous matrix containing interlacing fibres and cells. It also contains pigments and fat cells, scattered muscle fibres and branches of haemocoelomic channeles.

***4. Muscles.*** The musculature lies beneath the dermis and consists of thin layer of circular muscles and a thicker layer of longitudinal muscles. The longitudinal muscles are powerfully developed and their fibres converge into the two suckers. Circular muscles are arranged concentrically in the suckers. Between the circular and longitudinal muscles leeches have an additional double layer of oblique muscles directed spirally around the body like a coil. There are also dorsoventral muscles arranged segmentally throughout the body, they run from the dorsal to the ventral side in each segment. There are some radial muscles whose fibres radiate from the alimentary canal to the skin, they take the place of septa. The muscle fibres of leech have a characteristic structure, each fibre has an outer striated and contractile cortex or myoplasm, and an inner unmodified protoplasm called medulla or sarcoplasm.

***5. Botryoidal tissue.*** It is situated beneath the longitudinal muscles surrounding the alimentary canal. It consists of a network of large branching tubular cells arranged end to end. The walls of the cells are loaded with a dark brown pigment while their lumen or intracellular canals contain haemocoelomic fluid. It is probably excretory in function.

**Locomotion**

Leeches are capable of crawling as well as swimming.

1. ***Crawling movements.*** In crawling, the leech first fixes its posterior sucker to the substratum. This is followed by an elongation of the body on the substratum brought about by the contraction of the circular muscles and relaxation of the longitudinal muscles. The animal then fixes its anterior sucker to the substratum and draws forwards its posterior sucker and the rest of the body by the contraction of the longitudinal muscles and relaxation of the circular muscles. The posterior sucker is then fixed to the substratum close behind the anterior sucker, so that the body forms a loop. The posterior sucker now remain fixed and the rest of the body elongates by the mechanism described above.
2. ***Swimming.*** It is brought about by the undulatory motion of the body, which is in a state of dorso-ventral flattenning owing to the sustained contraction of dorso-ventral muscles. The undulatory movement is brought about by waves of contraction and relaxation passing along the longitudinal muscles.

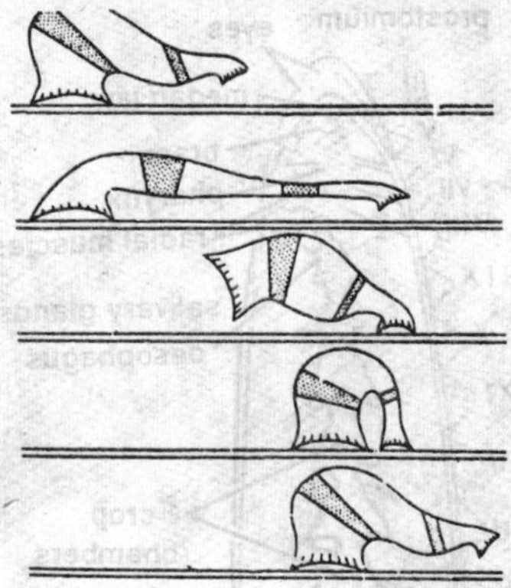

Fig. 17.5 *Hirudinaria*. Stages of locomotion

## Digestive System

The digestive system comprises the alimentary canal and glands associated with it.

***Alimentary canal.*** It is a straight tube of varying diameter running from the anterior to the posterior end. It is supported on all the sides by botryoidal tissue. It is differentiated into mouth, buccal cavity, pharynx, oesophagus, crop, stomach, intestine and rectum.

1. *Preoral chamber.* The cup-shaped depression of the anterior sucker represents the preoral chamber. Its base is perforated by triradiate mouth which opens into the buccal cavity. The preoral chamber and buccal cavity are separated from one another by a membranous partition, the *velum*. It consists of three flaps, which form three lips of the mouth.

2. *Buccal cavity.* The mouth leads into the buccal cavity, which is a short chamber behind the velum. In its mucous lining are embedded three crescentic jaws, each lodged in a depression. One jaw lies behind each ray or chink of the mouth, i.e. one is medio-dorsal and the other two ventro-lateral in position. Each jaw is a laterally compressed muscular cushion, covered with fine cuticle which is thickened at the free edge to form a ridge bearing minute teeth or denticles in a single row. Such jaws are termed *monostichodont*. The number of the denticles is 103-128 on the median jaw and 85-115 on lateral jaw. On each side of the jaw are about 42-54 button-shaped protuberances, the salivary papillae, each bearing numerous opening of the salivary glands.

3. ***Pharynx.*** The buccal cavity is followed by a highly muscular, oval sac, the pharynx. It extends from 5th to 8th segment. It is covered with

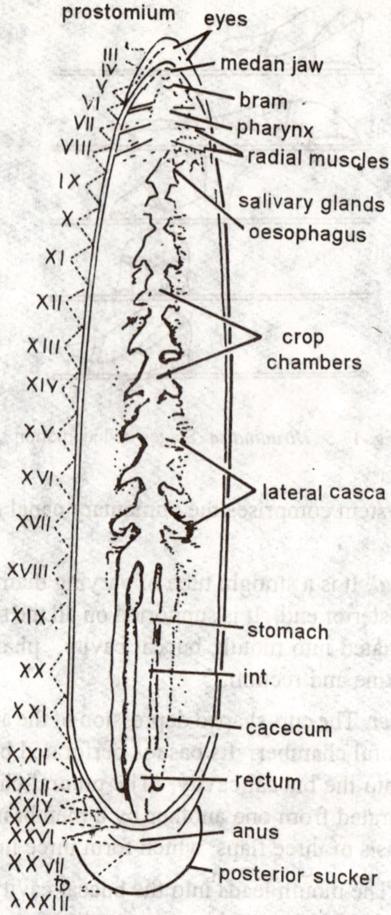


Fig. 17.6 *Hirudinaria*. Alimentry Canal.

unicellular salivary glands, whose ducts open into the salivary papillae of jaws. The entire space between the wall and the pharynx is filled up by salivary glands. The saliva secreted by salivary glands has a kind of substance called hirudin or anticoagulin, which does not allow the blood to coagulate during the act of sucking it from the prey.

4. *Oesophagus*. The pharynx opens into the oesophagus, which is a narand the epithelium is highly folded.

5. *Crop*. The crop is the largest region of the alimentary canal extending

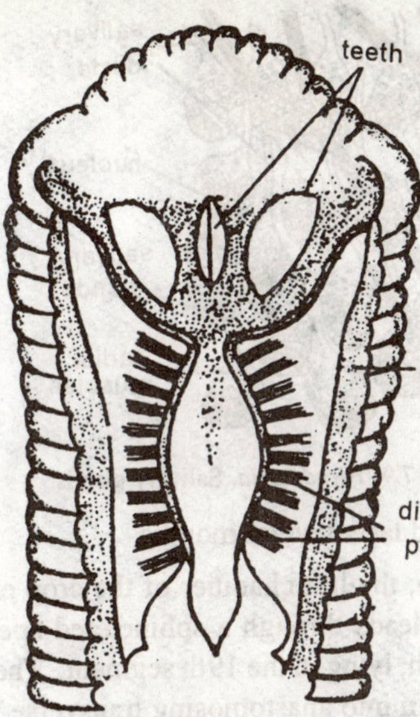


Fig. 17.7 *Hirudinaria*. Ventral dissection of anterior end showing there jaws.

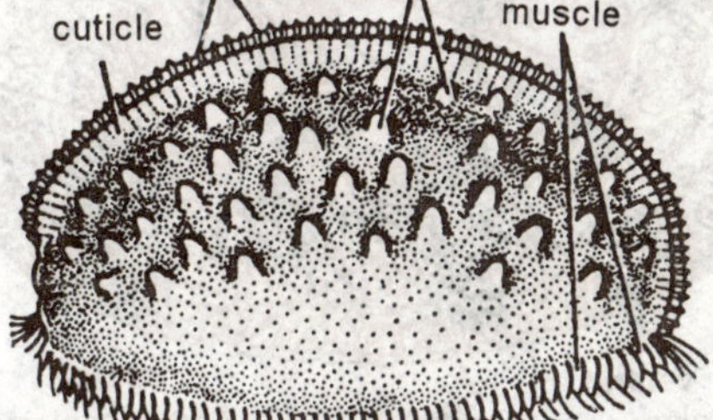


Fig. 17.8 *Hirudinaria*. A jaw.

from the ninth to the eighteenth segment. It has ten thin-walled chambers, one in each segment. Each chamber consists of a small anterior and a broad posterior part which is produced into a pair of lateral outgrowths the *caeca* or *diverticula*. The chambers and the caeca go on gradually increasing in size towards the posterior side. The last chamber of the crop is the largest and its caeca are prolonged backwards upto the twenty second segment. The crop is a very important part of the alimentary canal, as it is elastic and is used for storing blood,

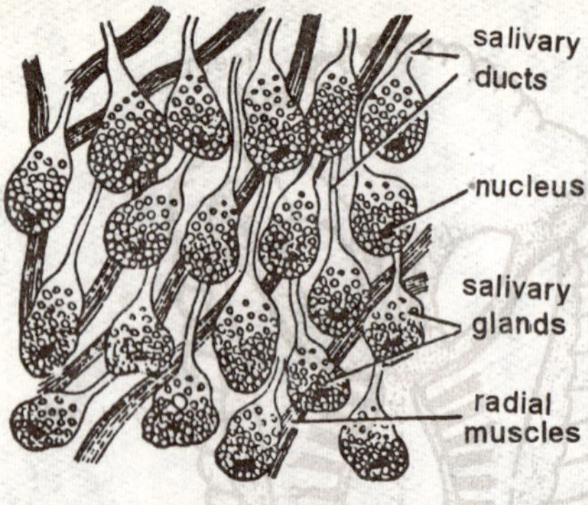


Fig. 17.9 *Hirudinaria*. Salivary glands.

one crop-ful of blood lasts several months.

6. *Stomach*. Posteriorly, the last chamber of the crop ends into a funnel-shaped tube, which leads through a sphinctered aperture into a small heart-shaped stomach, lying in the 19th segment. The mucous lining of the stomach is thrown into anastomosing transverse folds.

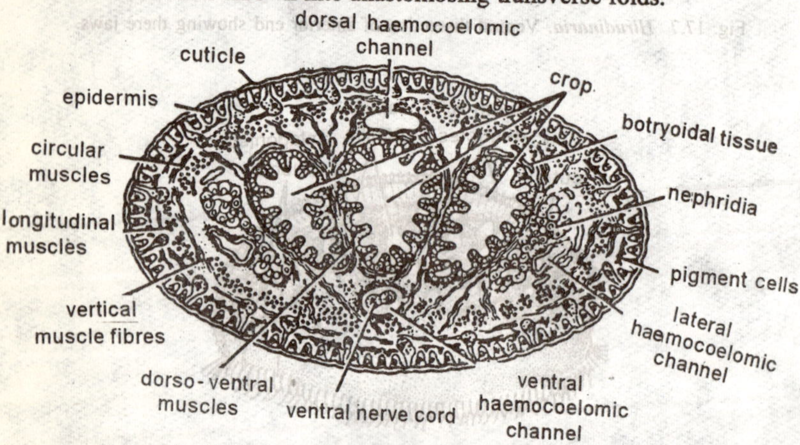


Fig. 17.10 *Hirudinaria*. T.S. body through crop and its diverticulum.

7. *Intestine*. It is a thin-walled narrow tube from 20th to 22nd segment. Its lining is also produced into spiral villi-like processes to increase the surface of absorption. The stomach and intestine are not marked externally.

8. *Rectum*. Behind the intestine lies the thin-walled wide rectum which extends upto 23rd-26th segment. There are no internal foldings in rectum. It opens out posteriorly by an anus which lies on the dorsal surface of 26th segment.

**Histology**

The pre-oral and buccal cavity are lined with cuticle continous with that of external surface. They form the stomodaeum. The rectum is also lined with cuticle and forms the proctodaeum. The remaining parts of alimentary canal are lined with endoderm and forms the midgut.

Fig. 17.11 *Hirudinaria* T.S. body throughthe region of stomach.

***Food.*** Food of these highly organised animals is usually blood and, because, they live on the blood of various animals, they are called blood-suckers. Some leeches are parasites on the blood of man, others live on horses and a few on other cattle. Some large leeches in Northern United States, feed voraciously on large snails.

***Feeding or ingestion.*** In feeding the animal attaches itself to its prey by the posterior sucker. At the other end the pre-oral chamber is flattened and applied to the host. The jaws are protruded through the mouth. Soon the back and forth movements of jaw brought about by special sets of muscles inserted over them work a triradiate or 'Y'-shaped incision in the skin of victim. The incision lays open the cutaneous vessles from which the blood is sucked with the help of pre-oral chamber, the buccal cavity and the pharynx that form highly efficient apparatus for the purpose. A leech can suck blood several times its own weight in a single meal. When the crops are full of blood it drops off and preers darker places and remains hidden till this meal exhausts. The salivary glands produce a substances, "hirudin" which prevents the coagulation of blood during their feeding time.

***Digestion.*** In the cm of the crops. Digestion of haemolysed blood takes place in the stomach by the proteolytic enzymes (*Abderhalden* and *Heise*, (1909). *Bupsing* and his colleagues discovered in 1953 that digestion in *Hirudo* occur entirely by the gut bacteria. The process of digestion is very slow and it takes 10-14 months to digest one crop-full of blood.

***Absorption.*** Digested food is absorbed in the intestine. The absorptive surface of intestine is increased by villi-like transverse and longitudinal folds.

***Egestion.*** Elemination of undigested food occurs through the anus.

**Coelom.**

Coelom, as seen in *Pheretima*, is not present around the alimentary canal of the leech. Actually, the coelom is present, although in a considerably reduced state, owing to the growth of botryoidal tissue. It is represented by the four longitudinal vessels, which are not true blood vessels, but shut off parts of the coelom, the branches of these vessels and a fairly large number of spaces.

The coelomic fluid, running through the channels and their branches, contains colourless amoeboid corpuscles and is coloured red due to haemoglobin dissolved in it. The blood-like coelomic fluid is called the haemocoelomic fluid and the channels are termed the haemocoelomic channels, which together form the haemocoelomic system.

Thus, in *Hirudinaria*, there are no true blood vessels, their function being taken up by the haemocoelomic system.

**Haemocoelomic system.**

It consists of following four longitudinal vessels:

1. A dorsal longitudinal channel.
2. A ventral longitudinal channel
3. Two lateral longitudinal channels.

**1. *Dorsal Haemocoelomic Channel.*** It is thin-walled non-contractile vessel running middorsally above the alimentary canal from one end of the body to the other. The haemocoelomic fluid flows from behind forward. The valves are absent. Anteriorly, it remifies in the 6th segment and supplies the anterior five segments, posteriorly, it bifurcates in the 22nd segment, the two branches run ventrally around the rectum to join ventral sinus.

It gives two types of branches:

(i) *Dorso-laterals.* Two pairs of dorsolateral branches arise from the ventrolateral side of the dorsal channel in each segment. Each branch runs outwards to form a capillary plexus in the dorsal and dorso-lateral regions of the body wall.

(ii) *Dorso-intestinals.* These are numerous small branches arising from the ventral side of the dorsal channels, all along its length. They supply the gut wall.

**2. *Ventral Haemocoelomic Channel.*** It runs below the alimentary canal in the mid-ventral region from anterior to the posterior end of the body. The flow of the blood is from anterior to the posterior end. It encloses the nerve-cord, cerebral ganglia, nerve-ring and sub-pharyngeal ganglion. It is wider than the dorsal sinus but there are no valves. All the arches are segmental.

(i) First pair of branches originate at the lavel of nerve-ganglion and give rise to two main branches: (a) *Ventral branches*, which is found on the

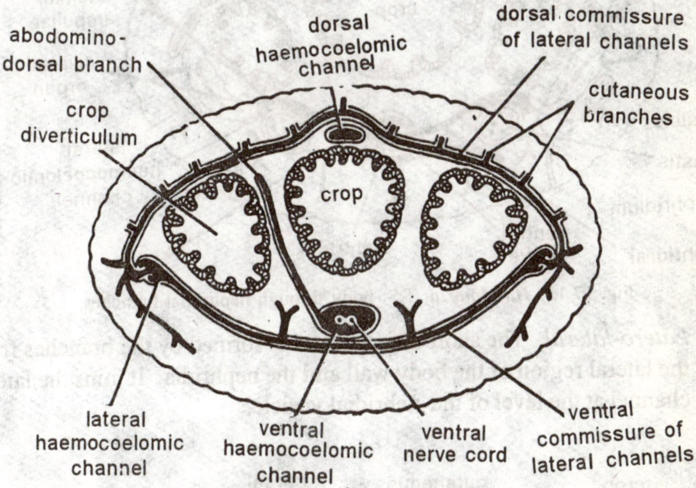


Fig. 17.12 *Hirudinaria*. T.S. body through dorsal and ventral commissures of the lateral haemocoelomic channels.

ventro-lateral side of the body-wall. (b) *Abdomino-dorsal branch* which runs upwards between the diverticula of crop and supplies dorsolateral part of the body-wall.

(ii) The second pair is of nephridial branches, which are found in 11 segments i.e. from 12th to 22nd segment, having testis-sacs. Each branch form 2 or 3 sets of saccules, the perinephrostomal ampullae, in which ciliated organs are enclosed and finally, passes out and gives off branches to lateral part of the body wall.

**3. *Lateral channels.*** There are two lateral haemocoelomic channels placed one on each side of the alimentary canal. They are wide and uniform in diameter. Their walls are thick, muscular and possess valves. The haemocoelomic fluid flows from posterior to anterior direction. They are collecting as well as distributing vessels.

Each lateral channel receives on its outer sides two branches, the latero-lateral and the latero-dorsal, and gives off on the inner side one branch, the latero-ventral, in each segment.

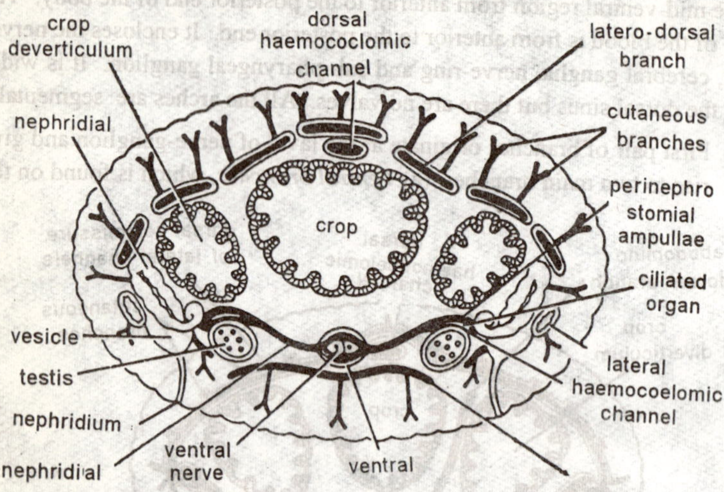


Fig. 17.13 *Hirudinaria.* T.S. body through nephridial branches.

(i) *Latero-lateral.* The short latero-lateral is formed by the branches from the lateral region of the body wall and the nephridia. It joins the lateral channel at the level of the nehridial vesicle.

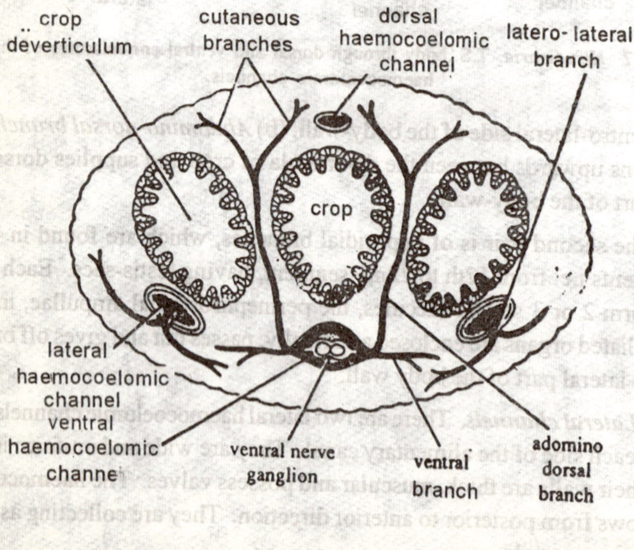


Fig. 17.14 *Hirudinaria.* T.S. body through dorso-abdominal branches.

(ii) *Latero-dorsal.* It opens into the lateral vessel at the level of nephridium proper. This is a large branch and collects blood from the dorsal and dorso-lateral branches of both the sides are connected by a dorsal commissure, which runs transversely above the dorsal canal in the segments from 6th to 22nd.

Latero-lateral and latero-dorsal branches of each side are connected by a lateral commissure in each segment. These collecting branches possess valves at their openings in the lateral channel.

(iii) *Latero-ventral.* It arises from the inner side of the lateral channel in each segment. Immediately after its emergence it gives off a small branch to the nephridium and the ventral and ventrolateral region of the body wall. Soon it divides into two branches. One of them is anterior branch, while the other is posterior branch. The two branches diverge from one another and unite with the corresponding branches of the other side below the ventral channel. This forms a characteristic rhomboid figure, the ventral commissure of laterals. Eighteen such rhomboids are found one in each of the segments from 6th to 23rd. The rhomboids of the adjacent segments are also joined together by three longitudinal intersegmental commissures. The latero-ventrals supply branches to the nephridia, alimentary canal, reproductive organs and ventral side of the body wall.

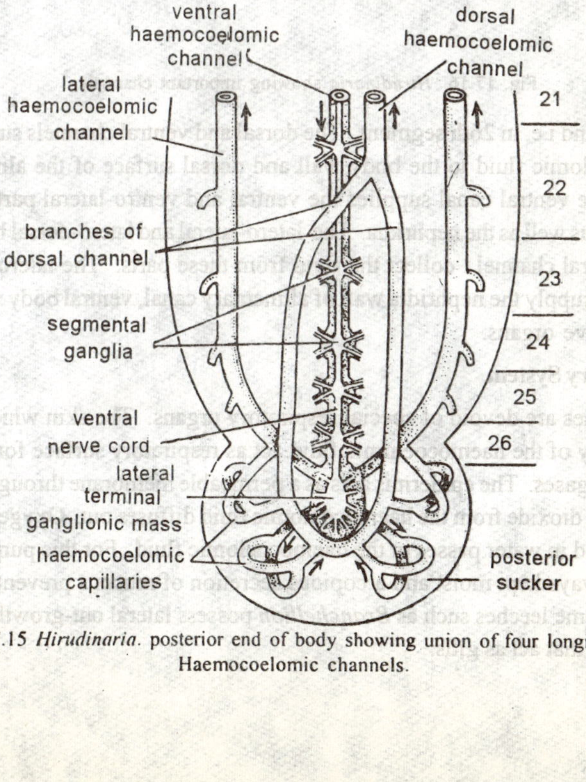


Fig. 17.15 *Hirudinaria.* posterior end of body showing union of four longitudinal Haemocoelomic channels.

***Course of circulation :*** Of the four longitudinal channels, the fluid in the dorsal and the two lateral channels move in anterior direction, while in ventral channel it moves in posterior direction. The dorsal and ventral are the distributing channels, while the laterals are collecting as well as distributing channels. All the channels are in direct communication with one another at the

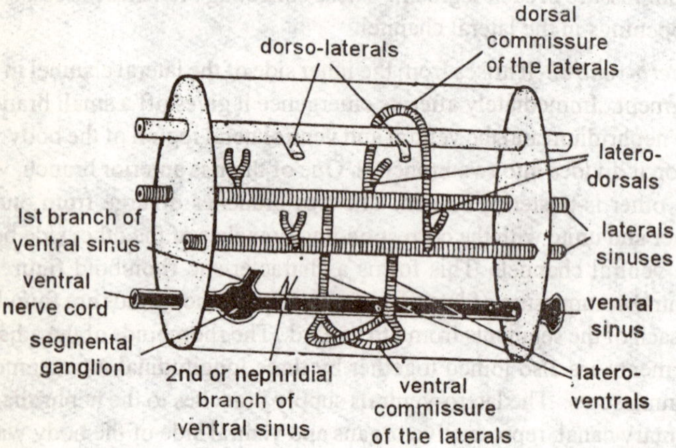


Fig. 17.16 *Hirudinaria* showing important channels.

posterior end i.e. in 26th segment. The dorsal and ventral channels supply the haemocoelomic fluid to the body wall and dorsal surface of the alimentary canal. The ventral canal supplies the ventral and ventro-lateral parts of the body wall as well as the nephridia. The latero-lateral and latero-dorsal branches of the lateral channel collect the fluid from these parts. The latero-ventral branches supply the nephridia wall of alimentary canal, ventral body wall and reproductive organs.

**Respiratory System**

Leeches are devoid of special respiratory organs. The skin which has a rich supply of the haemocoelomic fluid act as respiratory surface for the exchange of gases. The epidermis acts as a permeable membrane through which the carbon dioxide from the haemocoelomic fluid diffuses out. Oxygen which is dissolved in water passes in the haemocoelomic fluid. For this purpose the skin is always kept moist and a copious secretion of mucous prevent it from drying. Some leeches such as *Branchellion* possess lateral out-growths of the body wall that act as gills.

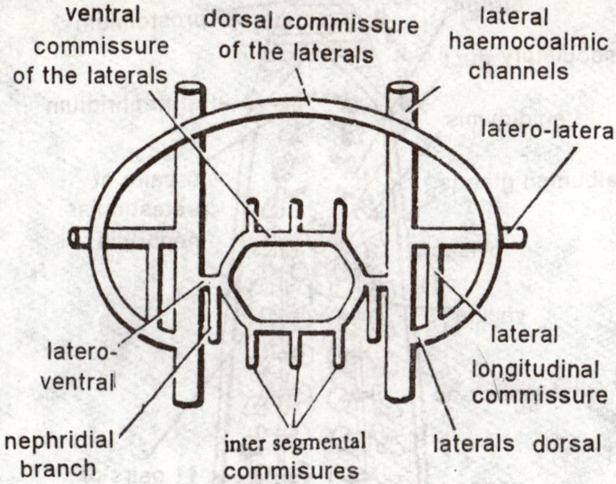


Fig. 17.17 *Hirudinaria.* Diagrammatic representation of the lateral channels and their branches in the segment in dorsal view.

## Excretory System

The excretory system of leech comprises 17 pairs of segmentally arranged nephridia, situated one pair in each of the segment 6th to 22nd. Of these the first 6 six pairs are located in segments 6th to 11th devoid of testis-sacs and are called the *pretesticular nephridia* and the remaining 11 pairs are the *testicular nephridia*. They remain in close association with the testis-sacs. The two types differ only in a minor respect.

***Testicular nephridium.*** This is regarded to be a typical nephridium. Its shape is just like that of a horse-shoe and consists of seven parts:

1. Main lobe
2. Vesicle duct
3. Vesicle
4. Apical lobe
5. Inner lobe
6. Initial lobe
7. Ciliated organ

1. *Main lobe.* The main lobe is actually just like a horse-shoe and is situated in the ventro-lateral position between the adjacent caeca of the crop. It forms the major part of the nephridium and consists of two unequal limbs, out of which anterior is long and the posterior limb is small.

2. *Vesicle duct.* The vesicle duct is a narrow tube that starts from the inner

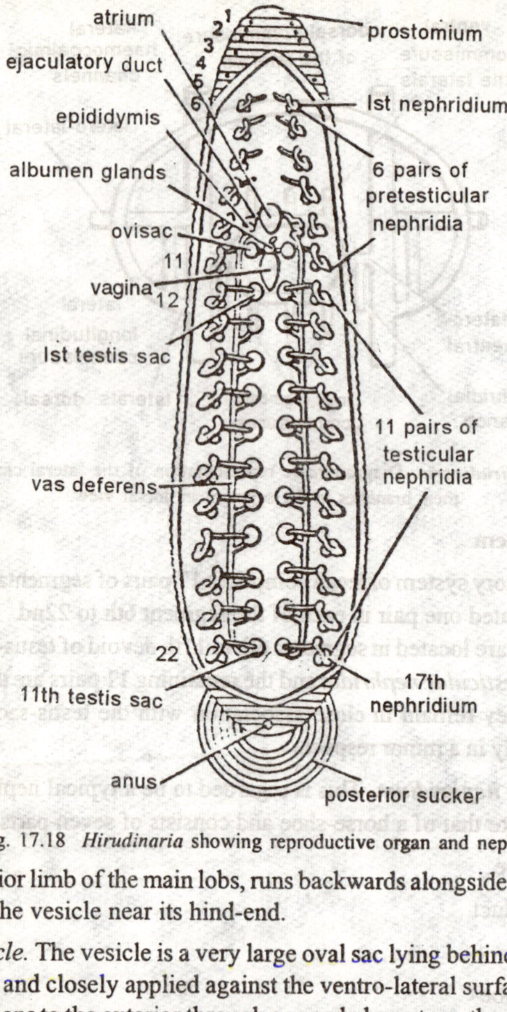


Fig. 17.18 *Hirudinaria* showing reproductive organ and nephridia.

end of anterior limb of the main lobs, runs backwards alongside the vesicle and opens into the vesicle near its hind-end.

3. *Vesicle.* The vesicle is a very large oval sac lying behind the rest of the nephridium and closely applied against the ventro-lateral surface of the body wall, and opens to the exterior through a rounded aperture, the *nephridiopore*.

4. *Apical lobe.* The posterior limb of main lobe is continued anteriorly into an apical lobe, situated antero-posteriorly beneath the crop. It is thick and stout and its anterior end is swollen and bent upon itself like the head of a walking stick. The cells lining the apical lobe are large with regular intracellular canals.

5. *Inner lobe.* Projecting from the anterior and the posterior limbs of the main lobe is found a short inner lobe situated in the concavity of the main lobe and running along the side of the apical-lobe for about half its length.

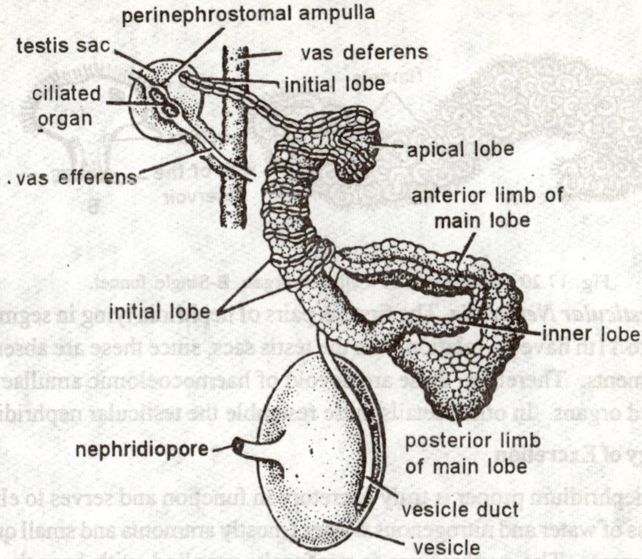


Fig. 17.19 *Hirudinaria*, A testicular nephridium.

6. *Initial lobe.* The testis-lobe or the initial lobe is a long, narrow, transparent and cord-like structure, formed of a single row of elongated tubular cells and closely twined around the apical lobe. Its posterior end joins the main lobe, while the anterior end runs inwards and reaches over the testis sac of its ownside, where it ends blindly close to the perinephrostomial ampullae. The intracellular canal of the initial lobe gives off many diverticula in each cell.

7. *Ciliated organ.* The ciliated organ corresponds to the funnel or nephrostome of a typical annelid nephridium and each consists‘ (i) a *central reservoir,* a perforated chamber, appearing more or less spongy, with an outer wall formed of a single-layer of cells and a central mass of connective tissue cells, the seat of manufacture of the coelomic corpuscles and (ii) *ciliated funnels*. Numerous funnels, each fitting into a pore of the reservoir by its narrow proximal part, the neck. They cover the reservoir from outside. Each funnel is like an ear-lobe, with about one-fourth of its margin incomplete along one side. The distal end is ciliated and broad.

In the adult *Hirudinaria* the ciliated organ has no connection whatever with the body of the nephridium although there is a distinct connection between the two in the embryo. In the adult it loses connection with nephridium, gives up excretory function and associated with the haemo-coelomic system, for which it manufactures coelomic corpuscles and drives them into the haemo-coelomic stream.

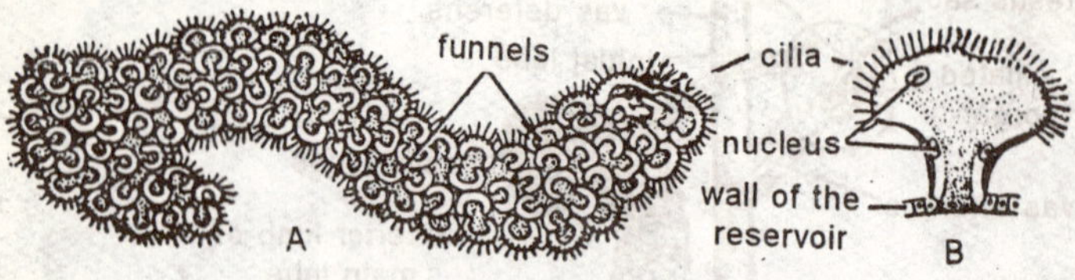


Fig. 17.20 *Hirudinaria*. A=Cilliated organ, B-Single funnel.

***Pretesticular Nephridia.*** The first six pairs of nephridia lying in segments from 6th to 11th have no relation with the testis sacs, since these are absent in these segments. Therefore, these are devoid of haemocoelomic amullae and the ciliated organs. In other details these resemble the testicular nephridia.

**Physiology of Excretion**

The nephridium proper is truly excretory in function and serves to eliminate excess of water and nitrogenous wastes,(mostly ammonia and small quantities of urea). The nephridium is profusely supplied with branches of haemocoelomic channels. Its gland cells separate the waste products from the haemocoelomic fluid. The excretory fluid is finally collected into the terminal vesicle to be discharged to outside through the nephridiopore.

Several workers have also attributed an excretory function to the botryoidal tissue, the intracellular capillaries of which communicate with the haemocoelomic vessels.

**Nervous System**

The nervous system satisfies the usual pattern of Annelida. It consists of three parts.

1. Central Nervous System,
2. Peripheral Nervous System, and
3. Sympathetic Nervous System.

*1.* ***Central Nervous System.*** The entire central nervous system lies within the ventral haemocoelomic channel and consists of five parts(i) cerebral ganglia, (ii) peripharyngeal connectives, (iii) sub-pharyngeal ganglionic mass, (iv) ventral nerve cord, and (v) terminal ganglionic mass.

(i) *Cerebral gonglia.* A pair of fused cerebral ganglia form a small brain situated above the pharynx just behind the median dorsal jaw in the 5th segment.

(ii) *Peripharyngeal connectives.* They start from the sides of the brain and pass downwards along the sides of pharynx to meet the sub-pharyngeal ganglionic mass.

(iii) *Sub-pharyngeal ganglionic mass.* Ventrally peripharyngeal connectives unite to form the subpharygeal ganglionic mass which lies below the pharynx exactly opposite to the cerebral ganglia. It is composite structure of more or less triangular shape, apex of the triangle pointing backwards formed by the fusion of four pairs of ganglia.

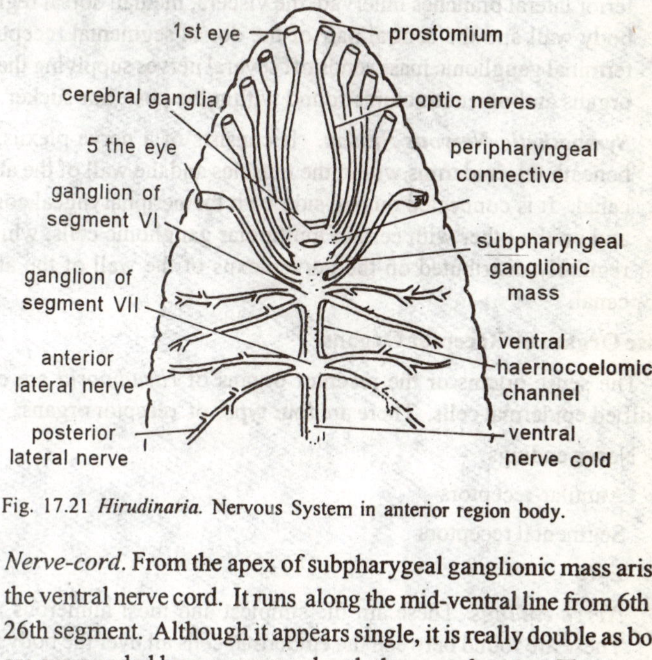


Fig. 17.21 *Hirudinaria.* Nervous System in anterior region body.

(iv) *Nerve-cord.* From the apex of subpharygeal ganglionic mass arises the ventral nerve cord. It runs along the mid-ventral line from 6th to 26th segment. Although it appears single, it is really double as both are surrounded by a common sheath the neurolemma. The ventral nerve cord carried 21 well formed ganglia, each located in the first annulus of its own segment Posteriorly, the ventral nerve cord ends in a large ovoid terminal ganglionic mass, situated within the posterior sucker. It is formed by the fusion of seven pairs of embryonic ganglia of the lasts seven segments that constitute the posterior sucker.

2. ***Peripheral Nervous System.*** It consists of several paired nerves which are originated from the central nervous system and innervate all parts of the body. A pair of stout nerves arise from the brain which run forward to supply the first pair of eyes, the prostomium and the roof of the buccal

chamber. The subpharyngeal ganglionic mass gives out four pairs of eyes and also to segmental receptors of the first five segments, roof of the buccal cavity and the muscles of body wall. From each ganglion of ventral nerve cord arise two pairs of stout nerves, the anterior laterals and posterior laterals. The anterior laterals arise from the anterior part of the ganglion, branch and innervate the vas deferens, nephridium, the nephridial vesicle, muscles of body wall, ventral nerve receptors and two outer dorsal receptors of their own side. The posterior laterals arise from the posterior part of ganglion just behind the anterior laterals. The posterior lateral branches innervate the viscera, median dorsal region of the body wall and the central pair of the dorsal segmental receptors. The terminal ganglionic mass sends off several nerves supplying the receptor organs and other structures found within the posterior sucker.

3. ***Sympathetic Nervous System.*** It consists of a nerve-plexus, situated beneath the epidermis within the muscles and the wall of the alimentary canal. It is connected on one side with the peripharyngeal connectives and on the other with certain multipolar ganglionic cells, which are irregularly distributed on the nerveplexus of the wall of the alimentary canal.

**Sense Organs or Receptor Organs.**

The sense-organs or the receptor organs of *Hirudinaria* are especially modified epidermal cells. There are four types of receptor organs:

1. Nerve endings
2. Annular receptors
3. Segmental receptors
4. Eyes.

*1.* ***Nerve endings.*** These are the simplest and most numerous receptors. These are found between the epidermal cells all over the body with their

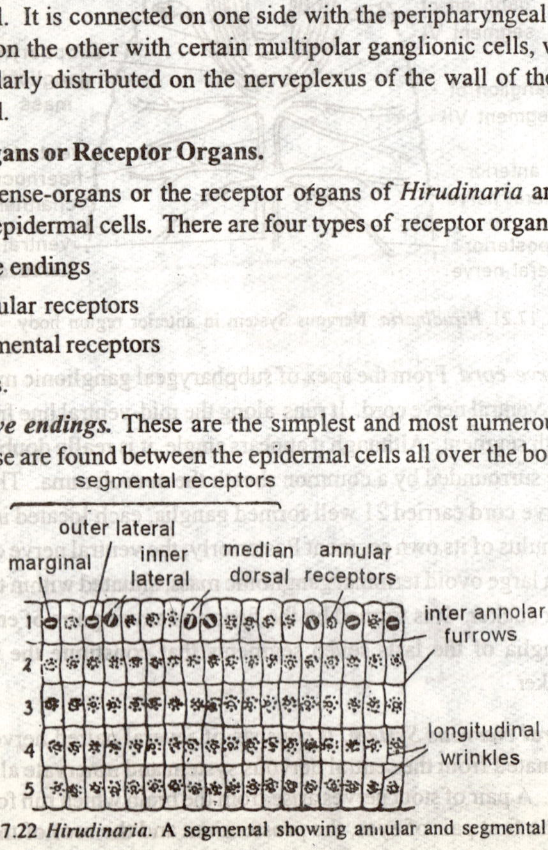


Fig. 17.22 *Hirudinaria.* A segmental showing annular and segmental receptor.

ganglionic cells situated beneath the epidermis. These are **simply the** ends of nerve fibres. These organs are probably sensitive to **chemical** stimuli i.e. changes in the surrounding water. They, thus , act as chaemoreceptors.

2. ***Annular receptors.*** These are in the form of minute papillae, formed of flattened overlapping epidermal cells. Their proximal ends are connected with the nerve fibres while the distal ends project out of the integument. About 36 annular receptors are found to be equally distributed on the dorsal and ventral sides in each annulus. These form transverse rows across the middle of the annulus. The annular receptors are tactile in function or tangoreceptors.

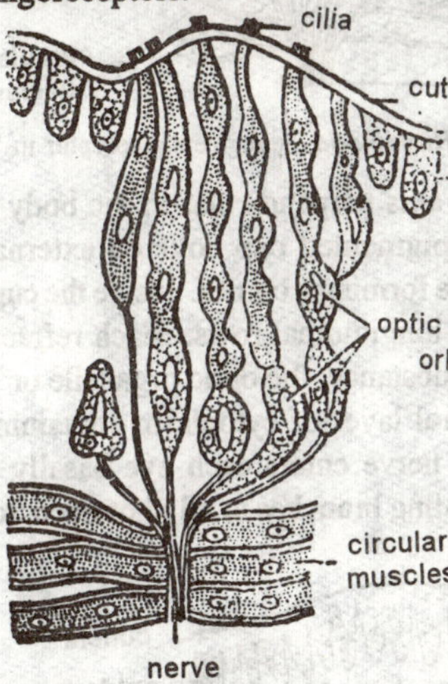


Fig. 17.23 *Hirudinaria.* An annular receptor in V.S.

3. ***Segmental receptors.*** These are small whitish elliptical papillae borne upon the first annulus of each body segment, four pairs dorsally and three pairs ventrally. Each receptor consists of two types of cells-tactile cells or tangoreceptors and light-sensitive cells or photoreceptors. There are 5 to 10 long, slender tactile cells, separated from one another and provided with hair-like processes at their outer free ends. The light-perceiving or the photoreceptor cells, found only in the dorsal receptors, contain a crescentic hyaline substance, the optic organelle or lens, in their cytoplasm. Each receptor receives a nerve branch and functions both as tangoreceptor and photoreceptor.

4. ***Eyes.*** There are five pairs of eyes lying in a semicircle of black spots on the dorsal surface of the anterior sucker, one pair in the first annulus of each of the first five segments. Each eye is **cylindrical or cup-like in**

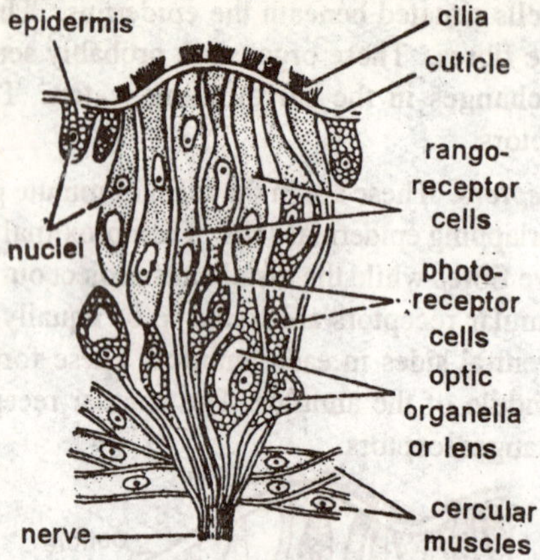


Fig. 17.24 *Hirudinaria*. A segmental receptor in V.S.

shape with its long axis perpendicular to the body surface. Each eye consists of a long pigmented cup covered externally by transparent epidermis and cuticle forming a cornea. Inside the cup are refractive cells arranged in several longitudinal rows. Each refractive cell contains a crescentic hyaline substance, the optic organelle or lens surrounded by a very thin peripheral layer of cytoplasm containing a small rounded nucleus. An optic nerve enters each eye basally and runs along its median axis distributing branches to all photoreceptor cells.

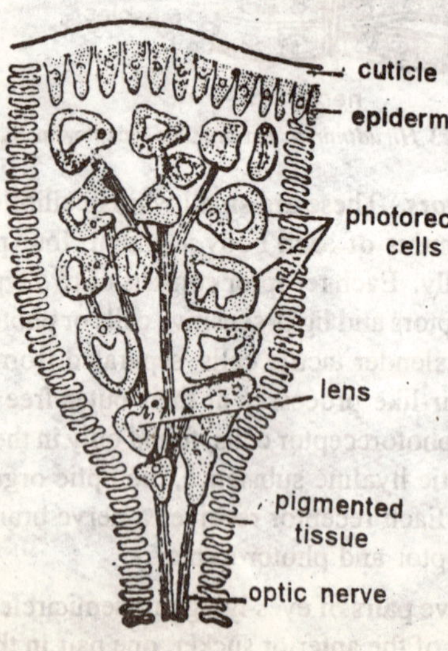


Fig. 17.25 *Hirudinaria*. V.S. of eye

The eyes are directed in different directions and each can receive light from only one direction. It is not yet known whether images of objects are formed in the eyes. However, it is presumed that eyes can only distinguish light from darkness and can locate the direction of the source of light.

On the basis of their metameric arrangement and histological structure, *Whiteman* regards the eyes to be serially homologous with the segmental sense organs. Moreover, the segments with eyes lack the usual segmental receptors.

**Reproductive System**

Leeches are hermaphrodite, i.e., male and female reproductive organs are found in the same individual.

***Male Reproductive Organs.*** Male reproductive or generative organs consist of the following structure:

1. Testis-sacs
2. Vasa efferentia
3. Vasa deferentia
4. Epididymis
5. Ejaculatory duct
6. Atrium.

1. *Testis sacs.* Testis sacs are eleven pairs of small, spherical or rounded vesicles found one pair in each of the segments from 12th to 22nd. These are located one on either side of the ventral nerve cord. From the wall of each testis sac spermatogonia are budded off which develop into spermatozoa.
2. *Vasa efferentia.* The spermatozoa pass from each testis sac into a short sinuous duct, the vas efferens, which arises from the postero-lateral border of the testis sac and runs outwards to join the common vas deferens of its side. All the vasa efferentia of one side open into the common vas deferens of that side.
3. *Vasa deferentia.* There are a pair of slender longitudinal ducts running forward, from twenty-second to the eleventh segment,lying along the ventral body wall on either side of nerve cord. Each vas deferens is enclosed within a tubular coelomic space which contains amoeboid corpuscles similar to those of the haemocoelomic fluid.
4. *Epididymis.* The distal end of each vas deferens in the 10th segment widens into a coiled mass which is termed as epididymis. This is said to store sperm broght from the testis sacs.

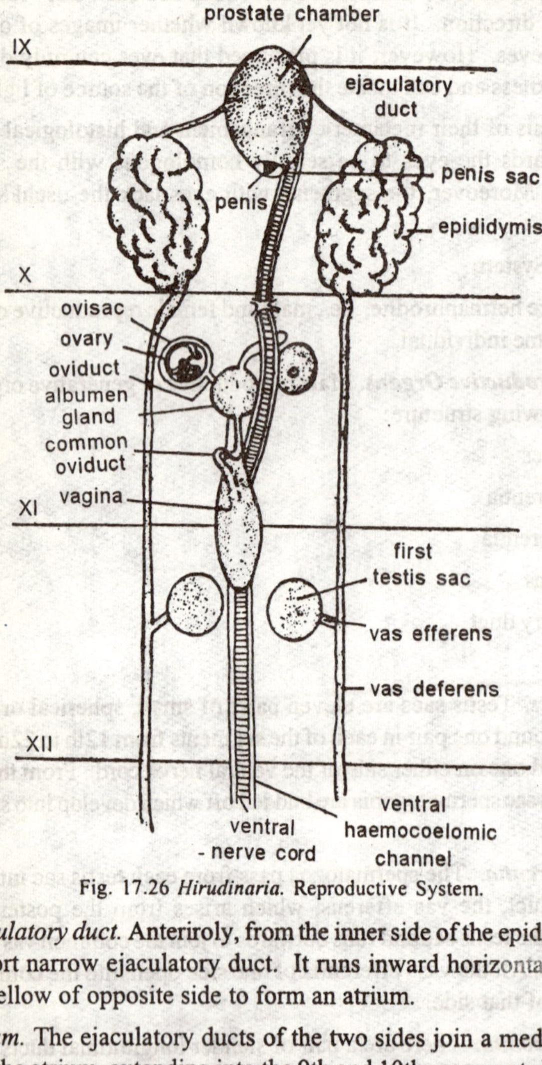


Fig. 17.26 *Hirudinaria.* Reproductive System.

5. *Ejaculatory duct.* Anteriroly, from the inner side of the epididymis arises a short narrow ejaculatory duct. It runs inward horizontally and joins the fellow of opposite side to form an atrium.

6. *Atrium.* The ejaculatory ducts of the two sides join a median pyriform sac, the atrium, extending into the 9th and 10th segments, and opening outside by the male genital pore. The atrium consists of two parts-a base-like broad anterior part called the prostate chamber and a neck-like narrow posteriorly directed part, the penis sac. The prostate chamber has thick muscular walls which are covered externally by numerous unicellular prostate glands.

The penis-sac is the posteriorly directed narrow neck-like highly muscular part of the atrium which contains a coiled filamentous tubular penis. It helps in copulation and transference of spermatophores into the vagina of other leech.

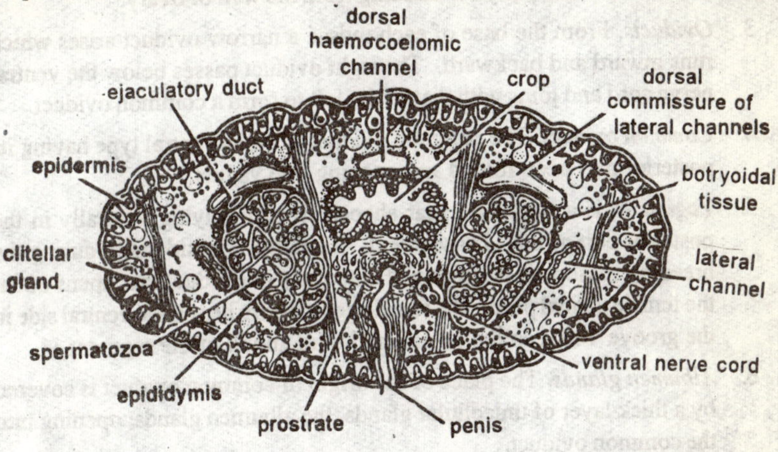


Fig. 17.27 *Hirudinaria*. T.S. body through epididymos, prostrate and penis sac.

***Female Reproductive Organs.*** The female reproductive system includes:

1. Ovisacs
2. Ovaries
3. Oviduct
4. Common Oviduct
5. Vagina
6. Albumen glands

1. *Ovisacs*. There is a pair of ovisacs situated ventro-laterally in the 11th segment, one on either side of ventral nerve cord. Each ovisac is a hollow structure filled with coelomic fluid, containing ana ovary.

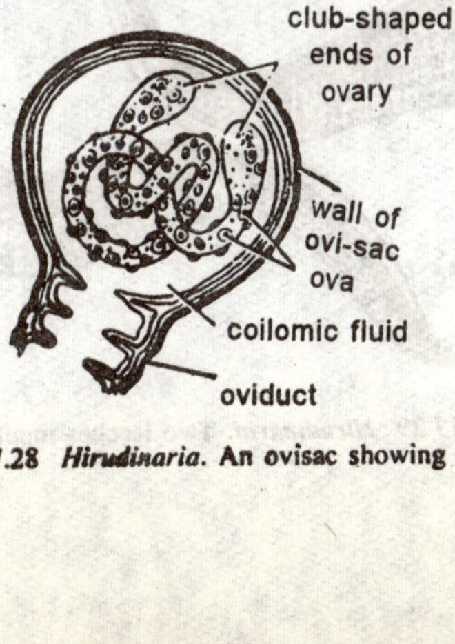


Fig. 17.28 *Hirudinaria*. An ovisac showing ovary.

2. *Ovaries*. A pair of ovary, each enclosed inside an ovisac. Each ovary is a delicate, coiled and filamentous structure which contains many nuclei. The ova are formed by the budding from the wall of ovary.
3. *Oviducts*. From the base of each ovisac a narrow oviduct arises which runs inward and backward. The right oviduct passes below the ventral nerve cord and joins with that of the left to form a common oviduct.
4. *Common oviduct*. The common oviduct is a mid-ventral type having its posterior part open folded and opening into vagina.
5. *Vagina*. It is a large and pear shaped structure, lying ventrally in the posterior part of the 11th segment. It enlarges considerably during the breeding season. It is continued anteriorly into a duct that opens out at the female genital pore. The female genital pore lies on the ventral side in the groove between the second and third annuli of the segment 11.
6. *Albumen glands*. The place of the origin of common oviduct is covered by a thick layer of unicellular glands, the albumen glands, opening into the common oviduct.

The ova, budded off from the ovaries, pass through the oviducts into the vagina, where they remain till fertilization. The ova are coated with the albumen from albumen glands. Fertilization takes place in vagina. Fertilized ova are discharged into a cocoon for further development.

**Copulation and Fertilization.**

Copulation takes place in March and April. It may occur in water or on land and lasts for about an hour. Two leeches come in contact by their ventral surfaces in head to tail position, so that the male genital pore of one lies against the female genital pore of the other. Each copulant protrudes its penis and passes it into the female genital pore of the other to shed sperms in the vagina. Thus cross-fertilization occurs.

**Fig. 17.29** *Hirudinaria*. Two leeches inpulation.

**Cocoon Formation.**

The cocoon is formed in the month of April to June after copulation. The cocoon is secreted in the form of snow-white, frothy girdle by the glandular skin of 9th, 10th and 11th segmengts. It is always formed during breeding season. After the cocoon is formed, clitellar glands secrete an albuminous fluid, which is deposited in the cocoon with the fertilized ova. The cocoon is then passed over the head of each individual. Leech then withdraws its anterior end backwards by rhythmic movements. When the prostomial region of

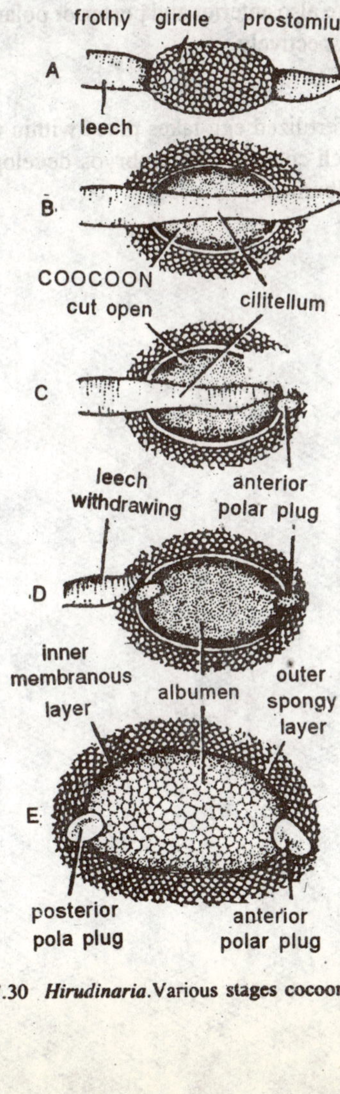


Fig. 17.30 *Hirudinaria.* Various stages cocoon formation.

the worm comes in contact with the anterior end of the cocoon, prostomial glands secrete a kind of plug which seals the anterior part of the cocoon and is called anterior polar plug. The leech goes on withdrawing its anterior end backwards till the prostomial region comes in contact with the posterior end of cocoon. There, it also secretes the posterior plug. After the posterior plug is completed, the leech shakes off the cocoon, which is laid in a moist place by the side of a pond or pool but never in water. Cocoon becomes hardened on exposure to the air. Its colour is light-yellow or amber and it is barrel-shaped. It measures about 25 mm to 30 mm. in length and 12 to 15 mm. in diameter. The wall of cocoon consists of an inner, thin, tough membrane and an outer thick, spongy layer. There are also anterior and posterior polar plugs on the anterior and posterior sides respectively.

**Development.**

Development of fertilized egg takes place within the cocoon and takes about 15 days. In each cocoon 1-24 embryos develop. The development occurs at the cost of albumen.

♦♦♦

# 18

# ARTHROPODA

## Characters and Classification

Phylum *Arthropoda* (Gr.*arthros* = joint; *pods* = foot) contains the great majority of the known animals, about one million species, and many of them are enormously abundant as individuals. It includes such common and well-known forms as the crabs, shrimps, insects, spiders, scorpions, ticks, centipedes, as well as a host of other less familiar forms. *Cuvier* (1800) placed all these animals together with Annelida under the group Articulata. *Van Siebold* (1845) separated Annelida from Articulata. Annelida were placed under class Vermes while the Crustacea, Arachnida and Insecta under an independent group, the *Arthropoda.*

## General Characters

1. *Arthoropoda* are triploblastic, bilaterally symmetrical, metamerically segmented animals.
2. Body is covered with a thick chitinous cuticle forming an exoskeleton.
3. Body segments usually bear paired lateral and jointed appendages.
4. Musculature is not continuous but comprises separate striped muscles.
5. Body cavity is haemocoel. The true coelom is reduced to the spaces of the organs.
6. Digestive system complete; mouthparts adapted for various modes of feeding.
7. Circulatory system open with a dorsal heart, arteries and blood sinuses.
8. Respiration by general body surface, gills, tracheae or book-lungs.
9. Excretory organs are *green glands or Malpighian tubules.*
10. Nervous system with a dorsal nerve ring and a double ventral nerve cord.
11. Cilia are entirely absent from all parts of the body.

12. Sexes are generally separate and sexual dimorphism is often exhibited by several forms.
13. Fertilization is internal. Development is usually indirect through larval stages.
14. Parental care is also often well marked in many arthropods.

Phylum Arthropoda is divided into following classes:

Class I-Crustacea

Class II-Myripoda

Class III-Insecta

Class IV-Arachnida

**CLASS-I. CRUSTACEA**

1. The body is divisible into head, thorax and abdomen, when head and thorax are fused they form cephalothorax.
2. The exoskeleton is chitinous.
3. The body is metamerically segmented, bilaterally symmetrical and triploblastic.
4. Each segment bears paired and jointed appendages.
5. The body cavity is filled with *haemolymph* and known as *Haemocoel.*
6. Respiration takes places either by the general body surface or by gills.
7. Excretion occurs by modified coelomoducts, which may take the form either of maxillary glands or antennary (green) glands.
8. These are unisexual except many Cirripedia and some Isopods.
9. The development is indirect.

Class Crustacea is divided into 7 subclasses

**Subclass-I. Branchiopoda**

1. Body has varying number of segments.
2. Antennules are small, uniramous and usually unsegmented.
3. Compound eyes are present.
4. Thorax bear four or more flat, leaf-like respiratory appendages also help in filter feeding and locomotion.
5. Abdomen without appendages.
6. Parthenogenesis is common.

Subclass-Branchiopoda is divided in three orders:

**Order (i) Anostraca**

1. The carapace is absent.
2. Eyes are stalked.
3. Antennae are uniramous, prehensile in male reduced in female.
4. Trunk elongated with 20 or more segments. Anterior 11-19 segment bear appendages.
5. The caudal styles are unjointed.

   Examples: *Branchipus, Streptocephalus, Artemia.*

**Order (ii) Notostraca**

1. A large dorsal shield-shaped carapace is present.
2. Eyes are sessile i.e. stalkless.
3. Anteenae are reduced.
4. Anterior half of the trunk bears 35-70 pairs of appendages.
5. The caudal styles are many jointed.

   Example: *Apus, Lapidurus.*

**Order (iii) Conchostraca**

1. The carapace is present but divided in two portions or valves.
2. Antennae biramous and large.
3. Eyes are sessile.
4. Caudal styles are in the form of unjointed, curved claws.

   Examples: *Limnetis, Estheria.*

**Order (iv) Cladocera**

1. The carapace bivalved, encloses the trunk but not the head and usually ends posteriorly in an apical spine.
2. Eyes are unstalked and united together.
3. Antennae biramous used for swimming.
4. Trunk segments 10-32, each with a pair of appendages.
5. Caudal styles are unjointed claw-like.

   Example: *Daphnia.*

**Subclass-2. Ostracoda**

1. Compound eyes may or may not be present.
2. Carapace is bivalved and compressed which encloses entire body.
3. Head forms about half of the body.

4. Thorax usually bears only two pairs of cylindrical appendages.
5. Abdomen is reduced, without appendages and ends in a pair of caudal styles.
6. Antennules and antennae locomotory.
7. Manibles have a palp.
8. Larva is *Nauplius.*
9. Parthenogenesis is common.

Subclass Ostracoda is divided into following orders:

***Order (i) Myodocopa***

1. Marine Crustaceans.
2. Carapace with antennal notches.
3. Antennae biramous.
4. Trunk appendages two pair.

Examples: *Cypridina, Conchoecia.*

***Order (ii) Cladocopa***

1. Marine forms.
2. Carapace unnotched.
3. Both pairs of antennae used in swimming.
4. Second pair of antennae with 2 branches.

Example: *Polycope.*

***Order (iii) Podocopa***

1. Marine or freshwater.
2. Carapace Unnotched.
3. Second antennae leg-like, clawed at tips.
4. Trunk appendages 2 pairs.

Examples: *Cypris, Entocythere, Cythereis.*

***Order (iv) Platycopa***

1. Marine forms
2. Carapace unnotched.
3. Both pairs of antennae large but not used for swimming, second pair flattened.
4. Trunk appendages single pair.

Example: ***Cytherella.***

**Subclass 3. Mystacocardia**

1. Primitive marine forms.
2. Size microscopic, cylindrical.
3. Antennules and antennae long, prominent.
4. Eyespots, compound eyes absent.
5. Thorax 4-segments, each with one pair of simple appendages.
6. Excretion by both antennal and maxillary glands.
7. Ventral nerve cord paired.
8. Sexes separate, earliest larva is *metanauplius*.

   Example: *Derocheilocarus*.

**Subclass 4. Copepoda**

1. Fresh water or marine, free-living, commensal or parasitic.
2. Body small, elongated, composed of head, thorax and abdomen.
3. Typically 10 free trunk somites, with 4 lacking appendages. Appendages mostly biramous.
4. Three ocelli often fused as median eye, no compound eyes.
5. Antennules and antennae well developed and used for swimming.
6. Abdomen without appendages, ends in a pair of caudal styles.
7. Excretion by maxillary gland.
8. Eggs on abdomen of female in 1 or 2 egg sacs.

   Subclass Copepoda is divided into 7 orders.

   Examples: *Cyclops, Salmincola, Caligus, Doropygus, Calanus, Monstrilla.*

**Subclass 5. Branchiura**

1. Ectoparasites of fresh water and marine fishes and some amphibians.
2. Body flat, carapace large, disk-like, covering head and thorax.
3. Abdomen small, bilobed.
4. First maxillae modified as suckers.
5. Compound eyes present, sessile.
6. Thoracic appendages 4 pairs, large, biramous, some of them are provided with flagella.

   Example: *Argulus*.

**Subclass-6. Cirripedia**

1. Marine forms.
2. Adults sessile, attached or parasitic; attached by cement gland on first antenna.
3. Carapace becomes calcareous and mantle surrounding body.
4. Six or fewer pairs of slender and bristly biramous appendages behind mouth used in food gathering.
5. Abdomen vestigeal, without appendages, ending in a pair of caudal styles.
6. Antennae and compound eyes disappear in adult.
7. Monoecious. Development includes free swimming *nauplius* which passes through a *cypris* stage which is enclosed in bivalve shell.

This subclass is divided into 5-orders :

***Order (i) Thoracica***

1. Non-parastic sedentary marine barnacles.
2. Body surrounded by calcareous shell.
3. Six pairs of biramous thoracic !imbs.
4. Alimentary canal is present.
5. Hermaphrodite.

Examples: *Balanus, Lepas.*

***Order (ii) Acrothoracica***

1. Parasitic, boring in the shells of mollusc.
2. Body is covered by large mantle. Calcareous shell is absent.
3. Less than 6 pairs of thoracic appendages.
4. Alimentary canal is present.
5. Chitinous disc is present for attachment.

Example: *Alcippe.*

***Order (iii) Ascothoracica***

1. Parasitizing corals and echinoderms.
2. Mouth appendages are modified into piercing and sucking organs.
3. Digestive tract with branches into mantle.
4. 6 pairs of trunk appendages.

Examples: *Laura, Petrarca.*

***Order (iv) Apoda***

1. Parasistic in mantle cavity of *Alepas*.
2. Body without mantle and thoracic appendages.
3. Body maggot-like.
4. Anus absent.
5. Hermaphrodite.

   Example: *Proteolepas*.

***Order (v) Rhizocephala***

1. Parasitic on crabs.
2. A mantle but no shell.
3. Appendages, gut are absent.
4. Body sac-like, with absorptive roots penetrating host.

   Example: *Sacculina*.

**Subclass-7-Malacostraca**

1. Mostly large sized crustaceans, marine and fresh water.
2. Body typically of 19 segments (5 head, 8 thorax and 6 abdomen).
3. Exoskeleton of head united with few or more thoracic segments to form cephalothoracic carapace.
4. Stalked compound eyes.
5. Abdomanal caudal style absent.
6. Development through zoaea stage. Nauplius stage rarely occurs.

   This subclass includes 2 series. :

**Series (A) Leptostraca**

1. Abdomen of eight somites.

***Order (i) Nabaliacea.***

1. Marine forms.
2. Body made up of 21 segments. Abdomen of 7 segments.
3. Thoracic appendages similar and foliaceous.
4. Telson has a pair of caudal styles.
5. Carapace bivaled. Eyes stalked.

   Example: *Nebalia*.

**Series (B). Eumalacostraca**

1. Abdomen of 6 or fewer somites.

*SUPER ORDER (A) SYNCARIDA*

1. Carapace is absent.

***Order(i) Anaspidacea***

1. Fresh waters of Australia.
2. First thoracic segment fused with head. Eyes stalked.

   Example: *Anaspides.*

*Order (ii) Bathynellacea*

1. Small animals present in central Europe.
2. Eyes absent. Pleopods reduced or absent.

   Example : *Bathynella.*

*SUPER ORDER (B) PERACARDIDA*

1. Carapace, when present, leaving last 4 or more thoracic somites distinct.
2. Females with thoracic brood pouch where young develop.
3. Development direct without larval stage.

***Order (i) Mysidacea***

1. Marine forms.
2. Carapace covers almost entire thorax.
3. Eyes stalked.
4. Thoracic appendages are biranmous.
5. Uropods form broad fan-like tail fin.

   Examples: *Mysis.*

***Order (ii) Cumacea***

1. Small, mostly marine, burrowing in surface sand or mud.
2. Carapace with 2 anterior extensions often joined over head.
3. Head and thorax greatly enlarged.
4. Abdomen slender, mobile.
5. Uropods slender.

   Example: *Diastyles, Cumopsis.*

***Order (iii) Tanaidacea***

1. Minute, marine, upto depths of 3650 meters in mud or tubes.
2. Carapace is small and fused to first 2 thoracic segments.

3. Second thoracic appendage with chela.
4. Telson unjointed fused with last abdominal segment.
5. Uropods slender.

   Examples: *Tanais, Apseudes.*

***Order (iv) Isopoda***

1. Mostly marine, some freshwater and few parasitic.
2. Body usually depressed dorsoventrally.
3. Carapace is absent. Abdomen short, partly or all fused.
4. Gills and heart abdominal.

   Examples: *Ligia, Asellus, Hemioniscus, Oniscus.*

***Order (v) Amphipoda***

1. Mostly marine, some freshwater and terrestrial.
2. Body often laterally compressed.
3. Carapace is absent.
4. Telson usually distinct.
5. Gills and heart thoracic.

   Example: *Gammarus, Caprella, Cyamus.*

*SUPER - ORDER (C) HOPLOCARIDA*

1. Head with twc movable anterior segment bearing eyes and antennules.

***Order (i) Stomatopoda***

1. Marine on bottom, in sand or cervices.
2. Antennae scales enlarged.
3. Gills on abdominal appendages modified for raptorial feeding.
4. Carapace small.

   Examples: *Squilla, Pseudosquilla, Chloridella.*

*SUPER ORDER (D) EUCARIDA*

1. Carapace is present, fused to and covering all the thorax.
2. Eyes stalked.
3. Gills thoracic.
4. Brood pouch absent.
5. Development usually indirect with a zoaea larva.

***Order (i) Euphausiacea***

1. Marine, small, pelagic forms.
2. Carapace does not cover sides of body and the gills.
3. Thoracic appendages all biramous and similar, none modified as maxillipeds.
4. Pleopods have setae, used for swimming.
5. Larva is *nauplius*.

   Examples: *Euphausia, Thysanopoda.*

***Order (ii) Decapoda***

1. Mostly marine forms.
2. Carapace covers the entire thorax.
3. Thoracic appendages are modified as 5 pairs of walking legs and three pairs of maxillipedes.
4. Thoracic appendages mostly uniramous.
5. Statocyst present.
6. Gills usually present in three series on the thorax.

Decapoda is divided into 3 suborders:

*Suborder (a) Macrura*

1. Abdomen is well developed and elongated.
2. The abdomen terminates in an expanded fan-like tail fin composed of telson and the uropods.
3. Eyes are not enclosed in orbits.

   Examples: *Lucifer, Palaemon, Homarus, Astacus, Cambarus.*

*Suborder (b) Anomura*

1. Abdomen is more or less reduced and flexed.
2. Tail fin is also reduced.

   Examples: *Eupagurus, Hippa.*

*Suborder (c) Brachyura*

1. Abdomen greatly reduced, hard, permanently flexed under cephalothorax.
2. Antennules and the eyes are both capable of being retracted into cavities.
3. Rostrum and uropods absent.

4. First legs usually large and chelate.

   Examples: *Cancer, Inachus, Uca, Libinia, Inachus*.

**CLASS-II MYRIPODA**

1. Terrestrial and air breathing arthropods.
2. Body is usually elongated with numerous segments, each bearing one or two pairs of appendages.
3. Head is distinct, bearing a pair of antennae, mandibles, eyes and maxillae.
4. Respiration by tracheae.
5. Excretory organs are one or two pairs of Malpighian tubules associated with hindgut.
6. Sexes are separate.

**Subclass-1 Progeneata**

1. The genital apertures are situated near the anterior end of the body.

***Order (i) Pauropoda***

1. The trunk segments are eleven and anal segment is also present.
2. There are 9 pairs of legs.
3. The head bears one pair of antennae, one pair of mandibles and one pair of maxillae.
4. These are without blood vascular system and trachea.

   Example: *Pauropus*.

***Order (ii) Diplopoda***

1. Body is usually cylindrical.
2. Body bears numerous segments each with two pairs of legs(expect first four).
3. Genital opening on the third segment behind the head.

   Example: *Julus, Millipede*.

***Order (iii) Symphyla***

1. There are not more than 12 segments in the trunk.
2. The head bears one pair of antennae, one pair of mandibles and 2 pairs of maxillae.
3. There is a single pair of tracheae.

   Example: *Scolopendrella, Scutigerella*

**Subclass-2. Opisthogoneata**

1. Genital apertures are situated at the posterior extremity of the body.

***Order (i) Chilopoda***

1. Body is usually dorsoventrally flattened.
2. Head bears many jointed antennae, a pair of mandibles and two pairs of maxillae.
3. Poison jaws are present.
4. Trunk bears 15-117 segments each with a single pair of legs.

   Examples: *Scoloplendra, Scutigera.*

**CLASS - III INSECTA**

1. These are air breathing, mostly terrestrial and rarely aquatic arthropods.
2. Body is divisible into head, thorax and abdomen.
3. Head bears a pair of antennae, compound eyes, mandibles and two pairs of maxillae.
4. Thorax consists of three segments bearing 3 pairs of legs and 2 pairs of wings.
5. Abdomen 7-11 segments which are without appendages in adult.
6. Liver absent but salivary glands are present.
7. Respiration by tracheae.
8. Excretion by Malpighian tubules.
9. Sexes separate.
10. Development is usually indirect, completed by complicated metamorphosis. Sometimes direct without larval stage.

    Class Insecta is divided into two subclasses:

    Subclass I Apterygota

    Subclass II Pterygota.

**Subclass-I-Apterygota.**

1. Wingless, primitive insects.
2. Metamorphosis is absent.

***Order (i) Protura***

1. Minute, elongate, whitish insects.
2. Antennae, eyes and ocelli are lacking.
3. Wings are absent.

4. Abdomen short consists of 12 segments. Bilateral styli on first three segments.
5. Tracheae lacking.
6. Malpighian tubules are small papillae.

   Example: *Acerentulus.*

***Order (ii) Collembola***

1. Minute, somewhat tubular or globose forms.
2. Eye patches consist of 1 to several lateral ocelli, dorsal ocelli lacking. Antennae short usually 4-segmented.
3. Abdomen consists of 6 segments, lobe-like organ on first segment. 4th segment bears furcula.
4. Mouth parts entognathus, chewing type.
5. The first segment bearing an adhesive tube-like organ called *Collophore.*

   Examples: *Podura, Isotoma, Sminthurus.*

***Order (iii) Archeognatha***

1. Small, elongate, with many scales on parts of body.
2. Mouth parts are ectognatous chewing type.
3. Antennae filiform, large compound eyes and 3 ocelli are present.
4. Abdomen small bilateral styli on venter of several segments.
5. Females with elongated jointed ovipositor.

   Example: *Machillis.*

***Order (iv) Thysanura***

1. Small, elongated insects. Usually covered with scales.
2. Compoud eyes usually present 0-3 ocelli.
3. Abdomen small, belateral style on venter of several segments.
4. Abdomen consists of 10 or 11 segments.

   Example : *Lepisma, Thermobia.*

**Subclass II Pterygota**

1. Adults are winged, some are wing-less.
2. Metamorphosis complete or incomplete.

**DIVISION (A) EXOPTERYGOTA (HETEROMETABOLA).**

1. Wings develop externally as buds.

2. Metamorphosis simple without plupal stage.

***Order (i) Odonata***

1. Medium to very large sized insects.
2. Head large, mobile bearing short antennae, large compound eyes and 3 dorsal ocelli.
3. Two pairs of membraneous wings, venation net-like.
4. Abdomen elongate. Adult males with gonopores on 9th segment and complex penis.
5. Metamorphosis incomplete. The nymphs aquatic, predatory with caudal tracheal gills.

   Examples : Dragon flies (*Lestes*), Damsel flies.

***Order (ii) Ephemeroptera***

1. Soft bodied, fragile, small to medium sized.
2. Mouth parts chewing type in nymph and vestigial in adults.
3. Antennae short. Compound eyes and 3 ocelli are present.
4. Two pairs of membranous wings. Fore-wings larger than hind-wings.
5. Abdomen with a pair of long cerci and often with a median caudal filament.
6. Metamorphoses incomplete. Nymph bears gills of the first 4-7 segments.

   Example : Mayflies (*Ephemera*).

***Order (iii) Placoptera***

1. Small to medium sized, soft-bodies, elongate, flattened insects.
2. Mouth parts chewing type, usually vestigial in adults.
3. Antennae long and tapering. Well developed compound eyes, usually 3 dorsal ocellia are present.
4. Two paris of membranous wings, males of some species apterous or brachypterous.
5. Metamorphosis incomplete. Nymphs aquatic with tracheal gills.

   Examples : Stoneflies (*Isoperla*, *Pteronarcys*).

***Order (iv) Embioptera***

1. Minute to small, elongated insects.
2. Head large bearing filiform antennae, compound eyes present, ocelli absent.

3. Mouth parts chewing type.
4. Thorax nearly as long as abdomen.
5. Some males with 2 pairs of membranous wings of nearly equal size and shape. females and some males wingless.
6. Metamorphosis gradual in male but absent in female.

   Example : Web-spinners (*Oligotoma*).

***Order (v) Phasmida***

1. Large size insects. Body elongate, cylindrical with short prothorax, large meso-and meta-thorax.
2. Head broad and bears a pair of long filiform or moniliform antennae. Compound eyes are present 2-3 or no ocelli.
3. Mouth parts biting type.
4. Wings present or absent.
5. Abdomen 11 segmental. Ovipositor small and somewhat concealed. Cerci short, unsegmented.

   *Examples : Phyllium, Carausius* (stick insect).

***Order (vi) Orthoptera***

1. Minute to very large sized insects.
2. Mouth parts chewing type. All gradation between hypognathous and opisthognathous.
3. A pair of antennae of variable structure.
4. Compound eyes and usually 2 or 3 ocelli are present.
5. Two pairs of wings are present. Forewings hard, narrow, parchment like opaque with indistinct venation forming tegmina (wing covers).
6. Abdomen 11 segmented. Females in several species with well developed ovipositors. Cerci variable, short, long, clasper-like, segmented or unsegmented.

   Examples : Grasshoppers (*Poecillocerus*), Locusts (*Schistocerca*), Crickets (*Gryllus*), Mole cricket (*Gryllotalpa*).

***Order (v) Dermaptera***

1. Small to medium sized insects.
2. Mouth parts chewing type.
3. Antennae long slender. Compound eyes present in most of the forms may be vestigial or absent in some forms.

4. Fore-wings short, leathery and veinless, hind-wings membraneous, broad, fan-like.
5. Abdomen 11 segmented bearing a pair of forceps-like cerci at the end.
6. Metamorphosis gradual.

   Example : *Earwigs* (*Forficula*).

***Order (vii) Grylloblattoidea***

1. Medium to large, somewhat elongate and slender.
2. Mouth-parts biting and chewing type.
3. Antennae filiform, compound eyes reduced or absent. Ocelli also absent.
4. Wings absent.
5. Abdomen well developed with sword-shaped ovipositor. Cerci long.
6. Metamorphosis simple.

   Example : *Grylloblatta*.

***Order (vii) Isoptera***

1. Social and polymorphic insects living in colonies under a caste system.
2. Mouth-parts chewing type.
3. Compound eyes are present in all winged forms; present or absent in apterous forms. O or 2 dorsal ocelli are present.
4. Antennae short, moniliform or filiform.
5. Wings when present, two pairs, similar in size, shape and venation.
6. Cerci short and simple.
7. Metamorphosis simple.

   Examples : Termites (*Zootermopsis Nasutitermes*).

***Order (ix) Blattaria***

1. Medium to large sized insects, some what dorsoventrally flattened.
2. Mouth parts chewing type.
3. Compound eyes well developed, 2 ocelli in some. Antennae long filiform.
4. Wings present or absent, when present forewings tegmina, hind wing membranous.
5. Pair of styli on 9th sternum of males. Ovipositor reduced and concealed by 7th abdominal sternum.

6. Metamorphosis simple.

   Example : *Periplaneta*, *Blattella*.

***Order (x) Mantodea***

1. Medium to large sized insects. Body usually elongate, somewhat cylindrical, with elongate prothorax.
2. Head small, triangular and bears large compound eyes with 3 or more ocelli.
3. Mouth parts chewing type.
4. Raptorial forelegs at anterior end of elongated prothorax.
5. Wings usually present in males, reduced or absent in females.
6. Abdomen comparatively short, multisegmented cerci, pair of styli usually associated with 9th sternum in males.
7. Metamorphosis simple.

   Example : Praying mantids (*Tenodera*).

***Order (xi) Zoraptera***

1. Minute sized insects.
2. Mouth parts chewing type.
3. Compound eyes and ocelli present in winged forms but absent in wingless forms.
4. Wings present or absent when present forewings larger than hindwings.
5. Abdomen short one segmented cerci.
6. Metamorphosis simple.

   Example : *Zorotypus*.

***Order (xii) Psocoptera***

1. Minute sized insects. Head capsule large compared to rest of body.
2. Mouth parts chewing type. Labial silk-glands are present.
3. Compound eyes weakly or strongly developed. Ocelli 3 in winged forms absent in wingless forms.
4. Wings when present, two pairs, membranous.
5. Cerci absent.
6. Metamorphosis simple.

   Examples : Book-lice (*Liposcelis*), *Peripsocus*.

***Order (xiii) Mallophaga***

1. Small, apterous, ectoparasitic insects.
2. Body is broad or elongated, flattened dorsally.
3. Head generally large. Compound eyes reduced. Ocelli absent.
4. Mouth parts biting type.
5. Cerci absent.
6. Metamorphosis absent.

   Examples : Chewing lice or biting lice (*Trichodectes*, *Menacanthus*).

***Order (xiv) Anoplura***

1. Minute, dorsoventrally flattened, extoparasitic.
2. Head narrower than thorax.
3. Mouth parts piercing and sucking, retracted into head when not feeding.
4. Compound eyes weakly developed or lacking, ocelli also absent.
5. Antennae short 3-5 segments, setiform.
6. Wings are absent.
7. Legs adapted for clinging to the hairs of the host and each terminating in a curved claw.
8. Metamorphosis absent.

   Examples : Human body-louse (*Pediculus*), *Phthirsu*.

***Order (xv) Thysanoptera***

1. Minute to small sized insects, Body slender.
2. Mouth parts rasping and sucking type : Asymmetrical, the right mandible reduced or absent.
3. Small, compound eyes; 3 dorsal ocelli in winged forms absent in wingles forms.
4. Antennae short 6-10 segments.
5. Wings two pairs, narrow fringed; coupted by basel hooks.
6. Abdomen 10-11 segments with an ovipositor. Cerci lacking.
7. Metamporphosis simple.

   Example : Thrips (*Heliothrips*).

***Order (xvi) Hemiptera***

1. Medium to very large sized insects.

2. Mouth parts piercing and sucking type.
3. Antennae 2-10 or rarely 25 segmented.
4. Eyes large with or without ocelli.
5. Wings when present, two pairs, fore wings usually thickened basally and membranous apically (suborder Heteroptera) and wholly membranous (suborder Homoptera).
6. A triangular plate is located between the bases of the wings.
7. Anal respirator filaments are present in some aquatic forms and anal filaments in male coccids.
8. Metamorphosis gradual.

   Examples : Gaint water bug (*Belostoma*), Bedbug (*Cimex*), Water strider (*Gerris*), Squash bug (*Anasa*), Leaf hopper (*Paraphlepsius*), Cicada (*Magicicada*).

*DIVISION (B) ENDOPTERYGOTA (HOLOMETABOLA)*

1. Wings develop internally.
2. Metamorphosis complete.

***Order (i) Coleoptera***

1. Minute to large sized insects.
2. Antennae variable in shape and size usually 11 segmented.
3. Compound eyes present or absent. Ocelli usually absent but present in some groups. Larva without compound eyes but lateral ocelli present.
4. Forewings sclerotized elytra which protect the membranous hindwings at rest.
5. Legs typically cursorial, some adapted for swimming, jumping or digging.
6. Abdomen usually 10 segmented. Cerci absent.
7. Mouth parts biting and chewing types.
8. Larvae diverse usually catterpillar-like or maggot-like.
9. Metamorphosis complex.

   Examples : Ground beetle (*Calosoma*), Blister beetle (*Epicauta*), Alfalfa beetle (*Hypnera*).

***Order (ii) Strepsiptera***

1. Minute sized mostly endoparasitic insects.
2. Mouthparts with reduced mandibles.

3. Compound eyes present in males and free-living females.
4. Antennae flabellate in males, absent in females.
5. Wings two pairs in males, forewings clublike, hindwings membranous and fan-like.
6. 3-5 genital pores in the abdomen of females

   Example : *Stylops*.

***Order (iii) Neuroptera***

1. Small to very large-sized, soft-bodied insects.
2. Mouth parts chewings type in adult and grasping-sucking type in larvae.
3. Compound eyes present. Ocelli present or absent.
4. Antennae filiform.
5. Wings two pairs, similar in size and appearence.
6. Cerci absent. variously modified ovipositors are present.
7. Metamorphosis complete.

   Examples : Green lacewing (*Chrysopa*), Snakefly (*Agulla*), Dobsonfly (*Corydalis*).

***Order (iv) Mecoptera***

1. Small to medium sized, slender, predaceous insects.
2. Head usually elongated. Mouth parts chewing type.
3. Compound eyes are present, Ocelli 0-3. Antennae long and filiform.
4. Wings 2 pairs, membranous, commonly with dark spots.
5. Cerci shorts, simple or two segmented.
6. Metamorphosis complete.

Examples : Scorpionfly *(Panorpa)*, Hanging Scorpionfly *Bittacus*.

***Order (v) Diptera***

1. Minute to medium sized, diurnal, nocturnal, terrestial or aquatic insects.
2. Mouth parts variously modified for sucking, lapping, piercing and sponging.
3. Compound eyes present. Most species with 3 ocelli.
4. Wings are present, fore wings developed for flight while hind wings are reduced to halters.

5. Cerci present or absent.
6. Metamorphosis complete.

Examples : Mosquitoes (*Aedes, Anopheles, Culex*), Black fly (*Simulium*), House fly (*Musca*), Fruit fly (*Drosophila*).

***Order (vi) Siphonoptera***

1. Minute, hard-bodied bilaterally compressed insects.
2. Mouth parts piercing and sucking type. In larvae chewing type.
3. Compound eyes lacking. Generally 2 ocelli.
4. Antennae short contained within antennae grooves.
5. Wing absent.
6. Male terminalia highly modified with one pair of 2 segemented claspers. Cerci lacking.
7. Metamorphosis complete.

Examples : Human fleas (*Pulex*), Polutary flea (*Echinophaga*), Rat flea (*Xenopsylla*), Dog and cat flea (*Ctenocephalides*).

***Order (vii) Trichoptera***

1. Small to medium sized insects.
2. Monthparts are non-feeding in adult stage. Adapted for inbibling fluid. Mandibles weakly developed or lacking. Larvae with chewing type.
3. Compound eyes present, O or 3 ocelli.
4. Antennae range from setaceous to filiform.
5. Wings 2 pairs, membranous, hind wings broader then forewings; covered with modified hair-like setae.
6. Larvae aquatic
7. Cerci one or two segmented.
8. Metamorphosis holometabolic.

Example : Caddis fly (*Philopotamus*).

***Order (viii) Lepidoptera***

1. Medium to large sized, social or parasitic insects.
2. Antennae 12 segmented in male and 13 segmented in females.
3. Mouth parts variable, chewing to lapping or sucking.
4. Compound eyes usually well developed. Commonly three dorsal ocelli.

5. Two pairs of membranous wings . Hindwings smaller than forewings, many wingless.
6. Abdomen with basal pedical.
7. Female with conspicuous ovipositor modified into saw, drell or sting.
8. Metamorphosis complete.

   Examples : Honey bees (*Apis*), Thief ant (*Solenopsis*), Wasp (*Vespa*), Ants etc.

**CLASS IV ARACHNIDA**

1. These are air breathing, mostly terrestrial arthropods.
2. Body is divisible into cephalothorax and abdomen.
3. Cephalothorax bears two pairs of jointed appendages the chelicerae and pedipalpi and four pairs of legs.
4. Antennae and true jaws are absent.
5. Eyes are sessile and simple.
6. respiration by tracheal or book lungs or book-gills.
7. Heart is dorsal and tubular.
8. Excretion by coxal glands or Malpighian tubules or both.
9. Sexes are separate with conspicuous sexual dimorphosim.
10. Development direct.

***Order (i) Eurypterida***

1. Extinct forms.
2. Exo-skeleton is characteristically sculptured.
3. Cephalothorax is relatively small and the abdomen bears 12 free segments and a telson.
4. Cephalothorax bears 6 pairs of appendages.

   Example : *Eurypterus.*

***Order (ii) Xiphosura***

1. Marine arachnids burrowing in the sand.
2. These animals have large subcircular, apparently unsegmented body with two body division-anterior large cephalothorax (prosoma) and posterior small abdomen (opisthosoma), the two are joined by a hinge.
3. The cephalothorax bears a pair of short chelate appendages and 5 pairs of legs.
4. Abdomen bears operculum.

5. Respiration takes place by lamelliform gills attached to the abdominal appendages.
6. Sexes separate. Development includes a trilobite larva.

   Example : King crab (*Limulus*).

***Order (iii) Scorpionida***

1. Terrestrial animals found under stones in tropical and sub-tropical regions.
2. Body is divisible into cephalothorax and abdomen.
3. Prosoma bears a pair of chelicerae, a pair of pedipalpi and four pairs of walking legs.
4. Respiration takes place by 4 pairs of book-lungs.

   Examples : *Palamnaeus, Buthus.*

***Order (iv) Pedipalpida***

1. Body consists of cephalothorax and abdomen.
2. The chelicerae are simple and non-chelate.
3. Respiratory organs are two pairs of book lungs.

   Example : *Phryniclus.*

***Order (v) Araneida***

1. Found under stones, bark leaves etc.
2. Body consists of an undivided cephalothorax and an unsegmented abdomen.
3. Cephalothorax bears 6 pairs of appendages.
4. Chelicerae are subchelate with poison glands. Pedipalpi ae simple and 6 jointed.
5. Eight eyes are arranged dorsally on the head.
6. Respiration by book-lungs or book-lung with tracheae.

   Example : Spider (*Aranea*), *Argiope.*

***Order (vi) Palpigradi***

1. Small arachnids.
2. Body consists of cephalothorax and an abdomen of 10 segments.
3. Chelicerae are chelate and pedipalpi leg-like.
4. Respiratory organs are 3 pair of book-lungs.

   Example : *Koenenia.*

***Order (vii) Solifugae***

1. Body divisible into prosome and opisthosoma.
2. Prosome consists of three fused and last three free segments.
3. Opisthosoma, composed of 10 segments, is not marked off from prosoma.
4. Chelicerae are large and chelate. Pedipalpi are elongated and leglike.
5. Respiration by tracheae. Poison gland are absent.

   Example : *Galeodes.*

***Order (viii) Pseudoscorpionida***

1. Minute arachnids having the appearance of scorpion, found under the tree barks.
2. Body consists of cephalothorax and an abdomen of 12 segments.
3. A pair of spinning glands are present.
4. Respiration by tracheae.

   Examples : *Chelifer.*

***Order (ix) Ricinulei***

1. Body consists of cephalothorax and an abdomen of 9 segments.
2. Cephalothorax and pedipalpi bear movable plates which are cucullus.
3. Respiratory organs are tracheae.

   Example : *Cryptocellus.*

***Order (x) Phalangida***

1. Body consists of unsegmented cephalothorax and an abdomen of 10 segments.
2. Chelicerae are chelate and pedipalpi leg - like.
3. Spinning glands are absent.
4. Respiratory organs are tracheae.

   Example : *Obligolophus.*

***Order (xi) Acarina***

1. Body is unsegmented.
2. Chelicerae and pedipalpi are usually small and associated with the mouth parts.
3. Mouth parts are piercing and sucking type.
4. Respiration by tracheae or skin.

   Example : Ticks (*Ixodes*), Mites (*Chorioptes*).

## PHYLUM ONYCHOPHORA

1. Aberrant forms as its several characters are of phylum Annelida and Arthropoda combined.
2. Body-soft, worm-like covered by papillae and hairy spines.
3. Cuticle is free, having transverse rings, metamerism not marked externally.
4. Jaws are only one pair.
5. Series of short, fleshy leg pairs provided with two claws.
6. Respiratory organs are tracheae.
7. Sexes are separate.
8. Regarded as living fossil.

Example : *Peripatus*.

## BRANCHIPUS

| | | |
|---|---|---|
| Phylum | - | Arthropoda |
| Class | - | Crustacea |
| Subclass | - | Branchiopoda |
| Order | - | Arostraca |
| Genus | - | *Branchipus* |

1. It lives in fresh water lakes, ponds etc.
2. It is commonly known as 'brine shrimp'.
3. Body is elongated pinkish, semitransparent and differentiated into head, thorax and abdomen.

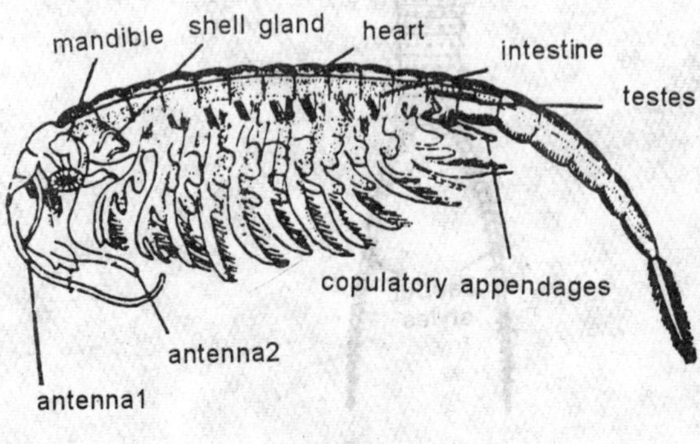


Fig. 18.1*Branchipus*

4. Carapace is absent.
5. Eyes are situated on unjointed stalks.
6. Antennae many and are modified into prehensile organs.
7. The caudal styles are unjointed.
8. Eggs are laid in mud during summer and hatch in the next summer.

**APUS**

| | | |
|---|---|---|
| Phylum | - | Arthropoda |
| Class | - | Crustacea |
| Subclass | - | Branchiopoda |
| Order | - | Notostraca |
| Genus | - | *Apus* |

1. It is.fresh water crustacean found in most parts of the world especially in U.S.A., Kansas, Texas, Mexico etc.
2. It is commonly called 'tadpole shrimp'.

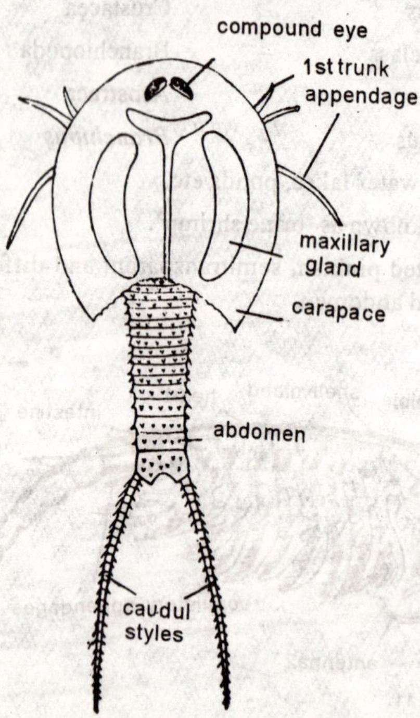


Fig. 18.2 *Apus*.

3. Body is elongated with broad shield-shaped carapace like horse-shoe.
4. Head is broad and depressed bears paired eyes, median eyes and dorsal organs above the antennules and antennae below.
5. Shell glands are present on the lateral surface of carapace.
6. The original segments of the body seem replaced by secondary annulation.
7. Caudal styles are long, filamentous and jointed.
8. Males are rare, hence reproduction commonly occurs by parthenogenesis.

## DAPHNIA

| | | |
|---|---|---|
| Phylum | - | Arthropoda |
| Class | - | Crustacea |
| Subclass | - | Branchiopoda |
| Order | - | Cladocera |
| Genus | - | *Daphnia* |

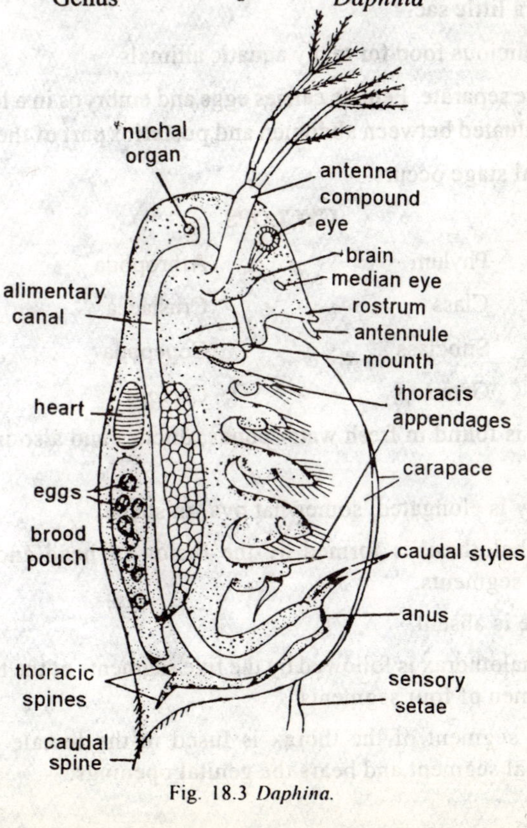


Fig. 18.3 *Daphina.*

1. It is a fresh water crustacean found in ponds, pools and ditches.
2. It is commonly known as 'water flea'.
3. The body except head is covered by folded carapace.
4. Head and abdomen are turned downward.
5. Head bears sessile eyes, nuchal sense organs above it, a pair of antennules, a pair of biramous antennae, large mandibles and two pairs of maxillae.
6. Large biramous antennae are the chief organ of locomotion.
7. Thorax bears usually five pairs of leaf like appendages.
8. Abdomen is devoid of appendages.
9. The female consists of a brood pouch between posterior part of carapace and abdomen in which eggs are stored.
10. The alimentary canal is simple.
11. Heart is a little sac.
12. It is a delicious food for many aquatic animals.
13. Sexes are separate. Female carries eggs and embryos in a large brood pouch situated between abdomen and posterior part of the carapace.
14. No larval stage occurs.

## CYCLOPS

| | | |
|---|---|---|
| Phylum | - | Athropoda |
| Class | - | Crustacea |
| Subclass | - | Copepoda |
| Genus | - | *Cyclops* |

1. *Cyclops* is found in fresh water ponds, ditches and also in brackish water.
2. The body is elongated, somewhat oval in shape.
3. The cephalothrax is formed by the fusion of head and the first thoracic segments.
4. Carapace is absent.
5. The cephalothorax is followed by the five segments of the thorax and an abdomen of four segments.
6. The last segment of the thorax is fused in the female with first abdominal segment and bears the genital openings.

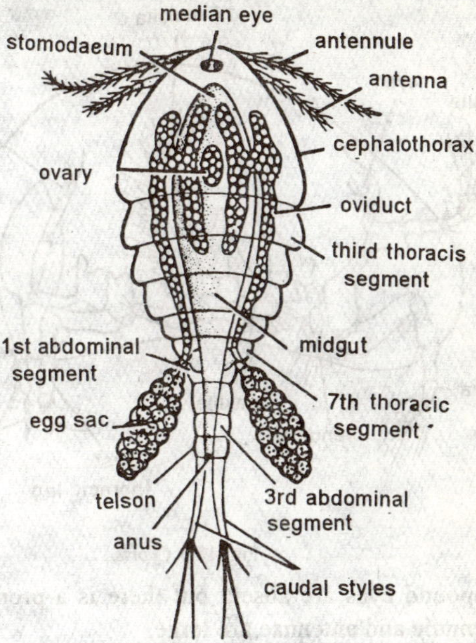


Fig. 18.4 *Cyclops*

7. The antennules are large, entire and segmented and they are meant for copulation in male.
8. The antennae are short and the mandibles have palps.
9. There are two pairs of maxillae following with are maxillipeds and which are the first thoracic appendages. Rest are used for swimming.
10. Abdomen is devoid of appendages except the caudal styles.
11. Sexes are separate. Mature female carries two egg sacs attached to the abdomen.

## CYPRIS

| | | |
|---|---|---|
| Phylum | - | Arthropoda |
| Class | - | Crustacea |
| Subclass | - | Ostracoda |
| Genus | - | *Cypris* |

1. It is free-swimming and occurs in fresh water stagnant ponds.
2. It has about 2 mm long, unsegmented, laterally compressed, clam-like body enclosed in a bivalved shell. The valves of the shell are closed by a transverse adductor muscle.

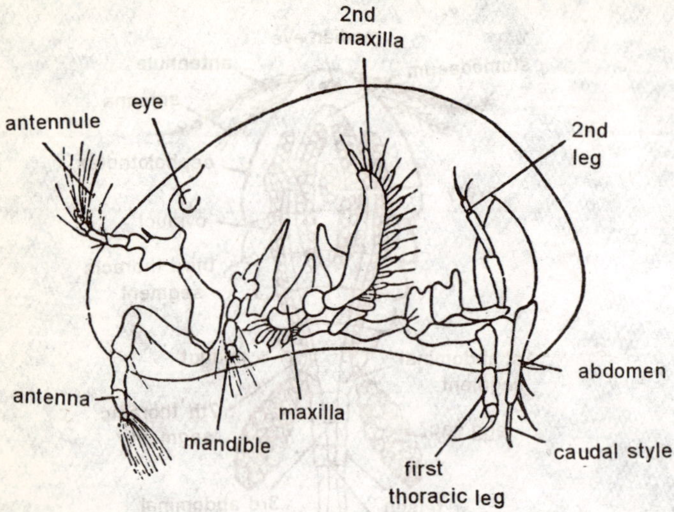


Fig. 18.5 *Cypris.*

3. Compound eyes are absent but there is a prominant median eye. Antennule and antennae are large.
4. It is a carnivorous filter feeder.
5. Abdomen is devoid of appendages and terminates into two caudal styles.
6. Respiration by integument.
7. Development by parthenogenesis.

## ARGULUS

| | | |
|---|---|---|
| Phylum | - | Arthropoda |
| Class | - | Crustacea |
| Subclass | - | Branchura |
| Genus | - | *Argulus* |

1. It is an ectoparasite on fresh water fishes and usually found in the walls of branchial cavity.
2. It is commonly known as 'carp louse'.
3. Body consists of an oval flattened cephalothorax and a small bilobed abdomen.
4. Mandibles and maxillae form piercing organs and are surrounded by a sucking tube called proboscis.
5. A poisonous spine, the stylet, lies in front of the proboscis.

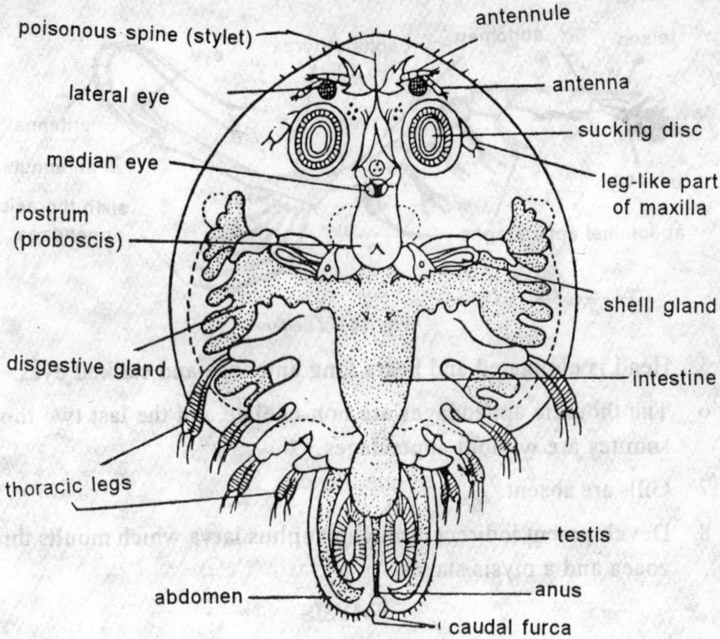


Fig. 18.6 *Argulus.*

6. Second maxillae are divided into two parts, an anterior sucking disc for attachment and a posterior leg like part.
7. A pair of large compound eyes and a median eye are present on the head.
8. Thorax bears 4 pairs of swimming appendages.
9. Gills are absent, respiration probably through the carapace surface.

**LUCIFER**

| | | |
|---|---|---|
| Phylum | - | Arthropoda |
| Class | - | Crustacea |
| Subclass | - | Malacostraca |
| Order | - | Decapoda |
| Genus | - | *Lucifer* |

1. It is a pelagic crustacean found in India, Asia and Europe.
2. The animal is minute, delicate and slender.
3. Body is divisible into cephalothorax and abdomen.
4. Cephalothorax is smaller than abdomen.

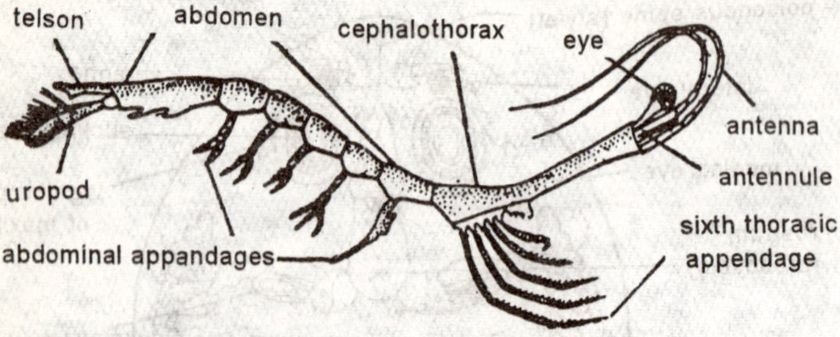


Fig. 18.7 *Lucifer*

5. Head is elongated and bears long antennae and stalked eyes.
6. The thoracic appendages are non-chelate and the last two thoracic somites are without appendages.
7. Gills are absent.
8. Development indirect through nauplius larva which moults through zoaea and a mysis stages.

## MYSIS

| | | |
|---|---|---|
| Phylum | - | Arthropoda |
| Class | - | Crustacea |
| Subclass | - | Malacostraca |
| Order | - | Mysidacea |
| Genus | - | *Mysis* |

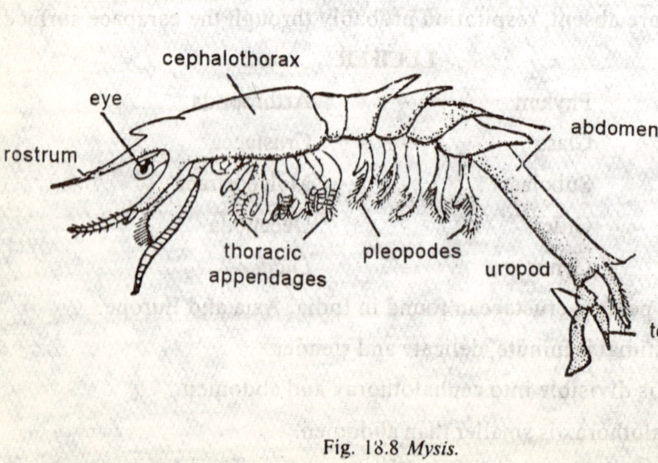


Fig. 18.8 *Mysis.*

1. It is found sometimes in sea water and some times in fresh water.
2. *Mysis* is a small transparent, shrimp-like form.
3. Body is bilaterally compressed and elongated measuring from 2-6 mm in length.
4. Carapace covers the entire thorax except the last two segments.
5. Head bears antennules, antennae and a pair of stalked eyes.
6. First pair of thoracic appendages are modifies as maxillipedes and the rest are biramous serving as swimming organs.
7. Brood pouch is attached to the posterior thoracic segments.
8. Development takes place within the brood pouch so there is no larval stage.

## GAMMARUS

| | | |
|---|---|---|
| Phylum | - | Arthropoda |
| Class | - | Crustacea |
| Subclass | - | Malacostraca |
| Order | - | Amphipoda |
| Genus | - | *Gammarus* |

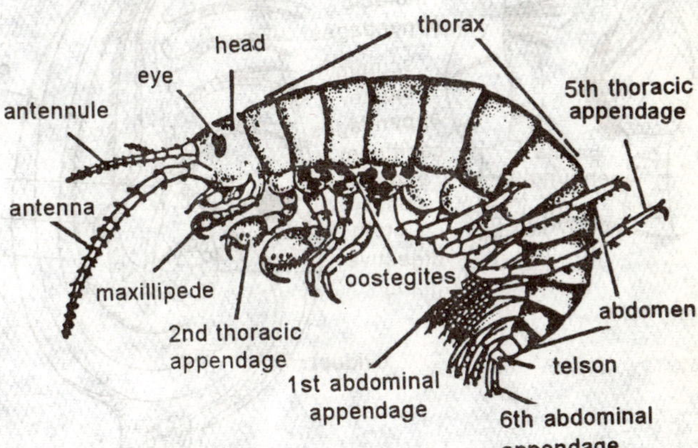


Fig. 18.9 *Grammarus.*

1. It is marine as well as found in fresh water. It is commonly called 'sand-flea'.
2. Body elongated and laterally compressed.

3. Cephalothorax is formed by the fusion of head and first thoracic segment.
4. Carapace lacking.
5. Head bears a pair of antennule, antennae and sessile eyes.
6. First thoracic segment bears maxillipeds and the remaining seven segments bear walking legs. First two pairs of legs are chelate.
7. The abdomen is of a considerable size and flexed ventrally, bears a bifid telson.
8. First three pairs of abdominal appendages are used for swimming and rest for jumping on the ground.
9. It feeds on dead and decaying organisms.

## LEPAS

| | | |
|---|---|---|
| Phylum | - | Arthropoda |
| Class | - | Crustacea |
| Subclass | - | Cirripedia |
| Order | - | Thoracica |
| Genus | - | *Lepas* |

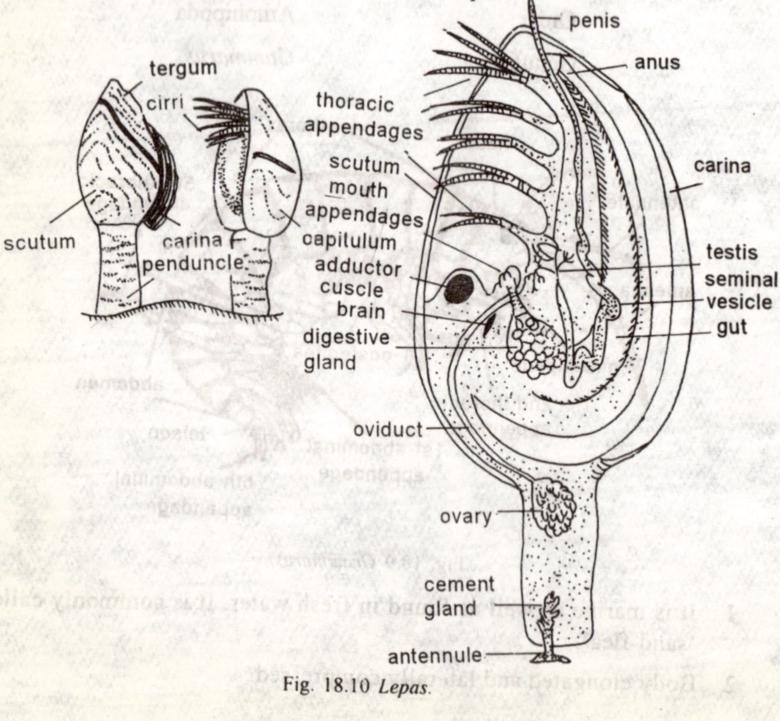


Fig. 18.10 *Lepas*.

1. It is sessile in habit and found attached in the clusters of floating objects like logs and the bottom of ships in the sea.
2. It has cosmopolitan distribution.
3. It is commonly called the 'goose barnacle'.
4. Body consists of a long stalk or pedicel and the body proper or capitulum.
5. Pedicle is covered by wrinkled skin.
6. The capitulum is enclosed in a bivalve carapace strengthened by 5 calcareous plates, 2 proximal scuta, 2 distal terga and a single dorsal carina.
7. The pedicle is really the highly modified anteriormost part of the head contains the vestige of the antennules, ovaries and the cement gland which provide the stickly secretion for attachment.
8. Antennae and paired eyes are absent.
9. Mouth is provided with a pair of mandibles and two pairs of maxillae.
10. Thorax bears 6 pairs of appendages.
11. Behind the appendages is a penis.
12. Abdomen is reduced to a process without any appendages.
13. It is bisexual, paired testes lie at the side of digestive tract and ovaries in the pedicle.
14. Development indirect through nauplius which moults into cypris stage which metamorphoses into adult.

## BALANUS

The systematic position is the same as the *Lepas*.

1. It is found attached to rocks and floating woods, shell of molluscs, carapace of turtle etc. in sea water.
2. It has cosmopolitan distribution.
3. It is commonly known as 'rock-barnacle' or 'acord-barnacle'.
4. It is sessible i.e. without pedicle, therefore, the body is directly attached to the rock or substratum.
5. Body is enclosed in a cylindrical case formed of six calcareous plates - unpaired carina and rostrum and two pairs of carinolaterals.
6. At the top of the shell there is an opening which is closed by two pairs of valves called opercular valves.
7. From the opening 6 pairs of thoracic feet which sweep in food particles of animal.

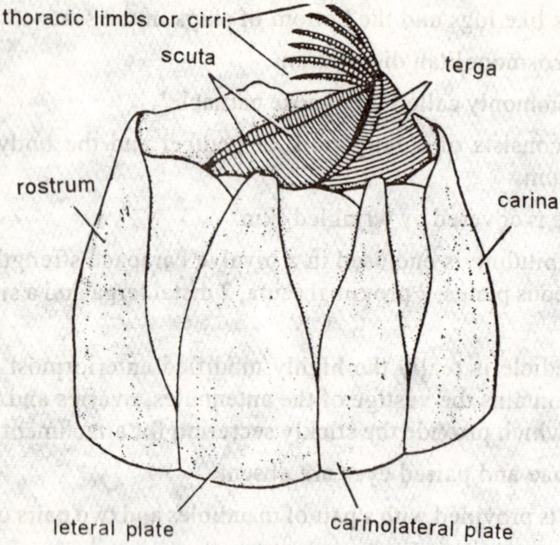


Fig. 18.11 *Balanus*.

8. Eyes and Entennae are absent.
9. Bisexual or hermaphrodite.
10. Development indirect and a nauplius larva is set free which moults into cypris stage and finally metamorphoses into adult.

## SACCULINA

| | | |
|---|---|---|
| Phylum | - | Arthropoda |
| Class | - | Crustacea |
| Subclass | - | Cirripedia |
| Order | - | Rhizocephala |
| Genus | - | *Sacculina* |

1. *Sacculina* is parasite on carbs found attached between thorax and abdomen.
2. It has world wide distribution.
3. It is commonly called 'root-headed barnacle'.
4. Adult is extremely degenerate and appears as a fleshy tumour on the host.
6. The peduncle gives out many delicate root-like process into the tissue of host through which it absorbs nourishment.

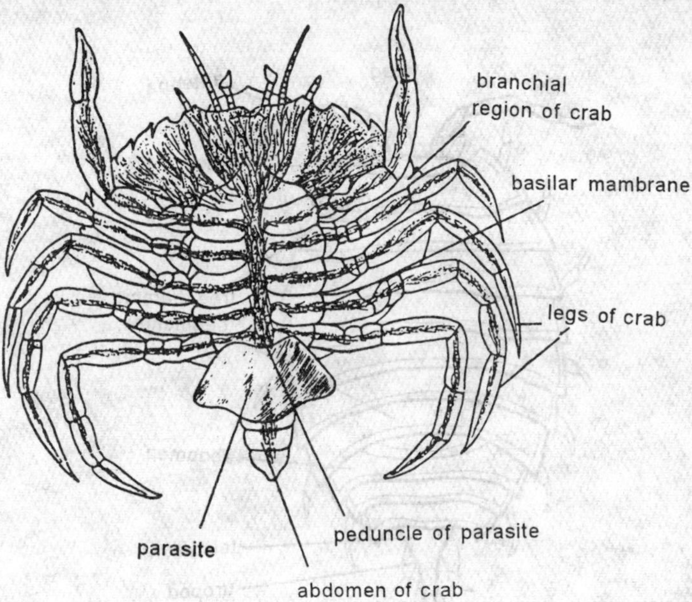


Fig. 18.12 *Sacculina* on the abdomen of crab.

7. Opening of mantle cavity at the posterior end of the body is known as cloacal aperture.
8. These are bisexual.
9. Development is indirect through cirripide nauplius, which transform into cypris. Cypris attaches itself to the crab by its antennules. The cypris moults and called kentrogen larva. It travels within the blood and attached to ventral side of intestine. Rootlets grow out from it and larva metamorphoses into adult.
10. It causes changes in the secondary sexual characters of host. Male crab develops female characters while female ovary degenerates. This is known as *Parasitic Castration.*

**ONISCUS**

| | | |
|---|---|---|
| Phylum | - | Arthropoda |
| Class | - | Crustacea |
| Subclass | - | Malacostraca |
| Order | - | Isopoda |
| Genus | - | *Oniscus* |

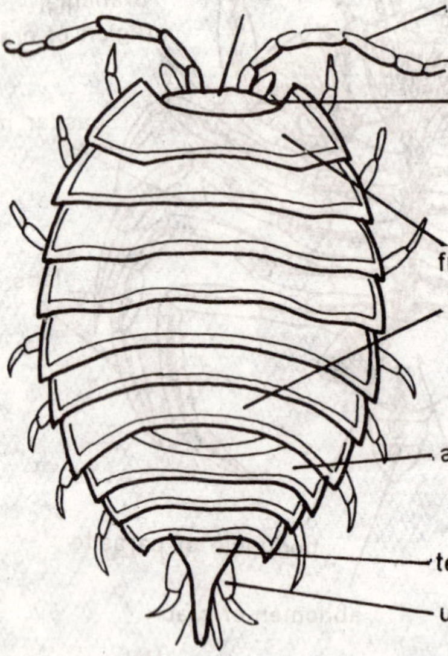


Fig. 18.13 *Oniscus*

1. It is found under stones, woods, etc., in damp soil. It is found in Europe.
2. It is commonly called as 'wood lice'.
3. The body is oval and dorso-ventrally flattened.
4. First thoracic segment fused with the head to form cephalothorax.
5. Unstalked compound eyes are present.
6. Antennules and antennae are uniramous.
7. The abdominal legs are adapted for aerial respiration and their endopodiates branchiae traversed by minute tube called pseudotracheae.
8. Abdomen short.
9. Youngs develop in a brood pouch, present on ventral side.

**SQUILLA**

| | | |
|---|---|---|
| Phylum | - | Arthropoda |
| Class | - | Crustacea |
| Subclass | - | Malacostraca |
| Order | - | Stomatopoda |
| Genus | - | *Squilla* |

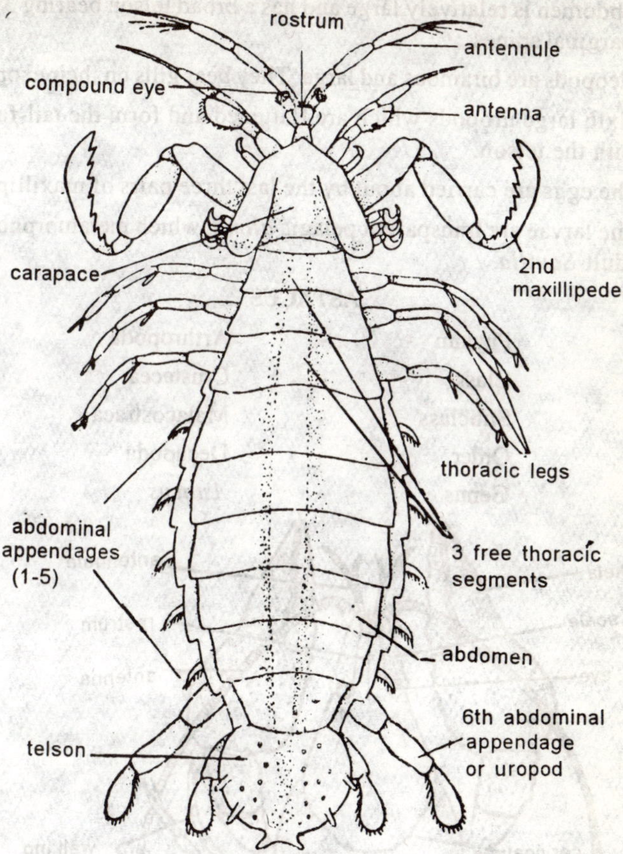


Fig. 18.14 *Squilla*

1. It is found in burrows of the sand or mud at the bottom of sea.
2. It is found in India, southern coast of Florida and Gulf of Mexico.
3. Body is elongated measuring 25 cm. in length, divisible into head, thorax and abdomen.
4. Carapace is thin and uncalcified. It is produced anteriorly into a movable rostrum.
5. Compound eyes are stalked.
6. Anterior 5 pairs of thoracic appendages are uniramous and directd forwards to act as maxillipedes. The remaining thoracic appendages are biramous and act as legs.
7. The 2nd pair of maxillipedes are very large and form powerful prehensile weapons.

8. Abdomen is relatively large and has a broad telson bearing 3 pairs of marginal spines.
9. Pleopods are biramous and large. They bear gills on their exopodities.
10. Sixth large uropods which are flattened and form the tail-fan along with the telson.
11. The eggs are carried about by the last three pairs of maxillipedes.
12. The larvae are transparent pelagic *alima*, which metamorphoses into adult *Squilla*.

## ASTACUS

| | | |
|---|---|---|
| Phylum | - | Arthropoda |
| Class | - | Crustacea |
| Subclass | - | Malacostraca |
| Order | - | Decapoda |
| Genus | - | *Astacus* |

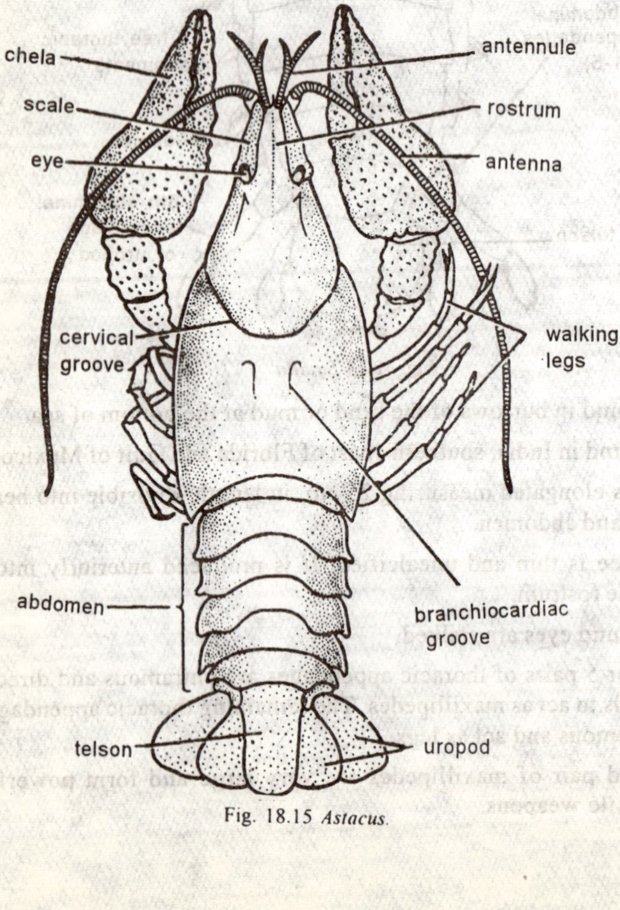


Fig. 18.15 *Astacus*.

1. *Astacus* is found in freshwater ponds, streams and lakes all over the world.
2. It is commonly called 'cray-fish'.
3. The body is long and robust, somewhat compressed.
4. Body is divisible into cephelothorax and abdomen.
5. The line of demarcation between the head and thorax is possibly indicated by cervical groove.
6. Rostrum is short and without serrated margins.
7. Head appendages 5 pairs. Eyes, antennules and antennae are present.
8. Thoracic appendages 8 pairs. First three pairs of thoracic legs are chelate and the first pair is largest.
9. Abdomen 6 segmented, with 6 pairs of appendages.
10. In females, eggs are carries on by the pleopods.
11. The development is direct.

## EUPAGURUS

| | | |
|---|---|---|
| Phylum | - | Arthropoda |
| Class | - | Crustacea |
| Subclass | - | Malacostraca |
| Order | - | Decapoda |
| Genus | - | *Eupagurus* |

1. It is found in the empty shells of gastropod molluscs.
2. It has cosmopolitan distribution.
3. It is commonly called 'hermit crab'.
4. The body is highly asymmetrical and consists of cephalothorax and abdomen.
5. Cephelothorax is broad and flattened bearings short antennules, antennae and stalked eyes.
6. Of the 5 pairs of thoracic legs, first are chelate, 2nd and 3rd are nonchelate, 4th and 5th are chelate but reduced.
7. Abdomen is soft, cylindrical, coiled and unsegmented. It bears reduced appendages on the left side and none on right side. Uropods are, however, well developed.
8. Swimmerets are found only in females for carrying eggs.

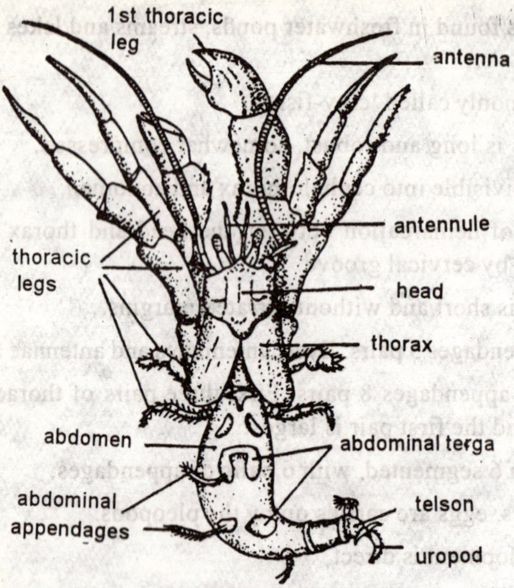


Fig. 18.16 *Eupagurus.*

9. They show *commensalism*. The close association between the hermit crabs and sea-anemone in which both are benefetted by each other.

**HIPPA**

| | | |
|---|---|---|
| Phylum | - | Arthropoda |
| Class | - | Crustacea |
| Subclass | - | Malacostraca |
| Order | - | Decapoda |
| Genus | - | *Hippa* |

1. It is found in the sand near the sea. It is commonly found on Pacific Coast, U.S.A. and Oregon to Mexico.
2. It is commonly known as 'Mole-crab'.
3. Body is ovate with large cephatothotrax.
4. Carapace is smooth.
5. Mouthparts are poorly developed.
6. Head bears long stalked eyes, a pair of biramous antennules and a pair of uniramous antennae.
7. Rostrum is simpe and pointed.
8. Thoracic appendages 7 pairs, walking legs 5 pairs adapted for digging.

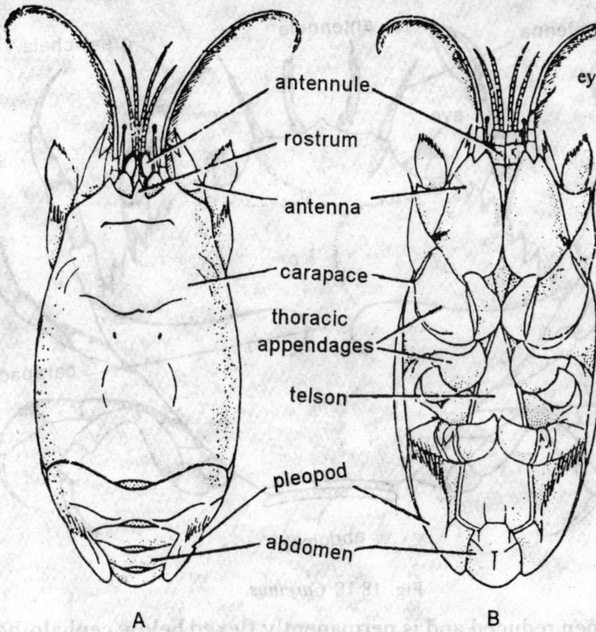


Fig. 18.17 *Hippa.* A= Dorsal view. B=Ventral view.

First two pairs of walking legs partially chelate.

9. Abdomen 6 segmented first three pairs of abdominal appendages modified for swimming and last three pairs are adapted for darting.
10. Feeds like an earthworm, swallowing sand.

## CARCINUS

| | | |
|---|---|---|
| Phylum | - | Arthropoda |
| Class | - | Malacostraca |
| Order | - | Decapoda |
| Genus | - | *Carcinus* |

1. It is commonly found buried among rocks and in sand near about the water. It has cosmopolitan distribution.
2. Cephelophorax broad and somewhat oval with a broad carapace.
3. The antennae and antennules are short.
4. The eyes are stalked and lodged in a pit of the cerapace.
5. Third maxillipedes broad, flat and plate-like.
6. Walking legs 5 pairs, the first pair is large chelate.

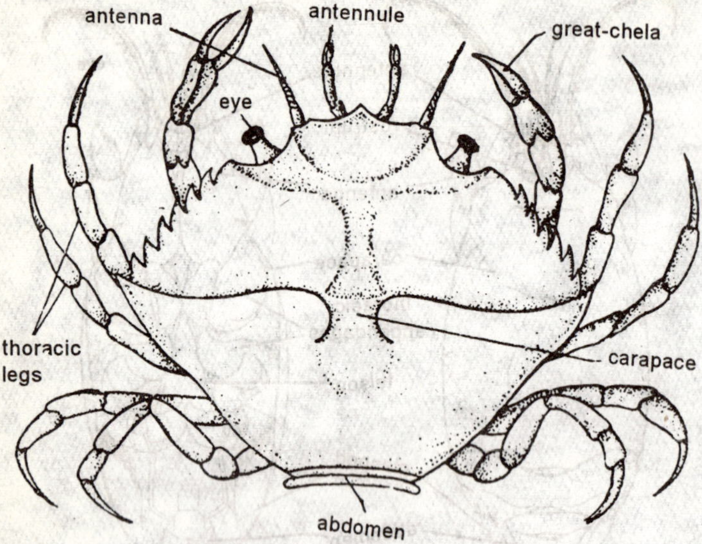


Fig. 18.18 *Carcinus.*

7. Abdomen reduced and is permanently flexed below cephalothorax.
8. Males with 2 pairs of pleopods, whereas female possesses 4 pairs of them. Uropods absent.
9. Eggs are carries by the females which are attached to the pleopods.
10. Development indirect. Egg develop into *zoaea* which passes into *megalopa* and metamorphoses into adult.

## JULUS

| | | |
|---|---|---|
| Phylum | - | Arthropoda |
| Class | - | Myripoda |
| Subclass | - | Progoneata |
| Order | - | Diplopoda |
| Genus | - | *Julus* |

1. It is found is damp and dark places under stones and decaying leaves. When disturbed usually roll up into tight coil and emits an unpleasing odour from segmental scent glands.
2. It has cosmopolitan distribution.
3. It is commonly called as 'Wire worm'.
4. Body cylindrical and metamerically segmented; differentiated into head and trunk.

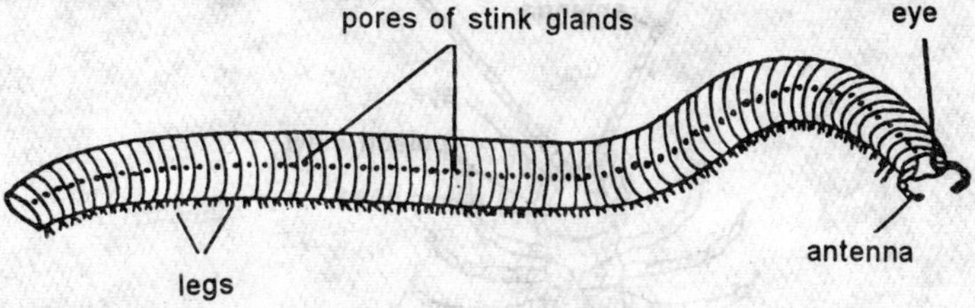


Fig. 18.19 *Julus*.

5. Head small; antennae short generally 7 segment. One pair of maxillae and mandibles forming gnathochilarium.
6. Eyes are minute.
7. Trunk cylindrical consisting of very closely ring like segments from 25 to more than 100 segments.
8. In each segment there are two pairs of legs. Since it has numerous legs it is called millipede.
9. Poison glands and poison claws absent.
10. Vascular system well developed.
11. Respiration by tracheae.
12. Sexes are separate.
13. Genital openings situated at the anterior end of body.
14. In male one or both pairs of legs of 7th segment are modified to form copulatory organs.
15. Herbivorous and even scavanger.

## SCOLOPENDRA

| | | |
|---|---|---|
| Phylum | - | Arthropoda |
| Class | - | Myripoda |
| Subclass | - | Opisthogoneata |
| Genus | - | *Scolopendra* |

1. It is found under stones, in cervices and other dark places. It is found in India and America.
2. It is commonly called 'Centipede'.
3. Body is elongated and dorsoventrally flattened with numerous segments.

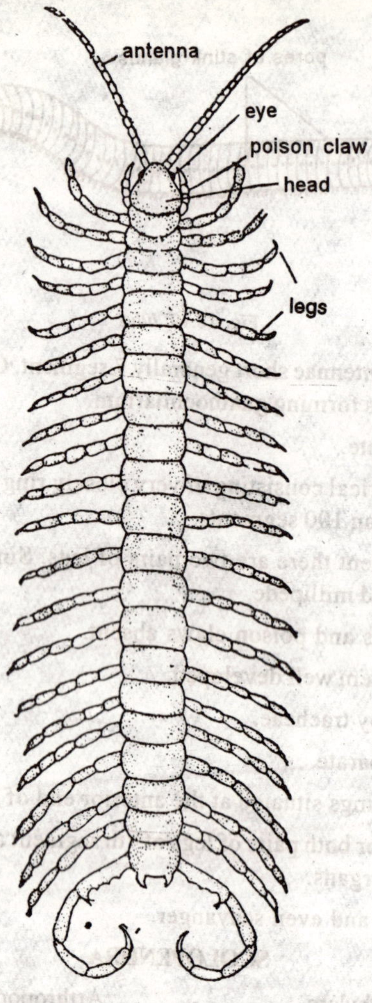


Fig. 18.20 *Scolopendra* - the centipede.

4. Head is distinct and bears a pair of antennae, a pair of mandibles and two pairs of maxillae.
5. Trunk segments numerous, each bearing a single pair of legs.
6. First pair of trunk appendages or maxillipedes bears a sharp claw connected with the poison gland.
7. The last pair of legs are usually longer than the rest.
8. The anus lies at the posterior end.

9. Respiration is affected by tracheal tube.
10. The vascular system well developed.
11. Sexes are separate.
12. The genital opening situated at the hind end of the body.
13. Bite is dangerous to man and sometime proves to be fatal.

## LEPISMA

| | | |
|---|---|---|
| Phylum | - | Arthropoda |
| Class | - | Insecta |
| Subclass | - | Apterygota |
| Order | - | Thysanura |
| Genus | - | *Lepisma* |

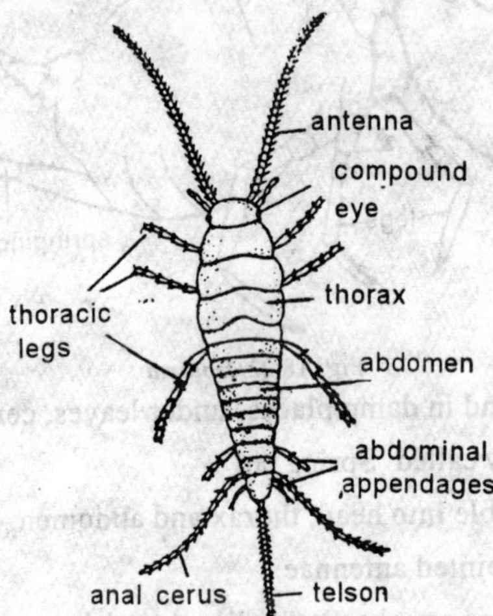


Fig. 18.21 *Lepsima*.

1. It is found in houses, in books, papers, clothes, under pictures in warm countries. It feeds on starch.
2. It is commonly known as 'silver fish'.
3. Body is flattened and covered with silvery, scales.
4. Body divisible into head, thorax and abdomen.
5. Head small with a pair of filiform antennae and eyes.
6. Mouth parts chewing type.
7. Thorax three segmented and bears 3 pairs of legs, wings are absent.

8. Abdomen 11 segmented with 3 anal cerci and a number of small delicate appendages.
9. Development direct without metamorphosis.

## ISOTOMA

| | | |
|---|---|---|
| Phylum | - | Arthropoda |
| Class | - | Insecta |
| Subclass | - | Apterygota |
| Order | - | Collembola |
| Genus | - | *Isotoma* |

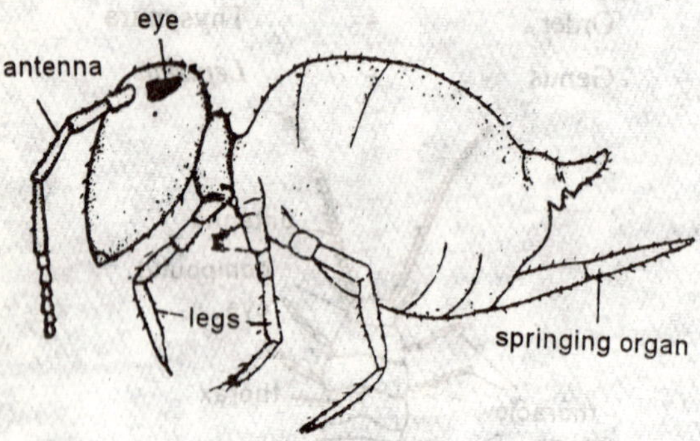


Fig. 18.22 *Isotoma*

1. These are found in damp places, under leaves, cerives of bark.
2. It is commonly called 'Spring tail'.
3. Body is divisible into head, thorax and abdomen.
4. Head with 4 jointed antennae.
5. Compound eyes are absent, ocelli scattered.
6. Thorax 3 segmented with 3 pairs of legs. Wings absent.
7. Abdomen 6 segmented. First segment with a tubular adhesive organs, *collophore*. 3rd segment with *tentaculum* and 4th with *furcula* (spring organ).
8. It secrets glue-like substance used sticking body to the object.

## SCHISTOCERCA

| | | |
|---|---|---|
| Phylum | - | Arthropoda |
| Class | - | Insecta |
| Subclass | - | Pterygota |
| Division | - | Exopterygota |

| | | |
|---|---|---|
| Order | - | Orthoptera |
| Genus | - | *Schistocerca* |
| Species | - | *gregaria* |

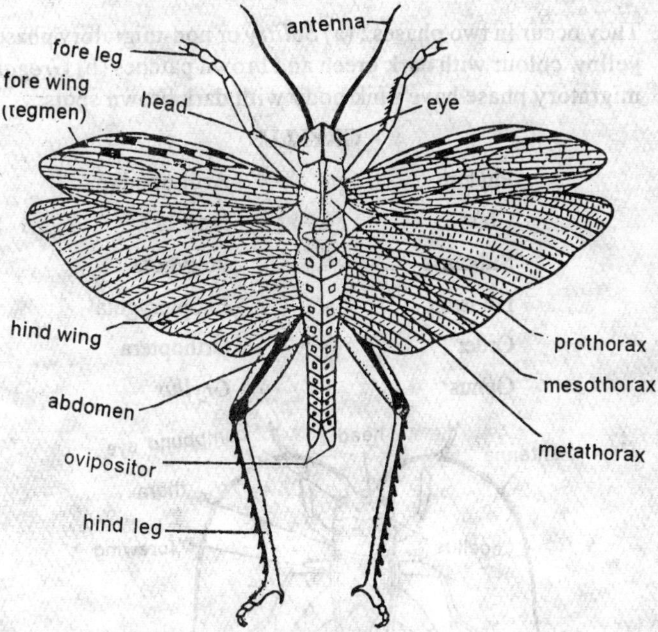


Fig. 18.23 *Schistocera.*

1. Gregarious herbivorous insect found in green vegetation.
2. Commonly found in Africa, Arabia, India, Japan, Mexico, U.S.A. and Afganisthan.
3. Body is divisible into head, thorax and abdomen.
4. Head with are pair of long antennae and well developed compound eyes.
5. Thorax 3 segmented. 3 pairs of legs and two pairs of wings are present. Wings are dissimilar and unequal.
6. The fore wings elongated, straight, thick, leathery with numerous brownish black sports.
7. The membranous hind wing large,plaietd anal area.
8. Legs well developed.
9. The hind legs modifies for jumping having long basally stout femora.
10. The abdomen has 10 evident and vestigial eleventh segments.

11. Mouth parts biting and chewing type.
12. Male with genital lamina incised posteriorly.
13. This is the most destructive of all locusts and its ravages were known in Biblical times.
14. They occur in two phases : (a) *Solitry* or non-migratory phase having yellow colour with dark green and brown patches (b) *Gregareous* or migratory phase have pink body with dark brown spots.

## GRYLLUS

| | | |
|---|---|---|
| Phylum | - | Arthropoda |
| Class | - | Insecta |
| Subclass | - | Pterygota |
| Division | - | Exopterygota |
| Order | - | Orthoptera |
| Genus | - | *Gryllus* |

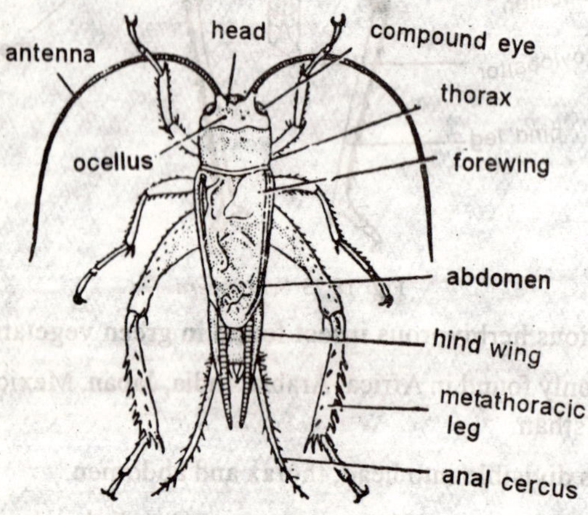


Fig. 18.24 *Gryllus*

1. It is a noctural and omnivorous insect found under various object, stones etc. during day time.
2. It is found in India, Sri Lanka, Asia, Burma, U.S.A., Europe and Canada.
3. It is commonly known as 'circket'.
4. Body divisible into head, thorax and abdomen.
5. Head bears compound eyes and a pair of filiform antennae.

6. Mouthparts chewing type.
7. Forewings hard, narrow and parchment like with pigmented spots (tegmina). Hindwings are membranous and folded.
8. Tibia of the fore-legs bear tympanic organs and the hind-legs are modified for jumping.
9. Well developed ovipositor in females.
10. Male circket chirps by rubbing from wings together during courtship.

## GRYLLOTALPA

The systematic position is same as that of *Gryllus*.

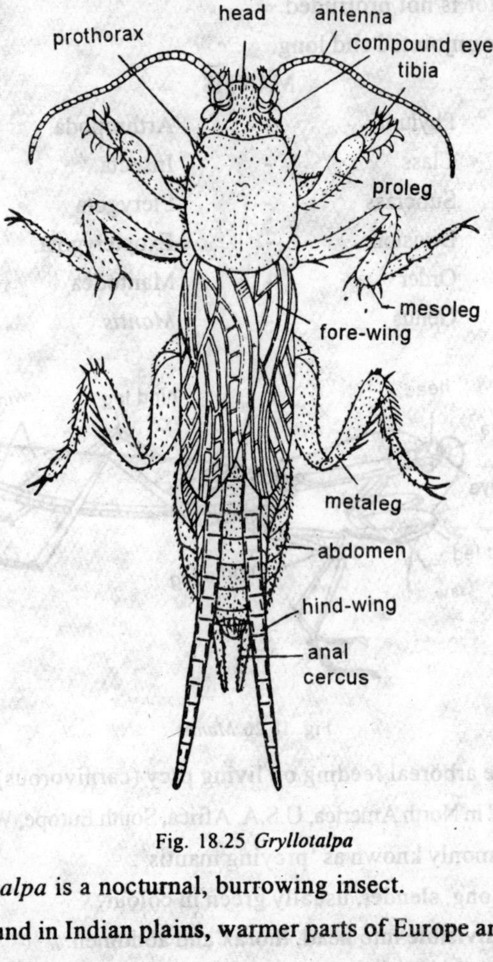


Fig. 18.25 *Gryllotalpa*

1. *Gryllotalpa* is a nocturnal, burrowing insect.
2. It is found in Indian plains, warmer parts of Europe and Asia.

3. It is Commonly known as 'mole-circket'.
4. Body is covered with short, fine hairs and is divisible into head, thorax and abdomen.
5. Head is well developed, bears a pair of antennae and commpound eyes.
6. Mouth parts are biting and chewing type.
7. Tibiae of the first pairs of legs are especially modified with stout spines for digging purpose.
8. Forewings are short and hard membranous.
9. Ovipositor is not protruded.
10. Cerci are unjointed and long.

## MANTIS

| | | |
|---|---|---|
| Phylum | - | Arthropoda |
| Class | - | Insecta |
| Subclass | - | Pterygota |
| Division | - | Exopterygota |
| Order | - | Mantodea |
| Genus | - | *Mantis* |

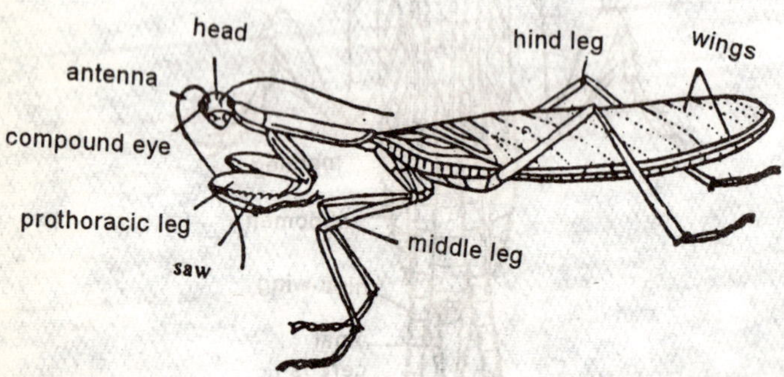


Fig. 18.26 *Mantis*

1. These are arboreal feeding on living prey (carnivorous).
2. It is found in North America, U.S.A. Africa, South Europe, Western Asia.
3. It is commonly known as 'preying mantis'.
4. Body is long, slender, usually green in colour.
5. Body is divisible into head, thorax and abdomen.

6. Head is small, triangular with vertical face and freely movable on slender neck.
7. Large compound eyes, three ocelli and long filiform antennae are present on head.
8. Mouthparts are chewing type.
9. Prothorax is much elongated and prothoracic legs are modifies for grasping and holding them prey.
10. Wings are well developed. These are folded flat and overlapping the sides of the body.
11. Abdomen 10 segment. Ovipositor is no exected.
12. Nineth segment in males bear a pair of styli.
13. Short flattened and segmental cerci present.
14. Female eats the male after nuptial. 200 or more eggs are enclosed in the ootheca formed of frothy quick drying material.

## PHYLLIUM

| | | |
|---|---|---|
| Phylum | - | Arthropoda |
| Class | - | Insecta |
| Subclass | - | Pterygota |
| Division | - | Exopterygota |
| Order | - | Phasmida |
| Genus | - | *Phyllium* |

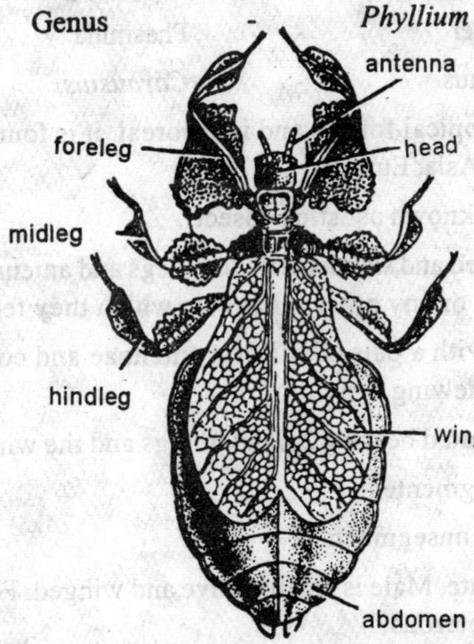


Fig. 18.27 *Phyllium*

1. It is found on the moist part of Oriental region, Forests of Assam etc.
2. It is comonly called 'leaf-insect'.
3. Body is broad and flattened
4. Body is divisible into head, thorax and abdomen.
5. Head is small bears small antennae.
6. Thorax is also small and bears 3-pairs of green coloured legs.
7. Tegmen or first pairs of wings are small than the other pair, having leaf-like colouration and veination.
8. Sexes distinct. Male with small tegmina, well developed wings and narrow bodies. Female with extensive tegmina produced into leaf like expansions.
9. It is an exanoke if mimicry or amouflage in which protective resemblance in both form and colouration, is similar to the objects in their surroundings.

## CARAUSIUS

| | | |
|---|---|---|
| Phylum | - | Arthropoda |
| Class | - | Insecta |
| Subclass | - | Pterygota |
| Division | - | Exopterygota |
| Order | - | Phasmida |
| Genus | - | *Carausius* |

1. It is found in tropical forests and is arboreal, it is found in India, Sri Lanka, Burma, Asia, Europe.
2. It is commonly known as 'stick-insect'.
3. Body is elongated and slender with long legs and antennae resembling a twing or stick or any other foliage on which they feed.
4. Head is small with a pair of filiform antennae and compound eyes. Mounthparts chewing type.
5. Thorax is elongated bearing 3 pairs of legs and the wings are absent.
6. Abdomen 10 segmented.
7. Cerci small and unsegmented.
8. Sexes are separate, Male is small, active and winged. Female is large, sluggish and apterous.
9. Exclusively parthenogenetic.

10. Power of regeneration or autotomy is very great.

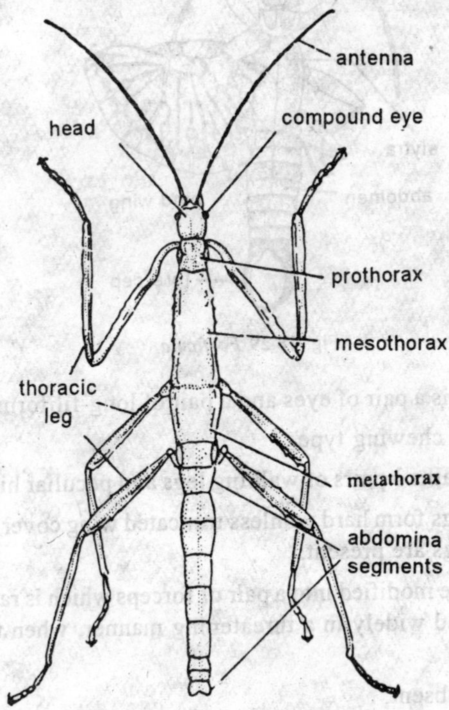


Fig. 18.28 *Carausius*

## FORFICULA

| | | |
|---|---|---|
| Phylum | - | Arthropoda |
| Class | - | Insecta |
| Subclass | - | Exopterygota |
| Order | - | Dermaptera |
| Genus | - | *Forficula* |

1. Found under stones, moist places under stones in decaying vegetable matter etc.
2. It is found in Europe, N. America, Mexico, Asia, Africa and on the coasts of Indian ocean.
3. It is commonly known as 'earwing'.
4. Small insect with elongate body.
5. Body is divisible into head, thorax and abdomen.

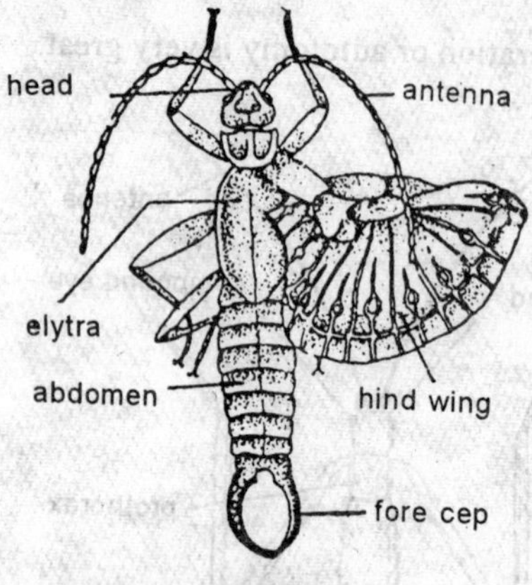


Fig. 18.29 *Forficula*

6. Head contains a pair of eyes and a pair of long-filiform antennae.
7. Mouth parts chewing type.
8. Thorax contains 3 pairs of walking legs and peculiar hindwings.
9. The forewings form hard veinless truncated wing cover below which the hindwings are present.
10. Anal cerci are modified into a pair of forceps which is raised upwards and is opened widely in a threatening manner, when the animal is disturbed.
11. Ovipositor absent.
12. Metamorphosis gradual and it exhibits parental care for eggs and young ones.

## PEDICULUS

| | | |
|---|---|---|
| Phylum | - | Arthropoda |
| Class | - | Insecta |
| Suborder | - | Pterygota |
| Division | - | Exopterygota |
| Order | - | Anoplura |
| Genus | - | *Pediculus* |

1. These are minute, flat, wingless and ectoparasite of man. It has cosmopolitan distribution.
2. It is commonly known as 'human louse'.
3. Body is dorso-ventrally flattened and pale in colour with dark marking along the sides.

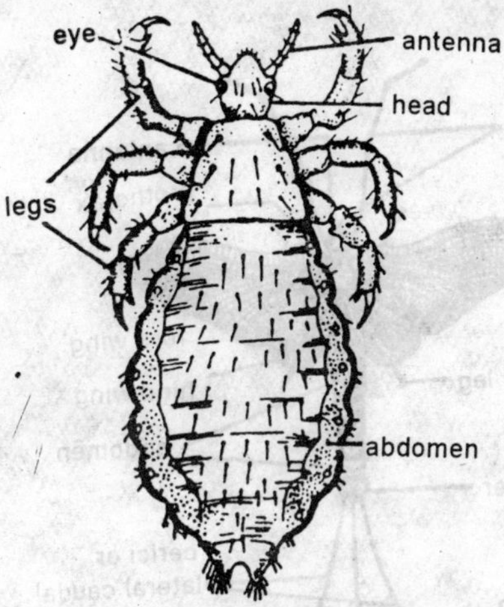


Fig. 18.30 *Pediculus*

4. Head bears a pairs of antennae which is 5 segmented, and a pair of compound eyes.
5. Mouthparts piercing and sucking type.
6. Thoracic segments are fused in one mass and contain 3 pairs of clawed legs, which form clinging organs with the tissue of the host.
7. Abdomen is 9 segmented.
8. Males are smaller than females and their posterior ends turned upward.
9. Metamorphosis is absent.
10. It is a major factor in transmitting relapsing fever, typhus and trench fever. It sucks the blood of man.

## EPHEMERA

| | | |
|---|---|---|
| Phylum | - | Arthropoda |
| Class | - | Insecta |
| Subclass | - | Pterygota |
| Division | - | Exopterygota |
| Order | - | Ephemerida |
| Genus | - | *Ephemera* |

1. It is a soft-bodies and most familiar insect found on the margins of streams, lacks etc.
2. It is found in U.S.A., India, Sri Lanka and Burma.

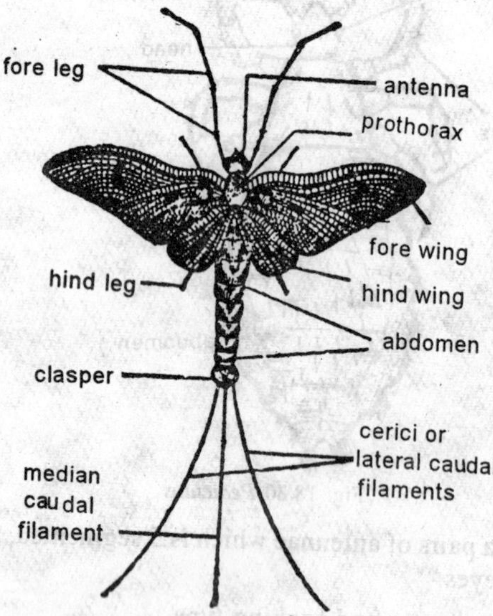


Fig. 18.31 *Ephemera*

3. It is commonly called as 'Mayfly',
4. Body is divisible into head, thorax and abdomen.
5. Head is small contain small setose antennae, compound eyes and 3 ocelli.
6. Mouthparts vestigial.
7. 3 pairs of inconstant walking legs.
8. Two pairs of membranous wings, forewings larger than hindwings.
9. Abdomen is 10 segmented with a reduced tergum. Last segment bears a pair of long many jointed cerci. Tergum is produced into a long caudal filament.
10. Male has 3 jointed, clasping organ and the female has no ovipositor.
11. Nymph or naiad lives for a long time upto 3 years in water, feeding on aquatic plants and animals. The imago moults prior to mating and is extremely short lived.

**DRAGONFLY**

| | | |
|---|---|---|
| Phylum | - | Arthropoda |
| Class | - | Insecta |
| Subclass | - | Pterygota |

| | | |
|---|---|---|
| Division | - | Exopterygota |
| Order | - | Odonata |
| Genus | - | *Libellula* |

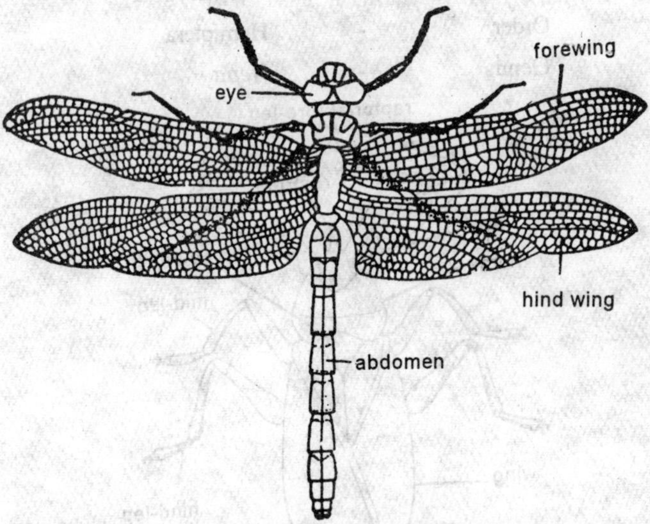


Fig. 18.32 *Libellula*

1. It is of world wide distribution and generally found darting balancing in the air.
2. Body is slender and divisbile into head, thorax and abdomen.
3. Head bears a pair of large compound eyes and small inconspicuous antenne.
4. Mouth parts chewing type.
5. Prothorax is small. Meso-and metathorax are fused. Legs are weak and used for resting.
6. Both pairs of wings are well developed and membranous. These are held in a horizontal position.
7. The abdomen in elongated often slender made up of 10 segments. The genital openings in both, male and female, lies at the hind end. Male copulatory,organs on the 2nd and 3rd sternite.
8. Nymphs are aquatic, breath by rectal gills and feeding upon aquatic insects and organisms.

## NEPA

| | | |
|---|---|---|
| Phylum | - | Arthropoda |
| Class | - | Insecta |
| Subclass | - | Pterygota |
| Division | - | Exopterygota |
| Order | - | Hemiptera |
| Genus | - | *Nepa* |

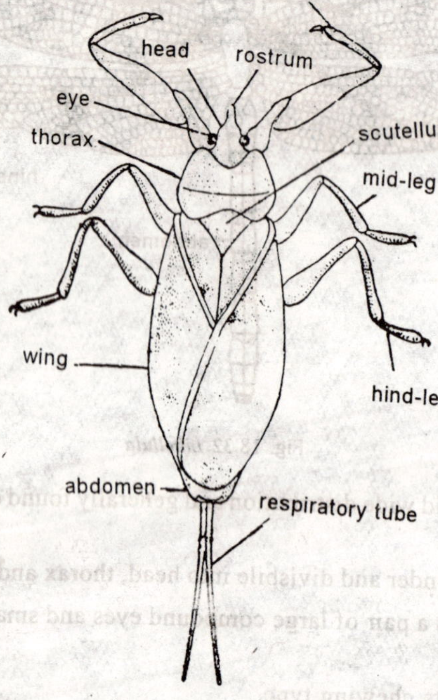


Fig. 18.33 *Nepa*

1. It is commonly found in shallow water. It has world wide distribution.
2. It is commonly called 'water scorpion'.
3. The body is dorsoventrally flattened and divisible into head, thorax and abdomen.
4. Head comprises a pair of 3 jointed antennae and compound eyes.
5. Mouth parts form a long rostrum adapted for piercing and sucking.
6. The respiratory tube is most peculiar. It consists of two spine-like process.

7. There are three pairs of false spiracles situated on 3rd, 4th, and 5th ventral abdominal segments respectively.
8. The false spiracles comprise sieve-like structures with the perforations occuleded by a delicate membrane.
9. Sexes are separate.
10. The female is provided with a pointed toothed ovipositor.
11. The eggs are deposited in chains. The ova adhere to one another by means of sevan long filaments radiating from one extremity.

## BELOSTOMA

The systematic position is the same as that of *Nepa*.

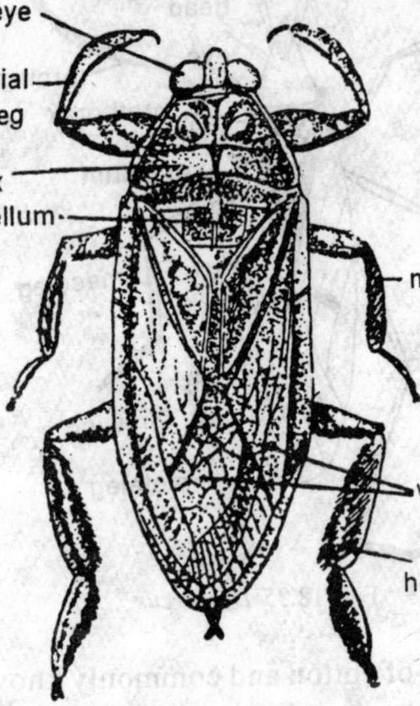


Fig. 18.34 *Belostoma*

1. *Belostoma* is the largest insect found in rivers, ponds, lakes, feeding on insects, tadpoles, snails etc.
2. It is found in India, U.K., North America and South Africa.
3. It is commonly called 'giant water bug'.
4. Body is dorsoventrally flattened and divisible into three parts, head, thorax and abdomen.
5. Head is small with a pair of large eyes and 4 jointed antennae. Mouth parts piercing and sucking type.
6. The thorax bears 3 pairs of legs. The midlegs and hindlegs are flattened and adapted for swimming, the forelegs raptorial.

7. First pair of wings hemielytra and second pair membranous.
8. The abdomen is provided with two retractile apical appendages.
9. Ovipositor well developed. Females of some species deposit their eggs on the back of male (parental care).

## DYSDERCUS

The systematic position is the same as that of *Nepa*.

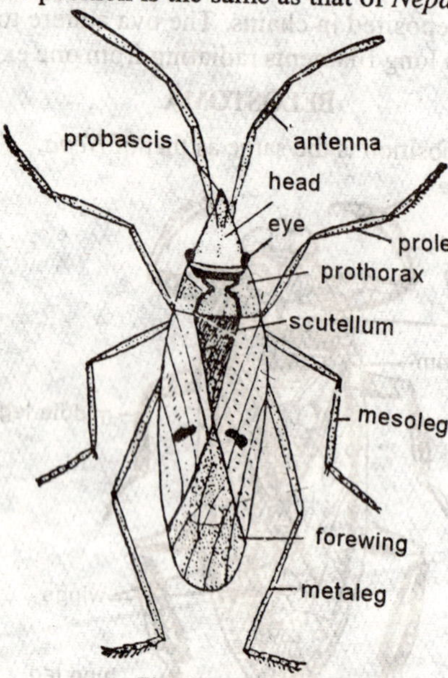


Fig. 18.35 *Dysdercus*

1. It is a common pest of cotton and commonly known as 'Red cotton-bug'.
2. Body is black and red in colour.
3. Body divisible into head thorax and abdomen.
4. Head bears a pair of compound eyes and 4 jointed antennae.
5. Mouth parts piercing and sucking type.
6. Wings two pairs, basal half of the forewings hard and leathery (hemielytra), hindwings soft and membranous.
7. Triangular scutellum present between the bases of wings.
8. Pronotum large.

## CICADA

The systematic position is same as that of *Nepa*.

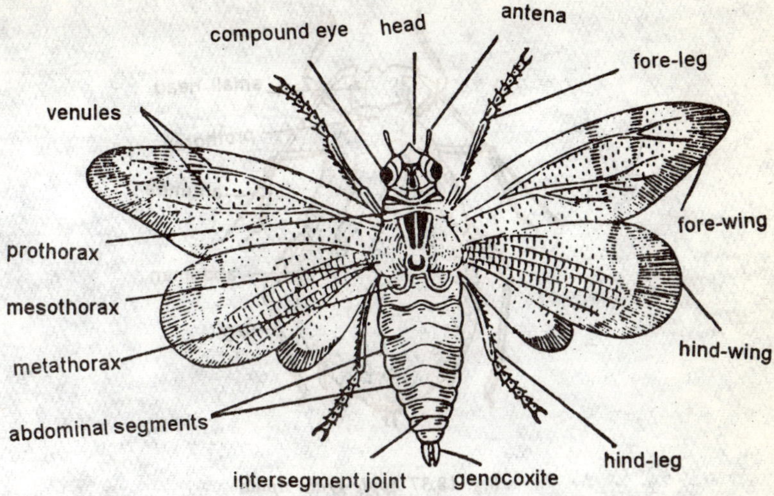


Fig. 18.36 *Cicada*

1. It is found in the forests and moist places, abundant in tropical, subtropical and temperate regions.
2. Body is divisible into head, thorax and abdomen.
3. The head is broad possessing two lateral compound eyes and two very short antennae.
4. The thorax comprises prothorax, mesothorax and metathorax.
5. There are two pairs of wings : fore-wings are larger and somewhat triangular and the hind-wings are oval.
6. The abdomen is elongated and consists of ten segments.
7. The sound production habit of this insect is very peculiar.
8. The sound is produced only by the male which possesses two sets of vibrating plates on the anterior part of abdomen.
9. Metamorphosis incomplete, egg hatches into nymph.
10. The life-history is very interesting as it takes about 13 to 17 years to complete.
11. Adults soon dies after mating.

**APHID**

The systematic position is the same as that *Nepa*.

1. It is an ectoparasite on plants, sucking their juice. It is found specially on he plants of mustard, wheat and cotton.

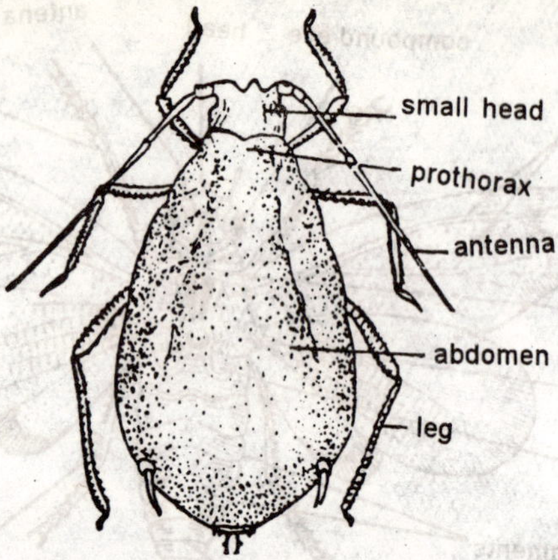


Fig. 18.37 *Apple-grain Aphid*

2. It is commonly called 'Plant-lice'.
3. Body is short divisible into head, thorax and abdomen.
4. Head is distinct with long straight antennae. Eyes are small.
5. Mouth parts are piercing and sucking type.
6. Thorax and abdomen robust. Abdomen is elongated.
7. Exhibit alternation of sexual and parthenogenetic phases, females always develop from parthenogenesis.
8. Both apterous and winged forms.

**DANAUS**

| | | |
|---|---|---|
| Phylum | - | Arthropoda |
| Class | - | Insecta |
| Subclass | - | Pterygota |
| Division | - | Endopterygota |
| Order | - | Lepidoptera |
| Genus | - | *Danaus* |

1. It is a flying, terrestrial, diurnel insect found in gardens.
2. Commonly called butter-flies.
3. Beautifully and brightly coloured insects.
4. Body is divisible into head, thorax and abdomen.
5. Head with a pair of compound eyes and club-shaped antennae.

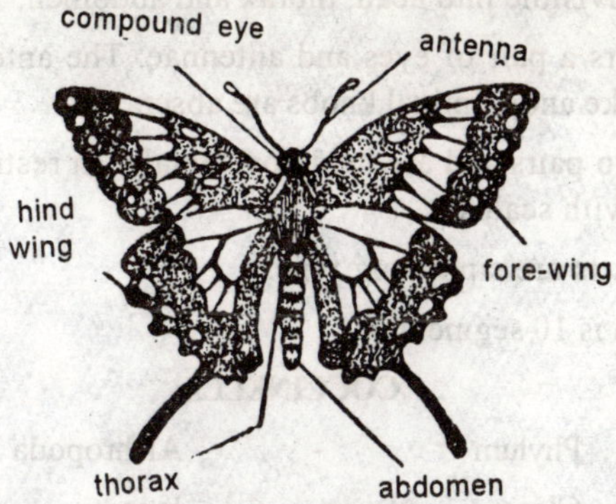


Fig. 18.38 *Danaus*

6. Mouth parts siphoning type. Maxillae are modified to form sucking tube.
7. Prothorax is fused with mesothorax.
8. Two pairs of wings which are typically covered with scales.
9. Abdomen is 10 segmented and covered with scales.
10. Cerci and ovipositor are lacking.
11. Development complete.

## MOTH

The systematic position is the same as that of *Danaus*.

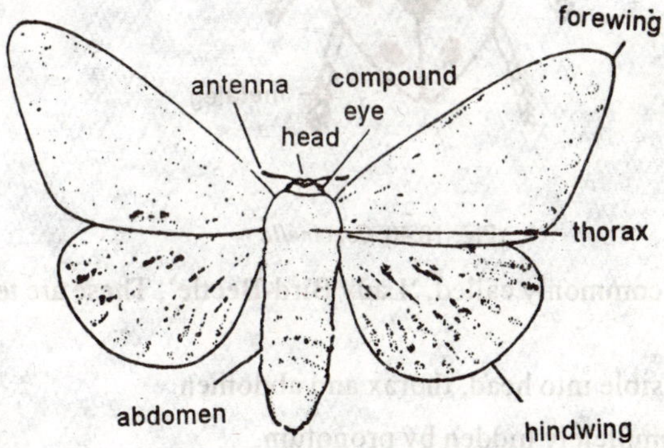


Fig. 18.39 *Moth*

1. It is found in garden among flowering plants. It is nocternal in habit.
2. Body is broad and stumpy.

3. Body is divisible into head, thorax and abdomen.
4. Head bears a pair of eyes and antennae. The antennae are usually feather-like and terminal knobs are absent.
5. Wings two pairs and are held horizontally in resting condition and covered with scales.
6. Mouth parts are siphoning type.
7. Abdomen is 10 segmented.

## COCCINELLA

| | | |
|---|---|---|
| Phylum | - | Arthropoda |
| Class | - | Insecta |
| Subclass | - | Pterygota |
| Division | - | Endopterygota |
| Order | - | Coleoptera |
| Genus | - | *Coccinella* |

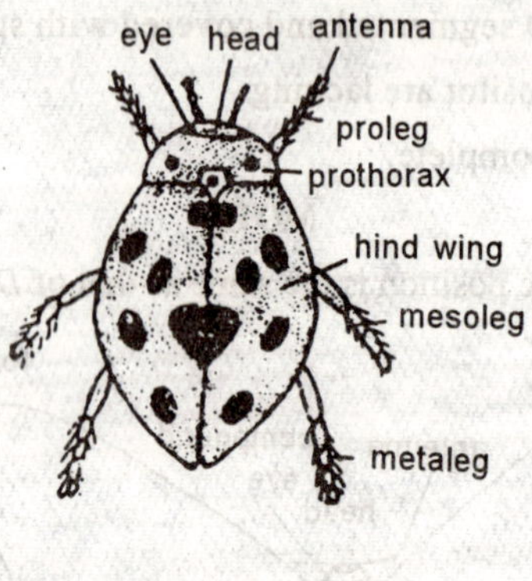


Fig. 18.40 *Coccinella*

1. These are commonly called, 'Lady Bird-Beetle'. These are terrestrial animal.
2. Body divisible into head, thorax and abdomen.
3. Head incompletely hidden by pronotum.
4. Head bears compound eyes and a pair of antennae. Mouth parts are chewing type.
5. Large number of species are carnivorous and predacious. Feeds on aphids, coccids and other soft bodied insects.

## VESPA

| | | |
|---|---|---|
| Phylum | - | Arthropoda |
| Class | - | Insecta |
| Subclass | - | Pterygota |
| Division | - | Endopterygota |
| Order | - | Hymenoptera |
| Genus | - | *Vespa* |

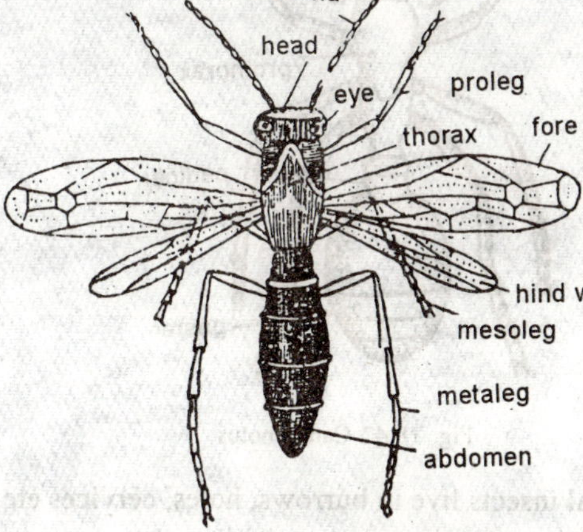


Fig. 18.41 Vespa

1. It is an Indian hornet, which is usually chestnut red and makes large nests of mud over walls and trees.
2. Body strongly buit, usually chestnut red in colour with yellow, transverse bars.
3. Body is divisible into head, thorax and abdomen.
4. Head is globular concave posteriosly with small ocelli and compound eyes.
5. Thorax deep and cubical wings are membranous. Hindwings short and bear hooks on their anterior margin which fit into a groove on the posterior margin of forewings.
6. The abdomen petiolate, large with pointed apex.
7. Modified ovipositor as sting is present. It stings painfully and causes local pain and swelling.

8. *Vespa orientalis* is a social colonial insect. Colonly consists of queen, males and workers. The impregnated female hibernates in winter and forms a new colony at the next spring.

## CAMPONOTUS

The systematic position is the same as that of *Vespa.*

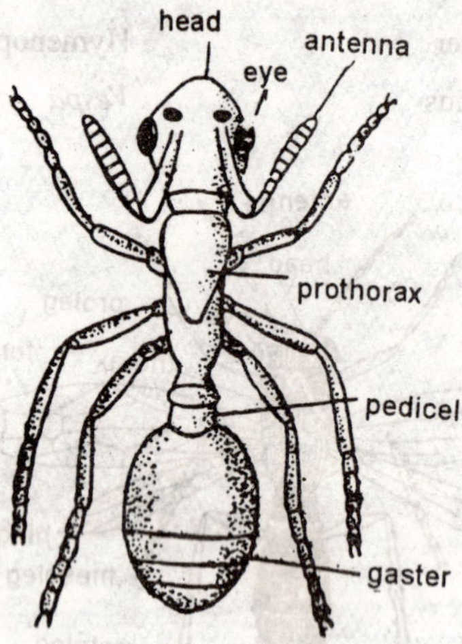


Fig. 18.42 Camponotus

1. Usually social insects live in burrows, holes, cervices etc.
2. Body divisible into head, thorax and abdomen.
3. Head varies in shape and mandibles are variable in forms.
4. Antennae consists of 4-13 segments. Male has one more segment than the female.
5. Compound eyes and ocelli are well developed in males where as reduced or vestigial in females and workers.
6. Stridulatory (sound producing) organ is present is abdomen.
7. Large and well developed stings are present in females and workers.
8. Mating takes place in winged stage during nuptial flight wings are then lost.
9. These are scavanger but are also harmful to man by destroying grain and seeds.
10. Other common Indian ants are : *Oceophylla smaragetina* (Weaver), *Holeomyrmex* (Harvester) and *Formica* (Red ant).

## XENOPSYLLA

| | | |
|---|---|---|
| Phylum | - | Arthoropoda |
| Class | - | Insecta |
| Subclass | - | Pterygota |
| Division | - | Endopterygota |
| Order | - | Siphonoptera |
| Genus | - | *Xenopsylla* |

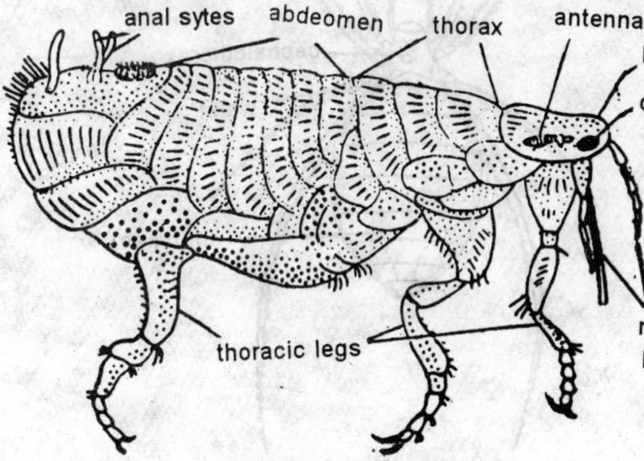


Fig. 18.43 Xenopsylla

1. It is a sanguivorous ectoparasite of rats, birds and man.
2. It is commonly found in India, Sri Lanka and Malaysia.
3. It is commonly called 'rat-flea'.
4. Body is laterally compressed without wings.
5. Integument is covered with spines and bristles.
6. Head broad jointed with small thorax. Abdomen is very large.
7. Antennae are clavate, short and concealed in grooves.
8. Compend eyes and ocelli are absent.
9. Mouthparts piercing and sucking type.
10. Walking legs are elongated, clawed and modified for jumping.
11. It transmits 'plague' from rat to man.

## ARANEAE

| | | |
|---|---|---|
| Phylum | - | Arthropoda |
| Class | - | Arachnida |

| | | |
|---|---|---|
| Order | - | Araneida |
| Genus | - | *Araneae* |

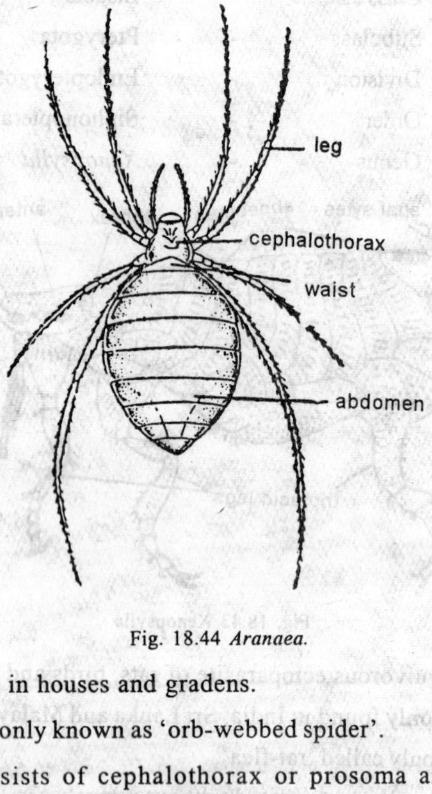


Fig. 18.44 *Aranaea.*

1. It is found in houses and gradens.
2. It is commonly known as 'orb-webbed spider'.
3. Body consists of cephalothorax or prosoma and abdomen or opisthosome, the two being connected by a narrow pedical.
4. Prosome with simple 8 eyes and 6 pairs of appendages. A pair of subchelate chelicerae, a pair of nonchelate pedipalps and 4 pairs of walking legs.
5. Opisthosoma is unsegmented, bears three pairs of spinnerets which produce thread for making web.
6. Poison glands open on the fangs of chelicerae.
7. Respiration by book-lung and tracheae.
8. Excretion by coxal gland and Malpighian tubules.
9. Sexes are separete. Female is larger than male and eats up after copulation.

## SARCOPTES

| | | |
|---|---|---|
| Phylum | - | Arthropoda |
| Class | - | Arachnida |
| Order | - | Acarina |
| Genus | - | *Sarcoptes* |

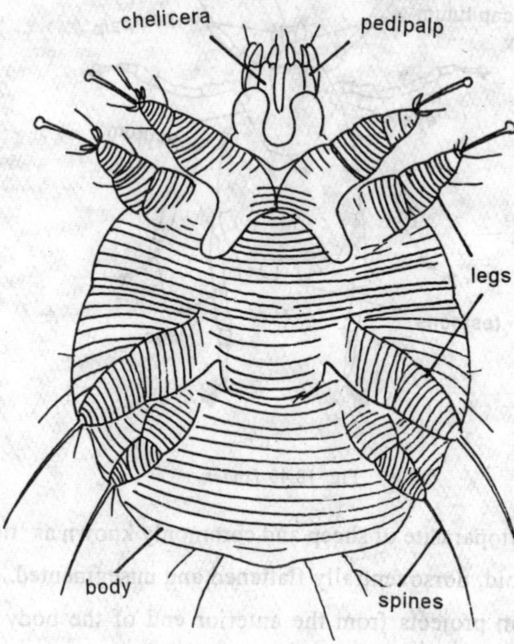


Fig. 18.45 Sarcoptes

1. It is an ectoparasite on human skin and causes horizontal burrows in region between fingers, elbow etc. Leads to echzema. Commonly known as 'itch-mite of human'
2. It is rounded or oval and dorsoventrally flattened with transverse striations and short spines.
3. Prosoma and opisthosoma not differentiated.

4. Four pairs of legs. Anterior two legs are stronger and project beyond the body and provided with terminal stalked suckers. Posterior two legs are shorter and attached more ventrally and carry long bristles.
5. Mouth parts borne on capitulum, chelicerae chelate.

## IXODES

The systematic position is the same as that of *Sarcoptes*.

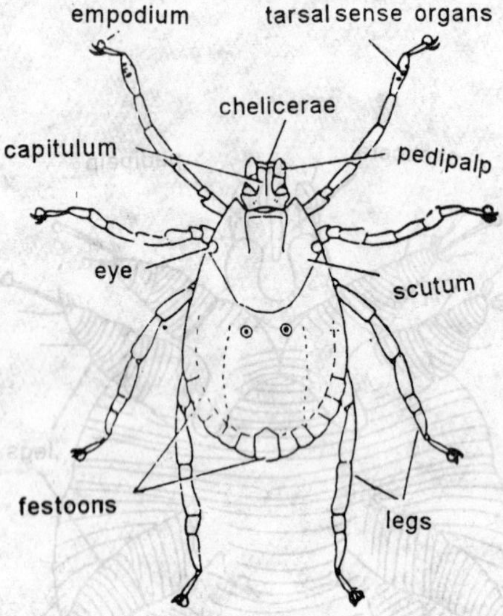


Fig. 18.46 *Ixodes.*

1. It is an ectoparasite of sheep and commonly known as 'tick'.
2. Body ovoid, dorsoventrally flattened and unsegmented.
3. Capitulum projects from the anterior end of the body and bears chelicerae and pedipalpi.
4. Dorsal surface of body is covered by shield-like scutum.
5. Eyes lateral and borne on scutum.
6. Walking legs 4 pairs with adhesive pads and claws.
7. Anus with adanal shield.
8. Female is larger than male.
9. They produce ixodin which acts as anesthetic, so that the host does not feel pains.

## CHELIFER

| | | |
|---|---|---|
| Phylum | - | Arthropoda |
| Class | - | Arachnida |
| Order | - | Pseudoscorpionida |
| Genus | - | *Chelifer* |

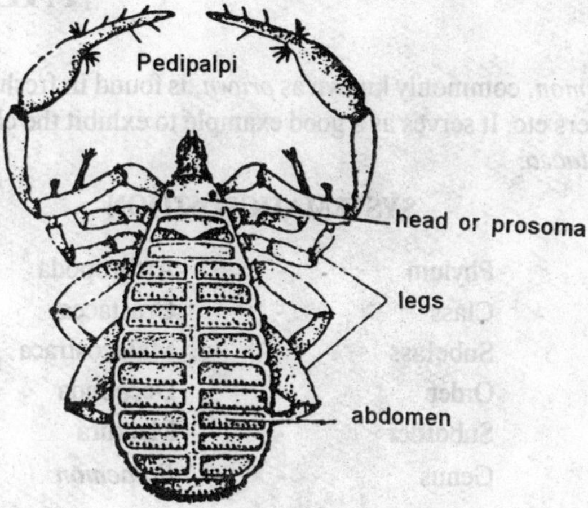


Fig. 18.47 *Chelifer*.

1. It is found commonly under the bark of trees.
2. Minute animals which have the appearance of scorpion but without tail hence commonly called 'pseudoscorpion'.
3. Body consists of prsoma and an opisthosoma.
4. Prosoma is broad and continuous bearing 6 pairs of appendages.
5. The chelicerae small 2 jointed and chelate.
6. Pedipalpi 5 jointed large and claw-like.
7. A pair of spining glands open on the tips of chelicerae.
8. Abdomen is 12 segmented.
9. Respiration by tracheae.
10. Males have a pair of reversible organ 'ram's horns, supposed to be copulatory in function.

♦♦♦

# 19

# PALAEMON

*Palaemon*, commonly known as *prawn*, is found in fresh water streams, ponds, rivers etc. It serves as a good example to exhibit the characteristics of class *Crustacea*.

### SYSTEMATIC POSITION

| | | |
|---|---|---|
| Phylum | - | Arthropoda |
| Class | - | Crustacea |
| Subclass | - | Malacostraca |
| Order | - | Decapoda |
| Suborder | - | Macrura |
| Genus | - | *Palaemon* |

### Habits and Habitat

A number of species of *Palaemon*, such as *P.carcinus, P.malcolmsonii, P.idae, P.rudis* etc., are found in the fresh water streams, ponds etc. of India, Bangla Desh and distant countries such as West Indies, Mexico and Australia. These animals generally like slow-running waters and are bottom-feeders. They are nocturnal in habit and come out in the night in search of food. The food is generally algae or other water plants and sometimes, insects. *P.malcolmsonii* breeds in May, June and July and the females carry eggs attached to their abdominal appendages.

### External Morphology

***Shape and Size.*** Prawn is a bilaterally symmetrical animal. The body is elongated and spindle-shaped and is well-adapted for swimming in water. The size of the adult palaemon varies considerably in different species. *P.carcinus* measures about 75 cm in length (found in Travancore). *P.malcolmsonii* from Central India, Tamilnadu and Bengal is about 25-30cm in length *P.lamarreii*, common in lakes and ponds is quite small about 2.5-5 cm. in length. *P.vulgaris*, a common prawn of America, measures 30 mm. in length.

***Colouration.*** The embryonic stages of prawn are usually translucent and whitish in colour. Adult animals of different species are of different colours. It is pale-yellow or greenish with a brown tinge or with organe-red patches or dull pale-blue.

***Division of body.*** The body is divisible into two district regions:

I. Anterior, immovable and unjointed cephabothorax.

II. Posterior, movable and jointed abdomen.

**I.** ***Cephalothorax.*** It is rigid and immovable. It is straight and oval in cross section. It is composed of 13 segments. In the embryonic condition, cephalothorax consists of first 14 fused segments and a pre-segmental region. Later on during development, first segment of the embryo disappears in the adult, while 13 segments remain. There are 13 pairs of appendages in these thirteen segments. The presegmental region is also found in adults and it possesses a pair of stalked compound eyes.

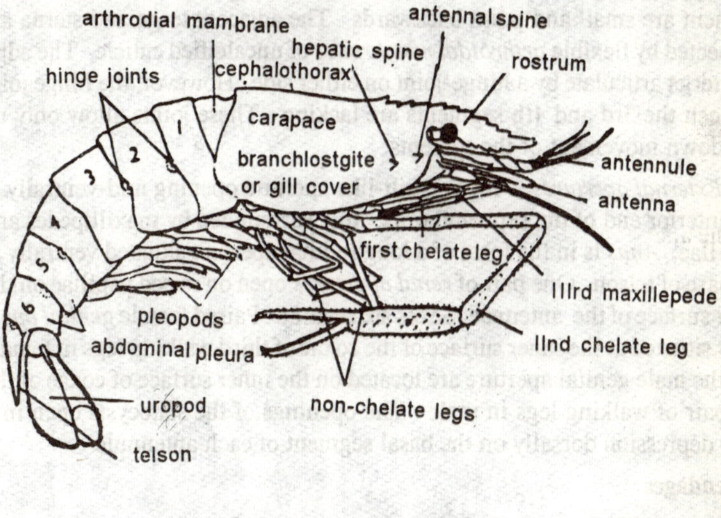


Fig.19.1 *Palaemon.* Lateral view

The cephalothorax is covered by a large chitinous shield, comprises two parts: dorsal shield and the sternal plates. The dorsal shield further consists of anterior dorsal plate and posterior *carapace* but there is no line of demarcation between two. The dorsal plate is trianagular and it extends forwardas as a laterally compressed process, the *rostrum*. The upper and lower margins of the rostrum are serrated and densely fringed with setae. At the base of rostrum, on either side, there is an *orbital notch* to which movable jointed stalk of the compound eyes is attached. Just below each orbital notch, the dorsal plate bears a prominant *antennal spine*. Behind each antennal spine there is a relatively small hepatic spine.

The carapace covers the thorax dorsally and laterally and represents the fused thoracic terga and pleura. On either sides of the body, the carapace hangs freely as a convex *gill cover or branchiostergite.* These enclose a narrow gill-chamber, which protects gills beneath them.

II. ***Abdomen.*** It consists of 6 posterior movable segments and a terminal conical structure, the *telson*. The telson articulates with the 6th abdominal segment by a pair of hinge-joints.

Each abdominal segment is covered dorsally by a convex sclerite, the *tergum* and ventrally by the *sternum*. The tergum extends downwards on either side as a free flap, the pleuron. The pleuron covers the base of the appendage of its side, and is connected with it by a small sclerite, the *epimeron*. *Pleura* of the 2nd abdominal segment are largest and differ from others in overlapping the pleura of both the Ist and the 3rd segments. Pleura of the 6th segment are small and point backwards. The adjacent terga and sterna are connected by flexible *arthroidal membranes* of uncalcified cuticle. The adjacent terga articulate by a hinge-joint on either side. However, the hinge joint between the 3rd and 4th segments are lacking. These joints allow only up and down movement of the sements.

*External apertures. Mouth* is silt-like aperture opening mid-ventrally at the anterior end of the cephelothorax. It is surrounded by maxillipedes and maxillae. *Anus* is in the form of a longitudinal aperture situated ventrally at the base of telson. One pair of *renal apertures* open on raised papillae on the inner surface of the antennae. Sexes are separate. Paired female *genital apertures* situated at the inner surface of the coxae of third walking legs in female and the male genital aperture are located on the inner surface of coxae of the 5th pair of walking legs in male. The openings of the statocysts open in a deep depression dorsally on the basal segment of each antennule.

**Appendages**

The appendages of prawn show considerable diversity in structure. In different regins of the body appendages are well-adopted for different functions. These appendages are homologous as all have the same typical structure and development.

Appendages appear very different from one another can be reduced to a general type of appendage . Typically each appendages consists of basal part or *protopodite*. The protopodite is a 2 jointed structure the proximal podomere with which the appendage is attached to the body is *coxopodite* or *coxa* and

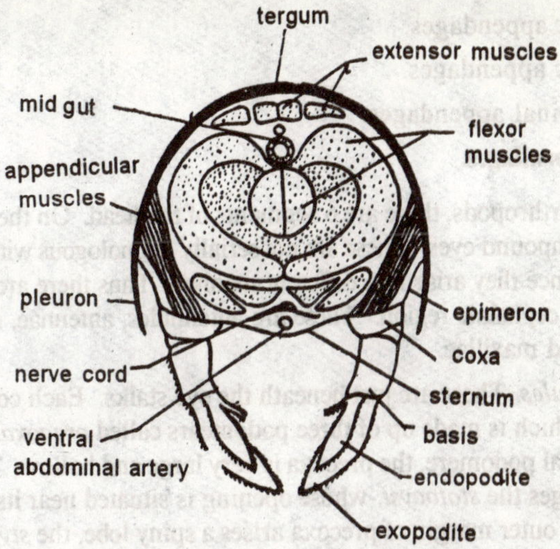


**Fig. 19.2** ***Palaemon*** **Section through abdominal region showing sclerites**

the distal joint is known as *basipodite* or *basis*. The basis bears at its tree end two ramii or branches. Rami are leaf like structures the outer one is called *exopodite* and inner one is *endopodite*. When there is only one ramus the appendage is called *uniramous* and when both are present it is a *biramous appendage*.

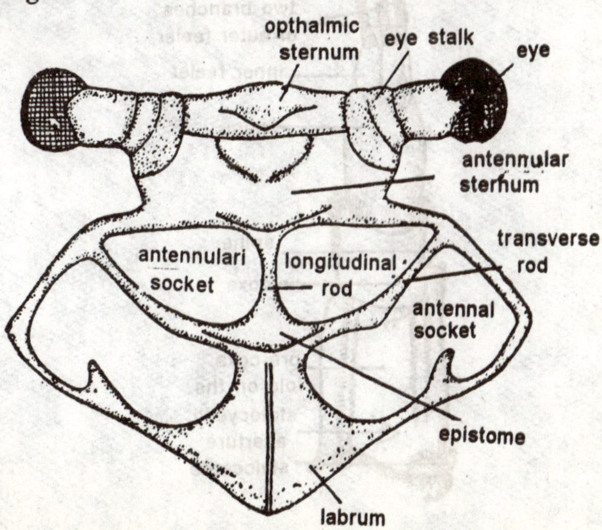


**Fig. 19.3** ***Palaemon.*** **Front view of preoral region**

There are 19 pairs of appendages which can be studies under 3 headings.

I. Cephalic appendages.

II. Thoracic appendages.

III. Abodominal appendages.

**I. Cephalic appendages.**

As in all arthropods, there are 6 segments of the head. On the first there is a pair of compound eyes. These are not serially homologous with the other appendages, since they arise in a different manner. Thus there are 5 pairs of appendages in cephalaic region. These are antennules, antennae, mandibles, maxillualae and maxillae.

1. *Antennules.* These are just beneath the eye stalks. Each consists of a protopodite, which is made up of three podomears called *precoxa, coxa* and *basis.* The basal podomere, the precoxa is very large and hollow. The cavity of precoxa lodges the *statocyst,* whose opening is situated near its proximal end. From the outer margin of precoxa arises a spiny lobe, the *stylosclerite.* The coxa is short and cylindrical. It bears setae on its inner and anterior borders. The basis is longer than the coxa and lacks setae. It bears a pair of long, slender, many-jointed,tapering processes, the feelers, with swollen bases. The inner feeler is unbranched but the outer one is divided into two unequal

**Fig. 19.4 *Palaemon,* Antennule**

branches, the inner smaller branch bearing special ***olfactory setae***. It is wrong to regard the two feelers as the two rami of biramous limb, perhaps they represent only the endopodite subdivided. The antennules are tectile in function.

2. *Antennae*. The antennae belong to segment 3, each with one feeler, situated immediately below and behind the antennules. The protopodite has two podomears, the coxa and basis. Each one is swollen at the bases as it lodges the *renal gland* within. The coxa is small in size and bears the renal aperture on its inner margin.The baiss bears two rami: a long, slender, many jointed *feeler* with a swollen basis and a broad leaflike *squama*. Of these the feeler corresponds to the endopodite and the squama the exopodite. Squama acts as a balancer in swimming, is fringed with setae along its inner border. Its outer border is smooth terminating in a small spine in front. The antennae are sensory, excretory and balancing in function.

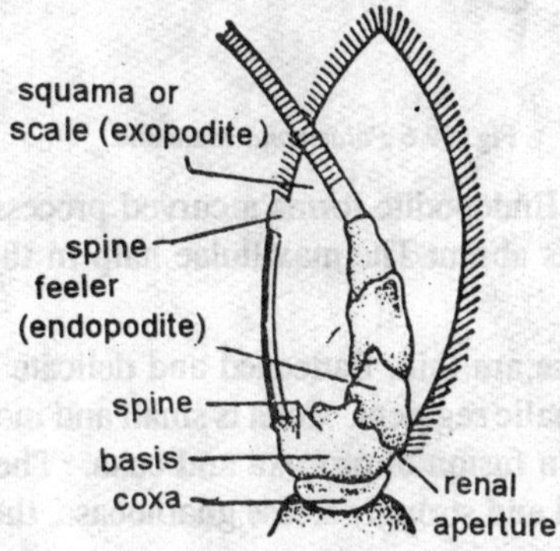


Fig. 19.5 *Palaemon* Antenna

3. *Mandibles*. The mandibles belong to the segment 4, lie on the sides of the mouth. They are very stout and highly calcified. The protopodite has two podomeres coxa and basis. The coxa is very large and can be divided into two parts: proximal, spoon-shaped hollow *apophysis* and the distal solid, *head*. The head is formed of two processes - a *molar process* having 5-6 yellow dental plates and a plate like *incisor process* ending in three teeth. Each mandible bears three segmented *mandibular palp* that arises from the outer margin of the head. The basal segment of the palp represents the basis of protopodite, the outer two segments represent the endopodite, and the exopodite is absent.The mandibles are masticatory in function.

4. *Maxillulae*. These are small thin, and leaf-like appendages of 5th segment. Coxa and basis are broad, and project inward as gnathobases or jaws,

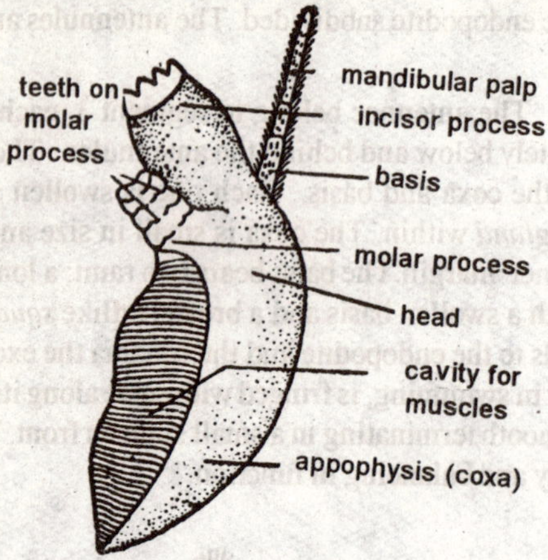


Fig. 19.6 *Palaemona* Mandible

fringed with bristles. Endopodite forms a curved process bifurcated at the apex. The exopodite is absent.The maxillulae halp in the manipulation of food.

5. *Maxillae.* These are thin, flatteened and delicate leaf-like structure present in the last cephalic segment. Coxa is small and incompletely divided into two representing a fusion of procoxa and coxa. The bais is large and produced into a forked and strongly setose gnathobase, the limbs of the fork

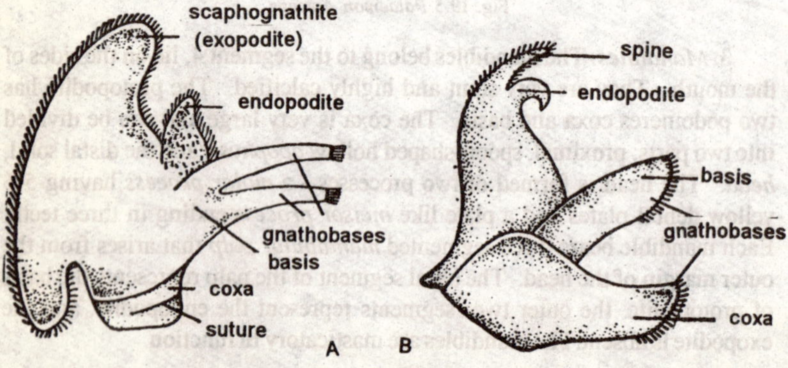


Fig. 19.7 *Palaemona* A = Maxilla B = Maxillula

are called endites. Endopodite is small and unjointed but the exopodite is large,flatteened and fan-like called *scaphognathite*.The movement of scaphognathite create a water current passing over the gills. It also bears sensory setae along its free margin. The maxillae help in manipulation and respiration.

**II. *Thoracic Appendages***

There are 8 pairs of thoracic appendages in prawn. Anterior 3 pairs are called *maxillipedes* are remaining 5 pairs of *palaeopods* or walking legs.

1. *First maxillipedes.* These are foliaceous. The coxa and basis of their protopodite extend inwards as two *gnathobases*, each bearing two rows of spines on the inner border. Coxa bears externally a bilobed *epipodite* or *primitive gill*.The endopodite and exopodite are unjointed and bear setae. The maxillipedes help in feeding and respiration.

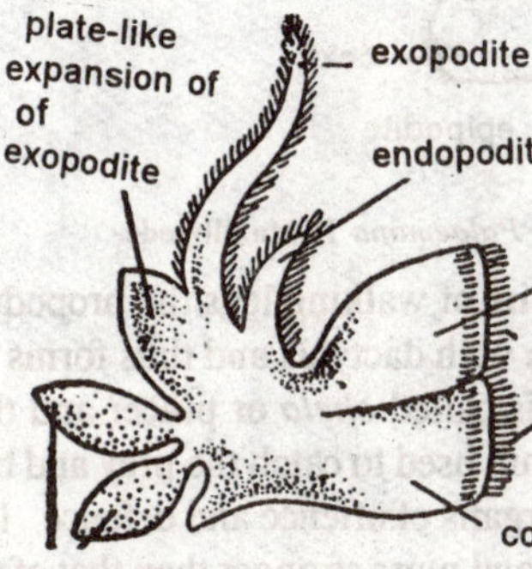


Fig.19.8 *Palaemona* I. Maxillipede

2. *Second maxillipedes.* The protopodite is represented by a coxa, which is short and bears setae on the inner side and a small epipodite and a gill on the outer border. Basis is short and is immovably joined with the endopodite which consists of five podomeres: (i) *ischium*, (ii) *merus*, (iii) *carpus*, (iv) *propodus* and (v) *dactylus*. Last two are bent backwards and inwards. Exopodite is long, slender, unjointed and is fringed with setae. The functions of second maxilipede are also the same as those of the first maxillipede i.e. to help feeding as well as respiration.

3. *Third maxillipedes.* These are leg-like in appearance but possess same parts as 2nd maxillipedes. The coxa bears an epipodite on the outer side and setae along the inner border. The basis carries outer unsegmented expodite and an inner three-jointed endopodite. The first podomere represents fused ischium and merus, the 2nd carpus and the third represents fused propodus and dactylus.

4. *Walking legs*. The 5 pairs of walking legs differ from the maxillipedes in the absence of exopodite and epipodites. Thus they have two segmented protopodite and a five segmented endopodite. These 7 segments are arranged in row. The fourth pair of legs represent the typical structure. It contains all the 7 podomeres as *coxa, basis, ischium, merus, carpus, propodus* and *dactylus.*

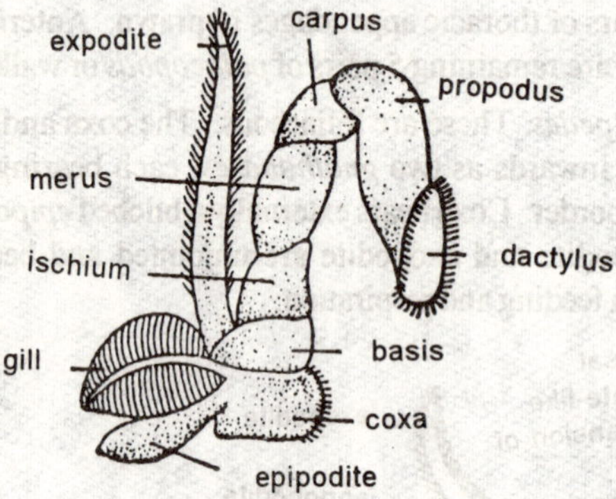


Fig. 19.9 *Palaemona* II Maxillipede

In the first and second pairs of walking legs the propodus is prolonged distally beyond its articulation with dactylus and thus forms a long forceps-like structure. The structure is called *chela* or pincer and the leg is called *chelate leg*. The chelate legs are used to catch the prey and transfer it to the mouth and also serve as the organs of offence and defence. In male the 2nd pairs of chelate legs are larger and more stronger than that of female.

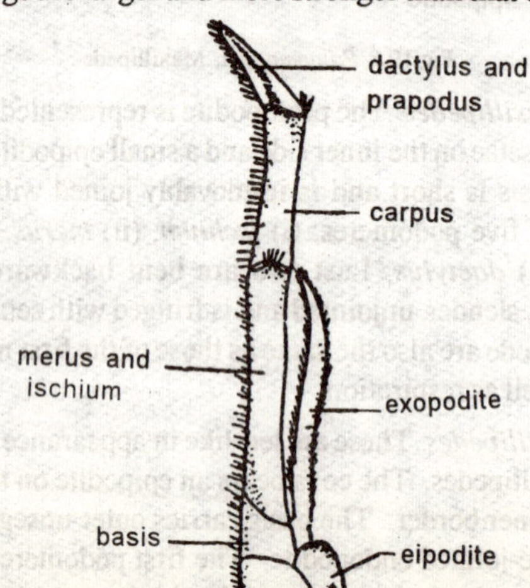


Fig. 19.10 *Palaemona.* III Maxillipede

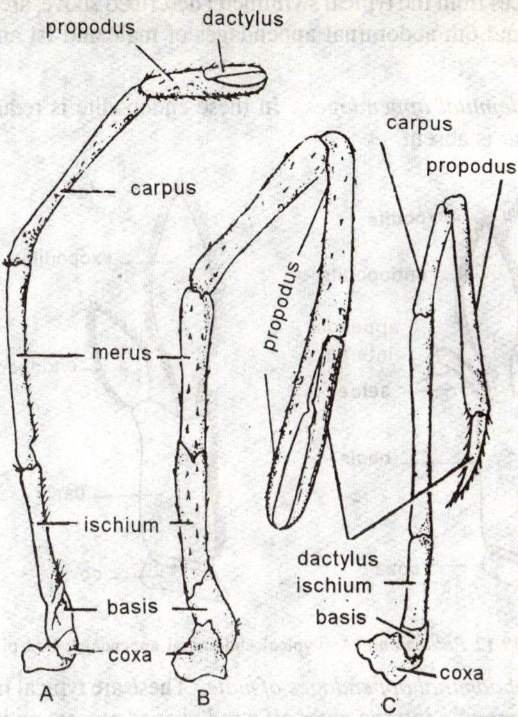


Fig. 19.11 *Palaemona.* A=walking leg, B=II walking leg, C=III walking leg.

The third, fourth and fifth pairs of walking legs are typical and non-chelate. In female, each of the third leg bears a crescentic female genital aperture on the inner side of its coxa, while in male, each of the fifth leg bears a slit-like male genital apeture on its inner side of the arthrodial membrane connecting the leg and thorax.

### *III. Abdominal appendages*

There are 6 pairs of abdominal appendages. These are collectively called *pleopods* or *swimmerets.* These are biramous appendages and help in swimming.

In a typical swimmeret (the 3rd abdominal appendage) the protopodite consists of two podomers, short ring-like segment devoid setae is known as the coxa and the basis which is cylindrical, bears setae on its surface and carries exopodite and endopodite. The endopodite is a leaf-like structure and gives of a small, rod-like structure, the *appendix interna*, with a knob-like head bearing many hook-like processes.

Differences from the typical swimmeret described above, are to be seen in the Ist, 2nd and 6th abdominal appendages of male and Ist and 6th of the female.

*First abdominal appendages.* In these endopodite is reduced and *appendix interna* is absent.

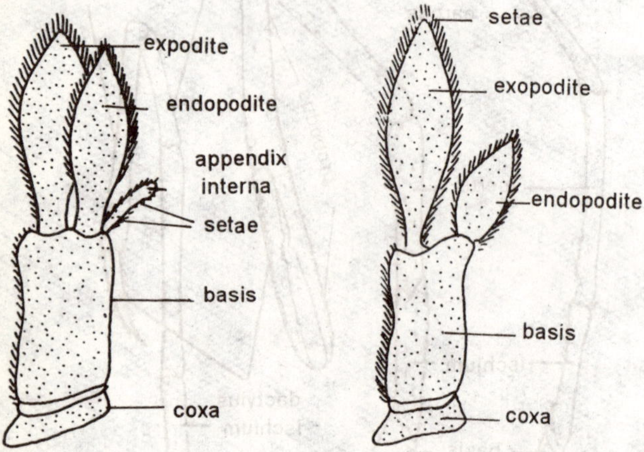


Fig. 19.12 *Palaemona.* A=Typical abdominal appendage, B=I pleopod

*Second abdominal appendages of male.* These are typical in female but in male the appendix interna gives off a rod-shaped process on its inner side called *appendix masculina.* It is covered with setae.

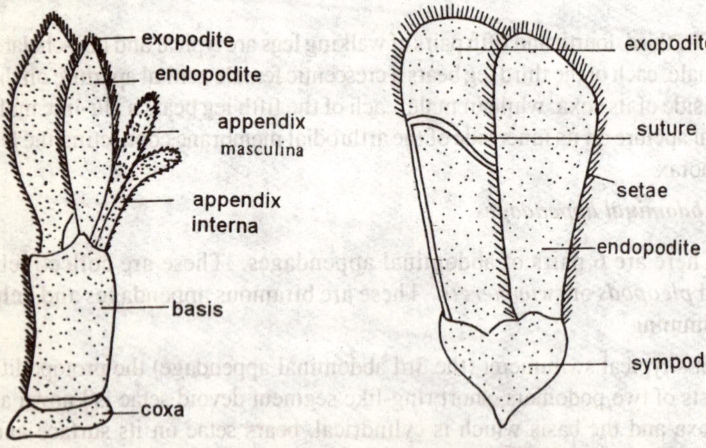


Fig. 19.13 *Palaemona.* A=II pleopod, B=Uropod

*Sixth abdominal appendages.* The sixth pair of abdominal appendages, called *uropods*, are large and lie one on either side of the telson forming tail-fin which enables the prawn to take a backward stearing in water. The coxa and basis fuse together to form a triangular sympod, bearing the oar-shaped endopodite and exopodite. The exopodite is bigger than the endopodite and incompletely divided in the middle by a transverse suture. Their margins, except the outer border of the exopodite, are fringed with numerous setae.

## Body wall

The body wall of *Palaemon* consists of an outer layer of cuticle a middle-single layered epidermis and an inner dermis.

***1. Cuticle.*** The cuticle of the body wall has become transofrmed into the exoskeleton which is thick and rigid at some places and thin and membranous at other. Cuticle is composed of two layers: (i) The outer layer is a thin non-chitinous *epicuticle*. It is yellow hyaline layer. It is further made up of two layers. Outer extremely thin made up of lipids and inner relatively thick protein layer. The protein layer is hardered and pigmented by its combination with an oxidised phenol. The epicuticle is secreted by the tegumental gland cells present is dermis. (ii) The *endocuticle* contains layers of chitin and has three successive layers : the outermost layer is a thin pigmented layer, below it is thick calcified layer and under it is an uncalcified layer. The pigmented layer is dark due to protein deposits and it becomes hardened by a process of tanning or sclerotization. In the calcified layers are deposits of carbonates and phosphates of calcium which make the integument hard. The chitin is an acetate of a polysaccharide containing glycosamine secreted by the underlying

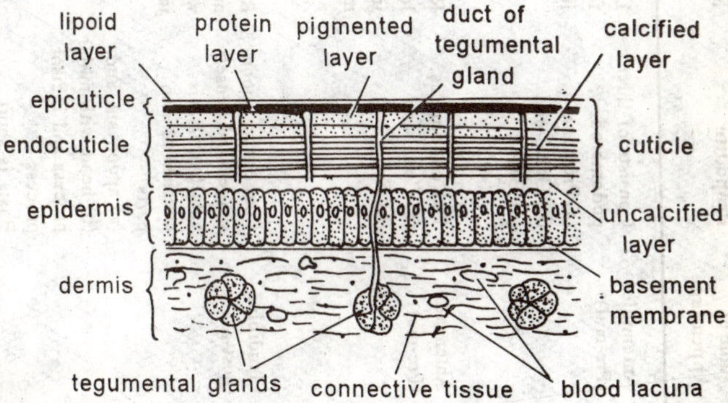


Fig. 19.14 *Palaemona* V. S. of Skin

## Table 19.1 Appendages of Prawn

| Body Region | No. | Name of Appendage | Propodite | Endopodite | Exopodite | Special features | Functions |
|---|---|---|---|---|---|---|---|
| | 1. | Antennules (Pre-oral) | Consists of three podomered, (a) Pre-coxa (b) Coxa, and (c) Basis. | Long, jointed feeler with swollen base. | Represented by an outer feeler which is further divided into an inner smaller branch and an outer large branch. | Procoxa has a cavity in which statocyst lodged. | Tactile, is olfactory and equilbrium librium |
| | 2. | Antennae (Pre-oral) | Two podomeres: (a) Coxa, and (b) Basis. | Represented by a longer feeler having three podomeres. | Flat, broad, leaf like squama or scale. | Both coxa and basis are swollen due to the excretory organ with them. Coxa possesses a renal opening. | Sensory, excretory and balancing. |
| Cephalic appendages (5) | 3. | Mandibles (post-oral) | Coxa large, spoon-shaped and densely calcified, Basis consists of two parts :- (a) apophysis, and (b) head with molar process and incisor process. Basis is short | Represented by a mandibular palp which is two jointed. | Absent. | Coxa-teeth are located at the side of the mouth. | Mastication of food |

| Body Region | No. | Name of Appendage | Propodite | Endopodite | Exopodite | Special features | Functions |
|---|---|---|---|---|---|---|---|
| | 4. | Maxillulae (Post-oral) | Coxa and basis both are leaf-like and project as gna thobases covered with pointed spines. | Bifurcated process on the outer side of the basis. | Absent. | Manipulation of food. | |
| | 5. | Maxillae (Post-oral) | Coxa is incomplete ly divided while basis from a bifur- | Reduced to a small process which is narrow and is for- | Flat, fan-shaped scaphognathite | Scaphognathite produces currents of water over gills. | Respiration and manipulation of food. |
| | 6. | 1st Maxilli-pede | Coxa and basis project like jaws | Endopodite is represented by a small, unjointed appendage. | Exopodite is also unjointed and a plate-like expansion. | Similar to a leaf in shape | To hold food and tactile. |
| | 7. | 2nd Maxilli pede. | Protopodite is repre-sented by coxa which is short and bears seate on the inner side and a small epipodite and gill on the outer border. Basis is short and is immovably joined with the endopodite. | It consists of 5 podomeres : 1. Ischium, 2. Merus. 3. Carpus. 4. propodus. 5. Dactylus. | Represented of long, slender, unjointed and Fringed seate. | | To help in feeding. |

| Body Region | No. | Name of Appendage | Propodite | Endopodite | Exopodite | Special features | Functions |
|---|---|---|---|---|---|---|---|
| Thoracic appendages (8) | 8. | 3rd Maxillipede. | Coxa is short and possesses epipodite on the inner side. Basis is short and supports exopodite and endopodite. | It consists of 3 joints :- 1. Fused ischium and merus. 2. Carpus. 3. Fused propodus and dactylus. | Unjointed, long, slender and thickly covered with setae. | Leg-like in appearance. | To hold the food in position and to help in feeding. |
| | 9. | Ist Chelate leg. | Coxa is short and ring-like, while basis is triangular. | 5. distinct podomeres, i.e., Ischium, merus, carpus, propodus and dactylus | Absent. | Chela or pincer is Present. | Catches and transfers the prey to mouth. |
| | 10. | 2nd Chelate leg. | " " | " " | " " | " " | Organ of offence and defence. Large in males. |
| | 11. | 3rd Non-chelate leg. | " " | Propodus and dactylus do not form chela. | | Chela absent. In female, coxa bears a female genital aperture | Walking |
| Thoracic appendages. | 12. | 4th Non- | " | " " | | Chela absent. | Walking. |
| | 13. | 5th walking leg. | " | " | " | In male, there is a male genital aper- | Walking |

| Body Region | No. | Name of Appendage | Propodite | Endopodite | Exopodite | Special features | Functions |
|---|---|---|---|---|---|---|---|
| | | | | | | ture between the leg and thorax. | |
| Abdominal appendages | 14. | Ist pleopod or swimmeret | Ring-like coxa and cyclindrical basis. Basis possesses two rami. | Small, leaf-like, possessing seatae. | Large, leaf-like. | | Swimming. |
| | 15. | 2nd pleopod. | " | In female, it help to carry the eggs with 3rd, 4th and 5th, appendix interma. | " | Appendix masculina is also present between endopodite and appendix. intema. | **Appendix interna is present on the inner side on the endopodite as a short, slender, slightly-curved rod.** Appendix |
| | 16. | 3rd pleopod. | " | " | | | intema is interlocked with its fellow of the opposite side to |
| | 17. | 4th pleopod. | " | " | " | " | from a basket-like structure. |
| | 18. | 5th pleopod. | " | " | " | " | |
| | 19. | Uropod or 6th swimmeret. | Coxa and basis fused to form a broad plate or sympod. | Flat, oval plate and provided with setae. | Large and divided into two pairs by a transverse suture. | With the telson form tail. | Swimming |

epidermis. The cuticle is relatively impermeable, except where it is thin and allows the passage of gases or absorption of water.

The seate and spines are the outgrowths of cuticle and have the same structure.

*2. Epidermis.* It comprises a single layer of glandular columnar epithelium resting on a thin basement membrane which is secreted by the underlying dermis. The epidermis is often called chitogenous eithelium as it secretes chitin.

*3. Dermis.* The dermis is thin and is made up of loose connective tissue beast with blood lacunae. It contains three types of tegumental glands that open out by fine ductules passing through overlying epidermis and the cuticle.

**Body cavity**

In *Palaemon* the body cavity is mostly occupied by musculature and organs (having interspaces containing blood) and all these together form a body cavity or *haemocoel*. However, true coelom is restricted to kidneys and gonads.

**Musculature**

The bulk of the body of the *Palaemon* consists mainly of muscles which are enclosed within the exoskeleton.These muscles are voluntary or striated type. The unstriated muscles are found in the organs such as blood vascular system and digestive system. There are two sets of muscles in prawn, the appendicular muscles that are segmentally and the abdominal muscles for straightening and flexure of the abdomen. The muscles are arranged in opposing series. By the contraction of one set the limbs are flexed (flexor muscles) and by the contraction of other, the limbs return to their normal position (extensor muscles). In the abdomen too such muscles are found. In the terminal sclerite of chelate leg, the flexor muscle become adductor, closing the chela, and the extensor, abductor.

The extensor of abdominal muscles include a pair of dorsal extensor, four pairs of inter-tergal abdominal muscles, a pair of intgral muscles and two pairs of extensor of the telson.

The flexors of abdominal muscles include 5 pairs of main flexor (first 5-segments), four pairs of accessory flexor muscles first 4 segments), a pair of flexor (between first abdominal segment and thorax), two pairs of lateral thoroaco-abdominal muscles, two pairs of flexor muscles of telson and a pair of ventral superficial flexor muscles.

All the appendages have their own muscles.

**Endoskeleton**

*Palaemon* has an endoskeleton in the cephalothorax fromed of a series of ingrowths of the exoskeleton called *apodemes* or *endophragms.* These ingrowths are connected together to form a regular framework that is known as *endophragmal skeleton.*

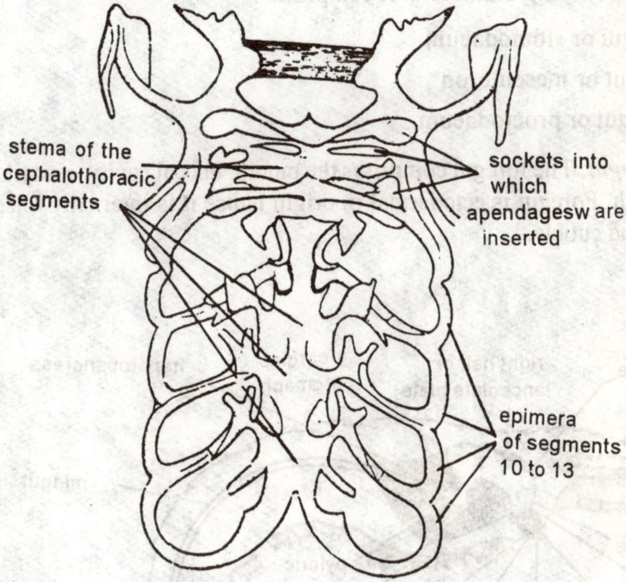


Fig. 19.15 *Palaemona* The stema and endophragmal skeleton (dorsal view)

The endophragmal skeleton is well developed in the segments 11-13, where, in each segment, it consists of an outgrowth from the anterolateral end of epimeron of the same sternal segment called *endopleurite posterior.*

The endopleurite posterior meets an ingrowth from the posterior-lateral end of the epimeron of the segment in front, called *endopleurite anterior.* These in turn meet respective ingrowths of the sternum of the same segment an endosternite posterior and an endosternite anterior. From the mesosternite, on each side arises an upwardly directed 'Y' shaped plate. The inner limb of the Y is called *mesophragm* and outer *paraphragm*. These apodemes provide for attachment of abdominal flexor muscles while other apodemes are for the attachment of appendicular muscles. A typical apodeme with complete number of endosternites, endopleurites and Y shaped structure occurs only between 11th, 12th and 13th segments. In other segments the apodeme is incomplete.

In mandibular region there is a H-shaped structure, the cephalic apodeme, for the insertion of muscles of branchiostegites, small gastric muscles and the muscles of mandibles.

**Digestive System**

The digestive system comprises, alimentary canal and digestive glands.

*1.* ***Alimentary Canal.*** The alimentary canal is straight tube-like structure of varying diameters. It comprises :

1. Foregut or stomodaeum.
2. Midgut or mesenteron
3. Hindgut or proctodaeum.

**1. *Foregut.*** The foregut comprises the mouth, buccal cavity, oesophagus and stomach. Foregut is ectodermal in origin hance it is internally lined by ectoderm and cuticle.

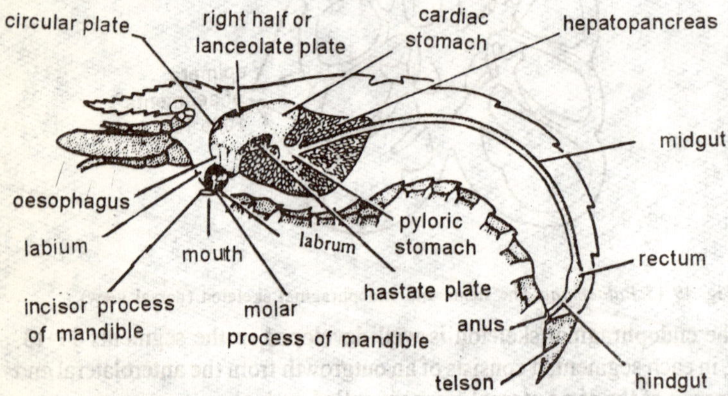


Fig. 19.16 *Palaemon* Alimentry canal in lateral view.

(i) *Mouth.* The mouth is large slit-like apertuse situated on the ventral side of the head below the anterior end. It is bounded anteriorly by a shield shaped labrum or upper lip. Laterally by the incisor processes of mandibles and posteriorly by a thin labium or lower lip which is bilobed. Mouth leads into the buccal cavity.

(ii) *Buccal cavity.* The buccal cavity is an antero-posteriorly compressed small chamber with irregularly folded thick chitinous lining. The molar processes of mandibles project into it and lie opposite each other.

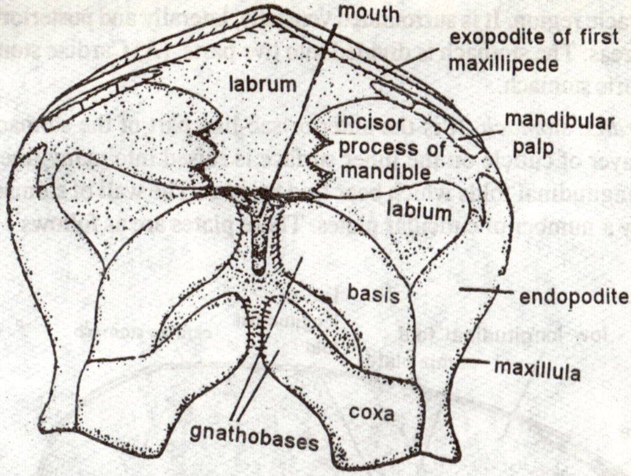


Fig. 19.17 *Palaemon* ventral view of oral region

(iii) *Oesophagus*. The oesophagus is a short, wide, vertical tube joined with the buccal cavity and stomach. Its inner wall is muscular and is produced into four longitudinal folds : an anterior fold which is short, a posterior fold and two lateral folds, which are large and more prominent.

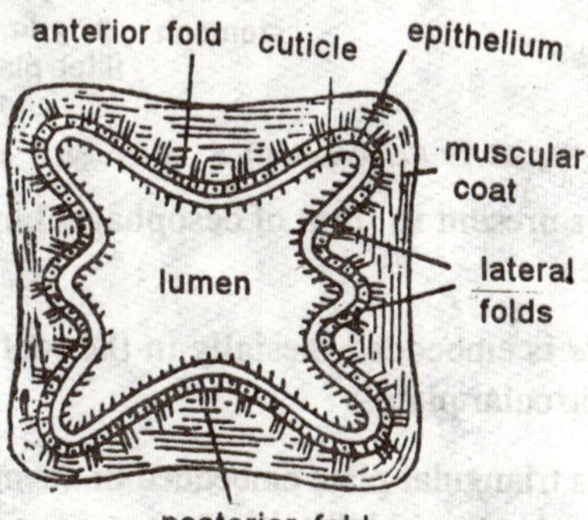


Fig. 19.18 *Palaemon* T.S. of oesophagus

The internal cuticle of oesophagus is covered with various bristles. Oesophagus opens into the stomach; the opening is provided with a valve like structure.

(iv) *Stomach*. It is a spacious chamber which occupies most of the cephalothoracic region. It is surrounded ventrally, laterally and posteriorly by hepatopancreas. The stomach is divided into two parts. (A) Cardiac stomach and (B) Pyloric stomach.

*(A) Cardiac stomach.* It is the anterior sac-like part of the stomach in which the layer of cuticle on the inner surface is raised into many fine and indistinct longitudinal folds which bear find bristles. The wall of stomach is supported by a number of cuticular plates. These plates are as follows :

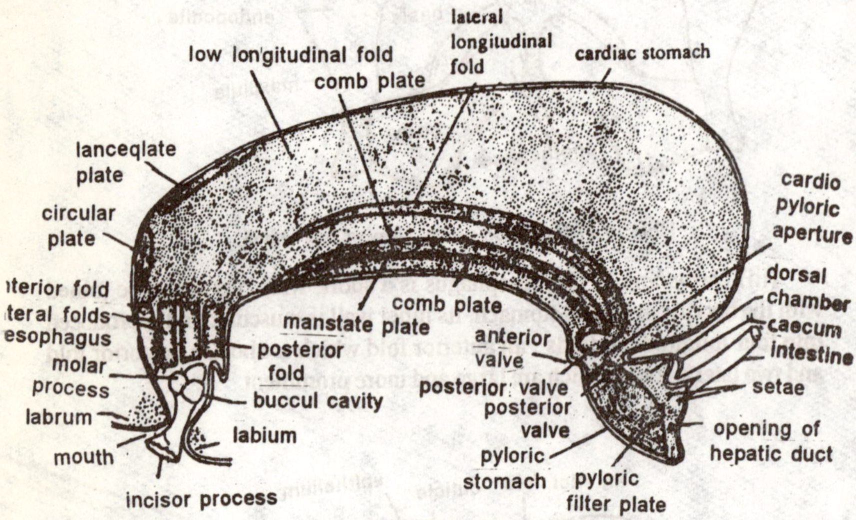


Fig. 19.19 *Palaemon* V.L.S. of foregut

(a) *Circular plate* is present in front of oesophageal aperture forming its anterior wall. .

(b) *Lanceolate plate* is embedded mesially in the roof of the stomach just behind the circular plate.

(c) *Hastate plate* is a triangular plate embedded in the middle of floor of stomach. It is raised in the middle with sloping sides. Its upper surface is covered with a thick growth of delicate setae. The posterior broad margin is depressed and fringed with setae to form the anterior valve of the cardio-pyloric aperture. The lateral margins of the hastate plate are supported by two circular supporting rods. A narrow lateral groove runs along the outer border of each supporting rod.

(d) *Groove plates* form the floor of lateral grooves and are in the form of open drain pipes.

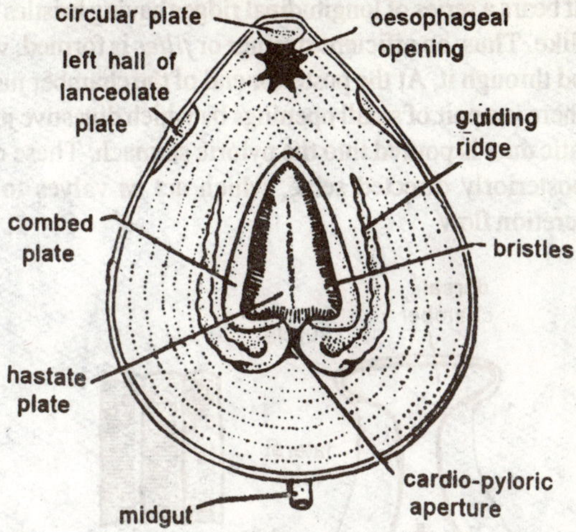


Fig. 19.20 ***Palaemon*** Floor of cardiac stomach

(e) *Combed plate*. Each lateral groove on its outer side is supported by a long ridge plate or combed plate. The two combed plates are united anteriorly but remain separate posteriorly by cardio-pyloric aperture. The inner border of each combed plate is beset with a row of long delicate bristles which bridge over the lateral groove and partially overlap the lateral margins of the hastate plate. The bristles remain in constant movement.

Just outside the combed plates, the wall of cardiac stomach is folded on each lateral side to form guiding ridges or lateral longitudinal folds. These ridges are low in front but gradually increase in high behind, but posteriorly the high walls of these folds bend inwards over the cardio-pyloric aperture, forming lateral valves. Thus these valves guard the passage of food towards the cardio-pyloric aperture.

The cardio-pyloric apertue is narrow x-shaped and is guarded by four valves, through which cardiac stomach opens into pyloric stomach.

*(B) Pyloric stomach*. Pyloric stomach is a short, narrow chamber. It is the posterior part of the fore-gut situated below the posterior end of the cardiac stomach. Its lateral walls are thick and muscular and project into the lumen as thick, longitudinal folds dividing it into a dorsal and a ventral chamber. These two chambers are connected by a narrow vertical passage. The floor of the

ventral chamber is elevated into a median longitudinal ridge so as to form two lateral compartments. The floor is provided with an '^' shaped *filter plate* or *pyloric filter*. It bears a series of longitudinal ridges having bristles and grooves, whch are felt-like. Thus, an efficient strainer or *filter* is formed, which allows only liquid food through it. At the posterior end of the chamber just behind the pyloric filter, there is a pair of small openings by which digestive juice from the hepatopancreatic duct is poured into the pyloric stomach. These openings are provided by posteriorly directed setae, which act as valves to control the direction of secretion flow.

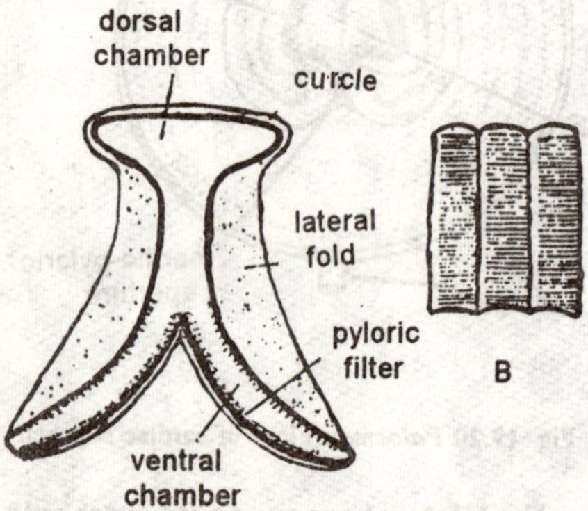


**Fig. 19.21** ***Palaemon*** **A=T.S. pyloric stomach B=Apart of pyloric stomach**

**The dorsal chamber gives out a small blind caecum dorsally and opens into the mid-gut. The junction of dorsal chamber and midgut is guarded by the median dorsal and two lateral groups of elongated setae. These setae control the flow of food material.**

***(v) Midgut or measenteron.*** **The midgut is a long, narrow, straight and slender tube running back along the median line above the mas of the ventral abdominal muscles upto the sixth abdominal segment. The internal lining of**

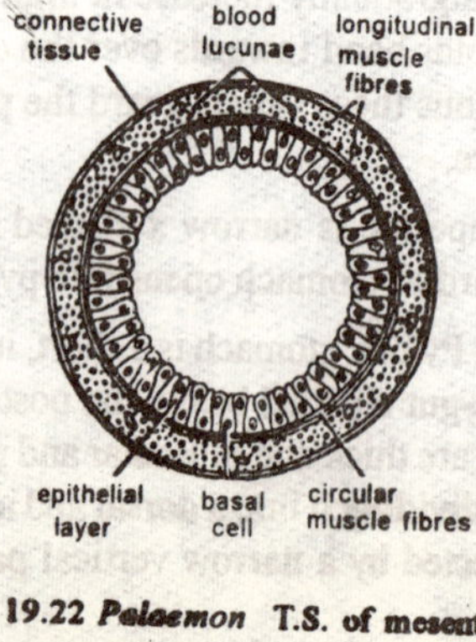


**Fig. 19.22** ***Palaemon*** **T.S. of mesenteron**

midgut is formed by epithelium which in the posterior part is thrown into many longitudinal folds thus greatly reducing its lumen.

***(vi) Hindgut or proctodaeum.*** It is the last part of alimentary canal which is internally lined with cuticle. Its anterior enlarged muscular part is called the ***intestinal bulb*** and the posterior less wide part is the ***rectum***. It opens out through the anus which is placed at the base of the telson.

***(vii) Anus.*** It is a longitudinal slit. It is surrounded by circular and radial muscle fibres. The former forms the sphincter and the later acts as dilator.

***Histology.*** The wall of midgut consists of 4 layers. Beginning from inside, these are (i) An epithelial layer of tall columnar cells resting on a basement membrane, having small basal or replacing cells (ii) A thin layer of circular muscle fibres (iii) A thick layer of longitudinal muscle fibres enclosing a plexus of blood lacunae (iv) A connective tissue sheath.

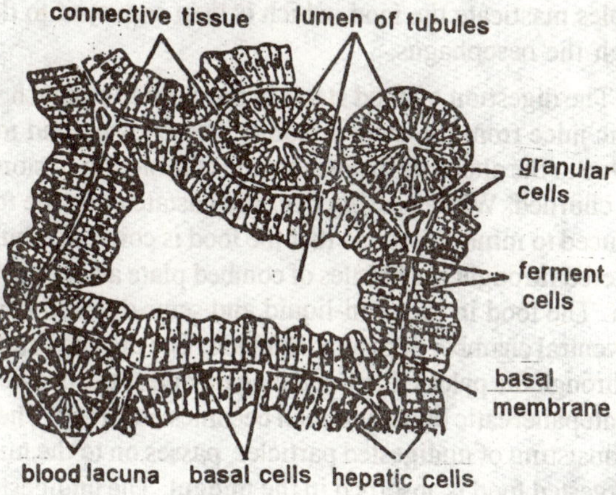


Fig. 19.23 *Palaemon* T.S. of hepatopancreas

The rectum also has the same layers except that its epithelial lining is covered by cuticle and there is an additional layer of longitudinal muscle fibres inside the circular muscle fibres.

***Digestive gland.*** The ***hepatopancreas*** is a large and physiologically very important gland. It consists of a pair of orange-red coloured lobes lying close to each other surrounding the cardiac stomach and has a massive appearance. It is a racemose gland which arises as a pair of tubular outgrowths from the midgut. These outgrowths grow, divide and redivide, thus forming a compact network which is held together by connective tissue. The wall of each tubule forming the gland is made up of four layers : the inner most layer of *epithelial cells*, a thin *basement membrane* on which epithelial cells rest and the outer most layer or *tunica properia* formed of a network of connective tissue and *muscle fibres*. The epithelial layer has four kinds of cells (i) *Hepatic cells* which are columnar shape and with their cytoplasm filled with fat droplets (ii) *Granular cells* (iii) Irregularly scattered *ferment cells* and (iv) Basal or replacing cells.

The hepatopancreas performs following functions :

1. It secretes enzyme for digestion of carbohydrates, fats and proteins.
2. It serves to store glycogen, calcium and fat.
3. It helps in absorption of digested food material.

***Food and feeding.*** *Palaemon* is omnivorous, feeding upon aquatic weeds, algae, mosses, insects and debris of bootom. It feeds at night. The chelate legs, aided by the third maxillipedes, capture and convey the food to the mouth. The coxae of the second maxillipedes hold the food, while the incisor processes of mandibles cut it into smaller pieces which are swallowed with the help of maxillipedes, maxillulae and maxillae. Inside the buccal cavity the molar processes of mandibles masticate the food, which is then conveyed to the cardiac stomach through the oesophagus.

***Digestion.*** The digestion of food starts in the cardiac stomach where the hepatopancreatic juice from the pyloric stomach ascends up and mixes well with the food. Due to the alternate expansion and contraction of stomach wall the food is well churned. While passing over the hastate plate the food particles are still reduced to minor pieces so that the food is converted into a paste. This paste is filtered throught the bristles of combed plate and is carried to the pyloric stomach. The food in the semi-liquid and semi-digestid condition is carried into the ventral chamber of pyloric stomach. The digested and liquified food is filtered through the pyloric filter and it again enters the hepatopancreas through the hepatopancreatic ducts and then becomes absorbed. The residual food material, consisting of undigested particles, passes on to the midgut and the remaining digested food is absorbed in the midgut. The undigested faecal matter pass on to the posterior side of the hindgut and ultimately expelled through the anus.

**Respiratory System**

The respiratory organs in prawn are developed and are located in the gill chambers, between the branchiostegites and the thorax. These are :

1. Branchiostegites or gill cover.
2. Three pairs of epipodites.
3. Eight pairs of gills or branchiae.

***1. Branchiostegite.*** The inner lining of branchiostegites is thin, membranous and highly vascular containing minute blood lacunae. These are constantly bathed in fresh water, thus forming the respiratory surface through which exchange of gases takes place between blood and water.

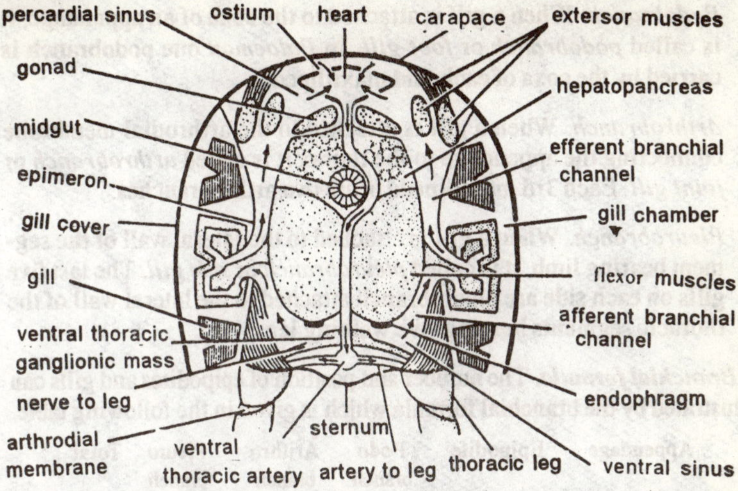


Fig. 19.24 *Palaemon* Section through cephalsthorax showing gill chamer

***2. Epipodites.*** These are thin, leaf-like processes, one attached to each maxillipede. These epipodites are full of blood lacunae and, infact, constitute a primitive type of gill. The epipodites lie in the anterior part of the gill-chamber below the scaphognathite. Exchange of gases between blood and lacunae takes place between water and blood in epipodites.

***3. Gills.*** There are 8 pairs of gills. These arise from the modification of epipodites and according to their position, they be distinguished into three types :

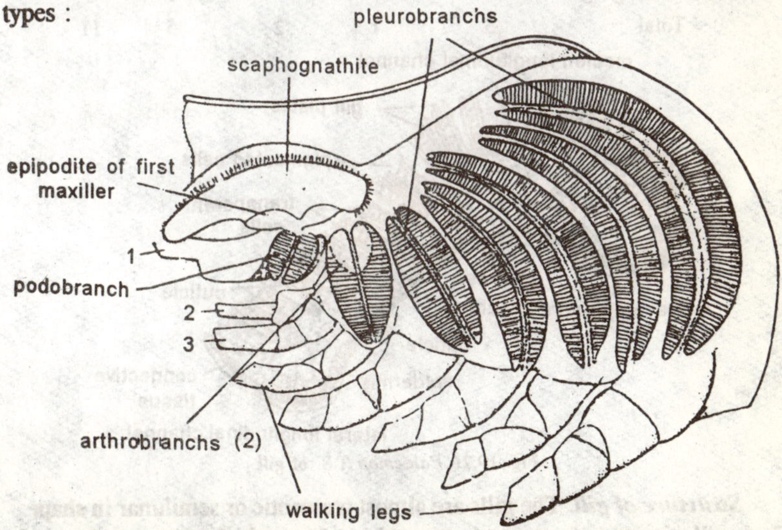


Fig. 19.25 *Palaemon* Gill Chamber exposed to show arrangement to gills

(i) *Podobranch.* When a gill is attached to the coxa of an appendage, it is called *podobranch* or *foot-gill*. In *Palaemon* one podobranch is carried by the coxa of each 2nd maxillipede.

(ii) *Arthrobranch.* When a gill is attached to the arthrodial membrane connecting the appendage to the thorax it is called *arthrobranch* or *joint gill*. Each 3rd maxillipede bears two arthrobranches.

(iii) *Pleurobranch.* When a gill is attached to the lateral wall of the segment bearing limb, it is called *pleurobranch* or *side gill.* The last five gills on each side are pleurobranch attached to the lateral wall of the thoracic segments bearing five walking legs.

***Branchial formula.*** The number and position of epipodites and gills can be illustrated by the branchial formula which is given in the following table.

| | Appendage | Epipodite | Podo branch | Arithro branch | Pleuro branch | Total |
|---|---|---|---|---|---|---|
| I | Maxillipede | 1 | - | - | - | 1 |
| II | Maxillipede | 1 | 1 | - | - | 2 |
| III | Maxillipede | 1 | - | 2 | - | 3 |
| I | Walking leg | - | - | - | 1 | 1 |
| II | Walking leg | - | - | - | 1 | 1 |
| III | Walking leg | - | - | - | 1 | 1 |
| IV | Walking leg | - | - | - | 1 | 1 |
| V | Walking leg | - | - | - | 1 | 1 |
| | Total | 3 | 1 | 2 | 5 | 11 |

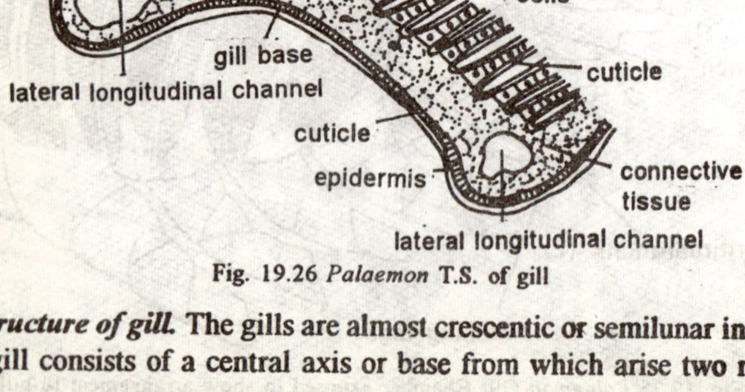


Fig. 19.26 *Palaemon* T.S. of gill

***Structure of gill.*** The gills are almost crescentic or semilunar in shape. Each gill consists of a central axis or base from which arise two rows of rhomboidal gill-plates or lamellae. Leaving a deep groove called *median longitudinal groove* between the two rows, such a gill is known as *phyllobranch*. The gill plates found in the middle of gills are large and gradually become smaller toward both the ends. In the middle of gill arises a small structure called *gill-root* by which it is attached to the thorax. Nerves and blood vessels enter or leave the gill through the gill roots.

The base of the gill appears triangular in a cross section. It is made up of loose conncetive tissue which is bounded by epidermis and cuticle. There is a row of epithelial cells in each gill-plates, outside which is a thin layer of cuticle. The cells are of two types viz. *pigmented* and *transparent* alterning with each other.

***Circulation of blood in gills.*** The gills are profusely supplied with blood. Three blood vessels run vertically in each axis - two *lateral longitudinal channals* along the two lateral margins and a *median longitudinal channel* in the middle. The lateral longitudinal channels are connected with each other by *transverse connective*, one of which, near the gill - root, communicates with the *afferent branchial channel* and receive blood from it. Each lateral channel gives a delicate marginal channel into each gill-plate which runs along the outer margin of the gill-plate and finally opens into the median longitudinal

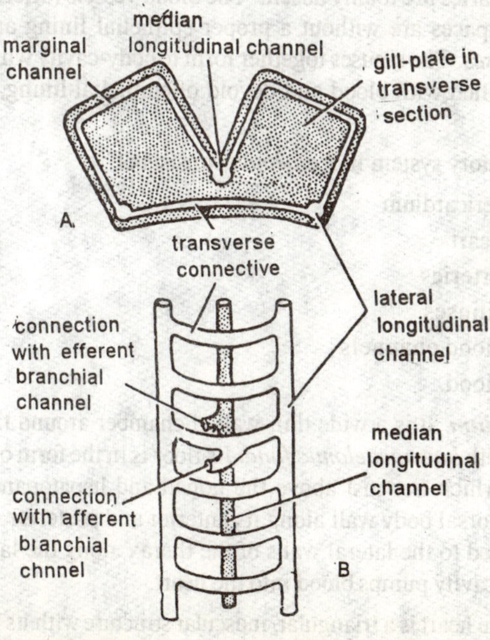


Fig. 19.27 *Palaemon* A=Blood supply in gill plates B=Blood channel in gill

channel. Blood from the afferent branchial channel passes into the transverse connective from where it goes into the lateral longitudinal channel, then through the marginal channel into the median longitudinal channel which finally pours the blood, at the gill-root, into the *efferent branchial channel* by a short connection.

***Mechanism of respiration.*** The exopodite of maxilla, called scaphognathite, is so situated that its anterior end penetrates into gill chamber and the posterior end lies outside. The gill chamber itself is covered dorsally and laterally but is open ventrally. The vibrating action of the scaphognathite bales out water from the anterior end of the chamber and simultaneously pushed a water current movement of the baler resuts in continuous entry of fresh water from behind and its subsequent exit from front. The incoming water current not only passes over the gills, epipodites but also washes the branchiostegites. The oxygen dissolved in water is taken into the blood and carbon dioxide passes out. The respiratory pigment, *haemocyanin* remains dissolved in blood. However, the haemocyanin content in the blood is low, therefore, the major part of oxygen is taken in as simple solution in plasma.

**Circulatory System**

In prawn the circulatory system is *open* or *lacuna*r type, in which the veins and capillaries are totally absent. The blood vessels (arteries) open into spaces. These spaces are without a proper epithelial lining and are called *lacunae* or *sinuses.* The sinuses together form te body-cavity which looks like a coelom, but filled with blood and devoid of epithelial lining, is known as *haemocoel.*

The circulatory system in *Palaemon* consists of :

1. Pericardium
2. Heart
3. Arteries
4. Sinuses
5. Blood channels
6. Blood.

***1. Pericardium.*** It is a wide thin-walled chamber around the heart and remains filled with *haemocoelomic fluid.* Its floor is in the form of a thin horizontal septum which lies just above the gonad and hepatopancreas and is attached to the dorsal body wall along its anterior and posterior aspects. The septum is attached to the lateral walls of the thorax along the lateral aspect. The pericardial cavity pumps blood into the heart.

***2. Heart.*** The heart is a triangular, muscular structure with its broader base facing posteriosly and the apex in front. It is found in the pericardial sinus in the median dorsal part of thorax. A median longitudinal strand, the *cardiopyloric*

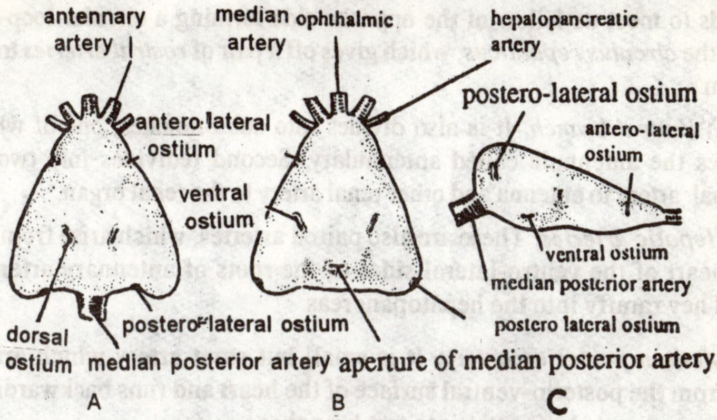


**Fig. 19.28 *Palaemon* Heart A=Dorsal view, B=ventral view C=lateral view**

***strand*, extending from the apex of the heart to the dorsal wall of the pyloric stomach, and two lateral strands, stratching from the lateral angle of the heart to the body wall, support the heart in the pericardium. The walls of the heart are very thick, muscular and extermely elastic. There are 5 pairs of valvular openings, the *ostia* in the wall of heart. The first pair of ostia is situated in the wall of heart mid-dorsally. The second part lie mid-ventrally. The third pair lies on the posterior border. The fourth pair lie anterio-laterally and the fifth pair posterio-laterally. The cavity of heart is not continuous but is traversed by a large number of interlacking muscle fibres.**

***3. Arteries.* The arteries are thick-walled muscular tubes, which arise from the heart and supply blood to different organs of the body. The main arteries are following :**

*(i)* *Median ophthalmic artery.* It arises as a single median cephalic or ophthalmic artery from the apex of the heart. It runs forward to the head region in the mid-dorsal line above the renal sac. Here it joins the two antennary arteries.

*(ii)* *Antennary arteries.* A pair of them originate from the apex of the heart one from either side of the median ophthalmic. Each runs obliquely forward along the outer border of the mandibular muscle. During its course it gives out the following branches :

(a) Pericardial branch to the pericardium,

(b) Gastric branch to the cardiac stomach, and

(c) Mandibular branch to the mandibular muscles.

**Eacn antennary artery passes forwards and divides into two sub-branches.**

(a) *Dorsal branch.* It sends an optic artery to the eye and then bend inwards to meet its fellow of the opposite side forming a circular loop-like artery the *circulus cephalicus*, which gives off a pair of *rostral arteries* to the rostrum.

(b) *Ventral branch.* It is also divides into sub-branches, one of which supplies the antennule called antennulary, Second redivides into two an antennal artery to antenna and other renal artery to the renal organ.

*(iii) Hepatic arteries.* These are also paired arteries, which arise from the heart of the ventro-lateral sides of the roots of antennary arteries. They ramify into the hepatopancreas.

*(iv) Median posterior artery.* It is small but stout artery which arises from the posterio-ventral surface of the heart and runs backwards. It immediately bifurcates into two branches.

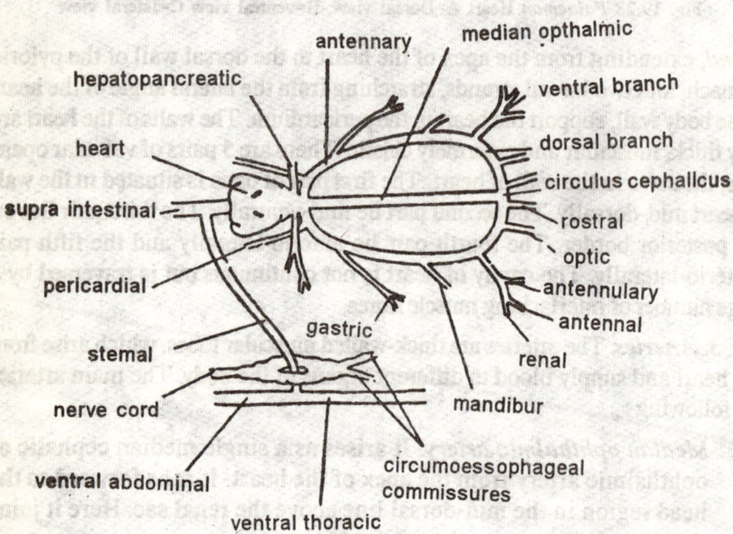


Fig. 19.29 *Palaemon* Showing heart and arteries

(a) *Supra-intestinal artery.* The supra-intestinal artery passes stright backwards along the dorsal surface of the intestine extending up to the hind-gut, while it divides into two branches supplying blood to the intestine and the dorsal abdominal muscles.

(b) *Sternal artery.* The sternal artery is the stoutest of all others. It runs downwards and forwards through the hepatopancreas and after passing through an aperture in the middle of the ventral thoracic ganglionic mass to reach the ventral side, it gets divided into two branches. The ventral thoracic branch runs anteriorly up to the mouth and supplies the sternal region, first three pair of walking legs, maxillae, and maxillulae, etc. The ventral abdominal branch runs posteriorly upto the anus and supplies blood to the ventral abdominal region, last two pairs of legs, pleopods and hind gut, etc.

***4. Blood Sinuses.*** The various arteries, on reaching the organs, branch repeatedly into minute branches which do not form capillaries but end into wide spaces, *blood sinuses or blood lacunae.* All the blood lacunae of the body collectively form a pair of elongated but ill-defined ventral sinuses. These lie below the hepathopancreas and above the floor of thorax. The ventral sinuses communicate with each other at various places.

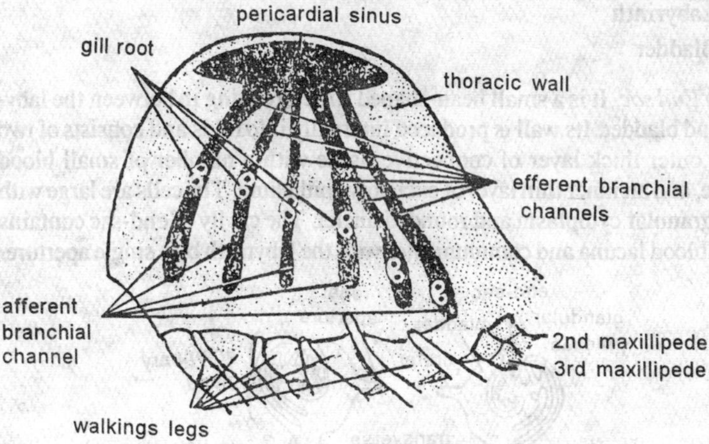


Fig. 19.30 *Palaemon* Branchial channels

***5. Blood Channels.*** In *Palaemon*, there are 6 pairs of affarent branchial channels and 6 pair of efferent branchial channels. These channels are lacunar in nature. The afferent branchial channels run along the inner surface of the thoracic wall and reach the gills through the gill-roots. Here each channel opens into the tranverse connective. The first afferent channel takes the blood to the podobranch and two arthrobranches. Remaining five channles supply the blood to the five pleurobranches. The blood is oxygnated while passing through the gills and this blood is carried to the pericardium by efferent branchaial channels.

***Course of blood circulation.*** It can be represented diagrammatically shown as below.

***6. Blood.*** The blood is a thin, watery, almost colourless liquid containing floating colourless amoebocytes. The colour of the blood is faintly bluish, when it is oxygenated. The colour is due to the presence of a respiratory pigment i.e. haemocyanin. It is a compound of copper and protein. There is also a lipochrome pigment called zoonerythin in the blood. The blood has the power of coagulation when it comes in contact with air.

**Excretory System**

The excretory system of adult *Palaemon* consists of a pair of antennary or green glands and a median sac termed a renal sac.

***1. Antennary or green glands.*** The antennary glands are opaque white organs of the size of a pea-seed, are enclosed inside the coxae of each antenna. Each gland consists of 3 parts :

(i) End Sac

(ii) Labyrinth

(iii) Bladder

(i) *End sac.* It is a small bean-shaped structure lying in between the labyrinth and bladder. Its wall is produced internally into folds and consists of two layers, outer thick layer of connective tissue with a number of small blood lacunae, and an inner thin layer of excretory epithelium. The cells are large with finely granular cytoplasm and rounded nuclei. The cavity of end-sac contains a large blood lacuna and communicates with the labyrinth by a single aperture.

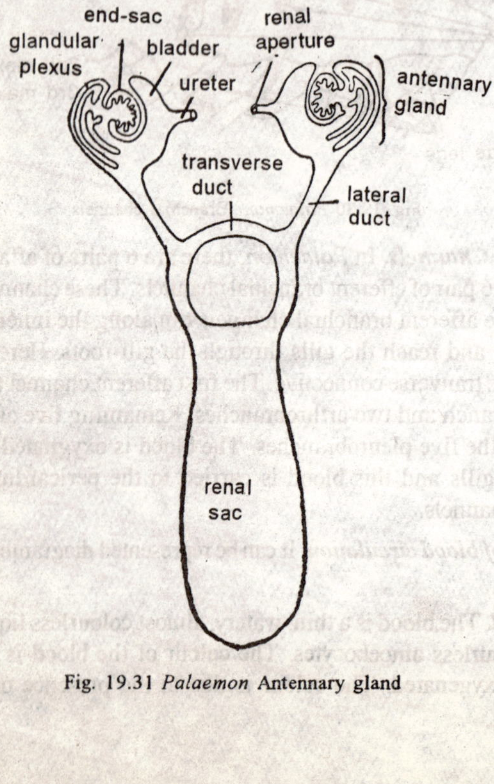


Fig. 19.31 *Palaemon* Antennary gland

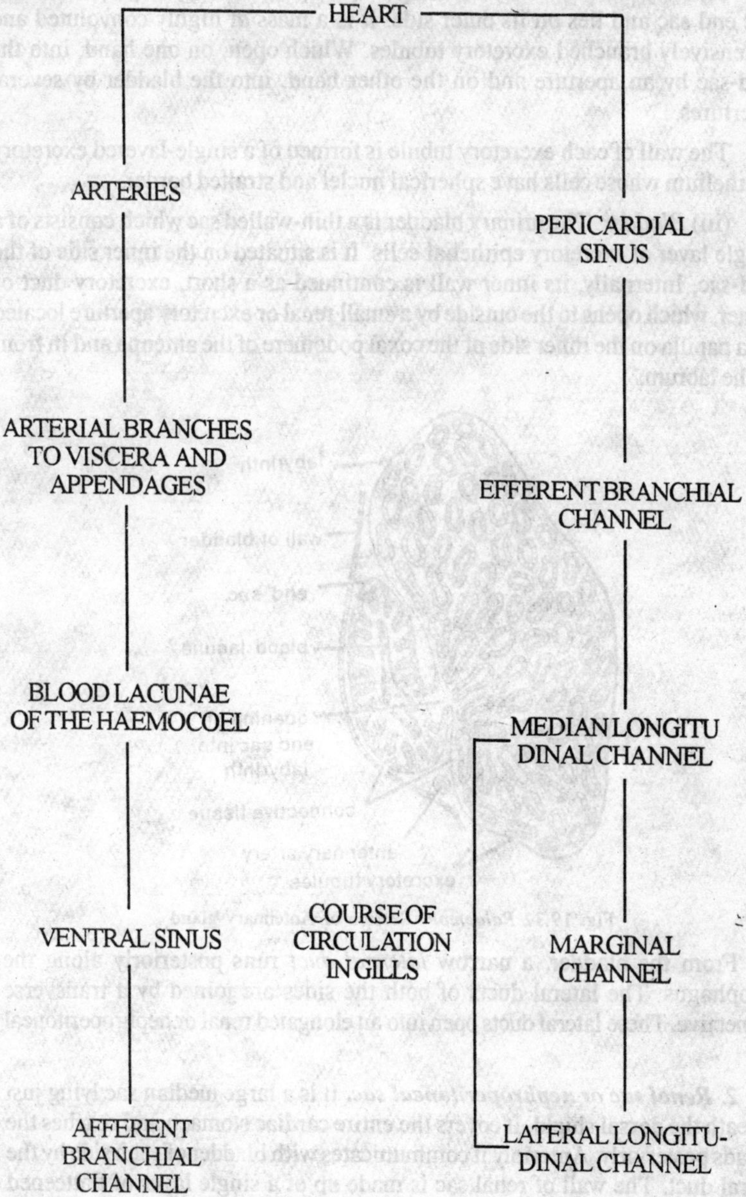


Chart Showing Course of Circulation in palaemon

(ii) *Labyrinth*. The labyrinth or glandular plexus is very larger in size then the end sac and lies on its outer side. It is a mass of highly convoluted and extensively branched excretory tubules. Which open, on one hand, into the end-sac by an aperture and on the other hand, into the bladder by several apertures.

The wall of each excretory tubule is formed of a single-layered excretory epithelium whose cells have spherical nuclei and straited border.

(iii) *Bladder*. The urinary bladder is a thin-walled sac which consists of a single layer of excretory epithelial cells. It is situated on the inner side of the end-sac, Internally, its inner wall is continued as a short, excretory duct or ureter, which opens to the outside by a small renal or excretory aperture located on a papilla on the inner side of the coxal podomere of the antenna and in front of the labrum.

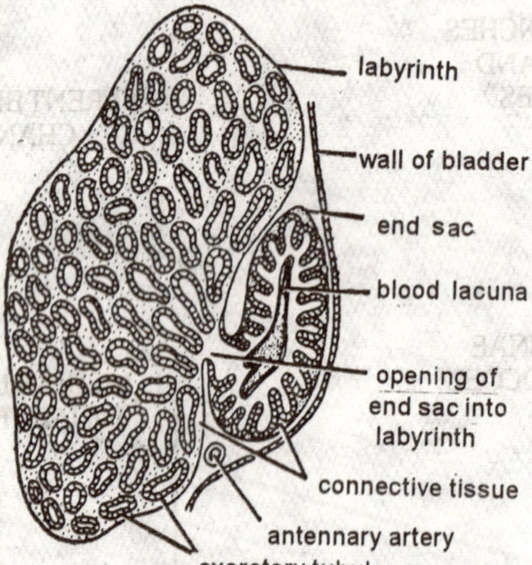


Fig. 19.32 *Palaemon* Section of Antennary gland

From the bladder, a narrow *laternal duct* runs posteriorly along the oesophagus. The lateral ducts of both the sides are joined by a transverse connective. These lateral ducts open into an elongated renal or nephroperitoneal sac.

***2. Renal sac or nephroperitoneal sac.*** It is a large median sac lying just beneath the dorsal shield. It covers the entire cardiac stomach and reaches the gonads posteriorly. Anterioly it communicates with bladder of each side by the lateral duct. The wall of renal sac is made up of a single layer of flatteened excretory epithelial cells.

***Physiology of excretion.*** The antennary glands perform two functions, *excretion* and *osmoregulation*. These glands are richly supplied with blood so that by ultrafilteration, the water and dissolved substances are passed into the end sac. The filterate passes into the labyrinths. Where selective reabsorption takes place. The useful substances are pass back into the blood. The remaining fluid, better to say the urine, flows into the bladders and the renal apertures. The urine contains mainly ammonium compounds, some *urea* and *amino acids*.

***Osmoregulation.*** *Palaemon* is a fresh water crustacean so that due to higher salt concentration of body the surrounding water continually diffuses into the blood through highly permeable gills. Excessive water is eleminated along with nitrogenous wastes that is why the urine of prawn is hypotonic.

***Integument.*** The integument is also regarded as an organ of excretion in young stages. The nitrogenous wastes are deposited over the integument, which is cast off now and then during ecdysis.

**Nervous System**

The nervous system of prawn is similar to that of earthworm in many respects. It consists of :

1. Central Nervous System
2. Peripheral Nervous System
3. Autonomic Nervous System.

**1. *Central Nervous System.*** The central nervous system includes a pair of cerebral ganglia, a pair of circum-oesophageal commissures and a double ventral ganglionated nerve cord.

(i) *Cerebral or supraoesophageal ganglia.* It is a bilobed structure situated at the base of rostrum in front of the oesophagus. It consists of three pairs of fused ganglia, as it appears from the three pairs of nerves originating from it.

(ii) *Circum-oesophageal commissure.* There are two stout nerves, which arise from the posterior end of brain and run backwards and downwards around the oesophagus. It unites ventrally with the sub-oesophageal ganglia which form the indistinguishable anterior part of the ventral thoracic ganglionic mass. Each commissure bears a small commissural ganglion near its anterior end, and gives off small nerve to the mandible of its side. The two circum-oesophageal commissures are connected with each other by a slender transverse commissure near the posterior end.

(iii) *Ventral nerve cord.* It lies in the mid-ventral line of the body just above the sternal plates. The ventral nerve cord is actually double but due to fusion it has lost its double nature. It bears 17 pairs of ganglia. first eleven pairs of ganglia belong to cephalothoracic re-

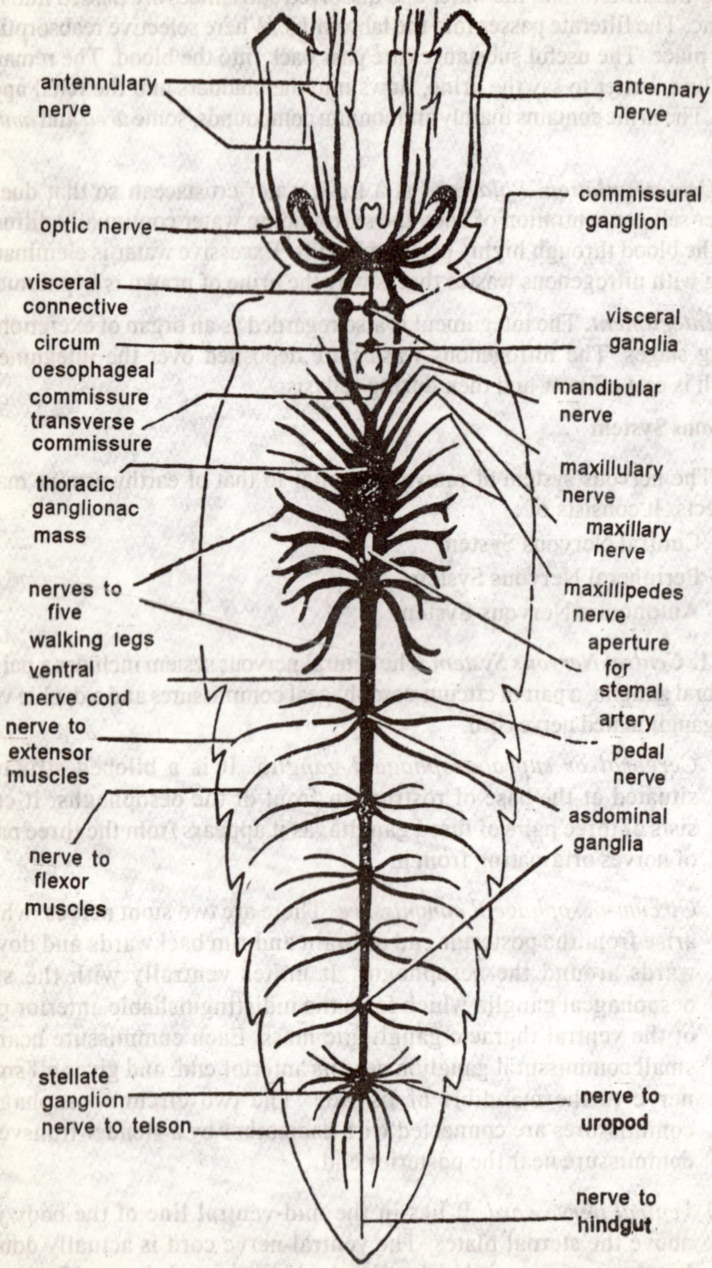


**Fig. 19.33** *Palaemon* **Nervous System**

gion. Thesae 11 pairs of ganglia are fused to from a single, large, oval mass called the *cephalothoracic ganglionc mass.* It is perforated by the sternal artery near its hind end. The posterior 6 pairs of ganglia of the nerve cord belong to the abdomen. The last abdominal ganglion is larger than others due to having several post-abdominal ganglia fused with it.

2. ***Peripheral Nervous System.*** It is formed by the paired nerves which arise from the various parts of central nervous system.

(A) Five pairs of nerves arise from brain are :

(i) *Antennulary nerves :* A pair of antennulary nerves arise from below the origin of optic nerves. Each nerve enters the antennule of its side into which it sends a statocystic branch to statocyst.

(ii) *Optic nerves.* A pair of stout optic nerves arise from the dorsal surface of the brain, one on each side, and supply the eye-stalks.

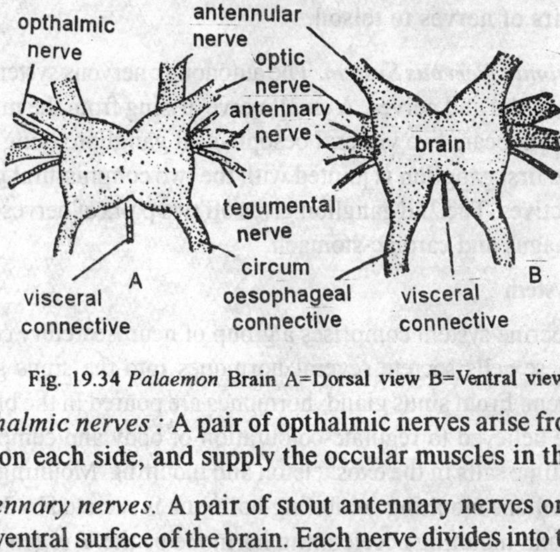


Fig. 19.34 *Palaemon* Brain A=Dorsal view B=Ventral view

(iii) *Opthalmic nerves :* A pair of opthalmic nerves arise from the brain, one on each side, and supply the occular muscles in the eye-stalks.

(iv) *Antennary nerves.* A pair of stout antennary nerves originate from the ventral surface of the brain. Each nerve divides into two branches, the outer innervating the squama and the inner innervating the feeler of antenna.

(v) *Tegumental nerves.* A pair of slender tegumental nerves arise just behind the origin of antennary nerves. These nerves innervate the labrum.

*(B) Mandibular nerve.* It arises from the outer side of each circum-oesophageal connective and supplies the mandible of its side.

*(C) Nerves from cephalothoracic ganglionic mass.* 11 pairs of nerves are given out from the mass.

(a) *Mandibular nerves for mandibles.*

(b) *Maxillulry nerves for maxillulae.*

(c) *Maxillipede nerves.* There are three pairs of maxillipede nerves which supply to maxillipedes.

(d) *Peraeopod nerves.* These are 5 pairs in number and supplying to the walking legs.

*(D) Nerves from abdominal ganglia.* Each of the first four abdominal ganglia gives off three pairs of nerves in each segment.

(a) *Pedal nerves.* A pair of pedal nerves to pleopods

(b) A pair of nerves to the extensor muscles.

(c) A pair of nerves to the flexor muscles of the succeeding segment.

(d) The Vth abdominal ganglion sends two pairs of nerves to the flexor muscles of sixth segment, two pairs of nerves to the uropods and two pairs of nerves to telson.

***3. Autonomic Nervous System.*** The autonomic nervous system comprises a few small ganglia and nerves. A small nerve arising from the mid-posterior part of the brain, bears two visceral oesophageal ganglion, lying one behind the other. The first ganglion is jointed with the two commissural ganglia by a pair of connectives. The 2nd ganglion gives off two pairs of nerves to the walls of the oesophagus and cardiac-stomach.

**Endocrine System**

The endocrine system comprises a group of neurosecretory cells in each eye-stalk. These cells secrete several hormones into the sinus gland lying adjacent to them. From sinus gland, hormones are poured in the blood. These hormones are believed to regulate-colouration of body and compound eyes; deposition of lime salts in the exoskeleton and moulting. Moulting is controlled by two hormons moulting inhibiting hormone is secreted by X-organs in the eyestalks and moulting accelerating hormone by the Y organ present beneath the adductor muscle of mandible.

**Sense Organs**

The sense organs of *Palaemon* are :

1. Statocyst.
2. Olfactory setae
3. Tectile setae
4. Compound eyes.

***1. Statocysts.*** The statocysts lie attached to the dorsal wall of the prexoxa of each antennule and open through apertures on the dorsal concave surface. Each aperture is covered by a fold of skin. A statocyst is a subspherical cuticular sac about 1-1.5 mm wide. Each statocyst is supplied by the statocystic branch of antennulary nerve.

The cavity of the capsule possesses a heap of sand particles lined by a number of delicate elongated statocystic setae. Each seta consists of a swollen base and a long shaft which is half bent upon itself and covered with fine bristles. The base movably articulates with the wall of the sac by means of an arthrodial membrane and receives a fine branch from the statocystic nerve.

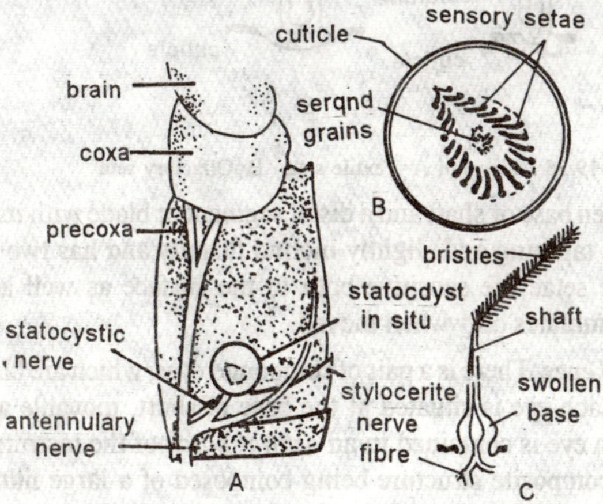


Fig. 19.35 *Palaemon* A=Statocyst in sity, B=T.S. C=a single receptor

The statocysts are the organs of orientation and wquilibrium. In the normal swimming position, the sand grains lie on the floor of the statocyst under the influence of the force of gravity. When the body of swimming prawn gets tilted to one side it causes the displacement of sand particles in the same direction. This presses the sensory setae and stimulates them. The stimulated setae transmit this feeling to the brain and the prawn knows about its position and balances itself accordingly.

***2. Olfactory setae.*** These are also known as *aesthetases*, occur on the middle small feeler of each antennule. Each setae consist of a basal segment or shaft and is attached segment of blade. Shaft is hollow and is attached to the integument while the later (blade) is free and bluntly rounded and is covered with a thin membrane. Small nerve fibres from the antennulary nerve supply these setae.

***3. Tactile setae.*** These are situated on the elongated feeler of the antennae and other parts of the body, especially on the appendages Each seta

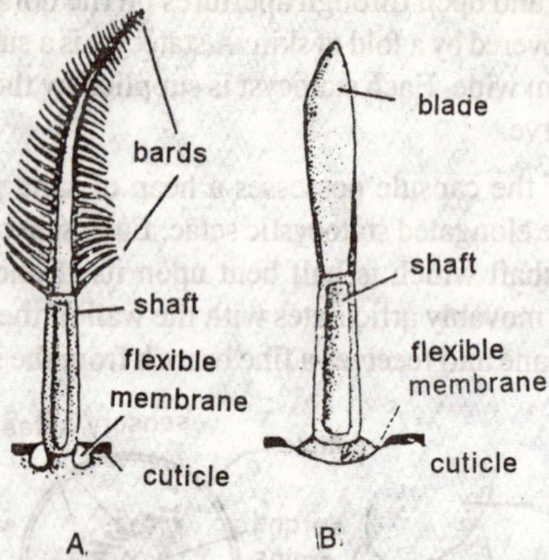


Fig. 19.36 *Palaemon* A=Textile seta, B=Olfactory seta

consists of a swollen base or shaft and a distal segment or blade with its central axis. The blade is tapering and slightly inclind distally and has two rows of barbs. The tactile setae are sensitive both to the surface as well as water current but get stimulates only when moved.

***4. Compound eyes.*** There is a pair of compound eyes, which are black and hemi-spherical. Each eye is situated at the base of short, movable and two jointed stalk. Each eye is contained in an *orbital notch* of the rostrum. Each eye is actually a composite structure being composed of a large number of visual units called as *ommatidia.* There are about 2500 such ommatidia in a compound eye. The surface of the eye is roughly hemi-spherical and is divided into a large number of small square facets. Each facet corresponds to a single

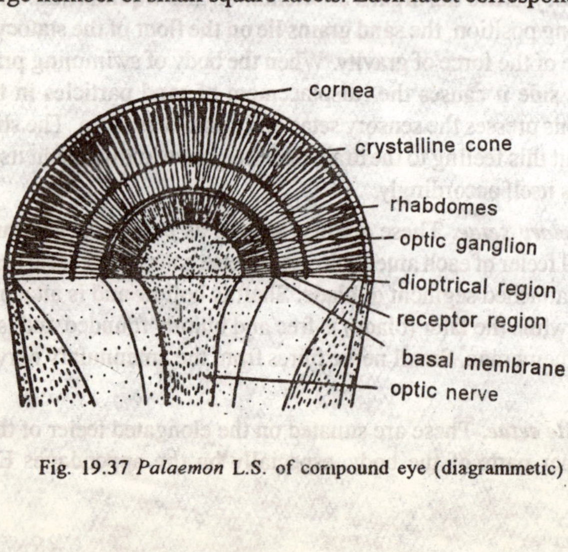


Fig. 19.37 *Palaemon* L.S. of compound eye (diagrammetic)

ommatidium. All the ommatidia are similar, displayed radially, lying side by side and separated by a dark pigment cells. Each ommatidium consists of following parts.

(i) *Cornea.* The eye is protected externally by a transparent covering of cuticle known as *cornea.* It is distinguished into facets like those of graph paper.

(ii) *Corneagen cells.* These are flat one pair and of epidermal in origin. They secrete a new corneal layer after ecdysis.

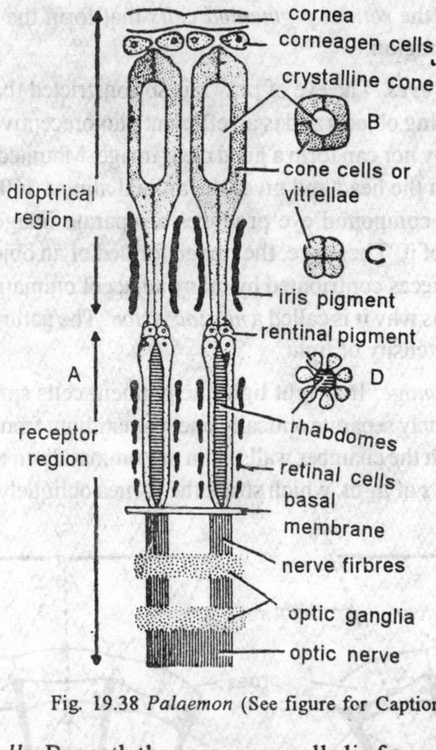


Fig. 19.38 *Palaemon* (See figure for Caption)

(iii) *Cone cells.* Beneath the corneagen cells lie four elongated cone cells or *vitrellae*, which constitute a transparent, homogenous crystelline cone. The inner ends of cone cells are long and tapering.

This part of the eye from the cornea to the posterior end of the cone cells is called *dioptrical region* as it is concerned with the focusing of the light on the underlying region known as receptor region.

(iv) *Rhabdome.* The inner ends of cone cells lie upon an elongated, spidle-shaped, transversely striated structure known as rhabdome. It is secreted and surrounded by a group of seven elongated *retinal cells*. The inner end of the rhabdome are supported by a basal membrane beyond which they are continuous with the nerve fibres. The rhabdome and retinal cells from the *receptor region*.

(v) *Nerve fibres.* The ommatidia are innervated by nerve fibres of the optic ganglion.

Each ommatidium is ensheathed by densely pigmented cells which optically separate from the adjacent ommatidia. There are two sets of these pigmented cells: the iris pigmented cells forming the iris sheath, round the dioptric region and the *retinal pigmented cells* that form the retinal sheath round the receptive region.

***Working of the eyes.*** The eye of prawn is so constricted that while it can swiftly spot out moving objects and is an efficient photoreceptive organ, it can neither focus properly nor can form a good clear image. Mounted on a movable stalk, it can move on the head and gives the animal almost $360^0$ vision. Each ommatidium of the compound eye produces a separate image of an object present just infront of it. Therefore, the image formed of an object by this eye consists of several pieces contributed by many adjacent ommatidia just like a mosaic art net. That is why it is called a *mosiac vision.* The nature of the image depends upon the intensity of light.

(a) *Apposition Image.* In bright light the pigment cells spread in such a way that they completely separate optically one ommatidium from adjacent. No light can pass through the chamber walls from one ommatidium to the other. In this condition the rays of light, which strike the cornea obliquely are absorbed

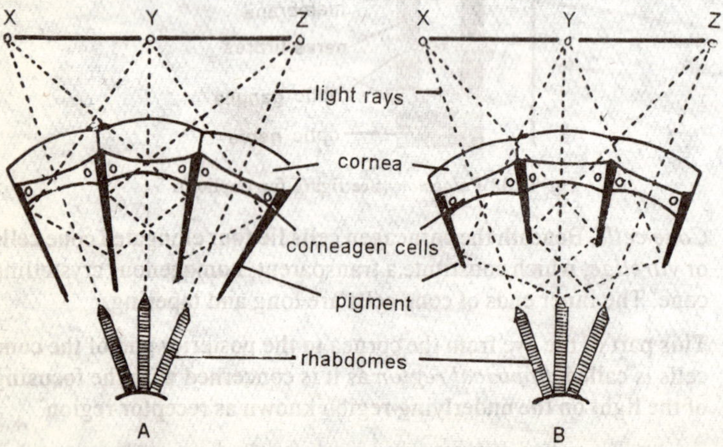


Fig. 19.39 *Palaemon* A=Apposition or mosaic vision, B=Super-position image

by the pigment cells without producing a visual effect. Only those rays of light, which pass directly through the centre of cornea, can traval through the ommatidium and reach the rhabdome to from an image of a part of object. As a result the complete image fromed is a mosaic of several components places in juxtaposition. Such as image is called *apposition image*.

(b) *Super Position Image*. In the weak light, the pigment cells migrate partly towards the outer and partly towards the basal ends of the ommatidia. When this is the condition, even the oblique rays of the source of light are capable of forming a point of image, after passing through a number of ommatidia. Thus, a combined image is thrown on the retinular layer. In this way, an erect superposition image is formed as the rays refracted by several crystalline cones are superposed on the focus of retina. The superposition image is less distinct in detail than the apposition image, but it is more sensitive to weak intensities of light.

**Reproductive System**

The sexes are spearate and there is a well marked sexual dimorphism. The males and females differ in several respects.

1. The males are larger than females.
2. Males have thoracic legs more closely arranged than in the females.
3. Males have narrower abdomens then those of the females.
4. In males the 2nd pair of chelate legs are longer and have a thicker covering of setae and spines.
5. Males have *appendix masculina* arising from the appendix interna of each pleopod of the second pair.
6. In males, the epimeres of abdominal segments are smaller in size than in the females, where they are bigger in size for carrying eggs.
7. In males, the paired genital and thoracic openings are situated on the arthrodial membrane between the 5th pair of the walking legs covered by a flap of integument, while, in the females, the genital apertures are situated on the coxae of 3rd pair of the walking legs.

***Male Reproductive Organs.*** The male reproductive system comprises:

1. Testes.
2. Vasa deferentia
3. Vesicula seminalis.

1. *Testes*. The testes are median dorsal structure situated in the posterior part of the thorax above the hepathopancreas and below the pericardium. They extend anteriorly upto the renal sac and posteriorly upto the first abdominal

segment. These are soft white structures which are fused anteriorly to from a single lobe, while at the posterior end these remain in close contact. The two testes enclose a central space or gap which gives way to the cardiopyloric strand connecting the heart to the cardiac stomach.

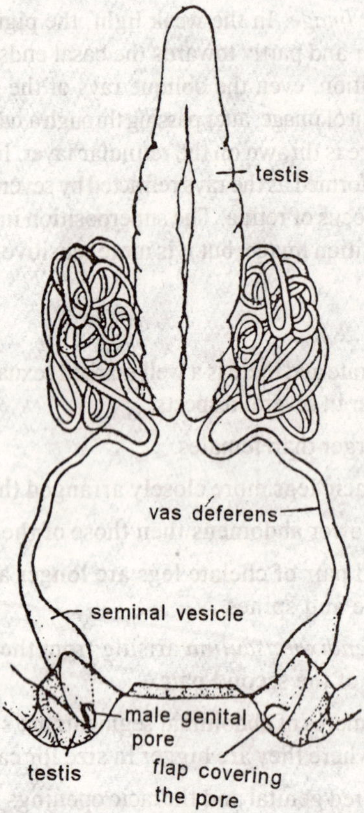


Fig. 19.40 *Palaemon* Male reproductive organ

Each testis consists of a large number of coiled, hollow thin walled seminiferous tubules buried in connective tissue. The cavity of each tubule is lines by a single layer of epithelium, the cells of which undergo spermatogenesis to form the spermatozoa. A mature sperm consists of a rounded cytoplasmic body, containing a large, dark crescentic nucleus, and a tail-like blunt process.

2. *Vasa deferentia.* From the outer surface of the posterior end of each testis arises a very long and narrow tube, the vas deferens. It is differentiated into two part : the proximal highly convoluted part which emerges out of the testis and the distal straight part which runs vertically downwards between the

abdominal flexor musceles and the thoracic wall. Each inwards to open into the vasicula seminalis.

3. *Semin·al vesicles.* The distal part of each vas deferens enlarges to form a seminal vesicle or vesicula seminalis near the coxa of 5th walking leg. It is club shaped structure which opens to the outside through the male genital pore on the inner surface or coxa of 5th leg. The male genital pore is covered by a small flap-like piece of skin. The vesicula seminalis stores the sperms.

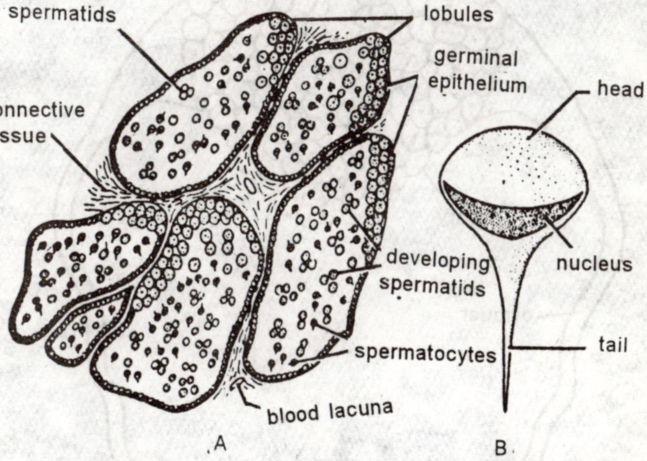


Fig. 19.41 *Palaemon* A=T.S. testis, B=A sperm

***Female Reproductive System.*** The female reproductive system comprises following organs :

1. Ovaries

2. Oviducts.

1. *Ovaries.* A pair of ovaries which are compact and sickle-shaped bodies touching each other at both the ends but leaving a gap in the middle for the passage of the cardiopyloric strand. The shape and size of the ovaries vary with age and the season of the year. Each ovary is enclosed within a membranous capsule and is made of numerous radial rows of ova in various stages of development. The immature ova lie towards the centre while the mature ova towards the surface of the ovary. The latter are large nucleated cells with planty of yolk material.

2. *Oviducts.* From the middle of the outer surface of each ovary arises a short thin-walled tube known as oviduct. Its anterior end is broad and forms the oviducal funnel. It lies opposed to the surface of ovary. The oviduct runs vertically downwards to open to the exterior by female genital pore. It is situated on the inner side of coxa of 3rd walking leg.

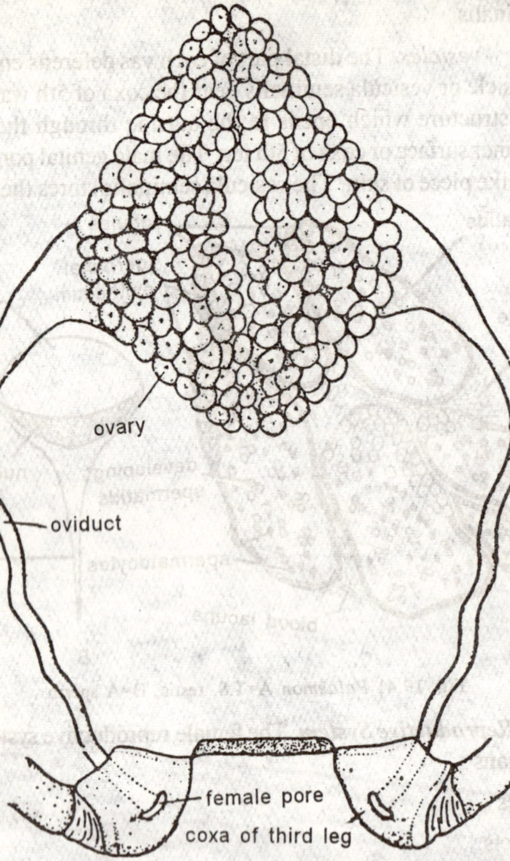


Fig. 19.42 *Palaemon* Female reproductive system

**Fertilization**

Breeding season in *Palaemon* is from May to July. During the season, the females carry eggs attached to their pleopods. Except in breeding season, the females are found without eggs. Fertilization is external. When ova are laid, they are attached to the setae on the pleopods with the help of a sticky secretion of glands situated both on the appendages and on the body segments. The male, later on, deposits sperms near the genital openings, where fertilization takes place. It is not known yet that how male deposits sepermatophores.

**Development**

In *Palaemon lamarrei, P. malcolmsonii* etc. The development is direct i.e. without larval stage. However, in other prawns (*Penaeus, Lacifer*) the development is indirect i.e. with larval stages.

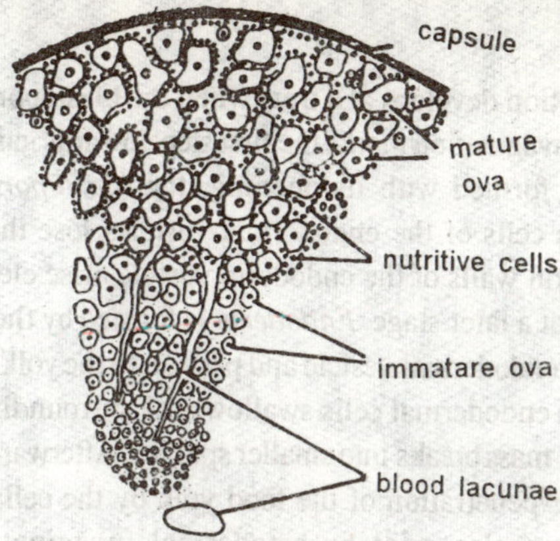


Fig. 19.43 *Palaemon* T.S. of ovary

The cleavage is superficial. The nucleus of fertilized egg, lying in the centre, divides and redivide forming a number of nuclei. These nuclei are later on enclosed by masses of protoplsm, but areas of the separate cells do not become marked off by furrows cutting right through the egg. As a result of segmentation, a blastoderm is formed, which appears at first on the ventral side of the egg. Its formation begins at one point on the surface and proceeds gradually from this point, which always represents the future ventral side of the egg. Gastrulation starts from the posterior end of the ventral side which is marked by the first formation of the blastoderm.

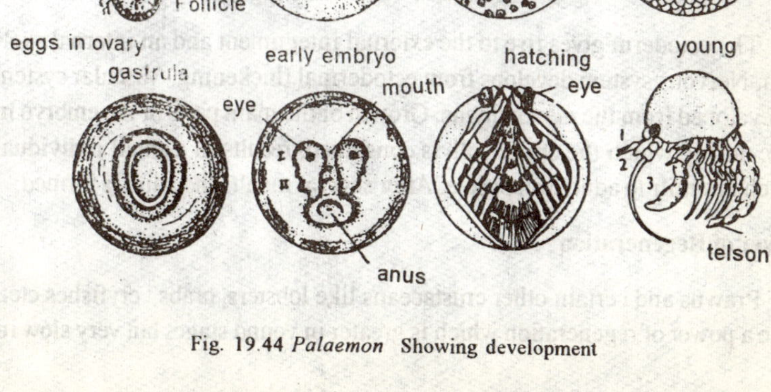


Fig. 19.44 *Palaemon* Showing development

**Gastrulation**

A small invagination develops at a time when the blastopore is not completely fromed. This invagination marks the first step towards gastrulation and small archenteron is formed with its opening, the *blastopore*. When the blastopore closes, the cells of the endodermal vesicles lose their epithelial nature. From the lateral walls of the endoderm vesicle arise elements which form the *mesoderma* at a later stage. *Endoderm* is formed by the cells, which arise from the floor of endodermal vesicle and pass on to the yolk and multiply therein. Each of these endodermal cells swallows the surrounding food yolk, with the result the yolk mass breaks into smaller spheres. Afterwards the midgut is formed by the inter-penetration of the food yolk by the cells of gastrula. Fore-gut and hind-gut also arise by ectodermal invagination forming *stomodaeum* and *proctodaeum*. Both reach the midgut and finally fuse.

When the mesoderm appears for the first time, it consists of a mass of cells near the point of origin. These cells multiply rapidly and spread out apparently in an irregulr manner between the ectoderm and the food yolk.

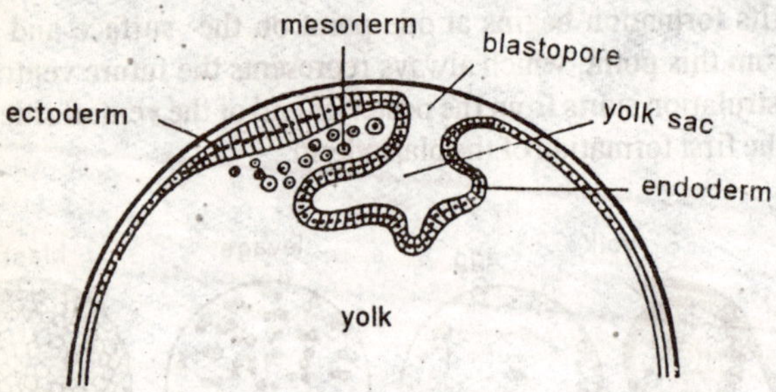


Fig. 19.45 *Palaemon* V.S. of early gagsgtrula

The ectoderm gives rise to the external integument and an internal skeleton. Nervous system develops from ectodermal thickening. Vascular system is developed from the mesenchyme. Growth of the main parts of the embryo in size, during which the entire yolk is consumed, results in a small individual which is similr to adult except size. After several moults the adult is formed.

**Power of Regeneration**

Prawns and certain other crustaceans like lobsters, crabs, cryfishes etc., have a power of regeneration which is greater in yound stages but very slow in adults for example, if the tip of eye is removed, normal regeneration takes place. If the entire eye stalk along with optic nerve is removed, an antenna like structure is fromed. Such type of regeneration in which normal organ is not developed, is called *heteromorphosis*.

Crustaceans have the power of *autotomy*. If a thoracic leg is broken in some mishap, its distal segments are spontaneously thrown off to feed the enemy.

♦♦♦

20

# CHARACTERS & CLASSIFICATION OF MOLLUSCA

Members of the phylum *Mollusca* are among the most conspicuous invertebrates and include such familiar forms as *clams, oysters, squids, octopods* and *snails* during the last 18th and 19th centuries, when natural history, occupied the time of many well to do gentlemen, shell collections were as popular as today such collections, often containing species gathered from various parts of the world, have contributed considerably to our knowledge of the phylum. In abundance of species, molluscs comprise the largest invertebrate phylum aside from the Arthropoda. Over 1,00,000 living species have been described. In addition, some 35,000 fossil species are known.

The term *Mollusca* (*L. mollis* = soft) mean soft bodied and for the first time coined by *Aristotle* in his famous Historia Animalium to describe various kinds of cephalopods and gastropods. The term *Mollusca* was used by *Jonston* (1650) to include cephalopods and barnacles.

The classification of *Mollusca* divided into the five classes-Amphineura, Gastropoda, Lamellibranchia and Cephalopoda, by *Laseneer* (1192) has remained standard ever since.

## CHARACTERS

1. Some are aquatic mostly marine, a few fresh water and some terrestrial forms.
2. Symmetry bilateral, viscera and shell coiled in Gastropoda and some Cephalopoda.
3. The body is triploblastic made up of 3 germ layers.
4. Body without any appendage and segmentation.
5. Body skin single layered mostly ciliated and with mucous glands.
6. Body usually small and soft and enclosed in an envelop, the *mantle* that secretes shell of various shape and sizes.
7. Generally the body is divided into 4 parts, the visceral mass carrying the vital organs of the body; the head, anterior most part bearing mouth,

eyes and tentacles but is absent in the *Pelecypoda* and *Scaphopoda*, the foot a muscular mass, the form and function of which varying in the different classes and lastly the mantle or pallium a thick muscular clock covering the entire animal.

8. Shell, when present, usually univalve, consituting an exoskeleton, internal in some.
9. Coelom reduced and represented mainly by pericardial cavity, gonadial cavity and kidney.
10. The alimentary tract is either straight or coiled. Jaws are present in many, especially in gastropods and cephalopods. In most of animals a rasping organ, the *radula* appears in the buccal cavity. the radula tears up the food.
11. Blood vascular system consists of dorsal heart with 1 or 2 auricles and 1 ventricle and is of open type.
12. Respiration by general body surface or mantle or by one to many gills or a lung in the mantle cavity.
13. Excretion by paired metanephridia (kidneys).
14. Nervous system of paired ganglia, connectives and nerves. Ganglia usually form a circumenteric ring.
15. Sense organs include eyes, statocysts and receptors for touch, smell and taste.
16. Sexes are usually separate (some hermaphrodite a few protandric); gonads 2 or 1 with ducts; fertilization external or internal. Mostly oviparous, few viviparous.
17. Development direct or interrupted with larval stage, the *veliger larva* of mollusca. The asexual reproduction is completely absent.

Phylum *Mollusca* is divided into six classes. Monoplacophora, Amphineura, Gastropoda, Scaphopoda, Pelecypoda and Cephalopoda.

## CLASS I. MONOPLACOPHORA

1. Abyssal or fossil, from lower Cambrain to Devonian, probably carnivorous.
2. Bilaterally symmetrical and metamerically segmented, but segmentation internal.
3. Shell dorsal and consisting of a single piece (hence Monoplacophora).
4. Mantle covering the dorsal surface of body.
5. Foot is ventral with a flat creeping sole.

6. Mouth is anterio-median and the anus is posterio-median.
7. Head is without eyes.
8. 5 or 6 pairs of gills externally and serially arranged.
9. Six pairs of nephridia and two pairs of gonads.
10. Stomach contains a crystelline style and the animal feeds on radiolarians.
11. Nervous system primitive, lacks ganglia.
12. Sexes are separate.
13. Development is indirect with a *trochophore larva*.

Example : *Neopilina galatheae.*

**CLASS II AMPHINEURA**

(Gr., *Amphi* = both; *neuron* = nerve)

1. Most primitive molluscs with either elongated and vermi-form or dorso-ventrally flattened body.
2. Head indistinct and without eyes and tentacles.
3. Foot broad, flat, sole-like adventral in position.
4. Shell and mantle covering the body from dorsal and lateral sides.
5. 6-60 pairs of gills situated laterally in the pallial cavity on either side of foot. They may be reduced to one pair.
6. Mouth and anus at the opposite ends of body.
7. Nervous system primitive and ladder-like with or without ganglia.
8. Heart with one or two auricles.
9. A pair of nephridia connected to the pericardium.
10. Sexes separate or united.

Class Amphineura is divided into two orders :

***Order - 1. Aplacophora***

1. The body is long, narrow, worm like.
2. The shell is absent.
3. Mantle contains tiny calcareous spicules.
4. Foot is reduced or absent.
5. Heart consists of a single auricle and a ventricle.
6. Gills are absent or reduced to a pair located in the cloacal cavity.
7. Alimentary canal is straight without radula.

8. Nervous system with a distinct brain and ganglia.
9. Sexes united.

Examples : *Chaetoderma, Neomenia.*

***Order - 2. Polyplacophora***

1. Body elliptical, convex dorsally and flattened ventrally.
2. Shell dorsal and formed of eight calcareous pieces.
3. Mantle with spicules disposed along the margins.
4. Gills 6 to 8 pairs, present in the pallial groove on the lateral sides of foot.
5. Head devoid of eyes and tentacles.
6. The whole ventral surface is occupied by the broad and flat foot.
7. Alimentary canal is coiled with well formed radula.
8. Heart with two auricles.
9. Nervous system without definite ganglia.
10. Sexes separate.

Examples : *Chiton, Chaetopleura.*

**CLASS.III. GASTROPODA.**

(Gr. *Gastros* = stomach; *podos* = food)

1. Body is asymmetrical with a mantle and shell.
2. The shell is univalved and spirally coiled.
3. Head is distinct bearing tentacles, eyes, and a mouth.
4. Foot is ventral and muscular.
5. Visceral mass usually coiled exhibiting various types of torsions.
6. Anus usually situated anteriorly close to the mouth.
7. Buccal cavity contains an *odontophore* with a radula.
8. Respiration is carried on by integument or gills or lungs.
9. Circulatory system open type with dorsal heart enclosed in pericardium.
10. Excretory system includes one or two kidneys or nephridia.
11. Nervous system contains distinct cerebral, pleural, pedal, parietal and visceral ganglia.
12. Sexes are separate. Mostly oviparous, a few viviparous.
13. Development indirect with *veliger* and *trochophore* larval stages.

Class Gastropoda is divided into three sub-classes. Prosobranchia, Opisthobranchia and Pulmonata

**Subclass (a) Prosopbranchia (Streptoneura)**

1. The visceral nerve commissures are twisted into a figure of 8, hence the name streptoneura (Gr. *streptos* = curved, *neuron* = nerve). This is due to torsion.
2. The mental cavity containing the pallial complex opens anteriorly.
3. The gills or ctenidia, if present, lie in front of the heart hence prosobranchia.
4. Head has a single pair of non-retractile tentacles.
5. Calcarious shell with an operculum.
6. Sexes separate.
7. Exclusively marine.

Subclass Prosobranchia is divided into three orders : Archaeogastropoda, Mesogastropoda and Neogastropoda.

***Order - 1. Archaeogastropoda (Aspidobranchia)***

1. Due to incomplete atrophy of organs, the kidneys, auricles, nephridia and ctenidia are paired.
2. Ctenidia are bipectinate or plume like.
3. Osphradium poorly developed.
4. The nervous system is little concentrated.
5. Gonad opens into the right kidney serving as gonoduct.
6. Fertilization external.

The order has been distinguished into the following two subroders :

***Surborder (i) Docoglossa***

1. Shell and visceral mass conical.
2. Ctenidium single or absent or replaced by secondary gills.
3. Radula with three marginals on either side.
4. Auricle single.

Examples : *Patella, Acmaea*

***Suborder (ii) Rhipidoglossa***

1. Shell spiral.
2. Ctenidia paired.
3. Radula with numerous marginals.

Examples : *Haliotis, Fissurella, Trochus.*

***Order - 2. Mesogastropoda (Pectinibranchia)***

1. These are advanced snails with complete atrophy of the left side so that usually unpaired auricle, nephridium, osphradium and ctenidium occur.
2. Ctenidium monopectinate or comb-like, attached to the mantle throughout its entire length.
3. Edge of the shell opening lacks a siphonal notch or canal.
4. Foot may be operculate.
5. Osphradium pectinate and well differentiated.
6. The nervous system is concentrated.
7. Gonad has its separate duct and opening.
8. Fertilization internal. Development is indirect with free-swimming veliger larva.

   Examples : *Pila, Triton, Valvata, Natica, Vermetus.*

***Order - 3. Neogastropoda (Stenoglossa)***

1. Carnivorous species with an eversible proboscis.
2. Osphradium is large and bipectinate.
3. Foot operculate.
4. One gill with filaments in one row, heart with one auricle, single nephridium present.
5. Nervous system is lightly concentrated.
6. Shell has a siphonal canal.
7. Embroys are intracapsular. Free swimming veliger suppressed.

   Examples : *Murex, Oliva, Magilus.*

**Subclass (b) Opisthobranchia (Euthyneura)**

1. Due to detorsion of visceral mass, visceral loop becomes untwisted hence named (Gr. *euthus* = straight, *neuron* = nerve).
2. Mantle cavity displaced posteriorly.
3. Head with two pairs of tentacles.
4. Shell and operculum, if present, reduced.
5. Usually one auricle and one nephridium present.
6. Ctenidia are replaced by secondary gills.
7. Hermaphroditism is universal. Larva is veliger.

8. Exclusively marine.

Subclass Opisthobranchia is divided into eight orders as follows :

**Order - 1. *Cephalaspidea***

1. These are borrowing forms having shield like head.
2. Shell and mantle cavity moderately developed.
3. Lateral parapodial lobes are prominent.

Examples : *Bulla, Acteon.*

**Order - 2. *Anaspidea***

1. These are crawling or burrowing forms having a pair of *rhinophores* on head.
2. Shell is small and internal.
3. Mantle cavity is reduced on right side.
4. Parapodial lobes are prominent.

Examples : *Aplysia, Akera.*

**Order - 3. *Thecostomata***

1. These are shelled pteropods.
2. Shell is spirally coiled or a non-spirally pseudoconch.
3. Mantle cavity is well developed.
4. Parapodial fins are large.

Examples : *Clione, Limacina.*

**Order - 4. *Gymnostomata***

1. These are naked pteropods and planktonic.
2. Shell and mantle cavity absent.
3. Small ventral parapodial fins are present.

Example : *Pneumoderma.*

**Order - 5. *Notaspidea***

1. Shell is external or reduced and internal.
2. Mantle cavity is absent, but a skirt-like projection of mantle covers gill on the right side.

Example : *Pleurobranchus.*

**Order - 6. *Acochlidiacea***

1. Small without shell and gill.

2. Naked visceral mass projecting behind the foot and covered with spiracles.

   Examples : *Acochlidium.*

***Order - 7. Saccoglossa***

1. These are herbivorous, with a modified redula and suctorial pharynx.
2. They are shelled or naked slug-like, with a gill.

   Examples : *Elysia, Oxynoe.*

***Order - 8. Nudibranchia (Acoela)***

1. They are naked bilaterally symmetrical sea-slugs without shell, mantle cavity, ctenidia and osphradium.
2. They have secondary gills around anus or surface outgrowths or cerata.

   Examples : *Doris, Aeolis, Tritonia.*

**Subclass (c) Pulmonata**

1. The ctenidia are absent and replaced by pulmonary sac or lung.
2. Shell is simple spiral or vestigial or absent and operculum never occur.
3. One or two pairs of tentacles and one pair of eyes are present.
4. The torsion takes place. The nervous system is secondary symmetrical due to shortening of connectives and concentration of ganglia into a circumoesophageal ganglionic complex.
5. The heart with a single auricle lying anterior to the ventricle.
6. The animals are bisexual. Gonad single.
7. Development direct without larval stage.

   The subclass is divided into two orders :

***Order - 1. Basommatophora***

1. One pair of tentacles with the eyes at the base.
2. Shell is delicate with a large aperture.
3. Some forms with secondary gills.
4. Male and female gonopores usually separate.

   Examples : *Lymnaea, Planorbis, Physa.*

***Order - 2. Stylommatophora***

1. Two pairs of retractile tentacles. Eyes are lodged on the tip of posterior pair of tentacle.

2. Shell is internal, reduced or absent.
3. Mostly terrstrial.
4. Male and female gonopores usually united.

   Examples : *Helix, Limax, Arion.*

**CLASS. IV. PELECYPODA**

1. The pelecypods or bivalvia are marine or fresh-water, bilaterally symmetrical molluscs.
2. Mantle consists of paired right and left lobes, which secrete a bivalved shell.
3. Devoid of head, eyes, tentacles and radula.
4. The muscular foot is usually wedge-shaped or tongue-shaped, which is adapte for burrowing.
5. Presence of two gills, one on either side of mantle cavity, the function of which is to produce respiratory and foot carrying currents of water.
6. Pharynx is absent.
7. Coelom is reduced to a dorsal placed pericardium.
8. Alimentary canal is coiled with a large paired digestive glands.
9. Heart is continued with pericardium and comprises two auricles and one ventricle.
10. The renal organs are a pair of coelomic kidneys, which lead into pericardium at one end and to outside at other.
11. The nervous system includes four pair of ganglion, viz., cerebral, pleural, pedal and visceral.
12. Sense organs are statocyst and osphradium.
13. Sexes may be separate or united. Development includes metamorphosis, followed by trochophore stage.

    Class - Pelecypoda is divided into following orders :

***Order - 1. Protobranchiata***

1. Presence of paired plume-shaped gills, each bears two rows of filaments.
2. The foot is flattened.
3. Presence of anterior and posterior adductor muscles.
4. Sexes separate, gonads lead into kidneys.

   Examples : *Nucula, Solenomya.*

**Order - 2. Filibranchiata**

1. Single pair of plate-like formed of distinct V-shaped filaments.
2. Inter-filamentar junctions are either absent or formed by groups of interlocking cilia.
3. Inter-lamellar junctions are either absent or non-vascular.
4. Two adductor muscles present, anterior may be reduced or absent.
5. Foot small or poorly developed.

   Example : *Mytilus, Arca, Noath.*

**Order - 3. Pseudolamellibranchiata**

1. The gills are plaited so as to present vertical folds.
2. Inter-filamentar junctions may be ciliary or vascular. Interlamellar junctions may be vascular or non-vascular.
3. Shell valves are frequently unequeal and single large adductor muscle present.
4. Foot is little developed or rudimentary.
5. Genital organs open into kidneys or near them.

   Examples : *Pecten, Ostrea, Pinna, Lima.*

**Order - 4. Eulamellibranchiata**

1. Gills are basket like, the gill-filaments are attached with vascular inter-filamentar and inter-lamellar junction.
2. Gills may be smooth or plaited.
3. Two equal adductor muscles.
4. Foot is well developed.
5. Siphon of small or large sized present.
6. Gonads open out by separate openings. They do not open into kidneys.

   Examples : *Unio, Teredo, Anodonta, Venus, Cardium, Mya.*

**Order - 5. Septibranchiata**

1. Gills replaced by horizontal muscular and perforated septa dividing the mantle cavity into upper and lower chambers.
2. Foot long and slender, byssus reduced or absent.
3. Adductor muscles two.
4. True ctenidia are absent.

   Examples : *Cuspidaria, Poromya.*

## CLASS .V. SCAPHOPODA

1. They are aberrant marine animals comprising only of three genera.
2. Body is elongated worm-like, enclosed in a tusk-like shell open at both ends.
3. Eyes, tentacles and gills are absent.
4. Mantle tubular completely enclosing the body.
5. Mouth surrounded by lobular processes or outgrowths.
6. Foot is reduced used for digging.
7. Heart rudimentary.
8. Sexes are separate.
9. Torsion takes place in some forms.

   Scaphopoda has two families :

### Family (i) Dentalidae

1. Foot is trilobed having two epipodial lobes and one hypopodial lobe.

   Examples : *Dentalium, Antalis.*

### Family (ii) Siphonodentalidae

1. Foot is elongated and capable of expansion into a terminal disc.

   Examples : *Pulsellum, Siphonodentalium.*

## CLASS .VI. CEPHALOPODA

1. Exclusively marine.
2. Shell spiral chambered or usually with or without shell embedded in mantle.
3. Body with head and trunk. Head bears eyes and mouth.
4. Trunk consists of symmetrical and uncoiled visceral mass.
5. Mantle encloses posteriorly and ventrally a large mantle cavity.
6. Foot altered into a series of sucker bearing arms or tentacles encircling the mouth.
7. Mouth bears jaws and radula.
8. Two or four pairs of bipactinate gills.
9. Circulatory system closed, heart with two or four auricles.
10. Excretory system comprises two or four pairs of nephridia.
11. Nervous system is highly developed and the principal ganglia are concentrated around the oesophagus.
12. Saxes are separate.

13. Development meroblastic without metamorphosis.
14. Most of the cephalopods have an *ink-gland* with a duct opening into the rectum.

Cephalopoda is divided into three subclasses :

**Subclass (a) Dibranchiata**

1. The foot modified into a circlet of eight to ten arms bearing suckers, around the mouth.
2. The funnel forms a complete tube.
3. The shell is generally internal.
4. Presence of paired gills and kidneys.
5. Presence of characteristic ink-gland and duct.

Dibranchiata is divided into two orders.

***Order - 1. Octopoda***

1. Body rounded or oval and with no lateral fins.
2. Arms eight, long arms or tentacles absent.
3. Suckers on the arms sessile and without horny rim.
4. Shell and nidamental glands absent.

Examples : *Octopus (*Devil-fish*), Agronauta.*

***Order - 2. Decapoda***

1. Body elongated and with no latèral fins.
2. Arms ten, eight short and 2 long and known as tentacles.
3. Suckers on the arms sessile and without horny rim.
4. Shell internal and well developed.

Examples : *Sepia, Loligo.*

**Subclass (b) Nautiloidea or Tetrabranchiata**

1. The foot divides into lobes bearing numerous tentacles.
2. The funnel is incomplete.
3. Shell is external which is divided into several compartments.
4. Kidneys and gills are four.
5. Devoid of ink gland.
6. Pericardium communicates to the exterior directly.
7. Eyes simple without crystelline lens.
8. Chromatophore, salivary glands absent.

Examples : *Nautilus.*

**Subclass (c) Ammonoidea**

1. All are extinct.
2. Shell variously modified in the form of straight spiral or turreted.
3. Siphon external, siphuncle simple and marginal.
4. Two pairs of ctenidia.

   Example : *Ammonities.*

**CHAETODERMA**

| | | |
|---|---|---|
| Phylum | - | Mollusca |
| Class | - | Amphineura |
| Subclass | - | Aplacophora |
| Genus | - | *Chaetoderma* |

Fig. 20.1 *Chaetoderma*

1. It is a most primitive mollusc. The body is worm-like cylindrical. It is found around North Atlantic.
2. The head is separate off from the body by a constriction.
3. Mantle covers the body completely.
4. Shell is absent but the body is covered by numerous calcareous spicules.
5. Rudimentary foot.
6. Gills reduced to a pair which are situated in the cloacal cavity.

7. Sexes are separated and each has one fused gonad.
8. A cavity is present at the posterior end of the body into which opens the anus.

## PATELLA

| | | |
|---|---|---|
| Phylum | - | Mollusca |
| Class | - | Gastropoda |
| Subclass | - | Prosobranchia |
| Order | - | Aspidobranchia |
| Suborder | - | Docoglossa |
| Genus | - | *Patella* |

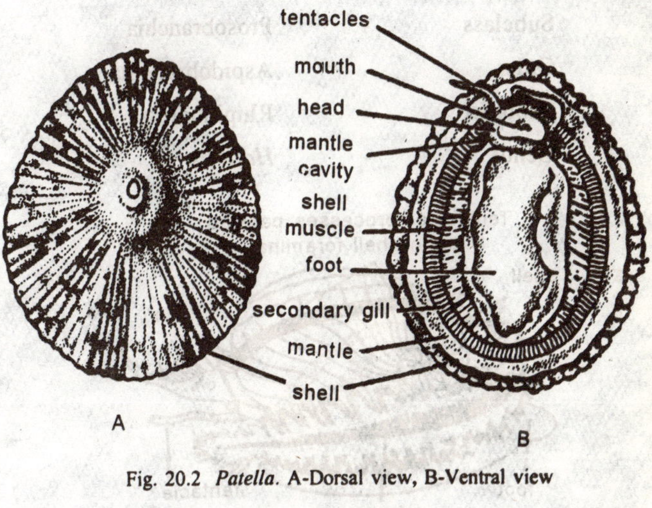


Fig. 20.2 *Patella*. A-Dorsal view, B-Ventral view

1. *Patella* or *limpet* is a littoral and is found attached to the rocks.
2. It is found along Pacific coast, Europe, U.S.A.
3. Body is oval with convex dorsal surface.
4. Shell is oval and core-like, with a conical projection from its dorsal surface.
5. Operculum is absent.
6. The head is laterally produced into a pair of small, stout, tectile tentacles.
7. Eyes are simple present at the base of tentacles.
8. Foot ventral and sole-like for creeping.

9. True mantle is restricted to the anterior and secondary mantle cavity is developed between foot and mantle.
10. Radula composed of very few, strong hooked teeth in each row.
11. All around the foot lies a secondary pallial cavity containing a series of secondary branchiae.
12. Heart with single auricle.
13. Herbivorous, feeding upon alage, sea-weeds and diatoms.
14. It is eaten by poor classes in country like France, Italy and Ireland.

## HALIOTIS

| | | |
|---|---|---|
| Phylum | - | Mollusca |
| Class | - | Gastropoda |
| Subclass | - | Prosobranchia |
| Order | - | Aspidobranchia |
| Suborder | - | Rhipidoglossa |
| Genus | - | *Haliotis* |

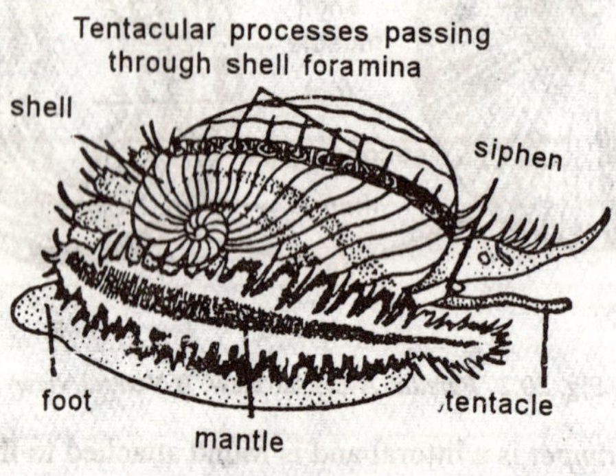


Fig. 20.3 *Haliotis.*

1. *Haliotis or 'Ear shell'* is marine and found attached to rocks and feeding on sea-weeds and other marine vegetation.
2. The animals are very common at and below low tide level.
3. *Haliotis or Abalone* has ear-shaped shell with small flattened spire.
4. The aperture is very large and without operculum.
5. The shell is perforated by a series of marginal slits through which project the tentacular processes of the mentle.

6. The foot is large and with epipoidia and the animal can move 2 to 4 meters per minute but clings very powerfully to the boulders.
7. Eyes are borne on stalks at the outer bases of the tentacles.
8. The mantle cavity is large, has bipectinate ctenidia or gill, the right being smaller.
9. The shell of the animal is of great economic importance, as beautiful pearly buttons are manufactures from the shell of *Haliotis* in Europe and U.S.A.
10. The animal is also eaten in countries like China, Japan and many cities along the Pacific coast.

**MUREX**

| | | |
|---|---|---|
| Phylum | - | Mollusca |
| Class | - | Gastropoda |
| Subclass | - | Prosobranchia |
| Order | - | Pectinibranchia |
| Genus | - | *Murex* |

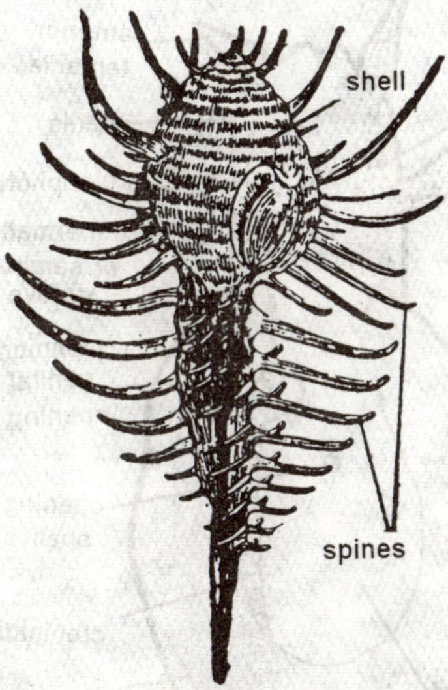


Fig. 20.4 *Murex*.

1. It is a marine animal found between low tide marks on stones and sea weeds.
2. It is founded on the Syrian coast, Italy, India, West Indies and U.S.A.

3. Shell is spirally coiled and with three or more rows of beautiful spines of different lengths.
4. It is carnivorous and highly predator feeding on living animals which it grasps by means of its foot. It is destructive to oyster bed.
5. A very prominent bipectinate osphradium perist.
6. Head is prolonged into a long proboscis which can be retracted within the proboscis-sheath.
7. Adrectal glands present which secrete a dye the Jyrain purple (Yellow in colour).
8. Sexes are separate.

## APLYSIA

| | | |
|---|---|---|
| Phylum | - | Mollusca |
| Class | - | Gastropoda |
| Subclass | - | Opisthobranchia |
| Order | - | Anaspidea |
| Genus | - | *Aplysia* |

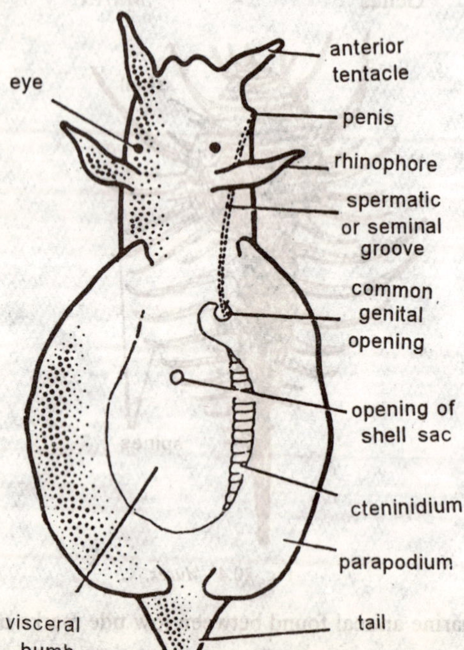


Fig. 20.5 *Aplysia*

1. It is a marine animal found among sea-weeds.
2. It is found in India, Asia, West Indies and on the Florida coast.
3. It is commonly called *'sea hare'*.
4. Body is soft, lumpy with a visceral hump.
5. Head with two pairs of tentcles. The anterior pair is large and ear-like, the posterior pair is olfactory and bear eyes at their base.
6. Shell thin, reduced and trnsparent, enclosed by mantle.
7. The foot is muscular and elongated pointed posteriorly with lateral out growths, the parapodia.
8. Mantle possesses unicellular ink glands secrete purple ink, used for defence.
9. Hermaphrodite. There is a single generative aperture and a single duct for the sperm and ova but a seminal groove runs from the aperture to the head and reciprocal fertilization is impossible.
10. Nervous system is well developed with perfectly symmetrical visceral loop.

**AEOLIS**

| | | |
|---|---|---|
| Phylum | - | Mollusca |
| Class | - | Gastropoda |
| Subclass | - | Opisthobranchia |
| Order | - | Nudibranchia |
| Genus | - | *Aeolis* |

1. It is a marine slug-like gastropod found in shallow waters crawling on under surface of sea weeds.
2. Found in U.S.A., Europe.
3. *Aeolis* is a small nudibranch 1 or 2 cm in length.
4. The soft body is covered with a transparent integument.
5. The shell, mantle and true gills are absent
6. The cerata are the respiratory organs.
7. There are cylindrical extensions of the integument, containing hollow extensions of the liver and nematocyst derived from ingested hydroids.
8. The cerata have characteristic arrangement, may occur in many transverse or in few longitudinal rows.
9. The head is small, bears tentacles, usually two pairs, one which may be retractile.

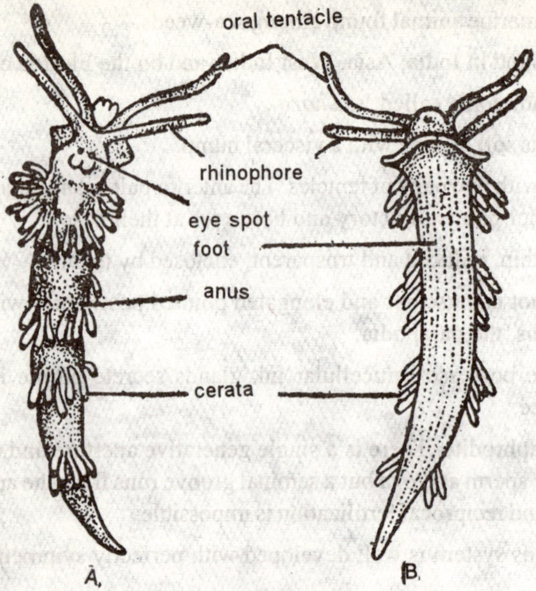


Fig. 20.6 *Eolis.*

10. The anterior or oral pair is the longer.
11. The posterior or dorsal pair, known as rhinophores bear simple sense organ with taste the surrounding water.
12. At the base of each rhinophore is an eyespot.
13. The mouth is prominent and can be most easily seen from the ventral view.
14. On the right side of the head is conspicuous genital opening.
15. Just posterior and dorsal to genital opening lies the small anus.
16. On the ventral side the animal has a highly muscular foot.
17. Proportion varies with the species.
18. Mouth is guarded by lips which cover both the radula and the horny jaws within the buccal cavity.
19. The powerful jaws cut the hydroid into two pieces. The *Aeolis* is hermaphroditic and gonad is protandric.

## DORIS

The systamatic position is same as that of *Aeolis*

1. It is a sluggish marine animal found under stones at low tide marks.

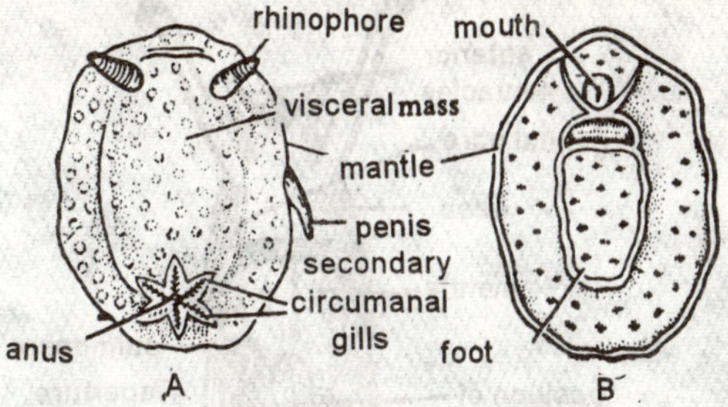


Fig. 20.7 *Doris.*

2. It is cosmopolitan in distribution.
3. It is commonly known as *'sea lemon'*.
4. The body is oval and depressed, concave above and flat below.
5. Head is indistnict, but there is a pair of short olfactory tentacles or rhinophores towards the anterior end.
6. Mantle is usually pigmented and contains calcareous spicules or dorsal tubercles.
7. Mouth is present on ventral side. Anus lies mid-posteriorly and is surrounded by a circlet of feathered retractile secondary branchiae.
8. The broad foot is on ventral side.
9. Hermaphrodite, gonopore is asymmetricelly placed on the right side of the body.
10. *Doris* has undergone detorsion, therefore, the shell, mantle cavity and primary gills are absent.

## LIMAX

| | | |
|---|---|---|
| Phylum | - | Mollusca |
| Class | - | Gastropoda |
| Subclass | - | Pulmonata |
| Order | - | Stylommatophora |
| Genus | - | *Limax* |

1. It is a terrestrial gastropod, found in gardens over damp soil.
2. It is cosmopolitin in distribution.
3. It is commonly known as *'gray slug.'*

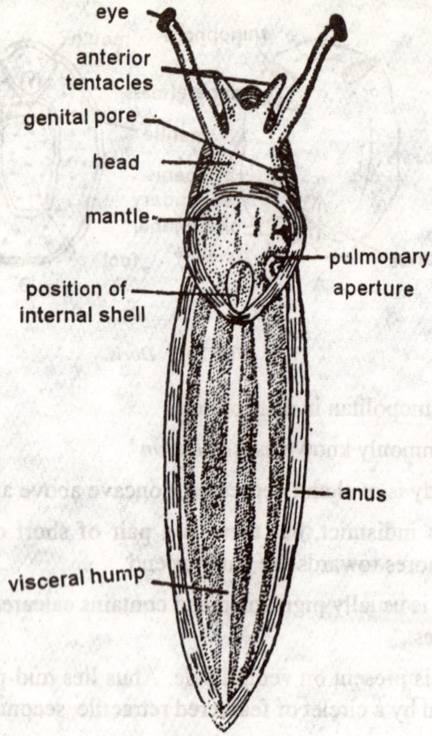


Fig. 20.8 *Limax*

4. **Body is elongated and tapering behind and is divisible into head, foot and visceral hump.**
5. **Two pairs of retractile tentacles are present on head. Eyes are present at the tips or posterior tentacles.**
6. **The mouth is on the ventral side of the anterior end and is bounded by a pair of lateral lips.**
7. **Mantle forms a shield like area in the dorsal side of the anterior part of the body. The mantle cavity acts as a lung for the breating, pneumostome, present on right side.**
8. **Shell is internal, thin and rudimentry, lies embedded in mantle.**
9. **Foot is on ventral side.**
10. **A pedal gland just behind the head for the mucous secretion.**
11. **Bisexual.**
12. **Nocturnal and herbivorous.**

## HELIX

The systematic position is same as that of *Limax*.

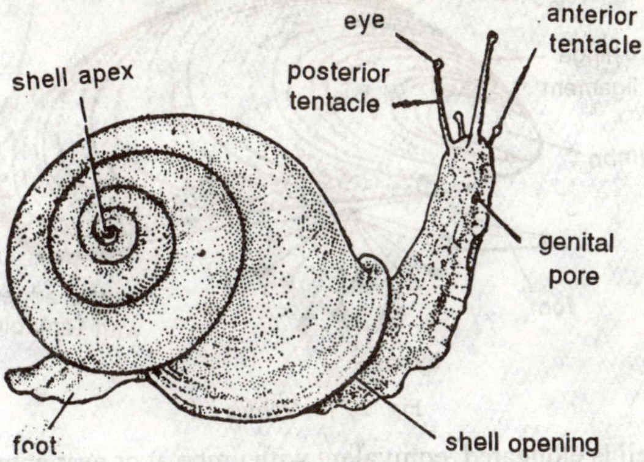


Fig. 20.9 *Helix*

1. It is a terrestrial animal found in damp places, under bark and fallen leaves.
2. It is found in Palearctic region.
3. It is commonly called *'garden snail'*.
4. The shell is thin, external and heliciform with a low conical spire, measuring 4 cm. in length.
5. Head bears two pairs of tentacles, first pair bears organs of small and second pair is large and bears a pair of simple eyes.
6. Foot possesses a flat ventral surface for creeping.
7. Respiration by pulmonary sac.
8. Gills are absent.
9. Hermaphrodite. The genital aperture opens above the right lateral lip. Gonads contain four angled dark and multifid digitiform glands.

## MYTILUS

| | | |
|---|---|---|
| Phylum | - | Mollusca |
| Class | - | Pelecypoda |
| Order | - | Filibranchia |
| Genus | - | *Mytilus* |

1. It is a marine, sedentary mollusc found at a depth of 2 or 3 fathoms in low tide mark, attached to rocks or wooden structure by its byssus threads.
2. It is commonly called as *'sea mussel'* and cosmopolitan in distribution.

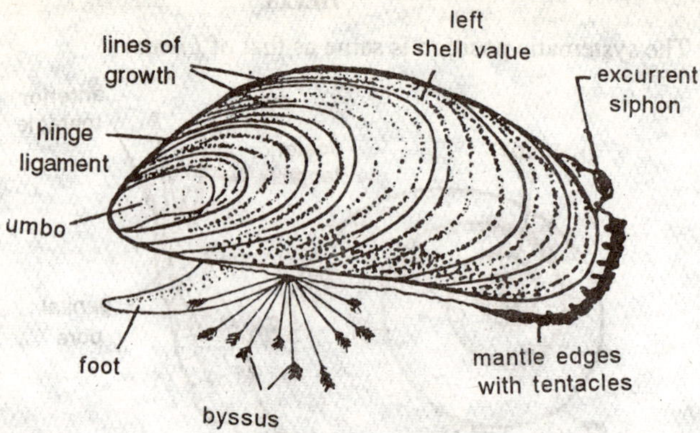


Fig. 20.10 *Mytilus*

3. The shell is elongated, equivalent with umbo at or near anterior end.
4. Hing toothless but may bear cerculations.
5. The foot is tongue-like and is modified into a bunch of byssal threads coming out from between the two shell valves, which are main anchoring organs.
6. The anterior adductor muscle is weaker where as posterior is strongly developed.
7. A pair of gills is present each gill is provided with gill- filaments. the gill are lamelliform.
8. The eyes are found anterior to the inner gill-lamella.
9. Sexes are separate. Gonads extend into the mantle.
10. It is used as food in Europe.

**PECTEN**

| | | |
|---|---|---|
| Phylum | - | Mollusca |
| Class | - | Pelecypoda |
| Order | - | Pseudolamellibranchia |
| Genus | - | *Pecten* |

1. It is free swimming marine mollusc lives on the sea bottom of 10 fathoms deep.
2. It is found in India and U.S.A.
3. It is commonly called as *scallop*.
4. Two shells are unequal. The right shell being larger and more convex and the animal rests on this valve.

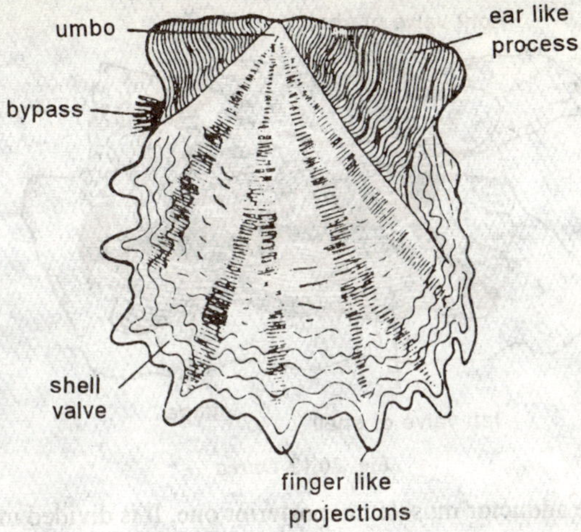


Fig. 20.11 *Pecten*

5. There is a single large adductor muscle. This is divided into two parts and the larger of these served for rapid contraction which cause swimming movements.
6. Foot is very much reduced but in larva very well developed and used for locomotion.
7. Two large and crescentric gills are present.
8. Mantle is tentaculiferous and it encloses viscera.
9. Large number of stalked eyes are present at regular intervals along the edges of the mantle.
10. These are hermaphrodite.
11. The ovary has pink colour when eggs are ripe.
12. The testis lies behind and is cream in colour.

**OSTREA**

The systematic position is same as that of *Pecten*.

1. It is a sedentary bivalve attached to rocks or other shells and found in shallow and brackish water.
2. It is found in Atlantic and Pacific coasts, India, Gulf of Mexico to Massachusetts.
3. It is commonly called *'edible oyster'*.
4. The shell valves are unequal the left is always larger and permanently attached to the rocks by byssus.

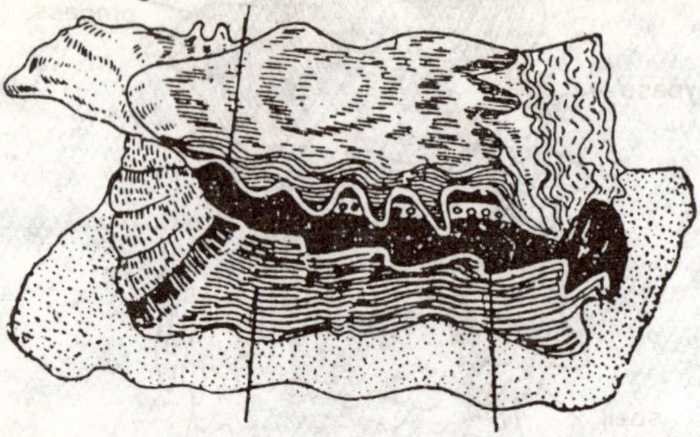


Fig. 20.12 *Ostrea*

5. Only one adductor muscle i.e. posterior one. It is divided into two parts one is straited and other is non-straited.
6. The foot is absent.
7. Heart with two fused auricles.
8. A pair of curved gills present.
9. It is peculiar in the sense as the individuals function alternately as males and females.
10. Spawning generally takes place at full moon. The eggs develop into veliger larva which metamorphoses into adult.

## TEREDO

| | | |
|---|---|---|
| Phylum | - | Mollusca |
| Class | - | Pelecypoda |
| Order | - | Eulamellibranchia |
| Genus | - | *Teredo* |

1. Marine pelecypod which burrows in wood of the boats, ships and other submerged wood.
2. It is found in India, Europe, Cape Cod to Florida and Massachusetts Bay to Florida.
3. It is commonly called *'shipworm'*.
4. Teredo has body extremely elongate and worm-like appearance.
5. Foot reduced.
6. Mantle tubular and opens anteriorly, shell is reduced and covers only the anterior part of the body.

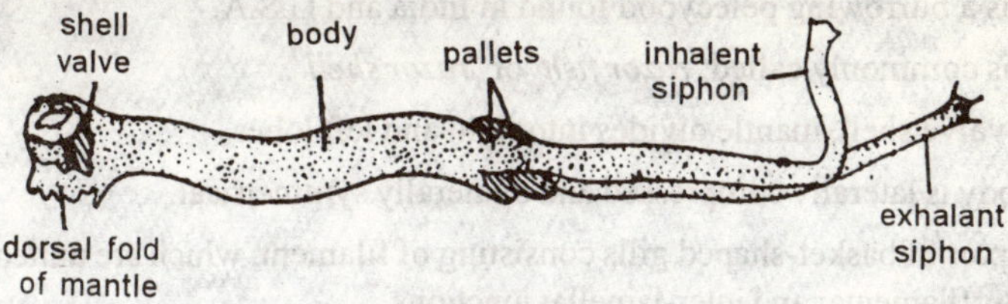


Fig. 20.13 *Teredo*

7. The mantle cavity and siphons are long and the latter unites to a large extend and provided with calcareous pellets for protection.
8. A constant current into and out of the mantle cavity is maintained by ciliary action.
9. The saw dust is swallowed by the animal and is digested by cellulose digesting enzymes.

## SOLENOCURTUS

| | | |
|---|---|---|
| Phylum | - | Mollusca |
| Class | - | Pelecypoda |
| Order | - | Eulamellibranchia |
| Genus | - | *Solenocurtus* |

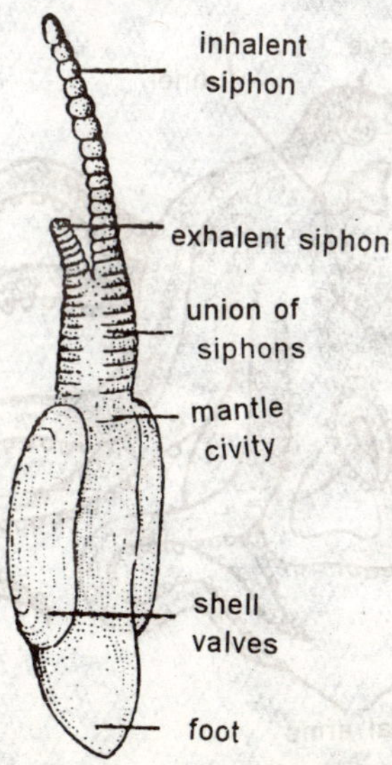


Fig. 20.14 *Solenocurtus* (Razor-fish)

1. It is a burrowing pelecypod found in India and U.S.A.
2. It is commonly called *'razor fish'* or *'razor shell'*.
3. Bivalve shell, mantle divides into right and left lobes.
4. Body is laterally compressed and bilaterally symmetrical.
5. A pair of basket-shaped gills consisting of filament, which are united by interfilamentar and inter-lamellar junctions.
6. Two equal-sized adductor muscles developed.
7. Presence of two elongated siphonal tubes which are united with each other.
8. The shell has the appearance of long scabbard.
9. The foot is long, cylindrical and enormously developed which is capable of penetrating into the loose sand.

## OCTOPUS

| | | |
|---|---|---|
| Phylum | - | Mollusca |
| Class | - | Cephalopoda |
| Subclass | - | Dibranchiata |
| Order | - | Octopoda |
| Genus | - | *Octopus* |

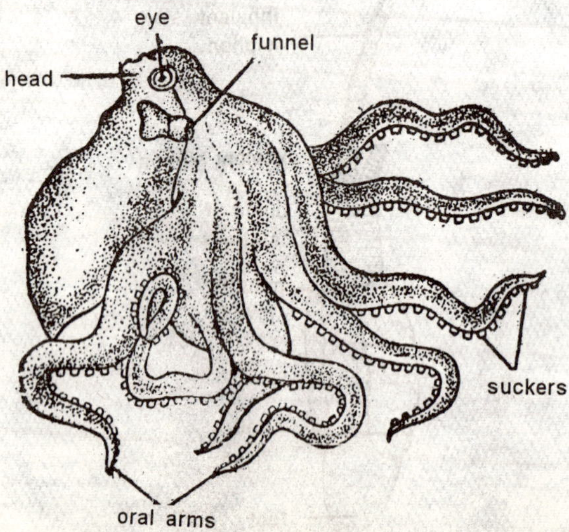


Fig. 20.15 *Octopus*

1. It is marine and cosmopolitan in distribution generally found on the Atlantic and Pacific coasts.
2. It is commonly called *devil fish*.
3. Body surrounded by mantle containing chromatophores. They can change their colour according to surrounding medium.
4. Foot modified into eight equal arms surrounding the head and siphon.
5. The arms are beset with suckers which are sessile and without horny rims.
6. Body is covered with tubercles, short, round aborally and without fins.
7. External shell is absent and body with vestiges of an internal shell.
8. Third arm on the right side in male modified to act interomittent organ or *hectocotylised*.
9. Like *Sepia* and *Loligo*, *Octopus* is also capable of producing inky liquids which form a smoky screen water.
10. Sexes are separate and there is no metamorphosis as the development is direct. Female cares the eggs.
11. Mouth with sharp and beak-like horny jaws.

## ARGONAUTA

| | | |
|---|---|---|
| Phylum | - | Mollusca |
| Class | - | Cephalopodata |
| Subclass | - | Dibranchia |
| Order | - | Octopoda |
| Genus | - | *Argonauta.* |

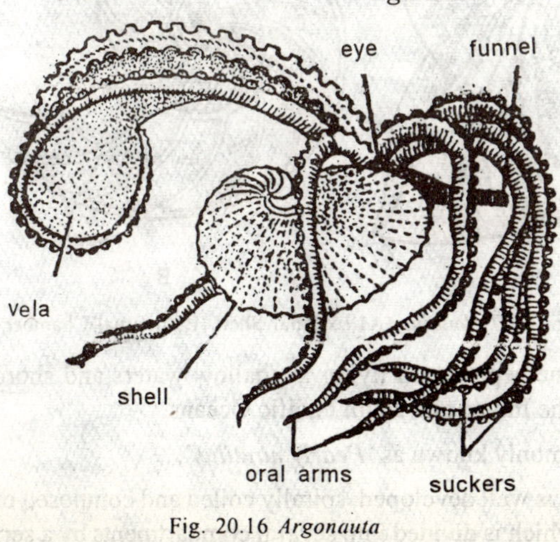


Fig. 20.16 *Argonauta*

1. It is a marine animal found in Atlantic and Indian oceans.
2. It is commonly called as '*paper nautilus*'.
3. Sexual dimorphism is remarkable and well-marked.
4. Male in 2.5 cm in length and without shell.
5. Female is 20 cm length and possesses a thin, transparent shell secreted by the extremities of two dorsal arms.
6. Shell is used for protection of eggs.
7. Third arm in male is hectocotylised.
8. Eyes and funnel and present.
9. Two plume-like ctenidia and two kidneys are present.

**NAUTILUS**

| | | |
|---|---|---|
| Phylum | - | Mollusca |
| Class | - | Cephalopoda |
| Subclass | - | Tetrabranchia |
| Genus | - | *Nautilus* |

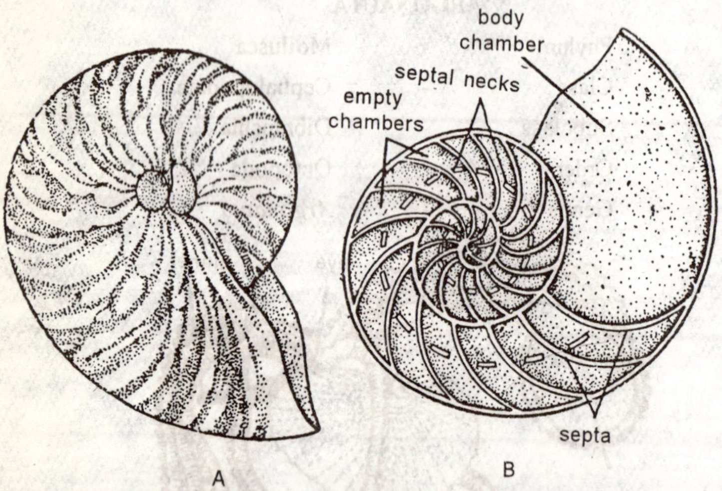


Fig. 20.17 *Nautilus* (A) External Shell (B) Internal Chamber

1. It a marine cephalopod living in shallow waters and shores and coral reefs in the Indian and south Pacific Oceans.
2. It is commonly known as '*Pearly nautilus*'.
3. The shell is well developed, spirally coiled and composed of calcareous matter, which is divided into several compartments by a series of septa.

4. Through the various compartments runs a tube called *siphuncle* the walls of which are formed of calcium carbonate. A narrow vascular prolongation of visceral region of the animal passes through this tube.
5. Head is well developed bearing the eyes.
6. The anterior part of food is modified into several arms provided with tentacles which are arranged in two series.
7. The posterior portion of the foot is modified into funnel situated below the head.
8. Buccal cavity with a pair of jaws and odontophore bearing several teeth.
9. Two pairs of gills and kidneys are present.
10. They are hyper-polyandrous i.e. male are abundant than female.
11. Presence of 60 tentacles in male and 90 in female.
12. Pericardium leads directly to exterior by two orifices located near the posterior renal aperture.

# 21

# PILA GLOBOSA

*Pila* is commonly called as *'Apple Snail'*. It comes under class Gastropoda. The gastropods have undergone an extensive adaptive radaition and have invaded various kinds of habitats.

**SYSTEMATIC POSITION**

| | | |
|---|---|---|
| Phylum | - | Mollusca |
| Class | - | Gastropoda |
| Subclass | - | Prosobranchia |
| Order | - | Pectinibranchiata |
| Genus | - | *Pila* |
| Species | - | *globosa* |

**Distribution**

The genus *Pila* is restricted to the Oriental and Ethiopian regions. In the Oriental region it occurs in India (except the province of Punjab), Burma, Ceylon, Thailand, Malaya, Vietnam, Indonesia and Philippines and in the Ethiopian region, in Africa, Arabia and Madagascar. *Pila globosa* is the common species found in North India except Punjab.

**Habits and Habitat**

These are largest known fresh water snails. They inhibit ponds, tank, paddy-fields and sometimes in fresh water rivers and in brackish water. They occur in those areas where there is a large amount of vegetation like *Vallisnqria or Pistia,* for food. They are amphibious being adapted for life in water as well as on land. They make long journeys on land in search of areas with more abundant food and water during rains. The animal creeps by its ventral muscular foot they pass over long periods of drought by remaining *torpid* with the aperture of shell tightly by the operculum.

**External Features**

***Shell.*** The body of *Pila* is enclosed in a calcareous shell, which is somewhat conical and is coiled around a central axis, the coils or whorls being marked off externally from each other by distinct suture.

The top of the shell is distinguished as the *apex*. Below this are distinguishable the successive lines called *sutures* running a spiral course. The lowermost whorl is considerably bigger than all the other whorls taken together, and ends in a wide opening, the *mouth* or *aperture*. In the living animal, the aperture is directed forwards and ventrally and the apex points backwards. The internal cavity contained in the shell, is a continuous one running through all the whorls. The margin of the aperture is named the *peristome*; its outer part is called the *outer lip* and the inner part, the *inner lip or* the *columellar lip*.

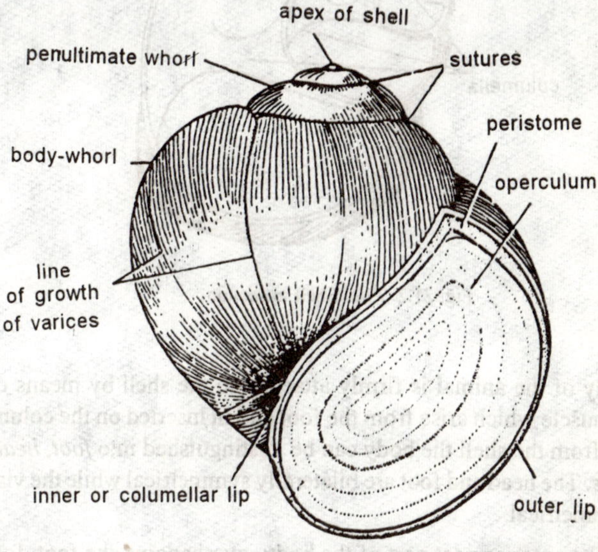


Fig. 21.1 *Pila*. Shell with operculum.

The axis of the shell or the *columella* is hollow and its cavity opens to outside by an opening called *umbilicus*, which lies close to the left corner of the inner margin of the peristome. Into the aperture fits an oblong flat plate, called *operculum* which is composed of material secreted by certain glandular cells of the foot. The inner and outer surface of the operculum bear characteristic markings. On its inner surfaces stands out a sharp elliptic impression, called the *boss* which is produced by the attachment of the *opercular muscle*. Round the boss runs a groove. The outer surface of the operculum is marked by a number of lines of growth running round a *nucleus* which is situate subcentrally near the inner margin.

*Microscopic structure of the shell.* The shell of *Pila* consists of three layers. The outerlayer is known as *periostracum*. It is greenish brown. The underlying two layers are made up of calcareous plates. In the *ostracum*, the middle layer, the calcareous plates run at right angles to the margin of the aperture of the shell. Ostracum is the thickest of the three layers. In *hypostracum*, the innermost layer, the plates are arranged parallel to the margin of aperture. The structure of the two inner layers is similar, but the calcareous plates in the two zones lie at right angles to one another.

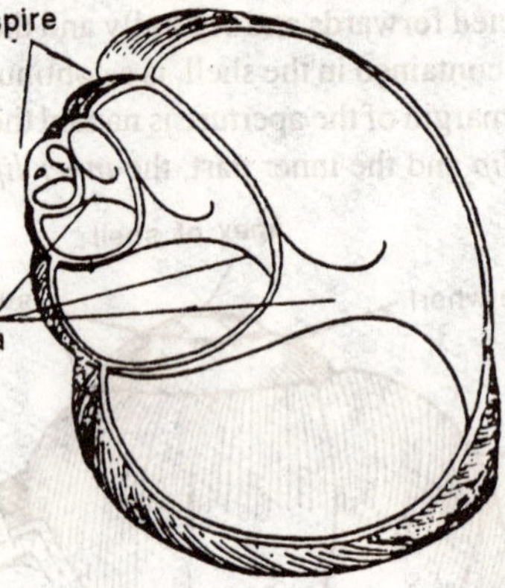


Fig. 21.2 *Pila.* Shell in section.

**Body**

The body of the animal is firmly attached to the shell by means of the columellar muscle which arise from the foot and in inserted on the columella. On removal from the shell the body can be distinguished into *foot*, *head* and *visceral mass*. The head and foot are bilaterally symmetrical while the visceral mass is asymmetrical.

***Head.*** The anteriormost part of the body, overhanging the foot, forms a definite head. It bears dorsally a pair of small tapering and contractile *first pair of tentacles* or *labial palps*. The snout bears the mouth in its anterior surface below the palps. Behind the labial palps are the longer, filament like *true tentacles* or *second pair of tentacles.* These are hollow and very contractile and become about 5 cm. long when fully extended, but only 1.5 cm. when contracted. The eyes are placed on tips of short stalks or *ommatophore* which arise from the outer sides of the tentacles near their bases. On the two sides of head connecting it and mantle, occur two fleshly structures, which are called *nuchal lobes* or *pseudepipodia*. The left nuchal lobe is better developed and forms the respiratory siphon.

***Foot.*** The large, strongly muscular, ventral part of the body forms the foot. When fully expanded, it is roughly triangular in shape, with the apex directed backwards. The foot bears the operculum dorsally on its posterior end, called the *operculiferous lobe*. When the foot is withdrawn, the operculum completely fits into the mouth of the shell and closes it.

The head-foot complex is attached to visceral mass by a short inconspicuous neck.

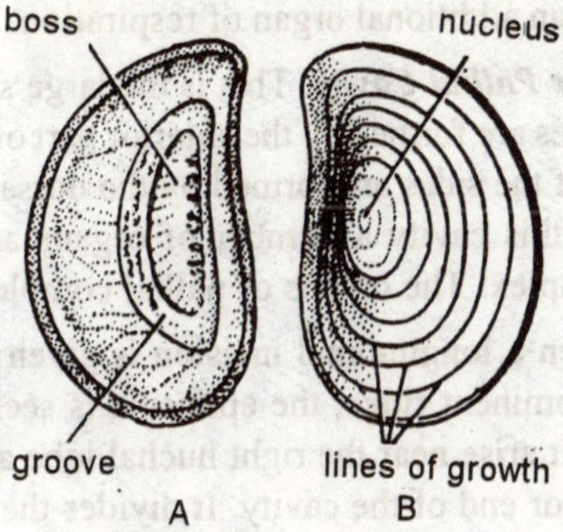


Fig. 21.3 *Pila.* Operculum A=inner view B=Outer view

***Visceral mass.*** The visceral mass is a spirally coiled hump-like structure on the dorsal side and contains all the vital organs of the body.

**Mantle**

The skin of the visceral mass form the *mantle* or the *pallium*, the characteristic molluscan structure. The free edge of the mantle is thickened and possesses the shell-secreting glands. The mantle keeps a regular check over the anterior part of animal. There is a long, narrow supramarginal groove along the thick free edge of the mantle. Behind the supramarginal groove, there is a slight, thickened band, which has the shell-secreting glands.

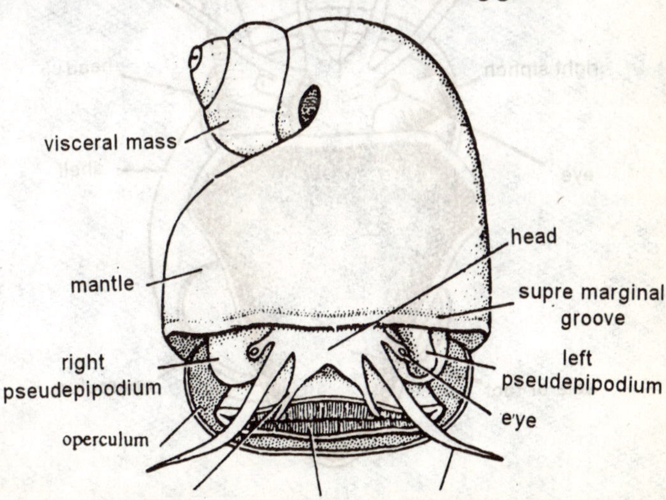


Fig. 21.4 *Pila.* Body after removing shell.

The shell is secreted by the mantle. The periostracum and the ostracum are secreted by the supra-marginal groove and the shell-gland respectively. The hypostracum is secreted by the general epithelial covering of the mantle, which also serves as an additional organ of respiration.

***Mantle Cavity or Pallial Cavity.*** This is the large space whose roof and greater part of the sides are formed by the anterior part of mantle and the floor and remaining part of the sides are formed by the dorsal surface of the body behind the head. In this cavity a number of organs are present known as organs of pallial complex. The organs of pallial complex are :

1. *Epitaenia.* When a longitudinal incision is given along the left of the ctenidium a prominent ridge, the epitaenia is seen on the floor of the mantle cavity. It arise near the right nuchal lobe and extends up to the extreme posterior end of the cavity. It divides the mantle cavity into a small right *branchial chamber* and large left *pulmonary chamber.*
2. *Ctenidium or gill.* It lies at the extreme right side of the mantle cavity hanging vertically downwards from its dorso-lateral wall. Being monopectinate (comb-like), its gill filaments or lamellae, which are somewhat triangular in shape, hang freely in the branchial chamber.

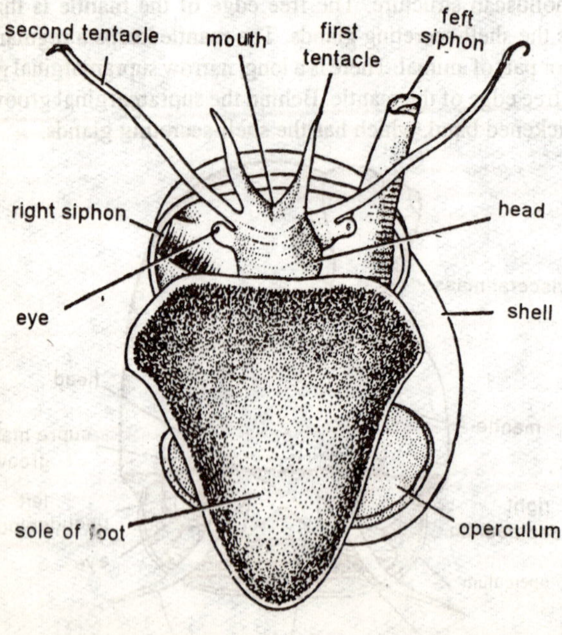


Fig. 21.5 *Pila.* Animal seen from the ventral side.

3. *Rectum.* It lies on the left of ctenidium on the floor of mantle cavity and opens to the exterior by anus, situated near the right pseudepipodium or right nuchal lobe.

4. *Genital duct.* The vas deferens in case of male or the oviduct in case of female lies in the branchial chamber in between the rectum and epitaenia. Its opening is situated close to the anus. A hook-like penis can be seen in front of the genital pore in cse of male specimen. At the base of penis is found the hypobranchial gland.

5. *Hypobranchial gland.* It is a glandular thickening at the base of the penis. Its secretion helps in the act of copulation.

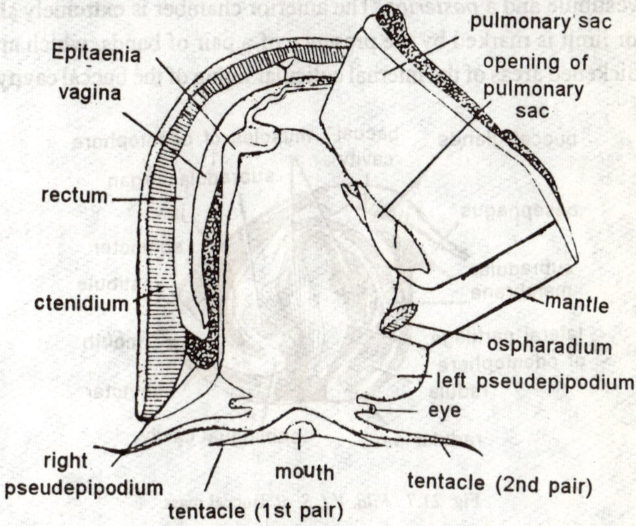


Fig. 21.6 *Pila.* Structure of mantle cavity.

6. *Renal opening.* The anterior chamber of the renal organ projects into the branchial chamber as a small area of reddish colour, near the posterior termination of the epitaenia. Its external opening is like a transversely oblique slit situated in a shallow depression.

7. *The pulmonary chamber.* It is a sac hanging in the mantle cavity from the roof of mantle. It opens into the mantle cavity by the two rows of lamellae. It is olfactory in function.

**Digestive System.**

The digestive system of *Pila* comprises :

I. Alimentary canal

II. Digestive glands.

*I.* ***Alimentary canal.*** The alimantary canal is tubular and peculiar due to the twisting of visceral mass. It is divisible into three parts, fore-gut, mid-gut and hind-gut.

1. *Fore gut.* It is also called *stomodaeum* and is ectodermal in origin. It is made up of buccal mass and the oesophagus.

*Buccal mass.* Mouth is a median vertical slit situated at the tip of snout and is surrounded by frilled folds of integument. True lips are absent. These frills serve as secondary lips. Mouth leads into the buccal cavity, which is the cavity of buccal mass. It is greatly obliterated and contains a pair of jaws and radula. The jaws divide the buccal cavity into two sections, the *anterior* also called vestibule and a *posterior*. The anterior chamber is extremely short. Its posterior limit is marked by the presence of a pair of bands, which are especially thickened areas of the internal cuticular lining of the buccal cavity. In the

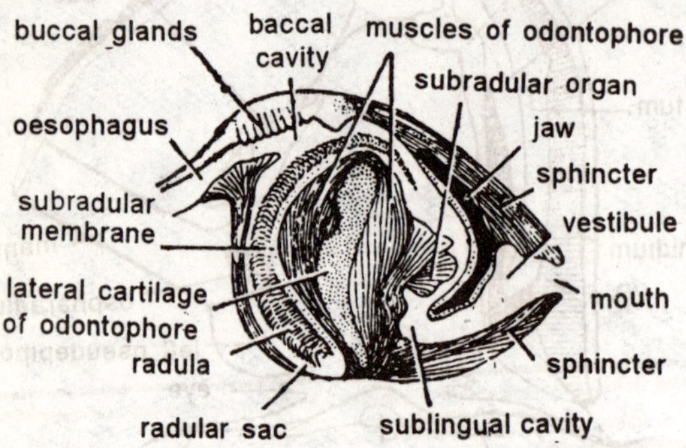


Fig. 21.7 *Pila. V.L.S. of* Buccal mass.

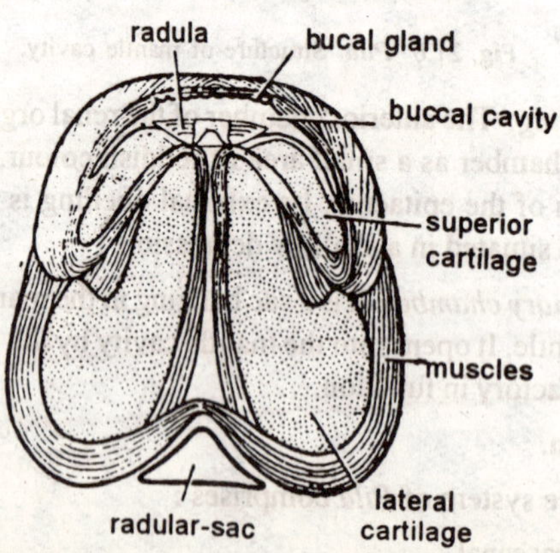


Fig. 21.8 *Pila.* Buccal mass in T.S.

posterior region of buccal cavity there lies an *odontophore* or *tongue mass* on the floor of the buccal cavity. On the roof of buccal cavity occur the dorsal *buccal glands.*

Fig. 21.9 *Pila.* Radula

The odontophore is supported by protractor and retractor muscles and two pairs of cartilages. A pair of large S-shaped lateral cartilages and a pair of triangular superior cartilages. The cartilages project into the buccal cavity.

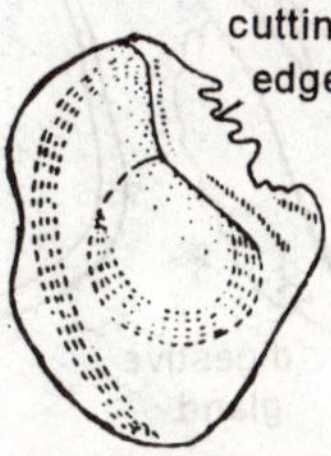


Fig. 21.10 *Pila.* A jaw

At the top of odontophore lies the *radula*, which is a*star-shaped* body and bears teeth on its upper surface which are arranged in transverse rows as in a file. Each transverse row of radula contains seven teeth including a large *central* or *median rachidian tooth* followed by one small *lateral* and two needle-like *marginals* on either side.

The teeth are made up of chitin reinforced by hardened protein. They bear sharp cutting edges meant for rasping vegetable. The old teeth are replaced by the new ones. The new teeth grow behind the radula and push it forward.

The number of radular teeth in a row and the size and shape of teeth vary in different genera and species.

The two chitinous jaws form the posterior limit of the vestibule. The jaws lying dorsolaterally in the region of the sphincter muscle of the buccal mass are connected together by a cuticular membrane. The anterior cutting edge of each is truncate and serrate and bears two or three large teeth. Their movements are brought about by the sphincter muscles of the buccal mass. They appear to be formed by the simple local thickenings of the cuticular lining of these areas.

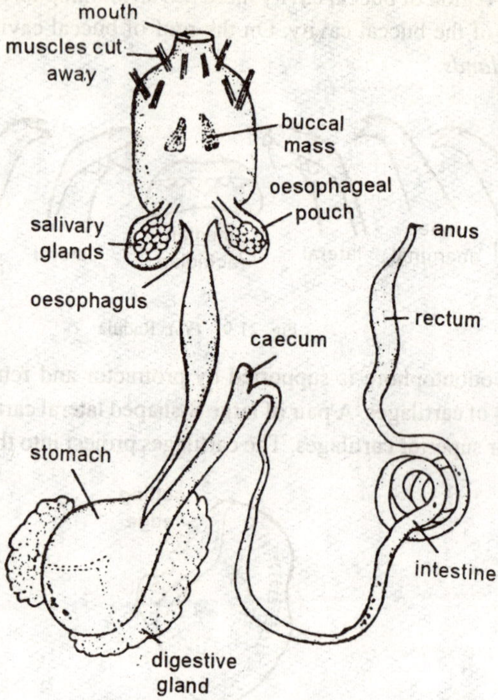

Fig. 21.11 *Pila.* Alimentry canal.

The buccal cavity lead into oesophagus.

*Oesophagus.* It is like a long and narrow tube. It starts from the dorsal part of buccal mass and extends backwards along the median line to open into stomach after the visceral mass. A pair of round coloured *oesophageal pouches* are found in the lateral sides of oesophagus. Each pouh open at the junction of the oesophegus and the buccal cavity by means of a narrow duct. The pouches are the projections of oesophageal wall but their function is not well known. Probably they secrete digestive enzymes. They are also used to store food for some times.

2. *Midgut.* It comprises stomach and intestine. it is endodermal in origin.

*Stomach.* It lies on the left side of visceral mass, below the pericardium, embedded in digestive glands. Its internal cavity is U-shaped in appearence. The stomach contains two parts or chambers : the *cardiac* and *pyloric*. The cardiac chamber is left or posterior part of stomach and forms its mainpart. It receives oesophagus. Its inner lining is thrown into longitudinal folds. The right or anterior limb of the stomach is pyloric chamber which is continued into the intestine. Its inner lining thrown into transverse folds. A bind pouch-like outgrowth the *caecum*, originates from the lower wall of pyloric stomach. It is devoid of crystalline cone which has been observed in many other gastropods.

The digestive gland opens into the stomach by means of a duct at the junction of cardiac and pyloric stomach.

*Intestine*. The pyloric stomach followed by a narrow coiled tube, the intestine. In the beginning the intestine runs backwards, then curves upward and takes 2½ or 3 coils between the gonad infront and the digestive gland behind, before joining the rectum.

3. *Hindgut.* The hind-gut or proctodaeum is ectodermal in origin, comprising rectum.

*Rectum.* It forms the terminal part of alimentary canal. It is thick walled tube running into the branchial chamber and opens to exterior by the anus lying about 0.6 mm. from the edge of the right nuchal lobe.

***II. Digestive glands.*** A number of digestive glands are associated with alimentary canal. These are :

1. *Salivary glands.* Salivary glands are paired, white branching glands lying one on either side of the oesophagus behind the buccal mass. From each salivary gland a duct enters the buccal mass to open into it in the region of the dorsal buccal glands.

They secrete mucus and enzyme. Mucus helps in the lubrication of the radula and the smooth transport of food. The enzyme helps in digestion of starch.

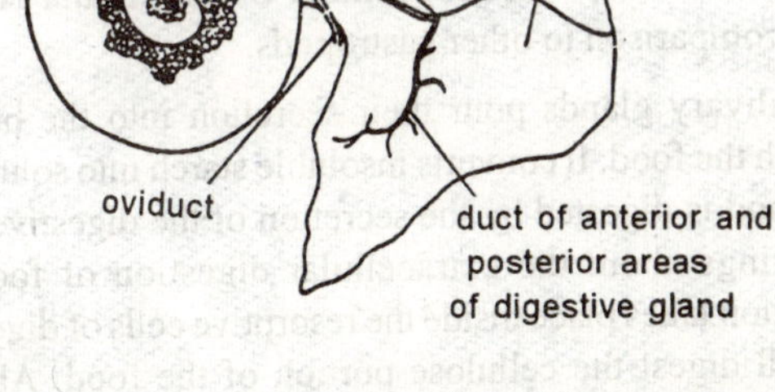


Fig. 21.12 *Pila.* Inner view of digestive gland.

2. *Digestive gland.* The digestive gland is large spirally coiled somewhat brownish or dirty-green in colour. It forms the greater part of the vis-

ceral mass. It consists or two lobes, the smaller is connected with the stomach, the bigger extends to apex of the spiral of the visceral mass. From each lobe arises a main duct, the two ducts unite to form a single duct before opening into the stomach. These ducts, into the digestive gland, branch repeatedly and end blindly in a very large number of small tubes, the *alveoli*. The alveoli are lined with a digestive epithelium made up of three types of cells - (i) *secretory cells* which secrete a cellulose digesting enzyme, (ii) *resorptive cells* which digest proteins intracellularly, and (iii) *lime cells* or *calciferous cells* which store calcium phosphate.Semi-digested food enters into these alveoli, where digestion of cellulose and proteins takes place.

This structure is regarded as liver by most of the zoologists but as it performs the functions of the various digestive glands and also is the main organ for the absorption of the digested food, therefore, *Dakin* has given the term digestive gland to this structure.

3. *Buccal glands.* A pair of elongated glands are present in the roof of buccal cavity. They secrete digestive enzymes.
4. *Oesophageal pouches.* As already described, they probably secrete digestive enzymes.

***Food and feeding.*** The food consists chiefly of aquatic succulent vegetation. It is also reported that sometimes *Pila* feeds on dead animal tissues too.

By the contraction of the sphincter and protractor muscles of the buccal mass, the jaws move to the mouth and cut thin pieces of plants. Meanwhile, the radula is brought forward and the pieces of aquatic plants caught by its teeth are carried in to the cavity of the buccal mass as the radula moves backwards. The chain saw movement (described by *Huxley*) of the radula i.e., to and fro movement of the radula over the cartilage just like a band over the pulley is very characteristic. The movement of the radula is very much limited in *Pila* in comparison to other gastropods.

***Digestion.*** The salivary glands pour their secretion into the buccal cavity where it mixes with the food. It converts insoluble starch into soluble sugar. In the stomach the food is digested by the secretion of the digestive gland. The digestive juice brings about the extracellular digestion of food, whereas intracellular digestion takes place inside the resorptive cells of digestive gland. The resorptive cell digest the cellulose portion of the food. Absorption of digested food takes place in the digestive gland and the intestine. The undigested food passes out from the rectum through the anus. From anus it comes into the branchial chamber and from there it comes to the exterior along with the out-going current of water.

**Respiratory System**

*Pila* is an *amphibious* animal so that it employes both aquatic and aerial methods of respiration.

***Organs of respiration.*** The organs of respiration consist of a pair of leafy expansions of mantle situated in the pallial cavity and called *ctenidia, pulmonary sac* and a pair of *nuchal lobes*.

1. *Ctenidia.* It consists of numerous thin lamellae attached to the mantle wall by their broad bases, but have their apices free in the branchial chamber. The line of attachment is known as *ctenidial axis*. A ctenidium with such an axis with a single row of lamellae is described as *monopectinate* gill. The lamellae present in the middle of the ctenidium are largest. They gradually become smaller towards the ends. Each lamella has a longer left side known as *efferent side* being on the side of efferent ctenidia sinus and the relatively smaller right or *afferent side* lying on the side of afferent ctenidial sinus.

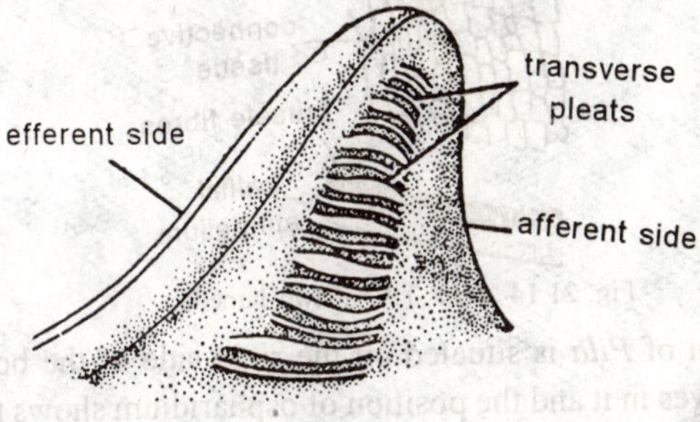


Fig. 21.13 *Pila.* A gill lamella

Shallow transverse ridges, called *pleats* are found in the anterior and posterior regions of each lamellae. Each lamella arises as a double-layer of epithelium internally supported by connective tissue and muscle fibres. The epithelial layer consists of three kinds of cells.

(i) Non-ciliated columnar cells

(ii) Ciliated columnar cells

(iii) Glandular cells.

Underlying the epithelium there is a thin basement membrane below which lies connective tissue layer consisting of cells with a few scattered nuclei and large number of obliquely running muscle fibres. After the muscle fibres is situated a layer of pallial epithelium.

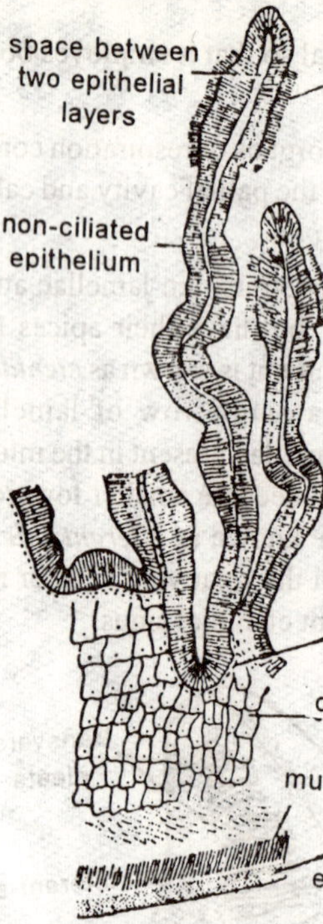


Fig. 21.14 *Pila*. T.S. of two lamella.

The ctenidium of *Pila* is situated on the right side of the body but the distribution of nerves in it and the position of ospharidium shows that gills of the left side has been pushed to the right side by extensive development of the pulmonary sac.

2. *Pulmonary Sac*. It is also known as *lung*. It is sac-like structure hanging from the roof of mantle cavity and is entirely closed except for a large opening with which it opens into the mantle cavity. It is developed from the mantle itself and consists of a densely pigmented dorsal wall and a creamy white ventral wall. The flaps on the two sides of the opening are of unequal size, the left being larger than the right and help in controlling the opening of pulmonary sac. The outer wall of the pulmonary sac consists of an external epithelial layer containing large numbers of fine black pigment granules. This is followed by a basement membrane below which lies a thick layer of connective tissue and finally the endothelium of the inner surface.

According to some workers the pulmonary sac is modified second ctenidium but this has not been proved to be correct.

3. *Nuchal lobes.* The mantle is provided with two fleshy and contractile nuchal lobes or pseudepipodia on the two sides of head over the foot. The left nuchal lobe is well developed whereas the right is less developed. They form two respiratory tubular funnels longer and smaller during respiration.

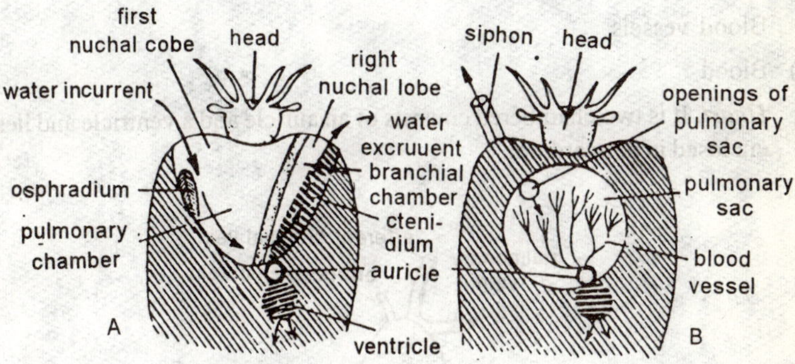


Fig. 21.15 *Pila.* Mechanism of respiration.
A=Water current in aquatic respiration. B= Air current in aerial respiration.

***Mechanism of respiration***

*Aquatic respiration.* Aquatic respiration usually takes place, when the animal lies submerged in the water, attached to the plants or floating in muddy water. During this process, the animal is fully expanded, and the two nuchal lobes, which are equally expanded, appear more or less like shallow channels. The left nuchal lobe after some time increases in size and takes the form of a gutter. The respiratory current of water enters through the gutter-like left nuchal lobe flowing beneath the osphradium and reaches the posterior part of the pulmonary chamber of the mantle cavity and then crossing the epitaenia it enters the branchial chamber, where it bathes the entire length of the gill allowing exchange of gases and, finally, passes outside through the right nuchal lobe.

*Aerial or pulmonary respiration.* The aerial respiration occurs either when the pond water is poor in oxygen or *Pila* is making excursion on land.

*Pila* moves towards the surface of water and thrusts out its left nuchal lobe which becomes elongated and tubular to form the respiratory siphon. The pulmonary chamber becomes greatly enlarged and is completely cut off from the branchial chamber. The air enters the pulmonary sac through respiratory siphon and the respiratory movements produced by the alternate contraction and expansion of the pulmonary sac help in inspiration.

While *Pila* is on land, the respiratory siphon is not formed and the pair enters pulmonary chamber directly through the expanded nuchal lobe.

**Blood Vascular System**

The blood vascular system is highly developed due to its amphibious nature. It consists of :

(I) Heart

(II) Blood vessels

(III) Blood

*I.* ***Heart.*** It is two chambered consists of an auricle and a ventricle and lies enclosed in pericardium.

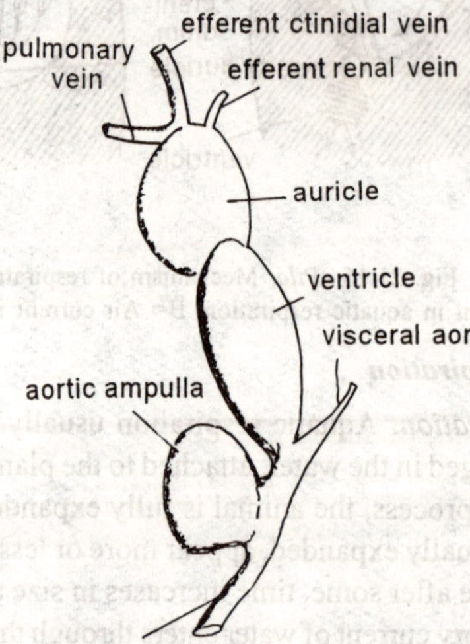


Fig. 21.16 *Pila.* Heart

*(i)* *Pericardium.* The pericardium is thin walled, oval chamber enclosing the heart and situated between the pulmonary chamber and the posterior renal sac on the left side of the body whorl. Posteriorly, it is surrounded by the stomach and the digestive gland. The renal sacs are present on its dorsal side while the oesophagus is found on its ventral side. The pericardial cavity is the true coelom as it communicates with renal organs and the reproductive organ develops from its wall in early stages of development. Pericardium communicates with the posterior renal chamber by reno-pericardial aperture.

(ii) *Auricle*. The auricle is a thin walled, roughly triangular sac highly contractile in nature. It is situated on the dorsal side of pericardium. It receives blood through three main veins - (a) efferent ctenidial vein (b) efferent renal vein and (c) pulmonary vein. At the lower end it is connected with ventricle through the auriculo-ventricular aperture. The aperture is guarded by two semilunar valves which allow the blood to flow from auricle to ventricle but not in reverse direction.

(iii) *Ventricle*. It is an ovoid sac lying below the auricle. It has thick, spongy and muscular walls and a reduced cavity due to a coarse meshwork of muscular strands. From the lower end of the ventricle arises a large artery, the aortic trunk. The opening between the ventricle and aorta is guarded by two *semi-lunar valves* which prevent the flow of blood back into the ventricle.

***II. Blood Vessels.*** It includes arteries, sinuses and veins.

*1.* ***Arteries.*** The aortic trunk divides immediately into two branches an anterior cephalic aorta and a posterior visceral aorta.

(a) *Cephalic aorta*. There is a thick-walled ampulla at the base of the cephalic aorta. It is highly elastic and helps in pumping the blood. The cephalic aorta, beyond the ampulla, gives off immediately three arteries on the outer side.

(i) A *cutaneous artery* which supplies the skin.

(ii) A pair of *oesophageal arteries* which supply the oesophagus.

(iii) A thick *pallial artery* which supplies the left side of mantle, left nuchal lobe and osphradium.

(iv) A *pericardial artery* arises from the inner side of the cephalic aorta. It supplied blood to the pericardium.

The main aortic trunk after giving off above branches, reaches the posterior renal chamber and supplies both the renal chambers by many branches, the *renal arteries*.

The main branch of the cephalic aorta now runs along the left side of the oesophagus for a short distance and then crosses over to its right. Here it appears as a large artery and gives off a number of small branches to the oesophagus and the floor of mantle cavity. On its right side it divides into three branches.

(i) *Right pallial artery* to the right side of the mantle.

(ii) The *right siphonal artery* to the right nuchal lobe.

(iii) The *penial artery* to the copulatory organs.

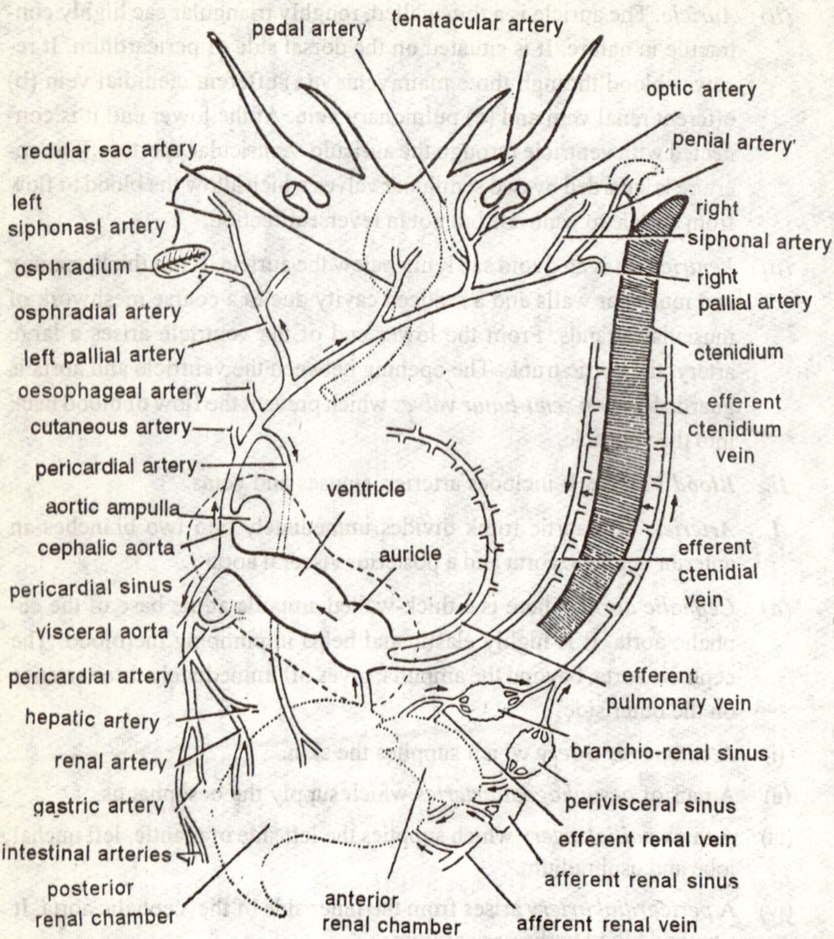


Fig. 21.17 *Pila.* Heart and Blood vessels

The main cephalic aorta again divides into following four branches :

(i) A *radular sac* artery to the radular sac.

(ii) A pair of *optic nerves* to the eyes.

(iii) A pair of *tentacular arteries* to the tentacles.

(iv) *Pedal arteries* which branches to form a network in foot.

*(b) Visceral aorta.* It passes through the visceral mass and supplies blood to the visceral organs by the following branches :

(i) A *pericardial branch*, which supplies the pericardium, the anterior part of the digestive gland and the skin.

(ii) A large *gastric branch*, which supplies the different parts of the stomach.

(iii) A number of *small intestinal arteries*, which supply the intestine.

(iv) A large *digestive branch*, which supplies the digestive gland and the gonad.

(v) Many *renal arteries* to the anterior and posterior renal chambers.

In the end the visceral aorta terminate into branches which supply the rectum and anus.

2. ***Sinuses.*** The ultimate branches of the arteries open into wide spaces without walls, called the blood sinuses. Some of them act as blood capillaries and contact the arteries with the veins. Other sinuses join together to form larger sinuses. In *Pila* there are four sinuses :

*(i)* *Perivisceral sinus.* It occupies the anterior part of the body above the foot and below the floor of mantle cavity. It surrounds the anterior part of alimentary canal.

*(ii)* *Peri-intestinal sinus.* The peri-intestinal sinus is confined around the coils of the intestine.

*(iii)* *Branchio-renal sinus.* It lies along the right side of the anterior renal chamber.

*(iv)* *Pulmonary sinus.* It is confined to the walls of the pulmonary sac.

3. ***Veins.*** The veins carry the venous blood from the various part of the body to the auricle directly or through the respiratory and excretory organs.

*(i)* *Afferent ctenidial vein.* It collects blood from the perivisceral sinus, branchio-renal sinus, rectum and the terminal part of the genital duct and carries it to the ctenidium, where it branches to supply blood to the branchial lamellae for purification.

*(ii)* *Efferent ctenidial vein.* It collects oxygenated blood from the lamellae of ctenidium, mantle and copulatory organ and carries to the auricle.

*(iii)* *Afferent renal vein.* It receives blood from the peri-intestinal sinus and empties into the posterior renal chamber.

*(iv)* *Efferent renal vein.* It originates from the posterior renal chamber and empties in to the auricle.

*(v)* *The pulmonary vein.* It collects blood from the pulmonary sac and carries it to the auricle.

*III.* **Blood.** The blood is colourless and contains colourless amoebocytes in the plasma. The blood of *Pila* has a light bluishtinge. A respiratory pigment, the haemocyanin is found dissolved in the plasma. The *haemocyanin* helps in respiration.

***Course of Circulation.*** The ventricle pumps the mixed blood through branches of cephalic aorta and visceral aorta to various parts of the body. The venous blood is collected from the various parts into the perivisceral and periintestinal sinuses. From these sinuses the blood is sent back to the auricle either directly through kidneys for the removal of nitrogenous waste products or indirectly to the gill and mantle during aquatic respiration. The auricle thus gets aerated and non-aerated blood from these organs and it is the mixed blood which is forced into the ventricle to be pumped again to various organs of the body.

**Excretory System**

In *Pila globosa,* there is a single large *renal organ or kidney or organ of Bojanus* lying behind the pericardium. The excretion is done by the left-kidney as the right kidney has modified into the gonoduct. On one side it opens into the pericardial cavity and on the other to the outside through the mantle cavity.

The renal organ is made up of two distinct parts : a right anterior renal chamber and a left posterior renal chamber.

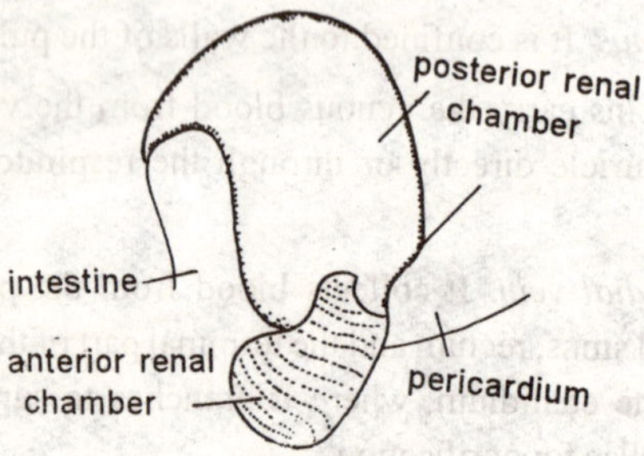


Fig. 21.18 *Pila.* Renal Chamber.

*1.* ***Anterior renal chamber.*** It is oval, reddish organ, lying on the dorsal side of the animal infront of the pericaridum and posterior renal chamber. The cavity of the anterior renal chamber is very small due to the presence of a number of triangular leaf-like processes or *lamellae* are arranged on either side along a thick median longitudinal axis, or the efferent renal sinus. On the floor of the chamber, the lamellae are arranged on either side of a median axis, the afferent renal sinus. The anterior renal chamber communicates with the posterior renal chamber through a small opening at one end and opens by an elongated opening into the mantle cavity lying to the right of the epitaenia, at the other end.

2. ***Posterior renal chamber.*** It is hook-like broad chamber brown or grey in colour situated behind the anterior renal chamber, in between the rectum on the right and the pericardium and the digestive gland on the left. Its large internal cavity encloses a part of the genital duct and a few coils of the intestine. At one end, it communicates with the anterior renal chamber through an aperture and at the other with the pericardium through an elongated slit-like reno-pericardial aperture, perforating a thin vertical reno-pericradial septum, separating the two. The afferent and efferent renal vessels profusely branch in the roof of this chamber.

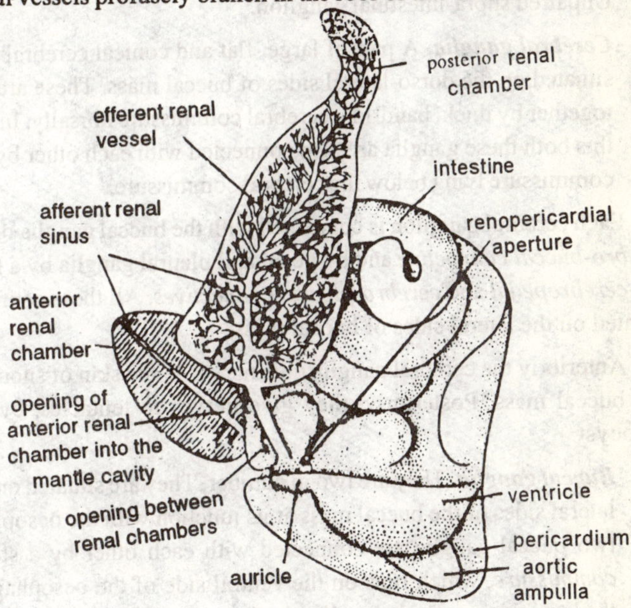


Fig. 21.19 Pila. Excretory System.

**Physiology of excretion.** The renal organ separates the nitrogenous waste from the blood. The excretory fluid passes from the posterior renal chamber to the anterior renal chamber from where it is discharged into the mantle cavity through the external renal aperture and finally it goes out with the outgoing current of water through right siphon. The excretory matter contains chiefly ammonia compounds, urea and uric acid. For conserving water, *Pila* converts ammonia into insoluble uric acid.

In most of the gastropods, the digestive gland also forms and organ of excretion. These are eleminated by way of stomach and intestine.

**Nervous System**

A well-developed nervous system. It possess two characteristics (i) except the visceral ganglia all other ganglia form a ring-like structure around the buccal mass (ii) due to torsion the visceral loop is twisted to form the shape of '8'. The nervous system of *Pila* includes the following ganglia, their commissures and connectives :

1. One pair of cerebral ganglia.
2. One pair of buccal ganglia.
3. Paired pleuro-pedal ganglionic mass.
4. One pair of visceral ganglia.
5. Unpaired supra-intestinal ganglion.

*1.* ***Cerebral ganglia.*** A pair of large, flat and conical cerebral ganglia are situated on the dorso-lateral sides of buccal mass. These are connected together by thick, band-like cerebral commissure dorsally. In addition to this both these ganglia are also connected with each other by thin labial commissure lying below the cerebral commissure.

Each cerebral ganglion is connected with the buccal ganglia of its side by *cerebro-buccal connective* and by pedal and pleural ganglia by a thick band-like *cerebropedal* and *cerebro-pleural connectives*. All these connectives are situated on the lateral sides of buccal mass.

Anteriorly the cerebral ganglia give nerves to the skin of snout, tentacles and buccal mass. Posteriorly, they give nerves to tentacles, eyes and the statocyst.

*2.* ***Buccal ganglia.*** They are two in number. They are situated on the dorso-lateral sides of the buccal mass at its junction with the oesophagus. The two buccal ganglia are connected with each other by a stout *buccal commissure*, which runs on the ventral side of the oesophagus behind the buccal mass. Each ganglion is also connected with the cerebral ganglion of its side by the *cerebro-buccal connective.*

The nerves are supplied from each buccal ganglion to the posterior part of buccal mass and radular sac.

*3.* ***Pleuro-pedal ganglia.*** A pair of large, somewhat triangular ganglionic masses present one on either ventro-lateral side of the buccal mass. Each one is formed by the fusion of an outer pleural and an inner pedal ganglion, separated by a faint notch. The right pleuropedal ganglionic mass also consists of the infra-intestinal ganglion fused with it.

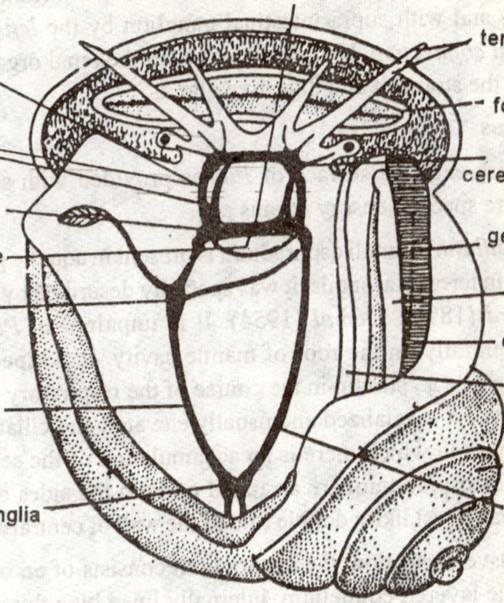


Fig. 21.20 *Pila.* Nervous System.

The two pedal ganglia are connected with each other by two *pedal commissures* which are placed one above the other. These commissures are situated below the buccal mass. Many nerves are given off to the foot from the anterior and posterior parts of the pedal ganglion. Behind the pedal commissures, the two pleural ganglia are connected with each other by infra intestinal nerve. The left pleural ganglion gives off nerves to the *mantle, osphradium, left nuchal lobe, columellar muscles* and the *anterior part of gill.* The right pleural ganglion supplies nerves to parietal wall, epitaenia, right nuchal lobe, copulatory organs, rectum, genital duct and columellar muscles.

4. ***Supra-intestinal ganglion.*** It is a fusiform ganglion situated on the left side 1/4th of an inch way from the pleuro-pedal mass of its side. It is connected anteriorly to the left pleural ganglion, posteriorly to the left visceral ganglion by thick nerves and on its inner side to the sub-intestinal ganglion of right side by a thin nerve. The thin nerve is known as *supra-intestinal nerve.* On its outer side the supra-intestinal nerve issues a short but thick nerve to the mantle and the anterior part of the ctenidium.

5. ***The visceral ganglion.*** It is formed by the fusion of two spindle-shaped

ganglia. It lies at the base of the visceral mass. It is connected with the right pleural ganglion by a thin nerve, the *infra-intestinal visceral connective* and with supra-intestinal ganglion by the *left supra-intestinal visceral connective*. It supplies nerves to the renal organ, genital organ and of the stomach and digestive gland.

## Sense Organs

The entire exposed surface of *Pila* is provided with sensory cells but followings are special sensory organs :

*1.* ***Osphradium.*** A small osphradium is present in aquatic gastropods while absent in terrestrial snails. it was specially described by *Spengel* (1881), *Bernard* (1890), *Demal* (1954). It is unpaired in *Pila* and situated dorsoventrally on the roof of mantle cavity. It is bipectinate structure hanging like a curtain in the course of the respiratory water current. It consists of a specialized and usually elevated and ciliated region of the epithelium in which there is an accumulation of the sensory cells. The leaves of osphradium are arranged on both the sides of a central axis. Each leaf is just like a double fold of the wall of central axis.

In a transverse section, the osphradium consists of an outermost covering of a single layered epithelium, internally lined by a thin basement membrane, the interior filled up with nerves, connective tissue and blood spaces. The epithelial cells are elongated possessing basal nuclei, and they are of three types (i) *sensory,* (ii) *ciliated* and (iii) *glandular*. The ciliated cells line the attached margin, while the sensory cells devoid of cilia cover the osphradium. The flash-shaped glandular cells are found scattered among the sensory cells.

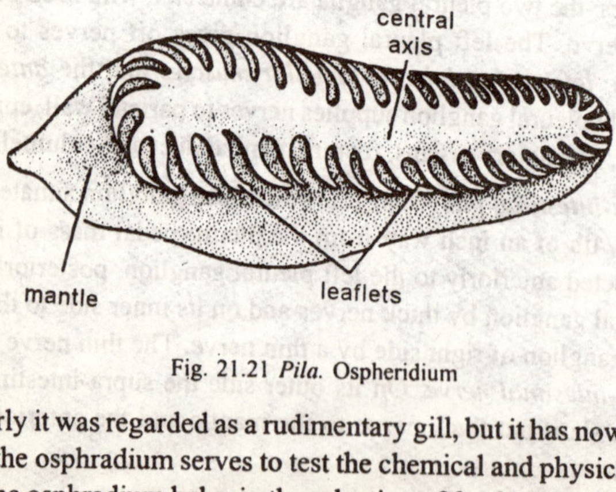


Fig. 21.21 *Pila*. Ospheridium

Formerly it was regarded as a rudimentary gill, but it has now been established that the osphradium serves to test the chemical and physical properties of water. The osphradium helps in the selection of food material (*Spengel*).

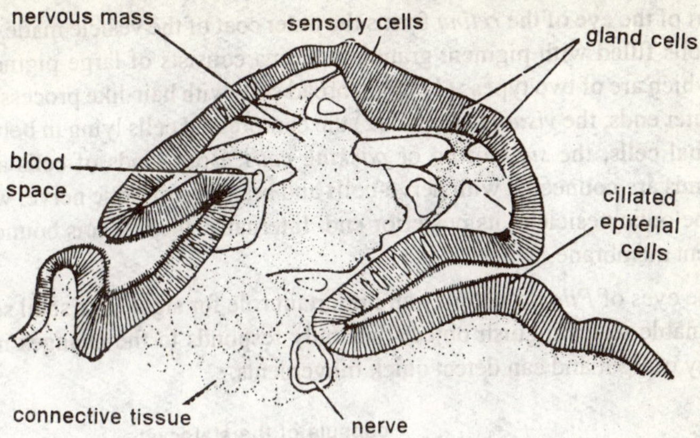


Fig. 21.22 *Pila.* T.S. of Ospheridium

2. ***Eyes.*** Two eyes are borne on short stalks called *ommatophores* situated behind the second pair of tentacles.

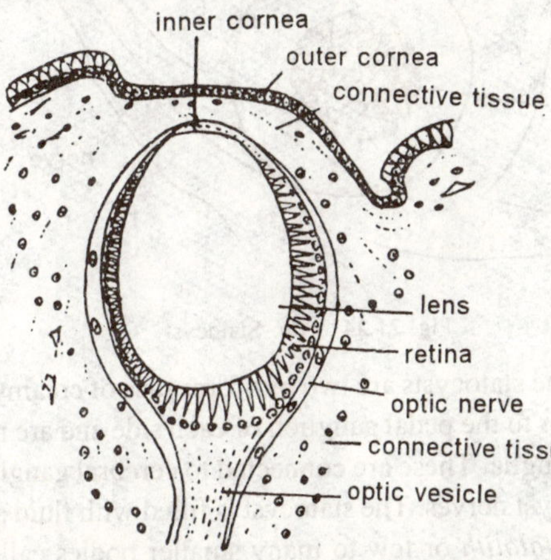


Fig. 21.23 *Pila.* H.L.S of eye

In section the eye is in the form of a pyriform cup or optic vesicle. The outer epithelium of the eye forming the *cornea* or *pellucida externa*, It consists of a row of squarish cells devoid of pigments. Lying below the cornea is thick layer of connective tissue made up of fibres and nuclei. This layer encloses the pyriform optic vesicles consisting of a transparent epithelium forming inner *cornea* or *pellucida interna* lying next to connective tissue. The ovoidal structureless lens occupies the central space of the vesicle. The sensitive part of the eye of the *retina* forms the outer coat of the vesicle made up of large cells filled with pigment granules. Retina consists of large pigmented cells, which are of two types : (1) large, broad cells with hair-like processes on their outer ends, the *visual cells* and (2) the compressed cells lying in between the visual cells, the *supporting* or *packing cells*. Both kinds of cells at the inner ends are connected with nerve-cells and branches of optic nerve, which enter the optic vesicle at its posterior end. Internally, the retina is bounded a basement membrane.

The eyes of *Pila* do not play an important role for sight. The snail seems to be unable to distinguish objects but only responds to the changes in the intensity of light and can detect quick movements.

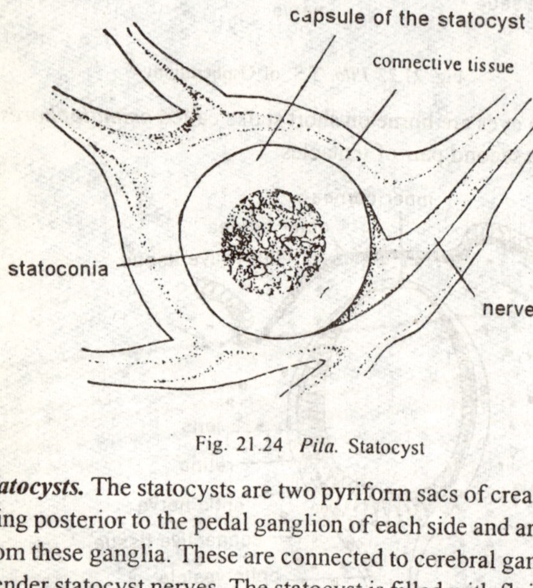


Fig. 21.24 *Pila.* Statocyst

3. ***Statocysts.*** The statocysts are two pyriform sacs of creamy white colour lying posterior to the pedal ganglion of each side and are not innervated from these ganglia. These are connected to cerebral ganglia by a pair of slender statocyst nerves. The statocyst is filled with fluid and contains a single large *statolith* or few to many smaller bodies called *statoconia* floating in the fluid. Formarly it was supposed to be the organ of hearing and hence it was called *otocyst* but it is proved that these are the organs of equilibrium (*Delage*).

4. ***Tentacles.*** There are two pairs of *cephalic tentacles* in *Pila*. The second pair of tentacles are *rhinophore or olfactory organs*.

**Reproductive System**

*Pila* is *unisexual*. A well marked sexual dimorphism is seen. The shell of male is smaller than that of female. In male penis is well developed and the female has a more swollen body short than the male.

***Male Reproductive System.*** The male reproductive organs comprises :

1. Testis
2. Vasa efferentia
3. Vas deferens
4. Seminal vesicle
5. Penis
6. Hypobranchial gland.

*1.* *Testis.* It is a single, flat, plate-like and roughly triangular whitish structure, occupying the upper part of the first two or three whorls of the shell. It lies in close association with the brownish or greenish digestive gland along its upper and inner or columellar side and appears quite distinct on account of its creamy colour. A thin cutaneous membrane separates the testis from the shell.

The germinal epithelium of testis produces two types of sperms:

(i) *Eupyrene sperms.* These are small and thread-like. Their body consists of a head and a tail. These are motile and fertilize the ova. They measure 25 long.

(ii) *Oligopyrene sperms.* They are large semicircular or spindle-shaped. The head is large and the tail bears 4 to 5 cilia. These are non-motile and incapable of fertilizing ova. They measure 32.5 in length.

*2.* *Vasa efferentia.* From the testis arise several delicate ducts called vasa efferentia. They unite towards the posterior edge of testis to form a common duct, the vas deferens.

*3.* *Vas deferens.* The vas deferens commences from the posterior part of the testis and passes under the skin and is differentiated into two parts :

(i) *Proximal tubular part.* It forms posterior part of testis under the skin along inner side of the digestive gland upto the posterior renal chamber and finally turns towards the left to approach the pericardium. Here it opens into the terminal glandular part of the vas deferens.

*(ii)* *Terminal glandular part.* It is the distal thickwalled glandular part of the vas deferens which runs along the left side of the rectum into the branchial chamber and opens into it by male genital pore. The male genital pore is situated a little behind the anus to a claw-like projection of genital papilla.

*4.* *Seminal vesicle.* It is situated on the right of the pericardium at the junction of the anterior and posterior renal chamber. It is swollen flask-shaped structure whose distal end is closed. The seminal vesicle stores the sperms and opens into the vas deferens at the junction of its tubular and glandular parts.

5. *Penis or Copulatory organs.* On the left side of the inner surface of the mantle, there is a thick glandular flap of yellowish colour. It is attached along its right side but its left free edge is rolled to form spout-like penis-sheath. Inside the sheath is the penis.

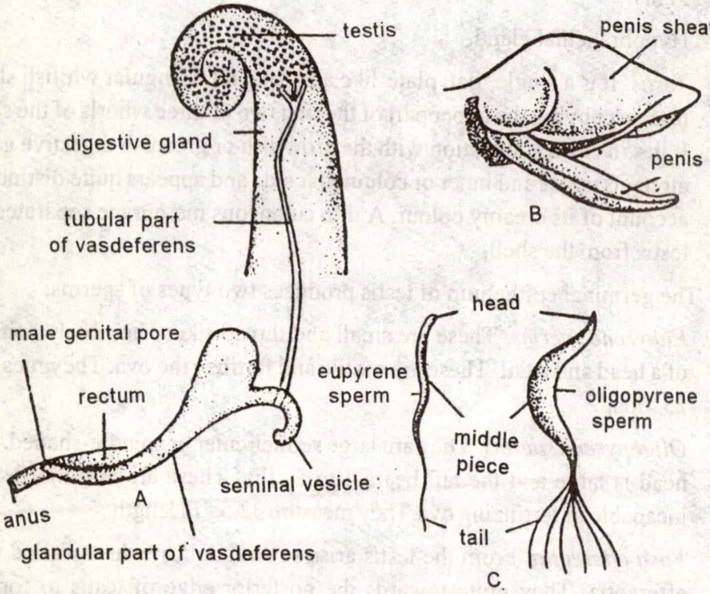


Fig. 21.25 *Pila*. A= Male reproductive organs, B=Copulatory organ, C=Sperms

The penis is about 1.25cm. long, provided with flagella arising from the roof of the mantle. It is slightly covered with swollen base and tapering end has a groove along its length on the inner side. It is capable of great extension.

6. *Hypobranchial gland.* It is a glandular structure with pleated surface at the base of penis-sheath. The gland is without duct and, therefore, its secretions are poured directly upon its surface.

*Female Reproductive Organs.* The female reproductive organs are following :

1. Ovary
2. Oviduct
3. Seminal recepticle
4. Uterus

5. Vagina
6. Copulatory organ
7. Hypobranchial gland.

1. *Ovary.* The ovary is present in the same position as the testis in the male. It is found in two or three whorls of the shell. It is orange coloured structure remains attached to the inner surface of the digestive gland. Branches of the ovary are single layered, flash-shaped structures with slender tubular necks uniting together to form minute ducts which open into oviduct.
2. *Oviduct.* The oviduct is a semi-transparent, narrow tube arising from the middle of the ovary and passing downwards along the margin of the digestive gland. Finally, it ascends upwards to receive the opening of receptaculum seminis.

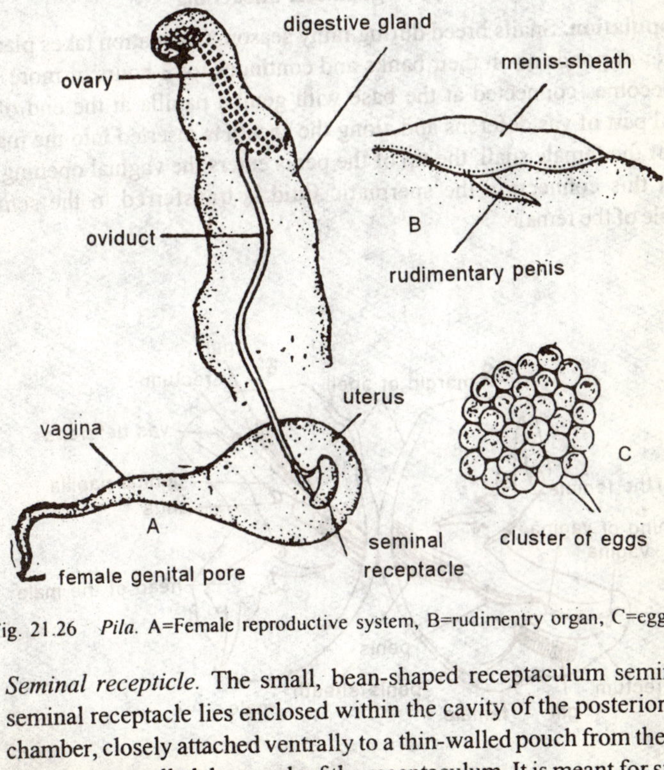


Fig. 21.26 *Pila.* A=Female reproductive system, B=rudimentry organ, C=eggs.

3. *Seminal recepticle.* The small, bean-shaped receptaculum seminis or seminal receptacle lies enclosed within the cavity of the posterior renal chamber, closely attached ventrally to a thin-walled pouch from the walls of the uterus, called the pouch of the receptaculum. It is meant for storing the sperms received from the male *Pila* during copulation.

4. *Uterus*. It is a large, yellow and pear-shaped structure situated below the intestine in the body whorl. It is differentiated into a broad rounded basal part and a narrow tubular apical part. Seminal recepticle opens into the basal part while apical part receives the vagina.

5. *Vagina*. The terminal narrow part of the oviduct is known as *vagina*. It is a cream-coloured tube which runs forward on the right side of rectum to open into the branchial chamber by female genital aperture situated behind the anus.

6. ***Copulatory Organ.*** The rudimentary penis is not enclosed within a well-developed penis-sheath, but lies beneath a glandular fold of the mantle. It is a thin, flageller structure nearly 6mm. long and uniform in thickness but pointed at the tip. It also bears a rudimentary groove along its inner surface, like the penis of the male.

7. ***Hypobranchial gland.*** It also represent the rudiment of the hypobranchial gland of male as an incipient glandular thickening.

**Copulation.** Snails breed during rainy season. Copulation takes place in the water of ponds or on their banks and continues for 3 hours or more. The penis becomes connected at the base with genital papilla at the end of the terminal part of vas deferens and along the sheath is inserted into the mantle cavity of the female snail, the tip of the penis enters the vaginal opening and through this connection the spermatic fluid is transferred to the seminal recepticle of the female.

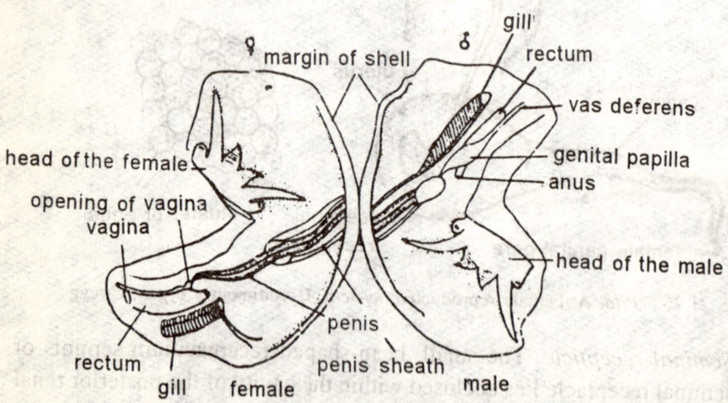


Fig. 21.27 *Pila*. Showing Copulation.

**Fertilization**. Fertilization is internal but the development of the embryo takes place outside the body of female. Ovulation takes place two days after copulation. Female lays 200-280 eggs at a time in moist earth in a sheltered cavity near ponds and lakes.

***Development***. The eggs are rounded and as big as the pea-seeds. Each egg consists of an outer whitish shell, enclosing a double shell membrane, a thick mass of solid white albumen and a little central mass of liquid albumen containing the embryo. During development the visceral mass and the shell of embryo become spirally coiled and a characteristic phenomenon of torsion takes place resulting in the asymmetry of the body. The young snails emerging from the eggs are similar to the adults.

☼☼

# 22
# UNIO

*Unio* comes under class Pelecypoda or Bivalvia in which an external and internal symmetry is seen unlike gastropods. The cephalic region is rudimentary and the body is laterally compressed.

**SYSTEMATIC POSITION**

| | | |
|---|---|---|
| Phylum | - | Mollusca |
| Class | - | Pelecypoda |
| Order | - | Eulamellibranchiata |
| Genus | - | *Unio or Anodonta* |

**Distribution**

The 'freshwater mussel' is a familiar representative of Phylum Mollusca. The family Unionidate to which *Unio* belongs is cosmopolitan in distribution. The family comprises several genera and about 1000 species of which a good number represented in India. *Anodonta* is the common representative of Britain, while in India *Unio* and *Lamellidents marginalis* are of common occurance.

**Habits and habitats**

It is found in rivers, streams, lakes and ponds. During day time it usually lives partly burried in mud or sand at the bottom, keeping its posterior end exposed. It comes out to shallow places in night. They may also occur wedged in between the rocks and stones, with the valves slightly spread and the two siphons exposed. It slowly creeps with a ploughshare-like muscular foot. It hardly covers 16 kilometres in a year. Sexes are separate but there is no sexual dimorphism externally. Development is indirect through a larval stage and takes place in brood chamber, which are enlarged water tubes of female's gills.

**External features**

***Shape and size.*** It is laterally compressed. It is elongated and enclosed and protected in a bivalve shell. The size varies from 2" to 4" in length.

## Shell

A typical bivalve shell consists of two similar more or less oval, usually convex valves which are attached and articulate dorsally with each other. Each valve bears a dorsal protuberance called the *umbo* which rises above the line of articulation. The umbo is the oldest part of the shell, and the concentric lines around it are lines of shell growth. The two valves are attached by an elastic protein band called the *hinge ligament*, which is covered above with the periotracum. The hinge ligament is so constructed that when the valves are closed, the dorsal part is stretched and the ventral or inner part is compressed. Thus, when adductor muscles relax, the ligament causes the valves to open. The inner surface of each valve possesses dorsally along the hinge line, small sharp ridges and teeth-like projections separated by grooves or sockets. These are called *hinge teeth*. The teeth of one valve fit into the corresponding sockets of the other to prevent lateral slipping. However, a dorsal fold of mantle tissue still lies between the tooth and socket of two opposing valves.

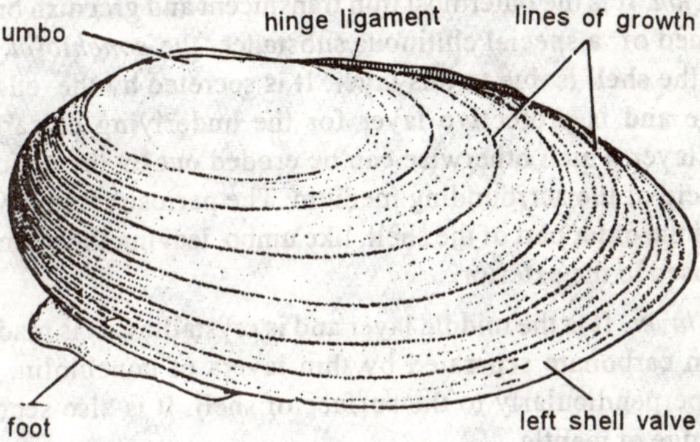


Fig. 22.1 *Unio* External feature

The valves of the shell are pulled together by two large dorsal muscles, called adductors. The posterior muscle being larger than that of the anterior. Near the impression of the anterior adductor muscles are two small impressions, the dorsal and posterior one left by *anterior retractor muscle* and a ventral and posterior by *protractor muscle.* A small impression of a *posterior retractor muscle* also lies dorsal to the impression of posterior adductor muscle.

Running parallel to the ventral margin of the valves from one adductor impression to the other is a thick line known as *pallial line*, which is caused due to the insertion of muscle fibres from the edges of the mantle.

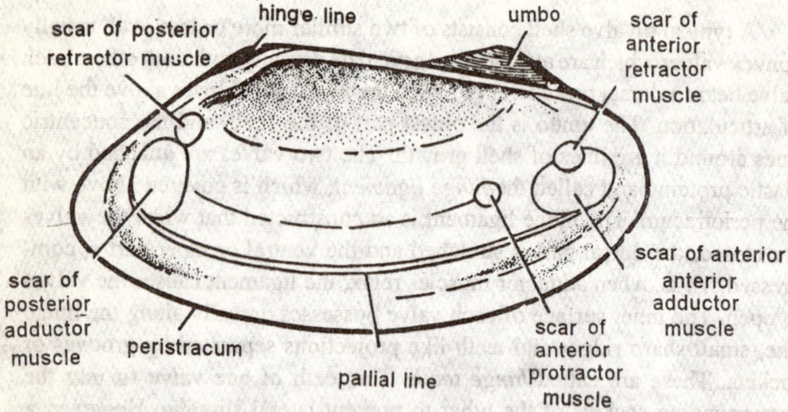


Fig. 22.2 *Unio* Inner view of left shell

***Microscopic structure of shell*** : The shell consists of three distinct layers

1. *Periostracum.* It is the outermost thin translucent and greenish brown layer formed of a special chitinous substance, the *conchiolin.* The colour of the shell is due to this layer. It is secreted by the edge of the mantle and is protective layer for the underlying calcareous prismatic layer which otherwise can be eroded out by the action of carbolic acid of the surrounding medium. The periostracum is worn away from the older part of the shell, like umbo, leaving the prismatic layer exposed to the exterior.

2. *Prismatic layer.* It is the middle layer and is crystalline. It is made up of calcium carbonate separated by thin layers of conchiolin. It is arranged perpendicularly to the surface of shell. It is also secreted from the edge of mantle.

3. *Nacreous layer or pearly layer.* It is the innermost layer of shell also known as the *mother of pearl or nacre,* since it presents a smooth, iridescent surface. It is secreted by the whole outer surface of mantle and is in the form of alternative transverse layers of $CaCO_3$ and conchiolin. The layer of both the types run parallel to the shell surface.

**Formation of shell.**

The shell is made up of calcium carbonate (89-99%), very little of phosphate of lime (1-2%) and organic base called the *conchiolin.* Two theories have been proposed regarding the formation of shell.

**1. *Bowerbank and Carpenter's Theory.*** According to *Bowerbank* and

*Carpenter*, the shell is formed by the growth of interstitial deposits like that of formation of teeth and bones in higher vertebrates.

**2. *Reaumur - Eisioz Theory.*** According to this theory the shell is formed by the cuticle-like deposition on the outside of skin. It is formed by minute calcareous particles which bound together Mantle takes important part in the deposition of shell. Every portion of mantle bears shell deposition cells. These cells separate the carbonate of lime from blood, which is the main constituent of shell. The umbo portion of shell is secreted first and subsequent layers are added in gradual succession forming the lines of growth.

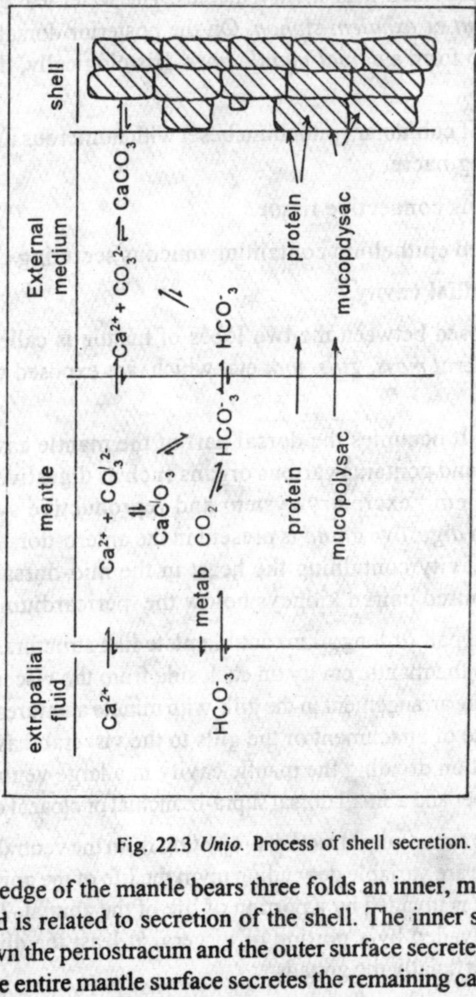


Fig. 22.3 *Unio*. Process of shell secretion.

The edge of the mantle bears three folds an inner, middle and outer. The outer fold is related to secretion of the shell. The inner surface of outer fold layer down the periostracum and the outer surface secretes the first calcareous layer. The entire mantle surface secretes the remaining calcareous cells.

**Mantle**

When the shell is removed, the soft body of the animal is exposed, covered in a thin, semi-transparent soft fold of skin, called *mantle* or *pallium*. It consists of two lateral halves or mantle lobes or folds, which correspond to the two valves of the shell. These lobes are thin sheets of tissues closely attached to the inner surface of the valve. The lobes come off from the dorsal part of the body-wall, ventrally and anteriorly, they are free, posteriorly, they are fused in the middle to from a *dorsal* and a *ventral siphon*. These siphons allow water in and out. Currents of water enter through the ventral or incurrent or inhalent siphon. The ventral siphon is formed by approximation of mantle lobes and its edge is provided with small tentacles. The water leaves through the dorsal or *excurrent* or *exhalent siphon*. On the posterior dorsal side, the two mantle lobes also form a *dorsal-mantle pore*. Histologically, the mantle consists of :

1. An outer layer of columnar epithelium beset with numerous unicellular glands secreting nacre.
2. A middle fibrous connective tissue.
3. An inner ciliated epithelium containing mucous secreting cells.

**Mantle cavity or Pallial cavity.**

The space enclosed between the two lobes of mantle is called ***mantle cavity***. It encloses *viseral mass, gills, foot* etc. which are exposed when one fold is removed.

**(i)** ***Visceral mass.*** It occupies the dorsal part of the mantle cavity. It is dark in colour and contains various organs such as digestive system, circulatory system., excretory system and reproductive system. A greenish-brown *digestive gland* is present in the antero-dorsal region. A pericardial cavity containing the heart in the mid-dorsal region, and a dark coloured paired kidneys below the pericardium.

**(ii)** ***Gills.*** Gills are a pair of long, thin, double plate like structures, which hang freely into the mantle cavity on each side from the visceral mass. There is sieve-like arrangement in the gills with minute apertures covered by cilia. The line of attachment of the gills to the visceral mass form a horizontal partition dividing the mantle cavity in a large ventral infra-branchial chamber and a small dorsal supra-branchial or cloacal chamber.

(iii) *Foot.* The foot is a muscular structure projecting from the ventral surface. Its size and form are variable depending upon the life of the animal. The mass of the foot is invaded by a portion of life of the animal. The mass of the foot is invaded by a portion of viscera, at least the alimentary canal, liver superficially the gonads.

(iv) ***Head.*** A distinct head is absent in *Unio*. The eyes and tentacles are also absent. The large mouth opens beneath the anterior adductor muscle. It is boarded by a pair of broad, lamellar *labial palps* on each side.

**Locomotion**

The locomotion is performed by muscular foot. It is affected by a combination of blood pressure and muscle action. Engorgement of the foot with blood, coupled with the action of pair of *pedal protractor muscles*, produces extension. The water in mantle cavity and the blood act as a hydrostatic skeleton. The pressure of these two fluids is elevated by the adducted valves. Blood from the visceral mass is forced down into the pedal haemocoel, causing the foot to dilate and anchor into the substratum.

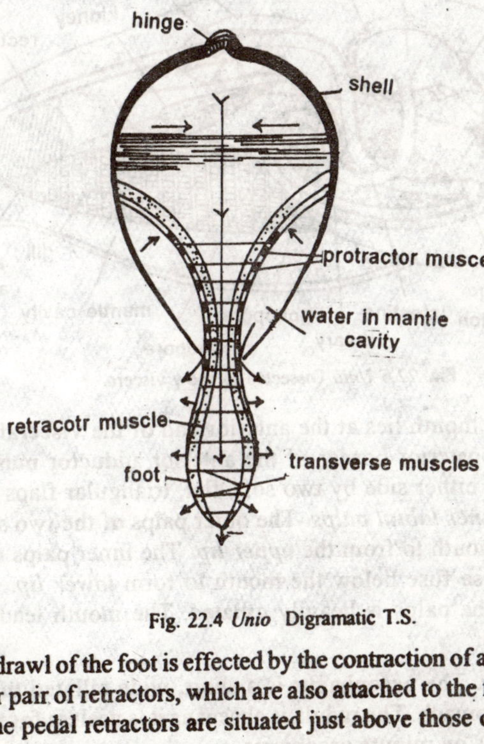


Fig. 22.4 *Unio* Digramatic T.S.

Withdrawl of the foot is effected by the contraction of an anterior pair and a posterior pair of retractors, which are also attached to the foot and shell. The scars of the pedal retractors are situated just above those of adductors.

**Coleom.**

Coelom is greatly reduced due to the development of connective tissue, unstriated muscle-fibres and blood-sinuses. Coelom is a ***haemocoel***, filled with blood. It is represented by three small cavities : (a) a single ovoidal chamber, called *pericardium*, which is situated dorsally and in it, heart is located, (b) the cavity of the kidney or *urocoel*, (c) *gonocoel* or cavity of gonad. Originally, all the three were intercommunicated but due to gradual separation, the cavities of the gonads are completely separated, while the cavities of kidney and pericardium are inter-communicated.

**Digestive System**

The digestive system comprises of alimentary canal and a paired digestive glands.

I. ***Alimentary canal.*** The alimentary canal is complete. It is a coiled tube of varying diameter and starts from mouth and terminates into an anus. Buccal cavity, jaws radula and salivary glands are absent.

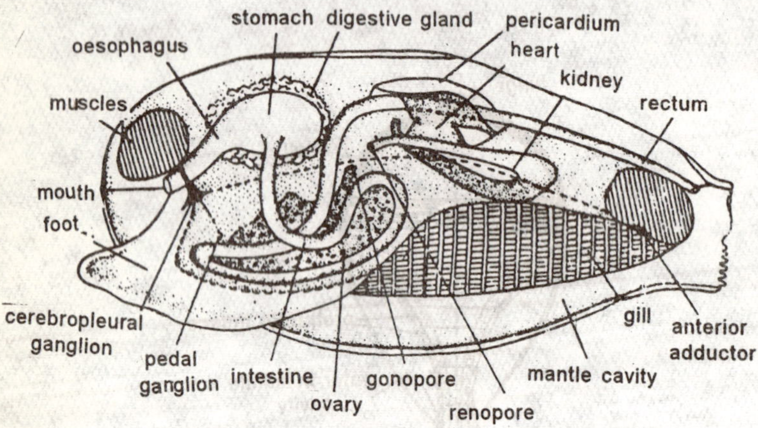


Fig. 22.5 *Unio* Dissected to show viscera.

1. *Mouth.* The mouth lies at the anterior end of the visceral mass just below the posterior border of the anterior adductor muscles. It is bounded on either side by two soft, flat, triangular flaps called the *outer* and *inner labial palps.* The outer palps of the two sides unite above the mouth to from the *upper lip.* The inner palps of the two sides likewise fuse below the mouth to form *lower lip.* The outer surface of the palps is heavily ciliated. The mouth leads into the oesophagus.

2. *Oesophagus.* The oesophagus is a short, wide ciliated tube, which leads into stomach. The radula is absent as it is useless for the animal since it feeds on minute organisms.

3. *Stomach.* The stomach is a wide thick-walled rounded sac, situated dorsally in the visceral mass and is surrounded by paired digestive

glands. The stomach is produced into a pouchlike out-growth known as *pyloric caecum*. It contains a transparent gelatinous flexible rod called the *crystalline style*. It has proteinous matrix deposited with amlyase which is a carbohydarate splitting enzyme. The style rotates due to cilia of the stomach and brings about the mixing of enzymes with food. The crystaliine style is secreted by the cells of the stomach and is contained in a style sac.

4. *Intestine*. From the posterior end of stomach arises the intestine and enters the substance of foot where it coils on itself and again ascent the place where the pericardial cavity starts. It then runs dorsally again and is continued into the rectum.

5. *Rectum*. It is the terminal part of intestine which starts from pericardium and runs for a short distance before opening finally through the anus in the cloacal chamber. Internally the rectum has a ridge, the *typhlosole*.

6. *Anus*. The anus is situated on an anal papilla just above the adductor muscle.

## II. *Digestive glands*

The digestive glands are a pair of large irregular green or dark-brown masses situated on the sides of the stomach. They lead into anterior part of stomach by many ducts. It is made up of highly branched tubules, line by columnar epithelium of unequal height containing basal nuclei. The cells of these glands also swallow the food particles and break the solid particles in addition to the secretion of enzymes. These cells intracellularly digest the protein and fats, and absorb the carbohydrates.

***Food and feeding.*** Freshwater mussel is a cliary feeder. If feeds upon micro-organisms and organic debris which is filtered from the respiratory nutritive current of water drawn into mantle cavity. A continuous current of water is maintained mainly by the action of lateral cilia of gill filaments. As the water current enters the mantle cavity through the inhalent siphalent siphon it slows down and the solid food particles settle down over the gill surface. These are removed to the edge of mantle where they are pushed back and are expelled out. The lighter food particles get entangled into the mucus secreted by the gills. By the beating of the cilia of labial palps, the mucus is drwan towards mouth. The particles of food which cannot be digested are dropped down into the mantle cavity, whereas the digestible particles are pushed into the mantle cavity, along the deep ciliated grooves between the labial palps.

**Digestion.**

The digestion is *intracellular*. Food from the stomach enters the digestive gland, the cells of which readily ingest and break down solid particles.

Protein and fat digestion is exclusively intracellular, and the cells of gland also absorb carbohydrates. The only extracellular enzyme is the carbohydrates digesting enzyme (*amylase*) released in the stomach by the dissolution of the crystelline style. The style pouch is lined with cilia, the beating of which rotates the style and moves it forward with the result that its free end constantly rubs against a particular portion of the stomach wall and its head is worn away getting mixed with stomach content. Digested food diffuses from the gland into the blood for distribution.

Undigested food particles are sent back into the stomach, whence they pass into the intestine and then into the rectum by ciliary action. Function of intestine or rectum is not propely understood. They may absorb water from faecal matters. From the rectum, faeces is passed through the anus into the cloaca, which it leaves with the outgoing current through exhalent siphon.

**Respiratory System**

The respiratory organs in *Unio* are a pair of *gills* or *ctenidia* and *mantle.*

***1. Ctenidia or gills.***

There are two gills that hang in the mantle cavity, one on either side between the mantle and visceral mass. The gills are plate-like hence they are called *lamellibranches*. Each gill has a central axis and plate like structure on either side. Hence the gills are called *bipectinate gills*.

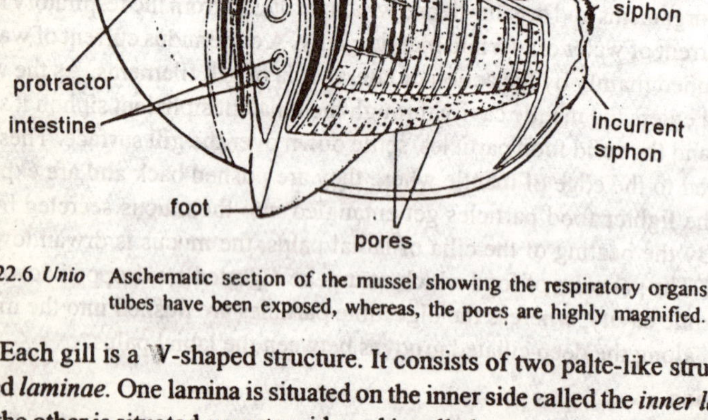


Fig. 22.6 *Unio* Aschematic section of the mussel showing the respiratory organs. Water tubes have been exposed, whereas, the pores are highly magnified.

Each gill is a V-shaped structure. It consists of two palte-like structures called *laminae*. One lamina is situated on the inner side called the *inner lamina* and the other is situated on outer side and is called *outer laminae*. Each lamina is V-shaped structure and formed of two vertical plates called *lamella*. The outer and inner lamellae are united together at their anterior, ventral and posterior edges, but are free dorsally. Each lamella is formed of a series of vertical filaments called *gill filaments*. The gill filaments of the two lamellae of a lamina are continuous at the free ventral side. The gill filaments are joined to one another by horizontal bars called *inter-lamellar junctions*. The inter-lamellar junctions divide the cavity lying between the lamellae into vertical chambers called *tubes*. Dorsally all the water tubes open into a large cavity called *supra-branchial chamber*. This chamber communicates to the exterior through the *exhalent siphon*. The portion of the mantle cavity lying below the gill is called *infrabranchial chamber* which communicates to the exterior through the inhalent siphon. The lamellae are perforated by minute openings, the *ostia*. They are located between the gill filaments and inter filamentar junctions. Through ostia water inters the mantle cavity.

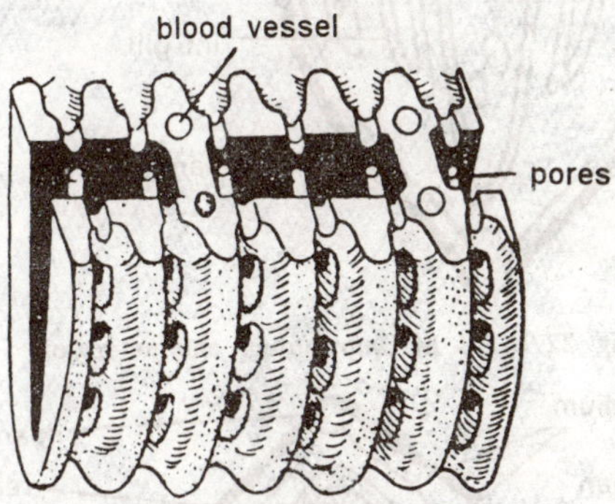


Fig. 22.7 *Unio*. Highly magnified gill

Each gill filament is covered by a layer of ciliated epithelial cells. Internally the gill filament is supported by chitinous rods.

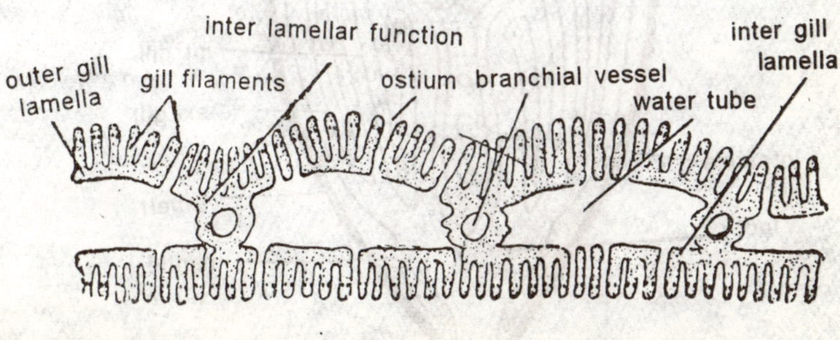


Fig. 22.8 *Unio*. T.S. of outer gill lamella.

**Attachment of gills.**

The mode of the attachment of the gills is very peculiar, as it determines

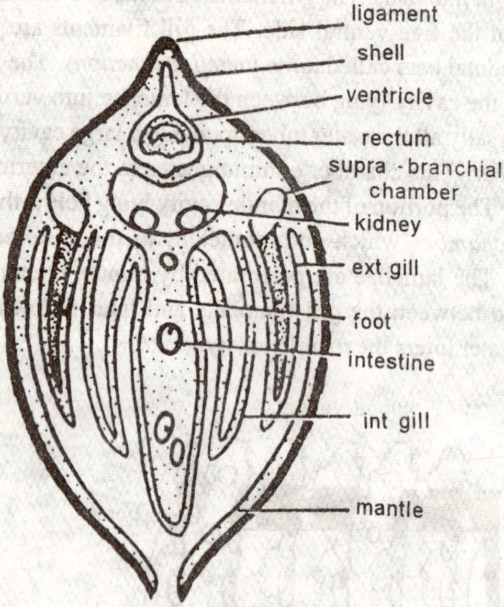


Fig. 22.9. T.S. of *Unio* through anterior region

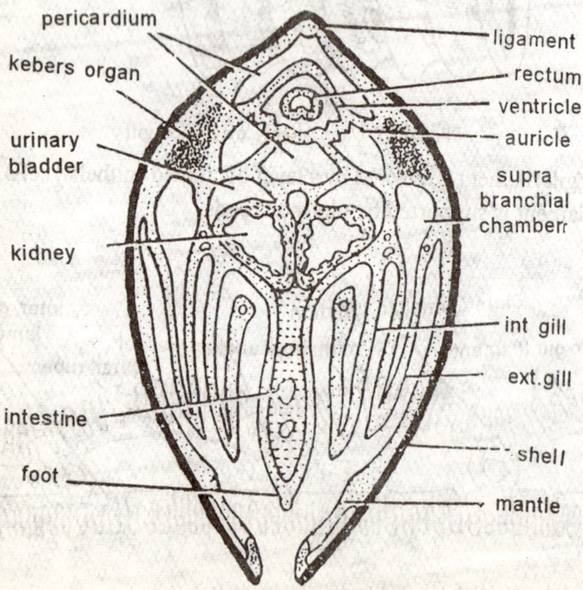


Fig. 22.10 T.S. of *Unio* through middle region.

the course of the water-current in the body. The outer lamella of the outer lamina is attached to the mantle along the whole length by its upper edge from anterior to posterior end. Inner lamella of the outer lamina and the outer lamella of the inner lamina are attached to the sides of the visceral mass a little below the origin of the mantle by dorsal edges. The inner lamella of the inner lamina is attached to the visceral mass in the anterior region, but is free in posterior region. Behind the posterior margin of the visceral mass inner lamella of inner lamina is also attached to its fellow of the opposite side. The inner laminae are attached together by their dorsal edge to the *posterior adductor muscles* on the sides of the visceral mass a little below the origin of the mantle. The outer lamella of the outer lamina is attached to the whole length of the mantle. The cavity between the upper edge of gill-lamella and the upper part of the mantle is called supra-branchial chamber, which continues posteriorly with the exhalent siphon.

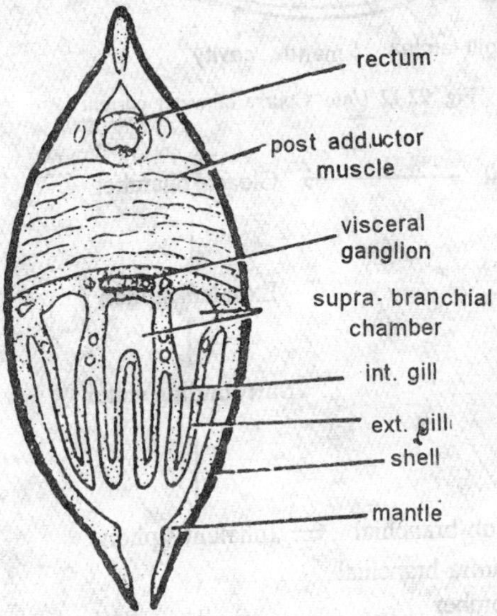


Fig. 22.11 T.S. of *Unio* through posterior region.

**Blood supply**

The gill receives venous blood from the kideny by the *afferent branchial vessels* and the oxygenated blood is conveyed to the heart by *efferent branchial vessels*.

**Course of Water current**

Beating of cilia of the gill filaments draws a continuous current of water into the infra-branchial chamber through the inhalent siphon. From here it enters the gills through ostia. Indise the gill, the water fills the watertubes and flows out into the *supra-branchial chamber* placed dorsal to each gill lamina and open posteriorly into the cloacal chamber. The water finally leaves the cloacal chamber through exhalent siphon which discharges water to the exterior.

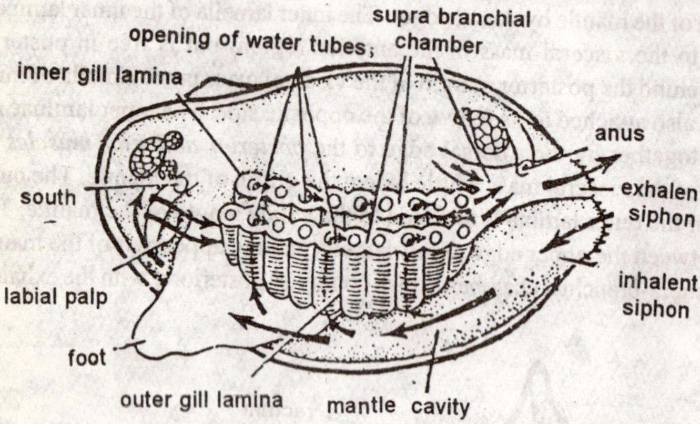


Fig. 22.12 *Unio*. Course of water current

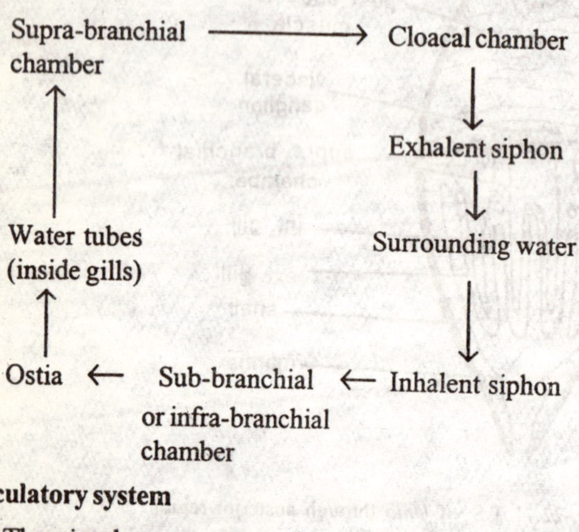


**Circulatory system**

The circulatory system of *Unio* is said to be *open type* because the capillaries are absent. The circulatory system comprises three parts.

1. Heart
2. Blood vessles
3. Blood

**1. *Heart.*** The heart is situated dorsally near the posterior adductor muscles. It is enclosed in *pericardium.* The pericardium is a thin walled triangular chamber. It is lined by epithelium and is filled with a fluid and represents a part of coelom. It is communicated with the supra-branchial chamber through the kidneys.

The heart is three chambered, having two *auricles or atria* and a *ventricle.* The auricles are thin-walled, highly distensible, triangular chambers one on either side of ventricle. They are attached by their broad bases to the pericardium and open dorsally into the ventricle. Each auriculo-ventricular *aperture* is provided with a valve that opens towards the ventricle and prevents the backward flow of blood. The ventricle has the form of a horizontal tube and surrounds a part of rectum.

The heart consists of smooth muscle-fibres covered externally by an epithelium called the *epicardium.* Endocardium is absent. The heart beats 20-100 times per minutes.

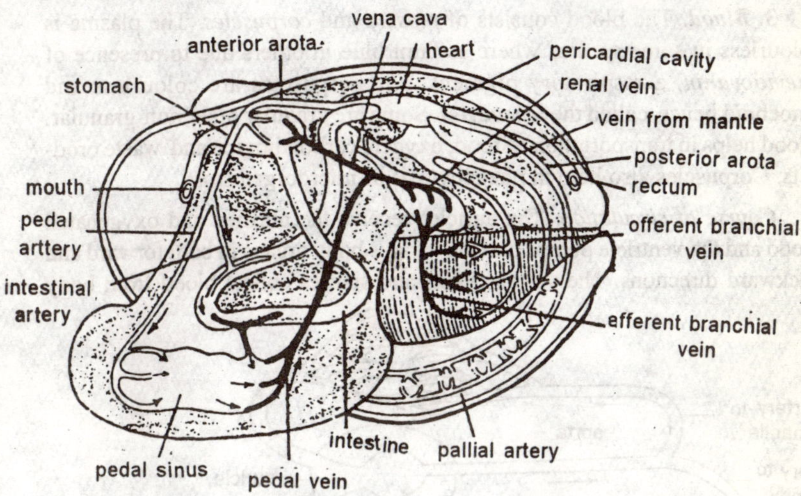


Fig. 22.13 *Unio* circulatory System

**2. *Blood vessels.*** The blood vessels comprise the (i) *arteries* (ii) sinuses (iii) veins,

(i) *Arteries.* From each end of ventricle arises, the *anterior aorta* passing above, and *posterior aorta* below the rectum. The anterior aorta supplies to the anterior parts of the body i.e. mantle, foot and visceral mass by *anterior pallial, pedal and visceral* arteries. The posterior aorta supplies to the posterior parts of the body i.e. rectum, mantle, nephidia, pericardium etc. by *rectal, posterior pallial, nephridial,* and *pericardial* arteries respectively.

(ii) *Sinuses.* The arteries ramify all the organs and finally empty into sinuses. From small sinuses blood is sent to larger sinues. The sinuses lack the epithelial lining of true blood vessels and connect directly with veins.

(iii) *Veins.* Blood from different parts of body is collected by many small veins, and in the end it reaches the large vein called *vena cava* which is placed longitudinally in the body. The vena cava collect blood from kidneys by the *renal vein,* here nitrogenous wastes are removed. Blood from kidneys is sent to the gills for aeration by *afferent branchial veins,* which gives off branches to all the gill-filaments. The oxygenated blood is sent to the auricles by *efferent branchaial veins.* Some deoxygenated blood is also sent directly to the auricles by vena cava. Blood from each mantle lobe is collected by a *pallial vein,* which also opens into the auricle of its side.

**3. *Blood.*** The blood consists of *plasma* and *corpuscles.* The plasma is colourless in some species where as faint blue in others due to presence of *haemocyanin,* a respiratory pigment. The corpuscles are colourless and amoeboid hence called the *leucocytes.* Some are granular other non-granular. Blood helps in transportation of food, oxygen, carbondioxide and waste products. Corpuscles also help in phagocytosis of microorganisms.

*Course of circulation.* The auricles receive the purifiea and oxygenated blood and the ventricle pumps it to the various body organs in both forward and backward directions. The various veins and sinuses collect blood from body

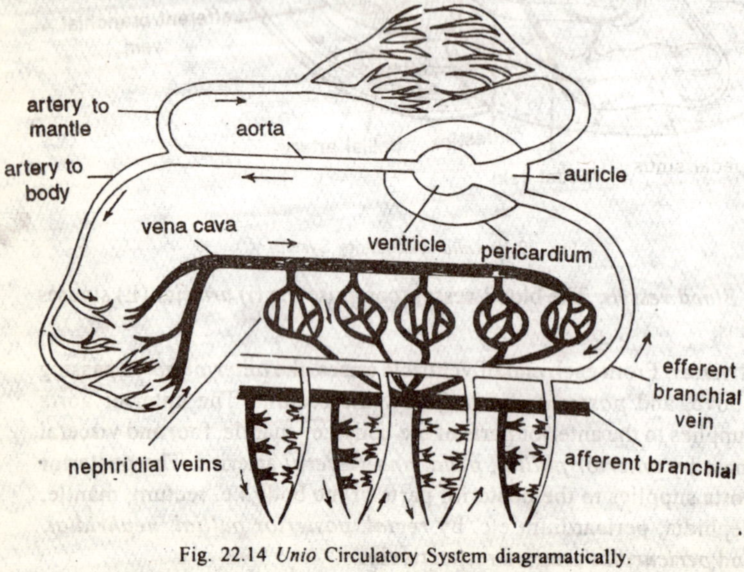


Fig. 22.14 *Unio* Circulatory System diagramatically.

organs and carry it to the kidney and gills for purification aeration. The course of blood is again carried to the auricles. The course of blood circulation can be shown as under :

Pallial Artery→Aorta→Blood→Capillaries ⟶ Smaller veins and sinuses
↓ ↓ ↓
Mantle Ventricle Vena cava
↓ ↓ ↓
Pallial veins Efferent Renal Veins
└⟶ Auricles←branchial←Gills←branchial←Kidneys
speaee vessels Afferent vessels

**Excretory system**

In *Unio* followings are the excretory organs.

1. A pair of kidneys or organs of Bojanus
2. Keber's organ

**1. Kidneys**

The kidneys or orgnas of *Bojanus* (after the name of their discoverer) are situated below the pericardial cavity on either side of vena cava. These are brownish-black in colour. Bojanus discovered these structures which are coelomic in origin. Each nephridium is a broad, U-shaped tube with its loop posterior, two ends anterior and two limbs situated parellel and one above the other. The lower limb of the tube is brown, spongy, thick walled, glandular in nature and forms the actual excretory part or the kidney proper. The upper limb

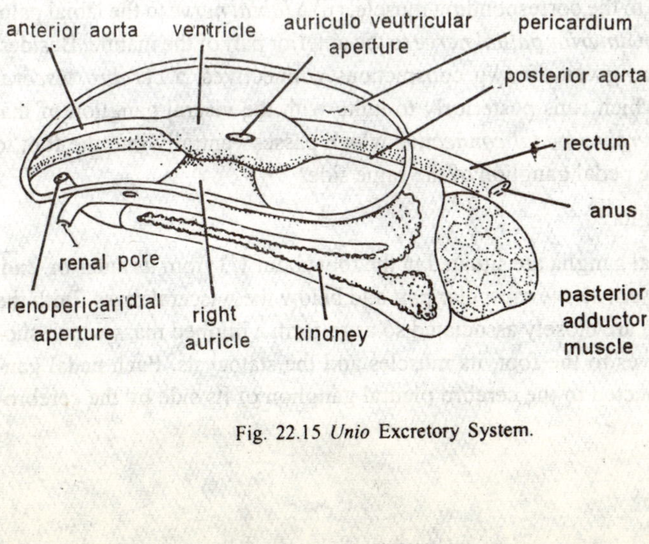


Fig. 22.15 *Unio* Excretory System.

is non-glandular lined by ciliated epithelium and is thin-walled and serves as ureter or urinary bladder. The kidney proper opens into the pericardium by a small, ciliated aperture known as reno-pericardial aperture. The ureter opens into the supra-branchial cavity by a small renal aperture between inner gill lamina and visceral mass.

The ventral glandular part of the kidney extracts nitrogenous products and guanin from the pericardial fluid and the blood. The ciliated epithelial lining of the bladder creates an outgoing current, which carries excretory fluid to the outside.

**2. Keber's Organ**

Keber's organ is a large reddish-brown gland situated in front of pericardium. It discharges waste products in the pericardium, from where it is eleminated out side via kidneys.

**Nervous system**

Due to sluggish, sedentary life, the nervous system of *Unio* is greatly reduced. It consists of three pairs of ganglia, connectives and commissures. These are :

**1. Cerebro-pleural ganglia**

At the base of the labial palps, just outside the corners of the mouth on either side, is placed a small yellowish, somewhat triangular *cerebropleural ganglion.* Each is formed by the fusion of two ganglia : a cerebral & a pleural. The ganglia of both the sides are equivalent of the brain. They are connected with each other by a thin transverse cerebral commissure, which passes over the oesophagus. Each cerebro-pleural ganglion gives out-(i) an *anterior adductor nerve* to the corresponding muscle, (ii) a *labial nerve* to the labial palp, and (iii) an *an anterior pallial nerve* to the anterior part of the mantle. Besides, each ganglion gives off two conspicuous connectives, a *cerebro visceral connective* which runs posteriorly to unite with the viseral ganglion of that side, and a *cerebro-pedal connective* which passes ventrally into the foot to unite with the pedal ganglion of the same side.

**2. Pedal ganglia**

The pedal ganglia are situated in the foot about 1/3 from its anterior end. They are present above muscular part and below the visceral mass. Both the pedal ganglia are closely associated so as to form a bilobed mass which supplies the nerves to the foot, its muscles and the statocysts. Each pedal ganglion is connected to the cerebro pleural ganglion of its side by the cerebro-pedal connective.

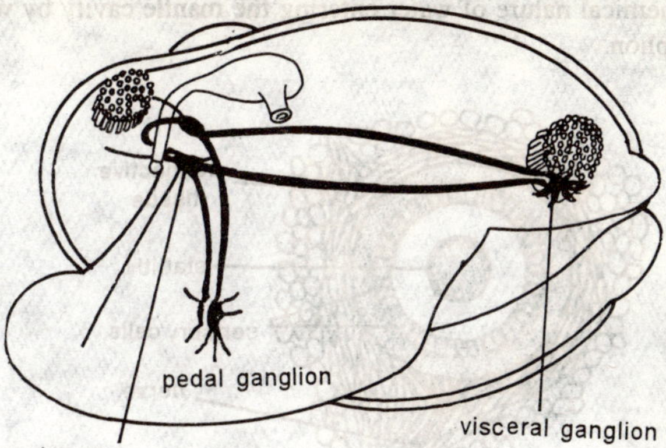


Fig. 22.16 *Unio*. Nervous System

### 3. Visceral ganglia

A pair of visceral ganglia are situated mid-ventrally on the posterior adductor muscles. The visceral ganglia are fused together to form a flattened, somewhat rectangular mass. On each side, the visceral ganglionic mass gives off (i) a dorsal pallial nerve, (ii) a posterior pallial nerve to the posterior part of the mantle, (iii) a posterior renal nerve to the kidney, (iv) branchial nerve to the gill. It is connected with the cerebropleural ganglion of its side by a thin cerebro-visceral connective which gives off several small nerves to the visceral mass.

Autonomous nervous system. Recently, *E.I.Thomas* (1974) has worked out anatomy of *Lamellidens* and has reported presence of minor ganglia in the chief muscles and has suggested that they are autonomous.

## Sense Organs

The sense organs are poolley developed due to sedentary mode of life. The main sensory structures are : (i) Statocysts (ii) Ospharidium and (ii) Sensory cells.

### (i) *Statocysts*

The foot contains a pair of minute hollow vesicles, the *statocysts,* one close to each pedal ganglion. They are innervated by the cerebropedal connectives. The statocyst is lined by sensory cells and contains a mass of lime, called *statolith,* the movement of which stimulates the sensory cells. The statocysts are thought to be organs of equilibrium.

### (ii) *Ospharidium*

A pair of dark yellow masses of sensory epithelial cells are found at the base of gills on the ventral surface of the posterior adductor muscles and above the visceral ganglionic mass. The ospharidium is a *chemoreceptor* as it tests the chemical nature of water entering the mantle cavity by way of the inhalent siphon.

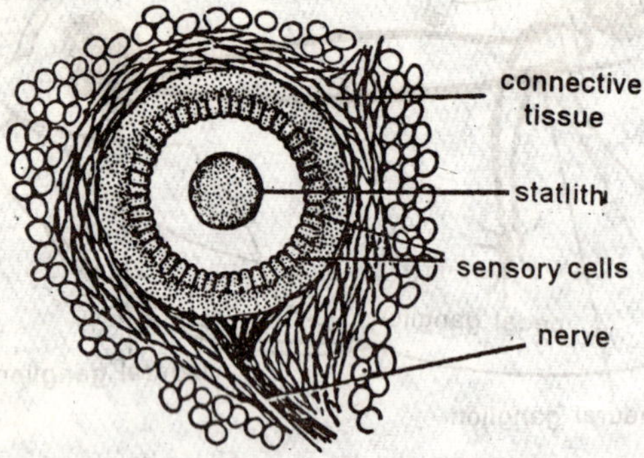


Fig. 22.17 *Unio*. Statocyst.

**(iii) *Sensory cells***

There are also other light responding devices found along the edges of mantle lobe and abundant on the inhalent siphon. They are regarded to be tactile organs and are also sensitive to light.

**Reproductive System**

The *Unio* is *dioecious* i.e. sexes are separate. The reproductive organs consist of the gonads and their ducts.

***Male reproductive organs.*** These consist of a pair of testes which are white-coloured, highly branched, glandular masses situated in the visceral mass between the loops of intestine. From each testis a *vas deferens* arises which opens into the supra-branchial chamber through male genital pore situated in front of renal aperture.

***Female reproductive organs.*** These are a pair of pink coloured glandular mass of branched tubules occupying the same position as testes in male. These become greatly enlarged during breeding season. Each ovary gives out an oviduct which opens through female genital aperture situated in front of renal opening. The accessory reproductive organs are absent.

**Fertilization**

The eggs are fertilized in the outer gill laminae. The eggs are collected in the supra-branchial chamber from here they reach the water tubes of the outer gill-laminae and are held there by the mucous. The sperms come out of the body of male body along with outgoing current of water. Some of these reach the gill laminae of the female through the incurrent siphon with water current and fertilize the eggs.

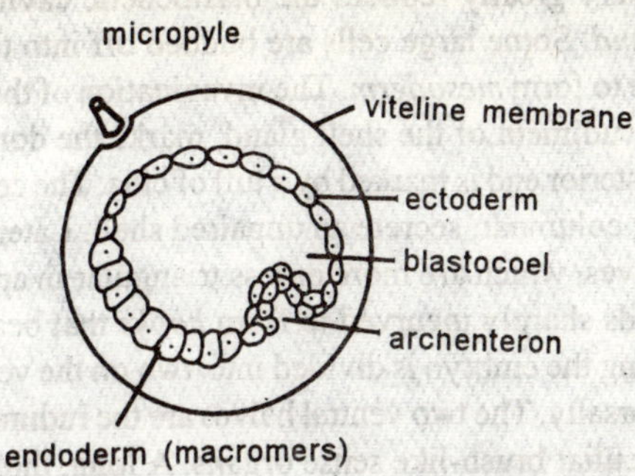


Fig. 22.18 *Unio*. A gastrula

## Development

The fertilized ovum undergoes segmentation. The cleavage is complete (holoblasitic) but unequal. A hollow, spherical embryo, the *blastula* with a fluid cavity the blastocoel is formed. The bigger blastomeres are called *macromeres* and the smaller ones, *micromeres*. In gastrulation some of the macromeres are invaginated, resulting in the formation of an exceedingly small cavity, the *archenteron*. Later on there is an extensive invagination of

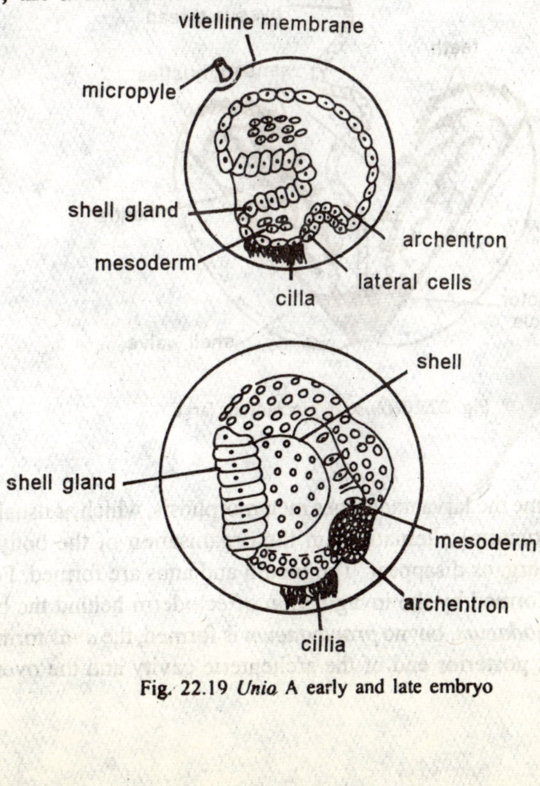


Fig. 22.19 *Unio* A early and late embryo

macromeres which very greatly reduces the blastocoelic cavity. This is the rudiment of *shell gland.* Some large cells are budded off into the blastocoel, where they aggregate to form *mesoderm.* The invagination of the macromeres which constitute the rudiment of the shell gland, marks the dorsal surface of embryo, while the posterior end is marked by a tuft of cilia. The cells of the shell gland, which are tall, columnar, secrete an unpaired shell. Later arises a shell consisting of two valves, which are more or less triangular in appearence and have their ventral ends sharply incurved to form *hooks* that bear *spines.* The mass of cells consisting the embryo is divided into two on the ventral side, but remains undivided dorsally. The two ventral halves are the rudiments of mantle lobes. These bear peculiar brush-like sense organs. A long, thread-like adhesive organ, the *provisional byssus* appears on the ventral surface of the body. The mesoderm produces a large *adductor muscle* that stretches between the two valves of the shell. While in the lamina, which serves as a *brood pouch* the larva, called *glochidium,* is nourished by the secretions of laminar wall but on escaping from the body of mother through the exhalent siphon, it starts leading an ectoparasitic life on the body of some fish, which is persisted till it is metamorphosed into adult. The glochidium is a minute larva 0.1-0.5 mm wide.

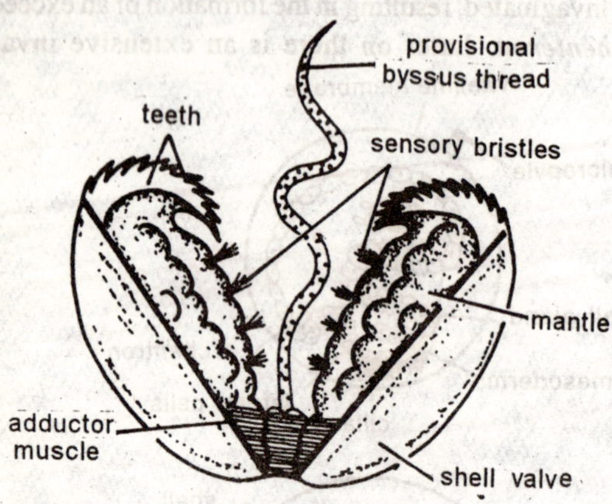


Fig. 22.20 *Unio.* A glochidium larva.

**Metamorphosis.**

After some time the larva undergoes metamorphosis, which as usual, consists of drastic structural alternations in the organisation of the body. The byssus and sense organs disappear. The mouth and anus are formed. Formation of mouth is formed by the invagination of ectoderm behind the byssus giving rise to *stomodaeum,* but no *proctodaeum* is formed, the *anus* formed by the perforation of posterior end of the archenteric cavity and the overlying ectoderm. The foot arises as a projection from the mid-ventral surface behind the mouth and no either side of it are formed the rudiments of the gills.

The glochindium continues to lead its ectoparasitic life to the end of its larval period, after the completion of metamosphosis it leaves the host and begins its independent life, growing on the adult state. The time taken by larva to metamorphose into little clams is about 10 weeks.

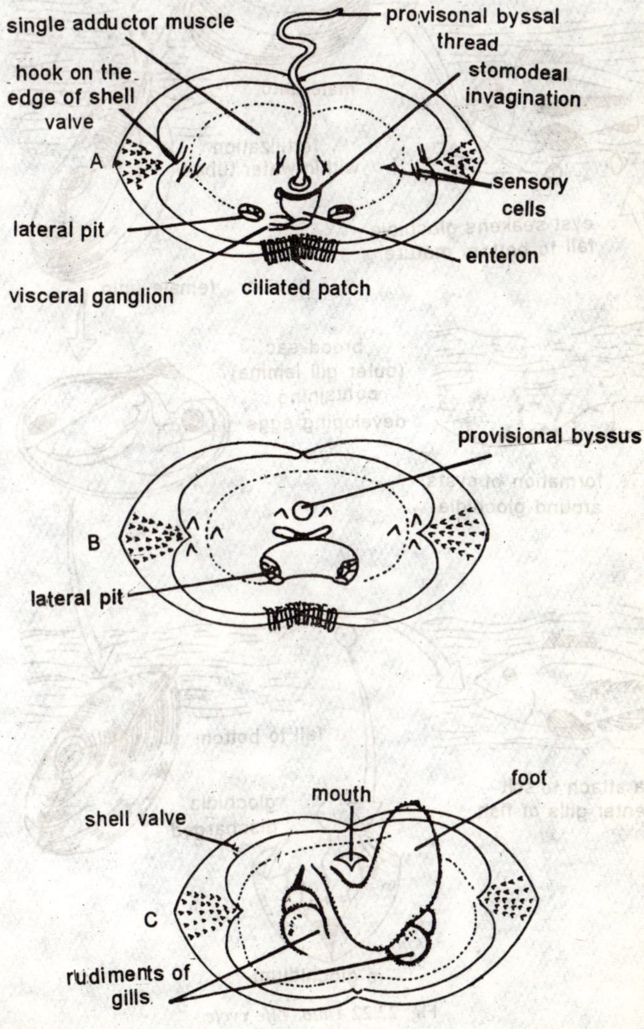


Fig. 22.21 *Unio*. The stages in metamorphosis

**Significance of glochidium**

Glochidium is an ectoparasete. It is found on fish that acts as its host. Fish helps the larve in dispersal and in capturing the food in addition to the protetion.

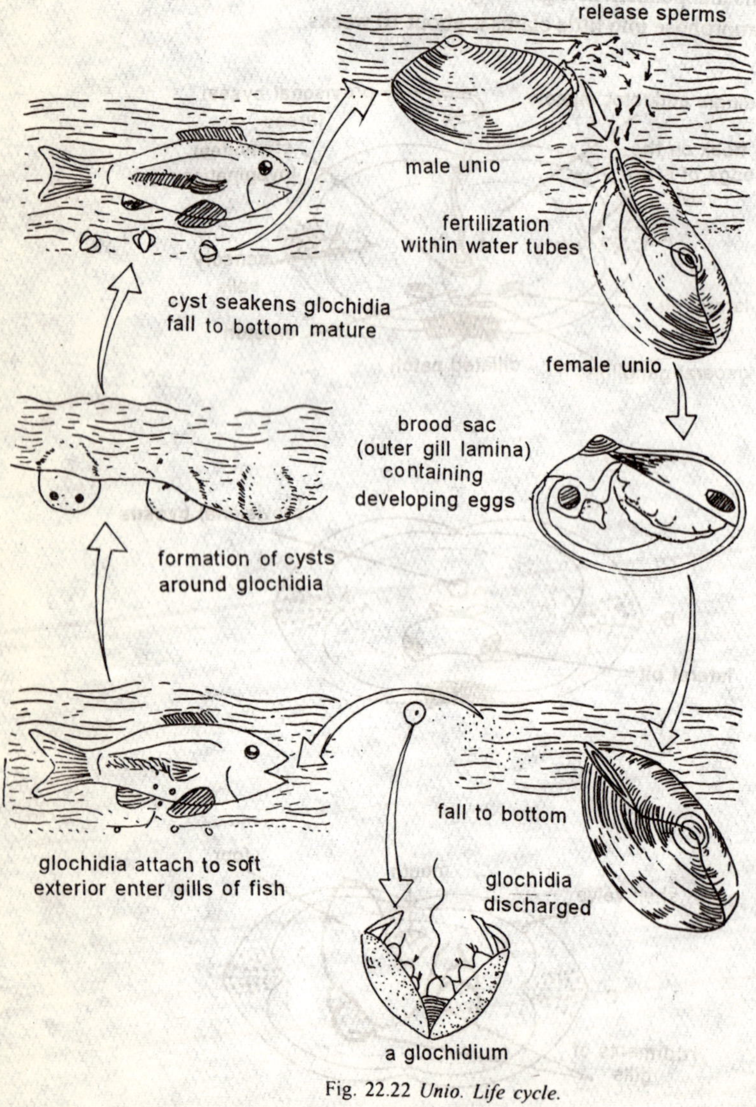


Fig. 22.22 *Unio. Life cycle.*

# CHARACTER AND CLASSIFICATION OF ECHINODERMATA

1. Small to large animals generally found in sea water.
2. Body colour often brilliant to different shades, few are almost transparent.
3. Body shape star-like, discoidal, circular or cylindrical with well differentiated oral-abroal surfaces.
4. Triploblastic, coelomate animal without well marked cephalization and segmentation.
5. Radial symmetrical in adult condition but the larva are bilaterally symmetrical.
6. The body wall is made up of three well defined layers, an outer *epidermis*, middle *dermis* and innermost *peritoneum*.
7. Endoskeleton in the form of calcareous, mesodermal plates and sometimes as spines hence the name echinodermata. (Gr., *echinus* = spiny; *derma* = skin).
8. An extensive body cavity or coelom surrounding the internal organs is present.
9. A part of embryonic coelom becomes modified into characteristic *ambulacral* or *water vascular system* or *hydrocoel* with many tube feet. This system is responsible for respiration and locomotion in animal.
10. Alimentary tract is usually coiled tube extending from the mouth located on the oral surface to the anus on the aboral or oral surface.
11. Circulatory or haemal or blood lacunar system is typically present.
12. Respiration occurs through a variety of structures, e.g. papulae in star-fishes, peristomial gills in sea urchins, genital bursae in brittle stars and cloacal respiratory trees in holothurians.
13. Excretory system is wanting.
14. Nervous system is primitive, consisting of networks concentrated into the radial ganglionated nerve cords.

15. Sense organs are poorly developed.
16. Sexes are usually separate (*dioecious*) with few exceptions. Gonads are simple with or without simple ducts.
17. Reproduction is usually sexual, few reproduce asexually or by regeneration.
18. Fertilization is external, while few echinoderms are viviparous.
19. Development is *indeterminate* including characteristic larva which undergo metamorphosis into the radially symmetrical adults.

Phylum *Echinodermata* has been divided into two subphyla;

I. Subphylum - Pelmatozoa

II. Subphylum - Eleutherozoa

**SUBPHYLUM I. PELMATOZOA**

1. It includes both extinct as well as living echinoderms usually attached by a stalk i.e. sedentary.
2. Entire structure pentamerous.
3. Mouth and anus on the oral surface.
4. Tube feet are absent and the tubular ciliated appendages when present are without suckers.
5. Theca is differentiated into aboral cup, calyx and oral roof.
6. Arms are movable with or without pinnules.
7. Main nervous system is aboral.

Subphylum Pelmatozoa is differentiated into 5 classes :

**Class - 1. Heterostelea**

1. Extinct echinoderms with laterally flattened non-porous theca attached horizontally by a stalk.
2. Mostly without arms extensions.

Example. *Trochocystites.*

**Class - 2. Blastoidea**

1. Extinct echinoderms. Test bud-like, short stemmed or stemless.
2. Theca is pentamerous, radial and made up of 13 plates arranged in three rows.
3. Ambulacra five.
4. Characteristic respiratory organs, the hydrospires, situated underneath the ambulacra.

Example. *Pentremites.*

**Class - 3. Cystodea**

1. Extinct animals of oval forms attached directly or by a stalk in upright direction.
2. Theca of rigid polygonal plates, of which some are porous.
3. Ambulacra vary from 2-5.

   Example : *Caryocrinites.*

**Class - 4. Edrioasteroidea**

1. Extinct, disc-shaped without stem or arms lying free or attached directly by the aboral surface.
2. Theca cushion-shaped with a star on top formed by five recumbent ambulacral grooves.
3. Anus and madreporite are situated on the oral surface.

   Examples : *Stromatocystites, Lepidodiscus.*

**Class - 5. Crinoidea**

1. It includes with living as well as non-living echinoderms usually attached by a stalk, but living members are stalkless and free living.
2. Entire structure strongly pentamerous.
3. Body consists of an aboral cup, the calyx and oral cover or roof, the tegmen.
4. Oral surface directed upwards.
5. Mouth usually central, anus usually excentric are present on the oral surface.
6. Arms movable, simple, mostly branched, usually five or ten in number with or without pinnules.
7. Ambulacral groves are open and extend along arms and pinnules to their tips.
8. Madreporite, spines and pedicellariae are present.
9. Sexes are separate. Development indirect through *doliolaria larva.*

   Class Crenoidea is divided into four orders :

***Order (i) Articulata***

1. Extinct as well as living forms.
2. Calyx pentamerous flexible and dicycle.

3. Mouth and ambulacral grooves are not covered by tegminal plates and remain exposed.
4. Tegmen leathery and contains small calcareous plates or minute ossicles.

   Example : *Metacrinus, Rhizocrinus, Antedon, Comatula.*

***Order (ii) Inadunata***

1. Extinct forms.
2. Calyx rigid and lower arm ossicles spring free from it.
3. Mouth covered over by tegminal plates but ambulacral grooves mostly exposed.
4. Arms three or more with or without pinnules.

   Examples : *Anartiocrimus, Hybocystites.*

***Order (iii) Flexibilia***

1. Extinct forms.
2. Calyx flexibel, dicyclic.
3. Arms branched, devoid of pinnules and typically curved inwards.
4. Mouth and ambulacra exposed.

   Example : *Forbesiocrinus.*

***Order (iv) Camerata***

1. Extinct forms.
2. Calyx rigid, mono-or dicyclic
3. Arms branched and pinnulate.
4. Mouth and ambulacra are covered.

   Example : *Xenocrinus.*

**SUBPHYLUM II. ELEUTHEROZOA**

1. Mostly living echinoderms.
2. Stem or stalk absent, usually free-living forms.
3. Body structure usually pentameous.
4. Oral surface bearing the mouth is downward or lying on one side.
5. Anus usually on the aboral surface.
6. Ambulacral grooves usually not for food gathering and the tube feet with suckers are chiefly locomotory organs.
7. Main nervous system is oral.

Subphylum Eluthrozoa is divided into four classes.

1. Class - Asteroidea
2. Class - Ophiuroidea
3. Class - Echinoidea
4. Class - Holothuroidea

**Class - 1. Asteroidea**

1. Body flattened and star-shaped with a central disc and five radiating arms which are not sharply marked off from the disc.
2. Arms hollow and contain gonads, prolongation of gut, coelom and other visceral organs.
3. Ambulacral grooves open in tube feet and extend upto the tip of arms.
4. Madreporite present and pedicellaria sessile or pedun-culate.

The modern classification distinguishes class Asteroidea into two fossil and there living orders :

***Order (i) Phanerozonia***

1. Arms short with broad bases and with two rows of large marginal plates.
2. Papulae limited to the aboral surface.
3. Pedicellariae generally sessile or absent.
4. Tube feet with or without suckers.

Examples : *Pentaceros*, *Luidia*, *Astropecten*.

***Order (ii) Spinulosa***

1. Arms generally without conspicuous marginal plates.
2. Pedicellariae rare.
3. Podia or tube feet are in two rows provided with suckers.
4. Aboral surface with low spines.
5. Ampullae single or bifurcated.

Examples : *Asterina*, *Solaster*, *Hymenaster*.

***Order (iii) Forcipulata***

1. No conspicuous marginal plates.
2. Aboral skeleton mostly reticulate with conspicuous spines.
3. Pedicellariae pedunculate and straight or crossed type.
4. Tube feet arranged in four rows provided with suckers.

5. Papulae on both surface.

   Example : *Asterias, Heliaster, Brisingaster.*

***Order (iv) Platyasteroida***

1. Extinct forms.
2. ambulacra widely open. Arms with series of long interambulacral ossicles.

   Example : *Platasterias.*

***Order (v) Hemizonida***

1. Extinct forms.
2. Ambulacral grooves are deep and ambulacral ossicles are insunk.

   Example : *Palasterina, Helianthaster.*

**Class-2. Ophiuroidea**

1. Marine usually deep-sea and free living echinoderms.
2. Body flattened with a pentamerous or rounded central disc.
3. Oral and aboral surfaces are distinct.
4. Arms usually five slender, jointed, solid, muscular, flexible rarely six or seven.
5. No ambulacral grooves.
6. Pedicellariae, skin gills and special sense organs are absent.
7. Madreporite opens on the oral surface.
8. Anus and alimentary canal are absent.
9. Tube feet arising ventrolaterally from arms in 2 row, ampullae and suckers are absent.
10. Sexes are separate, gonads pentamerous.
11. Development includes a free-swimming *plutus larva.*

    Class Ophiuroidea is divided into two orders :

***Order (i) Ophiurae***

1. Arms simple, mostly five in number, moving chiefly in transverse plane.
2. Arm ossicles articulated by pits and projections.
3. Disc and arms are usually covered with distinct shields or scales.
4. Arms spines are borne laterally and are directed outward or toward the arm tips, not downwards.

5. Single madreporite.

Examples : *Ophioderma, Ophioscolex, Ophiothrix, Ophiocoma.*

***Order (ii) Euryalae***

1. Arms simple or branched.
2. Skeleton of arms and disc poorly developed.
3. Arms move vertically and may be entwined around the objects.
4. It includes Brittle-stars.

Example : *Astrophyton.*

**Class-3. Echinoidea**

1. Body spherical disc-like, oval or heart shaped.
2. The body is lying enclosed in a shell formed by closely fitting calcareous plates.
3. Outer calcareous plates are distinguished into five alternating ambulacral and five inter-ambulacral areas.
4. The oral and aboral surfaces are distinct, the mouth is located on oral surface while the anus and madreporite are present on the aboral side.
5. Tube-feet come out from the pores of ambulacral plates and are locomotory organelles.
6. Pedicellariae are stalked and three jawed.
7. Sexes separate and, gonads 5 in number and inter-radial.
8. Development includes free-swimming *echinopluteus larva.*

Class Echinoidea is divided in three subclasses :

***Subclass (a) Bothriocidaroidea***

1. A single row of plates in each inter-ambulacral area.
2. Madreporite radial.
3. Lantern of Aristotle absent

Example : *Bothriocidaris (extinct).*

***Subclass (b) Regularia or Endocyclica***

1. It includes globular, circular and oval forms, which are pentamerously symmetrical with two rows of inter-ambulacral plates.
2. Mouth and anus lie at opposite pole.
3. An Aristotle's lantern is present.
4. Madreporite is ambulacral.

***Order(i) Lepidocentroida***

1. Test flexible with overlapping or separated plates.
2. Ambulacral plates continue upto mouth lips.

   Examples : *Palaeodiscus, Sperosoma.*

***Order (ii) Melonechinoida***

1. Test spherical, rigid.
2. Ambulacral plates continue upto mouth lip.
3. Interambulacral plates in 4 or more rows.

   Example : *Melonechinus.*

***Order (iii) Cidaroidea***

1. Test rigid and globular.
2. Two rows of long narrow ambulacral plates and two rows of inter-ambulacral plates present.
3. Ambulacral and inter-ambulacral plates continue up to mouth lip.
4. Gills and sphaeridia are absent.
5. Five bushy *Stewart's organs* are present appended to the lantern.

   Examples : *Cidaris, Notocidaris.*

***Order (iv) Centrechinoida***

1. Test rigid and rarely oval.
2. Peristome with 10 buccal plates.
3. Anus surrounded by a double circle of apical plates.
4. External gills and sphaeridia present.

   Examples : *Echinus, Solenia, Diadema, Arbacia.*

***Subclass- (c)- irregularia or Exocyclica***

1. Bilaterally symmetrical. Test mostly flattened, oval or circular.
2. Mouth centrally placed on the oral surface.
3. Anus is displaced posteriorly generally marginal at oral or aboral surface and lies outside the apical system of plates.
4. Podia or tube feet not locomotory.

***Order (i) Helectypoida***

1. Test regular with simple ambulacra and centrally located peristome and apical system.
2. Ambulacra narrow on oral surface.

3. Lantern teeth with lateral flanges.

   Example : *Helectypus (Extinct.)*

***Order (ii) Cassiduloida***

1. Aboral ambulacral area petaloid, forming a 5-armed figure.
2. Lantern absent.

   Example : *Cassidulus (Extinct.)*

***Order (iii) Clypeastroida***

1. Test flattened, oval or rounded in shape covered with small spines.
2. Mouth and apical system are usually central and oral in position.
3. Aboral ambulacral areas petaloid.
4. Aristotle's lantern present.
5. Gills present.

   Example : *Clypeaster, Laganum.*

***Order (iv) Spatangoida***

1. Test oval or heart-shaped with excentric mouth and anus.
2. Four aboral ambulacral areas petaloid, fifth not petaloid.
3. Lantern absent.
4. Gills absent.

   Examples : *Spatangus, Lovenia, Hemiaster.*

**Class-4. Holothuroidea**

1. Body bilaterally symmetrical, usually elongated in the oral-aboral axis having mouth at near one and anus at or near the other end.
2. Endoskeleton reduced to microscopic spicules or plates embedded in the body wall.
3. Mouth surrounded by a set of tentacles attached to water vascular system.
4. Podia or tube feet are usually present and locomotory.
5. Alimentary canal is long and coiled and cloaca usually with respiratory trees.
6. Sexes are usually separate and gonad single or paired tufts of tubules.
7. Development may be direct or indirect with a larva called *aūricularia*.

***Order (i) Aspidochirota***

1. Podia or tube feet numerous.

2. Mouth is surrounded by 10-30 mostly 20 peltate or branched oral tentacles.
3. Retractor muscles of pharynx absent.
4. A pair of well developed respiratory trees is present.

   Examples : *Holothuria, Stichopus, Mesothuria.*

***Order (ii) Dendrochirota***

1. Holothurians with numerous podia.
2. Oral tentacles dendroid
3. Retractor muscles and respiratory tree present.
4. Madreporite internal.

   Examples : *Cucumaria, Thyone.*

***Order (iii) Elasipoda***

1. Oral tentacles 10-20 peltate or shield shaped.
2. Tentacular ampullae, oral retractors and respiratory trees are absent.
3. Madreporite may be internal or external.
4. Body generally flat ventrally, with mouth usually ventral.
5. Tube feet few.

   Example : *Elipidia, Pelagothuria.*

***Order (iv) Molpadonia***

1. Oral tentacles unbranched and digitate or with minute branches at the tip.
2. Madreporite internal and respiratory tree present.
3. Tube feet absent and posterior end of body tapers to form caudal region

   Examples: *Molpadia, Caudina.*

***Order (v) Synaptida or Apoda***

1. Body vermiform having smooth or warty surface.
2. Podia or tube feet absent.
3. Oral tentacles are 10-20. Simple, digitate or pinnate.
4. Pharyngeal retractors are present in some forms.
5. Respiratory trees are absent.
6. Water vascular system greatly reduced.

   Examples : *Synapta, Chiridota.*

## METACRINUS

| | | |
|---|---|---|
| Phylum | - | Echinodermata |
| Subphylum | - | Pelmatozoa |
| Class | - | Crinoidea |
| Order | - | Articulata |
| Genus | - | *Metacrinus* |

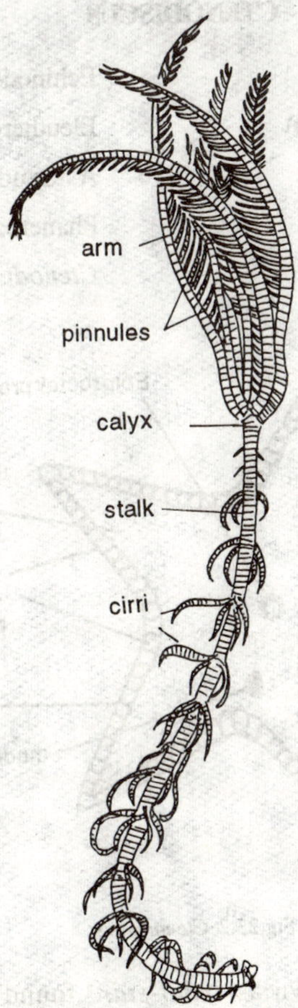


Fig.23.1 *Metacrinus*

1. It is commonly known as *'sea-lily'*.
2. It is a marine animal found at moderate depths. It is found in Japan, Malaya and Australian region.

3. The calyx is small and appears to be monocyclic.
4. It has four branchials between the radials and first branching.
5. The five arms are long, each dichotomously divided three times and with small uniserial pinnules.
6. Stalk is long, distinctly pentamerous and jointed.
7. Around the joints are present whorls of numerous slender, many jointed cirri, which are used for temporary attachment.

## CTENODISCUS

| | | |
|---|---|---|
| Phylum | - | Echinodermata |
| Subphylum | - | Eleutherozoa |
| Class | - | Asteroidea |
| Order | - | Phanerozonia |
| Genus | - | *Ctenodiscus* |

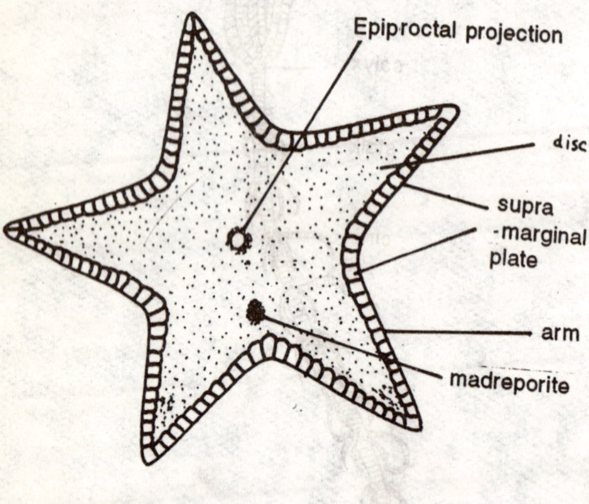


Fig.23.2 *Ctenodiscus*

1. It is commonly known as *'mud-star'* found in a shallow water circumpolar type ranging from cold northern waters to California, Japan and Cape Cod.
2. Body form varies from pentagonal to stellate with conspicuous marginal plates without spines.

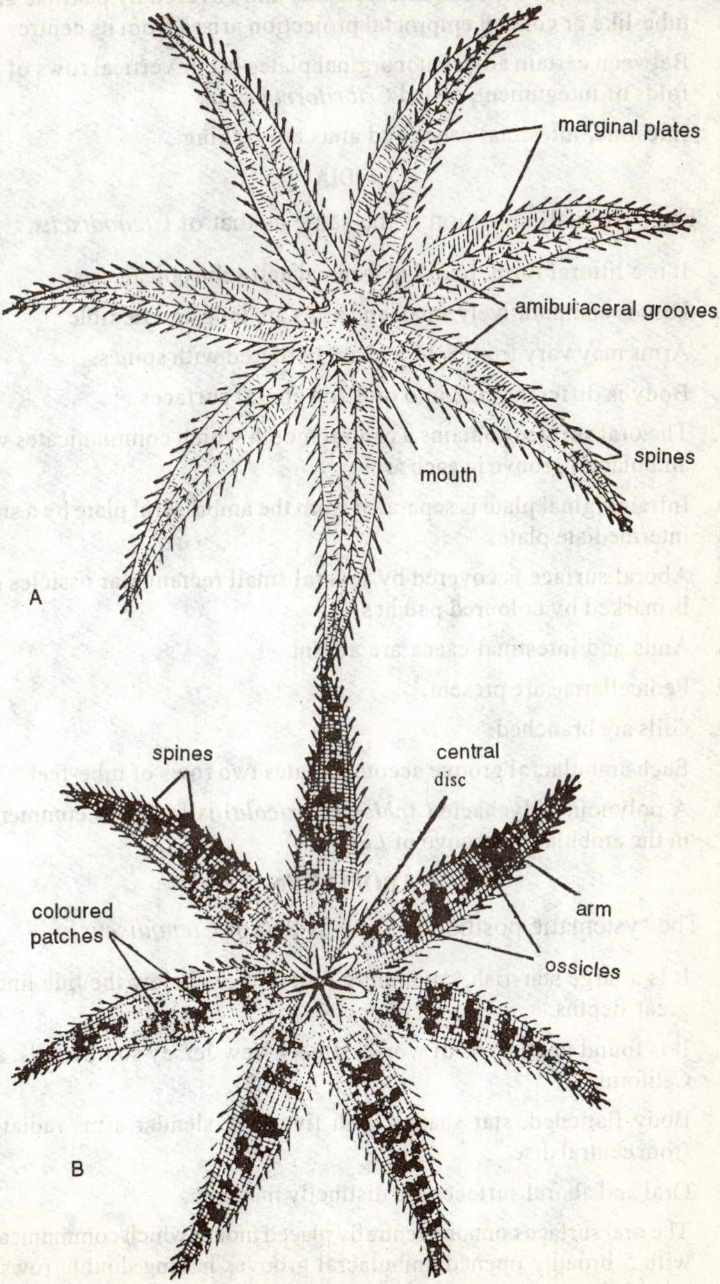


Fig.23.3 *Luidia* A= Oral view, B= Aboral view

3. The aboral wall is thin, membranous and covered by paxillae and a tube-like or conical epiproctal projection arises from its centre.
4. Between certain adjacent marginal plates occur vertical rows of thin folds of integument, called *Cribriform organ.*
5. Intestine, intestinal caeca and anus are lacking.

### LUIDIA

The systematic position is the same as that of *Ctenodiscus.*

1. It is a littoral form, found in tropical and subtropical seas.
2. Disc is comparatively small and arms are long and flexible.
3. Arms may vary from 5-11 and are bordered with spines.
4. Body is differentiated into oral and aboral surfaces.
5. The oral surface contains a central mouth which communicates with ambulacral groove in each arm.
6. Inframarginal plate is separated from the ambulacral plate by a small intermediate plate.
7. Aboral surface is covered by several small rectangular ossicles and is marked by coloured patches.
8. Anus and intestinal caeca are absent.
9. Pedicellariae are present.
10. Gills are branched.
11. Each ambulacral groove accommodates two rows of tube-feet.
12. A polynoid polychaete (*Achloe astericola)* is found as commensal in the ambulacral groove of *Luidia*.

### ASTROPECTEN

The systematic position is same as that of *Ctenodiscus.*

1. It is a large star-fish, commonly found in sea below the tide line to great depths.
2. It is found in all parts of world namely New Jersey southwards, and California.
3. Body flattened, star shaped with five long slender arms radiating from central disc.
4. Oral and aboral surfaces are distinctly marked.
5. The oral surface contains centrally placed mouth which communicates with 5 broadly opened ambulacral grooves having double rows of tube feet.

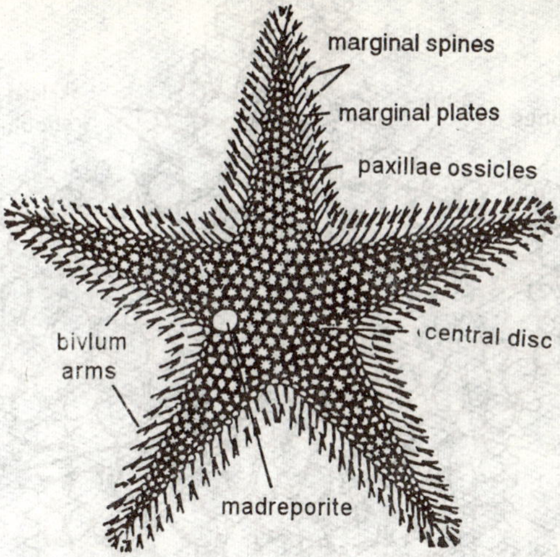


Fig.23.4 *Astropecian*

6. Tube feet without suckers.
7. The aboral surface has madreporite in one of its interradii.
8. Anus and pedicellariae absent.
9. The stone canal is extremely complicated due to internal development of ridge.
10. The optic cushion contains retinal cells.
11. Sexes are separate. The development includes branchiolaria larva.

**OREASTER (PENTACEROS)**

The systematic position is the same as that of *Ctenodiscus*.

1. It is most common starfish found in sea from shallow water to 1000 fathoms.
2. It is found in U.S.A., U.K., India and South America.
3. It is commonly called '*sea pentagon*'.
4. Body is enclosed in hard leathery integument which contains ossicles.
5. Central disc and arms are fused together.
6. Arms are five in number and arranged around a central disc in the form of star.
7. Body surface can be differentiated into oral and aboral surface.

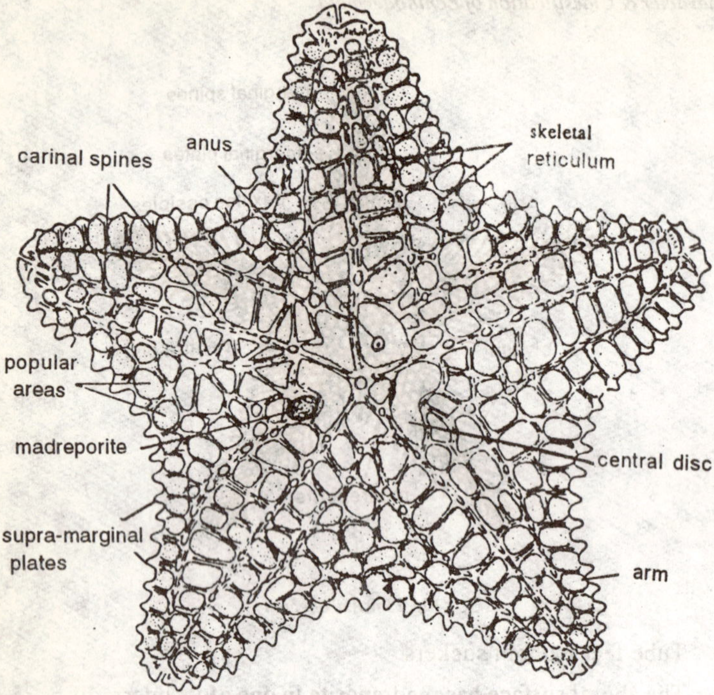


Fig. 23.5 *Oreaster* (Pentaceros)

8. Aboral surface is convex and brown in colour.
9. Oral surface is concave and dark brown in colour.
10. Mouth is present on the oral surface and is five angled groove up to the tip of an arm.
11. Madreporite plate is present in one of the inter-radii. It forms the first part of water vascular system.
12. The ambulacral grooves contain double row of locomotory organs or tube-feet.
13. Sexes are separate. Development is indirect through a bipinnaria larva.

**ANTHENEA**

The systematic position is the same as that of *Ctenodiscus.*

1. It is found in all the seas from shallow to deep waters.
2. It is most common in Australian sea.

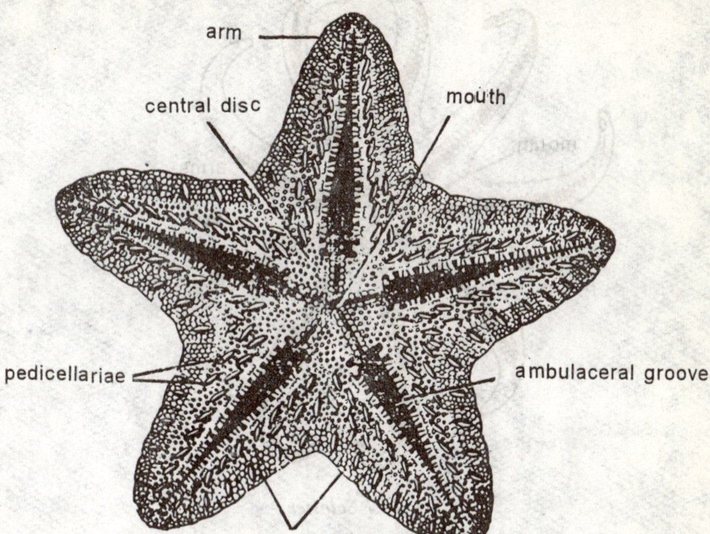


Fig. 23.6 *Anthenea*. Oral View

3. Body consists of relatively large central disc and five relatively short arms.
4. Body is differentiated into oral and aboral surfaces.
5. Aboral surface is convex, having interradial depression.
6. The oral surface is broad and flat containing a central mouth and five ambulacral grooves.
7. The flat surface between the ambulacral groove contains plate-like ossicles beset with a number of minute rounded tubercles and granules which some time assume the shape of spine.
8. Pedicellariae present on both the surfaces.
9. It possesses vertical calcareous and inter-radial partitions.

**SOLASTER**

| | | |
|---|---|---|
| Phylum | - | Echinodermata |
| Subphylum | - | Eleutherozoa |
| Class | - | Asteroidea |
| Order | - | Spinulosa |
| Genus | - | *Solaster* |

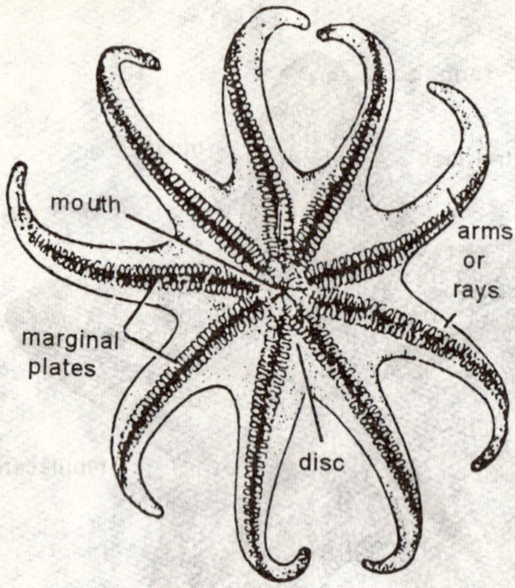


Fig. 23.7 *Solaster*

1. It is found in Northern waters.
2. Body has 9-14 arms with two rows of marginal plates.
3. The aboral skeleton is reticulate, the plates forming a fine network with samall meshes and carrying on a projecting tubercle, a crowded bundle of spinelets.
4. Pedicellariae are absent.
5. Carnivorous, feed on other echinoderms, particularly asteroids.
6. Anus is distinct.

## HELIASTER

| | | |
|---|---|---|
| Phylum | - | Echinodermata |
| Subphylum | - | Eleutherozoa |
| Class | - | Asteroidea |
| Order | - | Forcipulata |
| Genus | - | *Heliaster* |

1. It is found in shallow waters in Panamic region.
2. It has a broad disc with 40 or more short, tapering arms.
3. Skeleton is reticulate having both straight and crossed pedicellariae.
4. Madreporite is present.

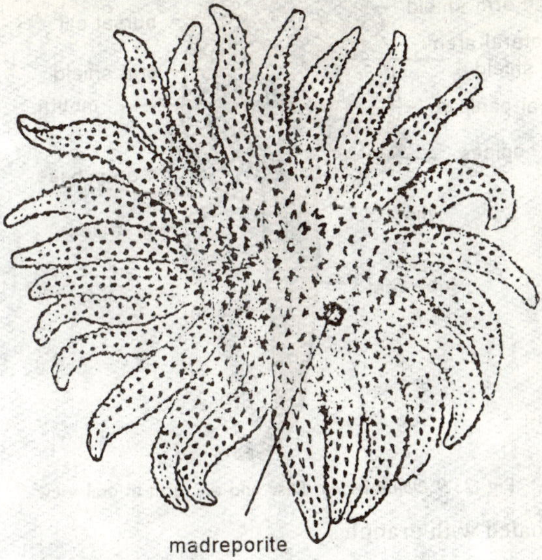


Fig. 23.8 *Heliaster*

5. Tube-feet arise in four rows.
6. Interbranchial septa are double.
7. It begins its life with five rays, but additional rays are added in inter-radii except that which contains the madreporite.

## OPHIODERMA

| | | |
|---|---|---|
| Phylum | - | Echinodermata |
| Subphylum | - | Eleutherozoa |
| Class | - | Ophiuroidea |
| Order | - | Ophiurea |
| Genus | - | *Ophioderma* |

1. *Ophioderma* is a *brittle star* of littoral zone tropical West Indian seas.
2. Found along the North Atlantic Mediterranean, littoral, tropical and West Indian seas.
3. Body with prominant pentagonal disc of brownish colour having five arms which are sharply distinct from disc and freely movable upon central disc.
4. Arms are slender, jointed,solid, cylindrical, muscular and flexible.

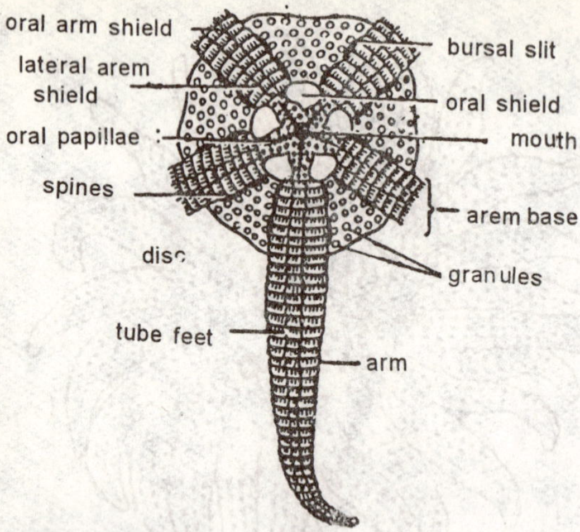


Fig. 23.9 *Ophioderma*. Disc and one arm in oral view.

5. Disc coated with granules.
6. Mouth fringed with numerous oral papillae.
7. Dental papillae are absent and teeth are arranged in a single row.
8. There are four bursal slits in each interradius, two located orally and two peripherally.
9. The spines of arms are very small so that the arms are smooth.
10. Sexes are separate. Development with free-swimming pluteus larva.

## OPHIOGLYPHA OR OPHIURA

The systematic position is the same as that *Ophioderma.*

1. *Ophioglypha* is a cosmopolitan brittle star found in almost all sea water.
2. It has cosmopolitan distribution.
3. Body with flat, pentagonal, central disc, covered with radial shields and numerous small scales.
4. Arms are five and markedly set off from central disc, which are of moderate length and covered with regular plates.
5. The arm plates of lateral side have closely lying spines.
6. Buccal podia of second pair open not into the mouth angles but on the oral surface.
7. Arms combs are present.

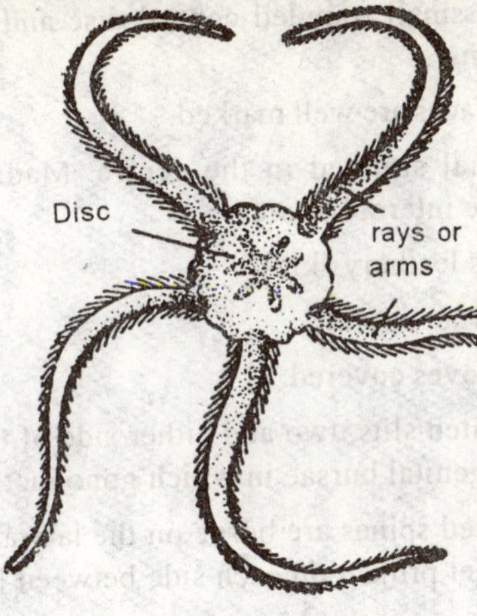


Fig. 23.10 *Ophioglypha*

8. Bursal slits are two in each inter-radius.
9. The prey is caught by coiling around the arms.
10. Male and female are separate, development indirect with ophiopluteus larva.

## OPHIOCOMA

The systematic postion of *Ophiocoma* is same as that of *Ophioderma*.

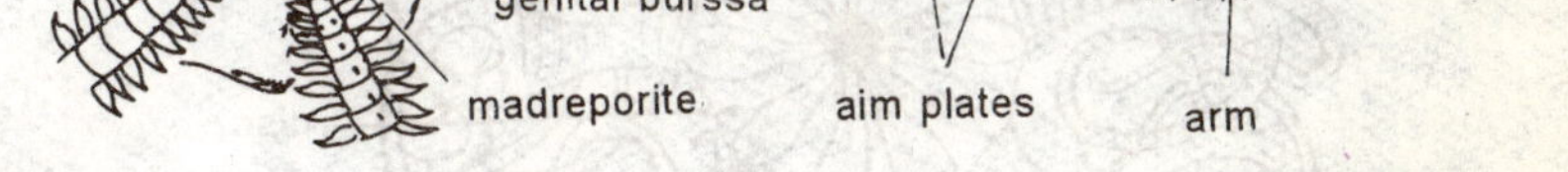


Fig. 23.11 *Ophiocoma*. Oral view.

1. It is commonly called *'brittle star'*. It is commonly found in shallow waters.
2. It is found along the Red Sea shores, West Indian region and Florida.

3. Body consists of a small rounded central disc and five long, slender tapering arms.
4. Oral and aboral surfaces are well marked.
5. Mouth is pentagonal situated in the centre. Madreporite is present in one of the interradial areas.
6. The oral surface has leathery skin.
7. Anus and pedicellariae absent.
8. The ambulacral grooves covered.
9. There are 10 elongated slits, two and either side of the base of each arm called as genital bursae in which gonoducts open.
10. Three rows of pointed spines are borne on the lateral plates. A row of short tube feet project on each side between lateral and lower plates.
11. The tube feet are devoid of suckers they are only respiratory and sensory.

## ASTROPHYTON

| | | |
|---|---|---|
| Phylum | - | Echinodermata |
| Subphylum | - | Eleutherozoa |
| Class | - | Ophiuroidea |
| Genus | - | *Astrophyton* |

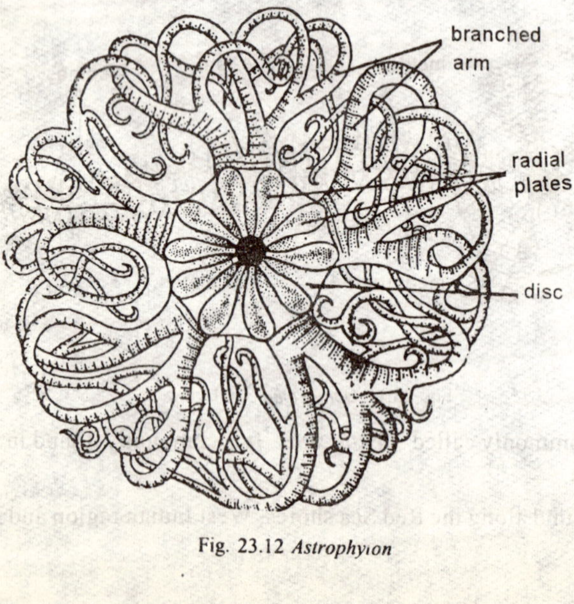


Fig. 23.12 *Astrophyton*

1. It is commonly known as *'basket star'*.
2. It is found in deep sea often rotting on sea bottom.
3. Body surface is soft and devoid of spines.
4. Pentagonal disc is very large and circular.
5. Aboral surface bears spoke-like radial shields which radiate from the centre to the periphery.
6. Fine arms are long, slender and repeatedly branched with tendrillar tip which assist in clinging the objects.
8. Development occurs passing through ohiopluteus larva.

## GORGONOCEPHALUS

The systematic position is the same as that of *Astrophyton*.

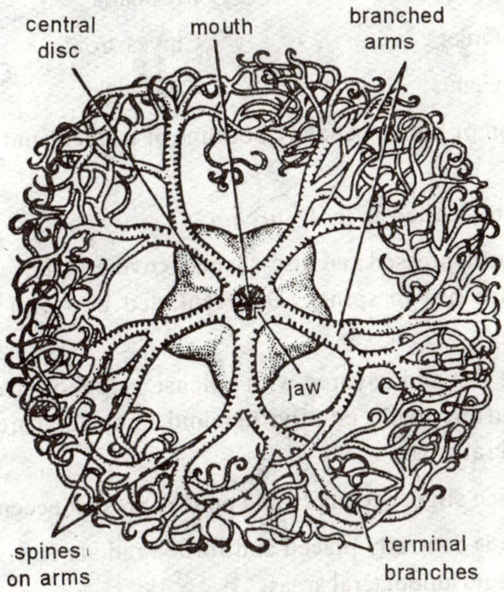


Fig. 23.13 *Gorgonocephalus*. Oral View.

1. It is another deep sea water echinoderm and commonly called *'basket star'*.
2. The disc is either pentagonal or circular and large.
3. The aboral surface of disc shown long and spoke-like radial shields, radiating from the centre of the disc to the margin.
4. Disc margin between the radial shields bear a series of plates.

5. The five arms are long, slender and flexible and repeatedly branching dichotomously.
6. One madreporite is present in each inter-radius.
7. Bursal slits are short and fusion of bursae form large spaces so that true coelom is greatly reduced.
8. Development occurs through ophiopluteus larva.

## CLYPEASTER

| | | |
|---|---|---|
| Phylum | - | Echinodermata |
| Subphylum | - | Eleutherozoa |
| Class | - | Echinoidea |
| Subclass | - | Irregularia |
| Order | - | Clpyeastroida |
| Genus | - | *Clpyeaster* |

1. It is found on or buries in sandy bottoms of tropical and subtropical seas.
2. It is commonly called as *'cake-urchin'*.
3. The corona is depressed and disc-shaped, covered all over with short stiff and dark brown spines. Differentiated into oral and aboral surfaces.
4. Oral surface flat and covered with a dense velvet of short and stiff spines and differentiates into five petaloid ambulacral areas and five interambulacral areas.
5. Aboral surface slightly convex and anus is present eccentrically.
6. Madreporite is centrally placed and form a radiate five wide, well-marked petaloid ambulacral areas.
7. Tube feet on the aboral ambulacral areas are flattened and respiratory.
8. Arms and ambulacral grooves are absent.
9. Pedicellariae are present.
10. Aristotle's lantern with teeth developed which is of simple type.
11. Genital plates are fused with the central pentagonal plate. Radial oscular plates are distinct.
12. Development includes *echinopluteus larva*.

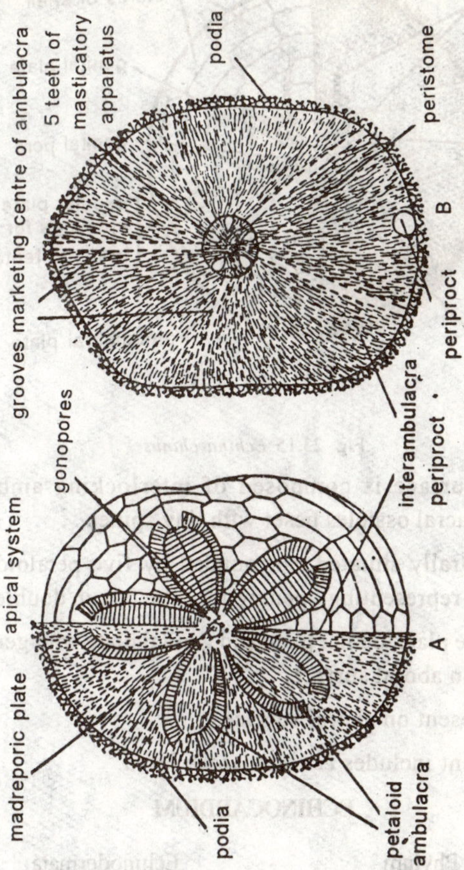


Fig. 23.14. *Clypeaster* A=Aboral view, B= Oral view

## ECHINARACHINUS

The systematic position is the same as that of *Clypeaster*.

1. It is commonly found about 800 fathoms deep, lying on sand or partly buried.
2. It is found along W.Coast of North America, Pacific Coast and U.S.A.
3. It is commonly called *'sand dollar'*.
4. Body is greatly flattened in the oral-aboral direction and covered with soft fur of delicate spines.

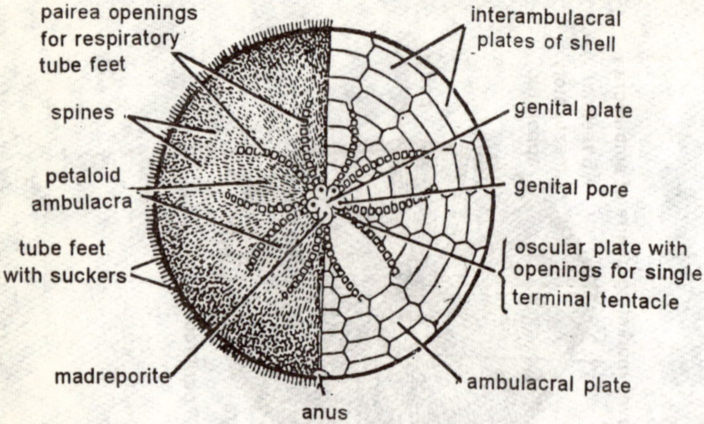


Fig. 23.15 *Echinarchinus*

5. The oral surface is composed of interlocking ambulacral and interambulacral ossicles beset with tiny spines.
6. Mouth centrally situated, surrounded by five petaloid ambulacra. Ambulacra representing opening for tube feet in double row.
7. Madreporite plate, ocular plate and genital plate with genital opening are found on aboral surface.
8. Anus is present on aboral surface.
9. Development includes *echinopluteus larva.*

## ECHINOCARDIUM

| | | |
|---|---|---|
| Phylum | - | Echinodermata |
| Subphylum | - | Eleutherozoa |
| Class | - | Echinoidea |
| Subclass | - | Irregularia |
| Order | - | Spatangoida |
| Genus | - | *Echinocardium* |

1. It burrrows in sand. It has world wide distribution.
2. It is commonly called *'heart-urchin'*, as the shape resembles a heart.
3. Body outline cardiform generally light brown in colour.
4. Four petaloids well formed which are flush with test.

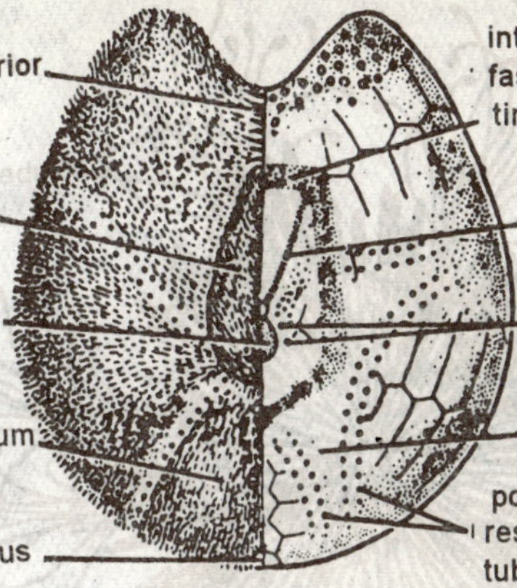


Fig. 23.16 *Echinocardium*

5. Mouth eccentric, lies near the anterior end with transversely elongated peristome.
6. Anus situated at the posterior end with periproct.
7. Ambulacra five, petaloid and aborally situated.
8. The Aristotle's lantern is absent.
9. Pedicellariae are of two types with only 2 jaws and scattered among the spines. Poisonous pedicellariae are absent.

**STRONGYLOCENTROTUS**

| | | |
|---|---|---|
| Phylum | - | Echinodermata |
| Subphylum | - | Eleutherozoa |
| Class | - | Echinoidea |
| Subclass | - | Regularia |
| Order | - | Camarodonta |
| Genus | - | *Strongylocentrotus* |

1. It is commonly called as 'sea-urchin', found in littoral sea.
2. Body is slender in out line and bears slender and short spines.
3. Mouth and anus lie at opposite poles.
4. Body is differentiated into ambulacral and inter-ambulacral areas.
5. Each ambulacral plate bears 7-10 pairs of pores.
6. Tubercles are unequal in size and arranged in numerous series and often crowded.

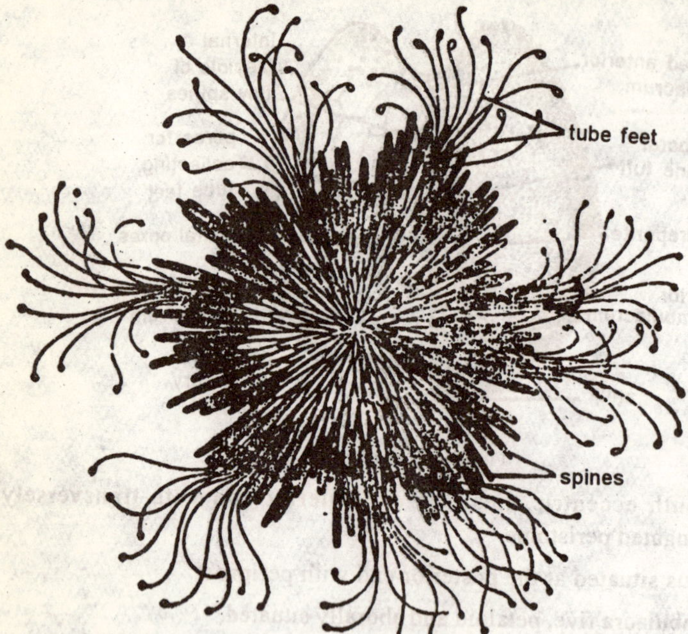


Fig. 23.17 *Strongylocentrotus*

8. Pedicellariae are gemmiform
9. Gill-clefts are shallow.

## CUCUMARIA

| | | |
|---|---|---|
| Phylum | - | Echinodermata |
| Subphylum | - | Eleutherozoa |
| Class | - | Holothuroidea |
| Order | - | Dendrochirota |
| Genus | - | *Cucumaria* |

1. *Cucumaria* is a very common sea cucumber, benthonic and sluggish is habits.
2. Generally feeds on small organisms. Cosmopolitan in distribution and specially found in Europe, U.S.A. and India.
3. Body is cucumber-like and is generally cylindrical.
4. Mouth is at broad oral end and the anus at the aboral end which is somewhat narrow.

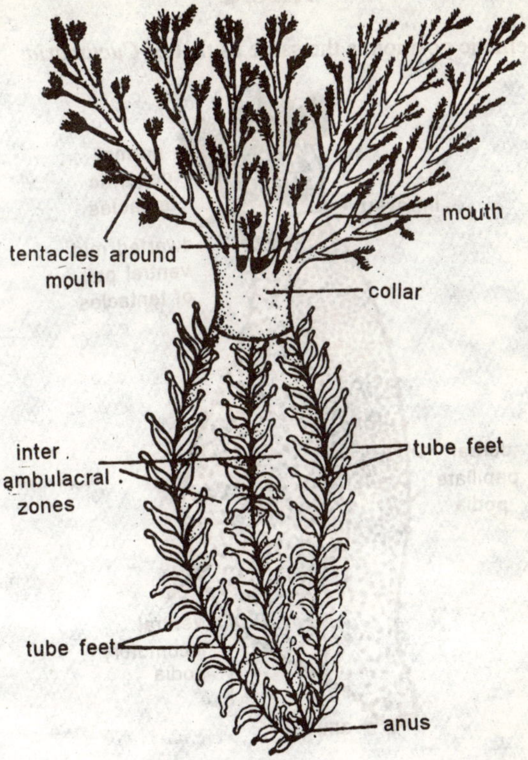


Fig 23.18 *Cucumaria*

5. Body is pentamerous, having five longitudinal bands or ambulacra, each provides with locomotory podia.
6. Mouth is surrounded by a lip and peristomial membrane having circular row often dendritic or tree-like tentacles.
7. This collar-like region at the base of tentacles is known as introvert.
8. The genital aperture is situated mid-dorsally in between the bases of two adjacent bivial ambualacral tentacles.
9. Body without spines and pedicellariae but highly muscular, leathery and slimy.
10. The exoskeleton in the form of calcareous spincules.
11. Sexes are separate.
12. Development with *auricularia larva* stage.

## THYONE

The systematic position is the same as that of *Cucumaria.*

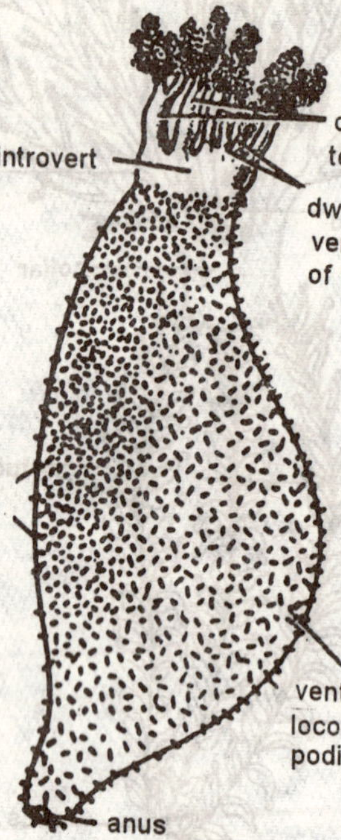


Fig. 23.19 *Thynoe*

1. It is free swimming holothurian generally found buried in muddy and sandy bottoms of sea.
2. It is found in U.S.A.
3. Podia or tube feet are distributed over the entire body surface.
4. Mouth is at the oral end and surrounded by introvert and ten dendritic tentacles which can be extended to great length.
5. Two tentacles attached to the mid-ventral ambulacral area, are much smaller than others.
6. Anus is present on aboral end.
7. *Thyone* is famous for its great power of regeneration and autonomy.
8. Sexes are separate.
9. **Development indirect with auricularia larva.**

## ACTINOPYGA

| | | |
|---|---|---|
| Phylum | - | Echinodermata |
| Subphylum | - | Eleutherozoa |
| Class | - | Holothuroidea |
| Order | - | Aspidochirota |
| Genus | - | *Actinopyga* |

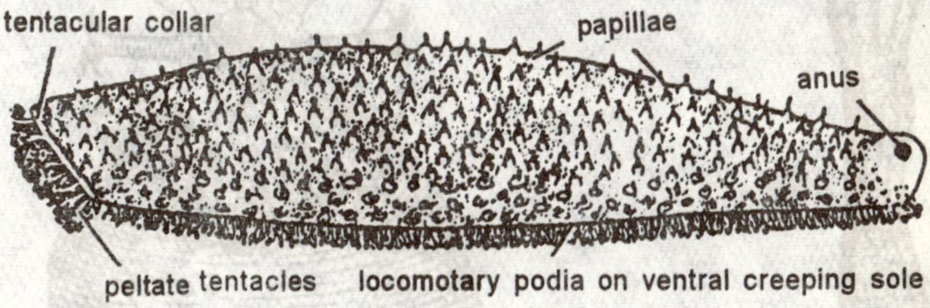


Fig. 23.20 *Actinopyga*

1. It is found in all the sea waters and commonly called *sea cucumber*.
2. Body is large cylindrical.
3. The entire ventral creeping sole is crowded with numerous locomotory tube-feet.
4. The warts covering the dorsal surface also lack any definite arrangement.
5. The anus is placed at the extreme aboral end. Five calcareous teeth are present within the rim of anus.
6. From the base of left respiratory tree arise a few or numerous pink *Cuvierian tubules* which are emitted from anus, when stimulated.
7. The tubules are toxic to fishes and other animals.

## SYNAPTA

| | | |
|---|---|---|
| Phylum | - | Echinodermata |
| Subphylum | - | Eleutherozoa |
| Class | - | Holothuroidea |
| Order | - | Apoda |
| Genus | - | *Synapta* |

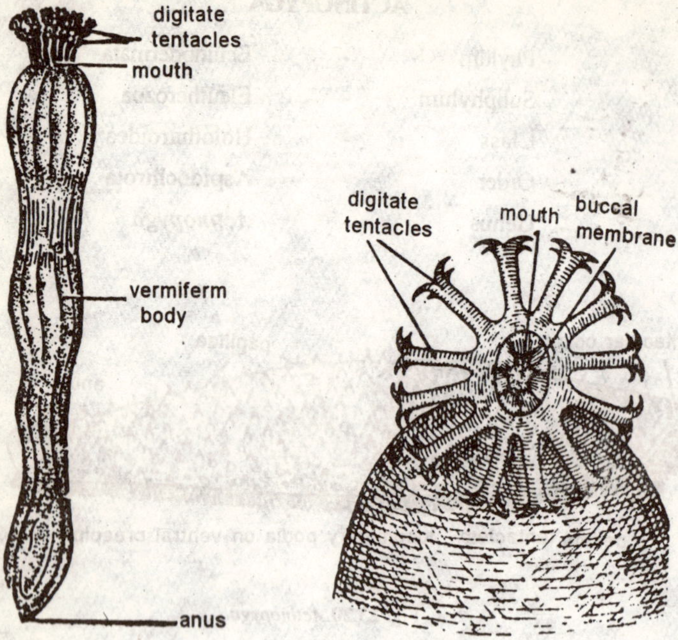


Fig. 23.21 *Synapta*. A= Entire, B= Oral

1. *Synapta* is generally found completely buried in soft, sandy soil of the sea bottom.
2. Found in Europe, Pacific coast and Bermuda.
3. Body is elongated, vermiform with rough or warty surface.
4. Tube feet are absent
5. Mouth is surrounded by 15 short, digitate tentacles.
6. Anus situated at the aboral end.
7. Respiratory tree and Cuvierian organ absent.
8. Exoskeleton of calcareous spicules are completely absent but with arches and anchor plates.
9. Hermaphrodite. Gondas consist of one pair of branching tubules.
10. Development indirect.

◆◆◆

# 24

# ASTERIAS

*Asterias* is commonly called 'starfish'. The name starfish is somewhat misleading suggesting an organism to be star-like and fish but it lacks in both the characters. However, it is appropriate to call it as *sea-star* as it is star-like in shape and occurs in sea. The geneus *Asterias* includes about 150 species.

**SYSTEMATIC POSITION**

| | | |
|---|---|---|
| Phylum | - | Echinodermata |
| Subphylum | - | Eleutherozoa |
| Class | - | Asteroidea |
| Order | - | Forcipulata |
| Genus | - | *Asterias* |

**Distribution**

There are about 1000 species of sea stars and genus *Asterias* includes about 150 species. They have different geographical distribution *Asterias rubens* occurs on the English and North European coasts, *A. vulgaris* on the North Atlantic coast of N. America, *A. forbesi* on the eastern shore from the Maine to the Gulf of Mexico, *A. amurensis* from the Behring sea, Japan, Korea Northward and *A. tenera* occurs on the sea shore from North Scotia to New Jersey.

**Habits and Habitat**

*Asterias* are free-living and marine echinoderms, crawling about slowly in shallow water to very great depths. It prefers rocky areas where locomotion and concealment are easier. However, the species that live more or less buried necessarily occur on the sandy bottom. The bottom living animals are known as *benthonic*. The most species of *Asterias* are generally solitary but many individuals gather at some place for protection from the direct rays of sun and from excessive drying. Most of them are nocturnal, remain quiet in day time and become active during night. It creeps slowly along the sea bottom. Inspite of having a hard integument, it can bend and twist in a variety of ways. They are carnivorous and feed voraciously. They exhibit remarkable power of autotomy and regeneration.

**External Morphology**

***Shape, Size and Colouration.*** The body is flattened in oral-aboral axis and is generally pentamerous in arrangement. The body consist of a central, pentagonal central disc from which radiate out five elongated, tapering, symmetrical spaced rays or arms. 6,7 or 8 rayed specimens are rare and probably result due to regeneration after injury. The body has two distinct surfaces, *oral* and *aboral*. The oral surface is flat, directed downwards having a centrally placed mouth. The aboral surface is convex, directed upwards and much darker in colour. The oral and aboral surfaces are not the ventral and dorsal surfaces but correspond to the left and right sides of the bilaterally symmetrical larva. The axis occupied by the arms are known as *radii* and the regions of the central disc between the arms are *interradii*. A well defined head is entirely absent.

The size varies from 10-20 cm in diameter. The smallest sea-star known is about 1 cm diameter and the largest 80 inches.

They are usually brightly coloured yellow, brown or orange and purple. The colouration is due to carotenoid pigments.

***Oral or Actinal surface***

It is the lower flat surface of the body which bears a mouth and usually applied to the substratum. It bears following structures :

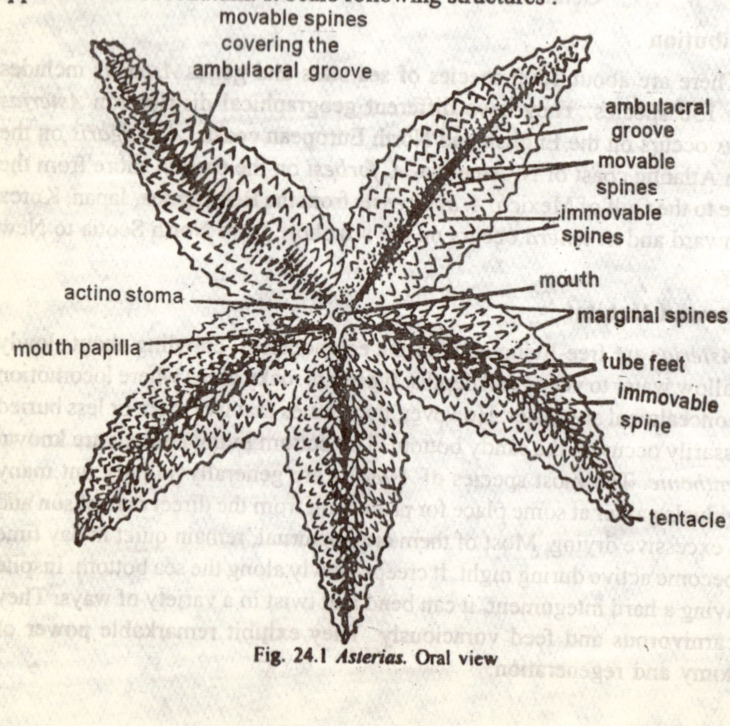


Fig. 24.1 *Asterias.* Oral view

1. ***Mouth*.** On the oral surface, in the centre of the pentagonal central disc is an aperture, the *actinostome* or *mouth*. It is a pentagonal aperture with five angles, each directed towards an arm. The mouth is surrounded by a soft and delicate membrane, the *peristomial membrane* or *peristome* and is guarded by five groups of oral spines or mouth papillae.

2. *Ambulacral grooves*. Ambulacral grooves radiate out from the five corners of the actinostome and extend one to each arm along its middle upto the distal end. Each groove is bordered by two rows of calcareous ambulacral spines. Their bases have muscles, which can bring about the movement of the spines overlapping the tube feet to protect them from injury. External to the ambulacral spine are three rows of large stout immovable spines.

3. *Tube feet*. The tube feet are soft, extensile, tubular processes arranged in 4 radial rows in each ambulacral groove. Each tube foot is provided with terminal disc or sucker. The sucker works as suction cup to afford a firm attachment on the surface to which it is applied. These are multipurpose organs as they help in locomotion, food capturing, respiration and as a sensory organs.

4. *Ambulacral spines*. The spines are present in a definite arrangement. Each ambulacral groove is bordered and guarded laterally by 2 or 3 rows of movable calcareous adambulacral spines. They can be brought together to form a lattice-work over the tube feet retracted into the ambulacral groove. At each angle of mouth, these spines become larger, stouter and grouped together to form a mouth papilla. Outside the adambulacral spines are three rows of immovable spines. There is a row of similar spines along the borders of the arms demarcating the oral from aboral surface.

5. *Sense organs*. These include five unpaired tentacles and five eye spots. The tip of each arm bears a small median non-retractile and sucker-less terminal tentacle. It acts as a tectile and olfactory organ. At the base of each tentacle, there is a small, light sensitive, bright-red spot called the *eye*. It is made up of several *ocelli*.

***Aboral or abactinal Surface***

**The aboral side remains directed upward and usually it is orange or purplish in colour. It bears following structures :**

1. *Anus*. **It is a small aperture hardly visible with the naked eye. It is situated very nearly, though not exactly, in the centre of the aboral surface, in an interradius next to that occupied by the madreporite.**

2. ***Madreporite*. The aboral surface is marked by the presence of an asym-**

metrical subcircular plate known as *madreporite*. It is present in the central disc in one of the interradii close to the base of two arms. The two rays between which the *madreporite* is placed are termed the *bivium*, remaining three are called *trivium*. Madreporite introduces bilateral symmetry, because it is only through one vertical plane, which passes through the middle of the madreporite as well as the middle of the opposite arm, that the *Asterias* can be divided into two equal, right and left, halves. The surface of madreporite bears a number of radiating, narrow straight or wavy grooves with pores in them.

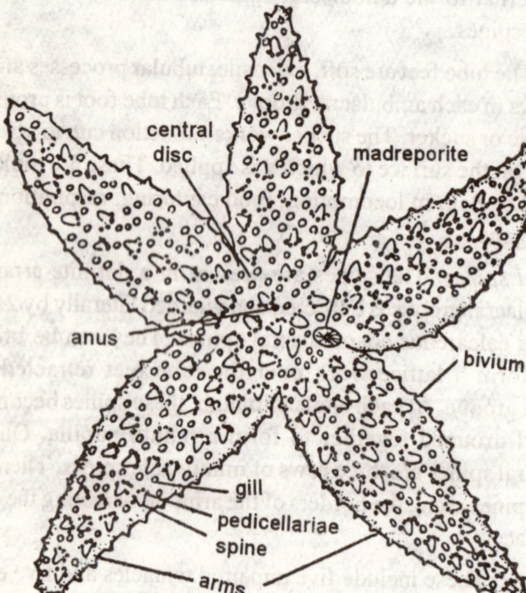


Fig. 24.2 Asterias Aboral view

3. *Spines*. The entire aboral surface is also covered with stout, blunt calcareous spines or tubercles. They arise from the *calcareous plates* or *ossicles*. The ossicles are embedded in the body wall and covered by epidermis forming the endoskeleton. The spines are arranged in irregular rows parallel with the long axis of the rays.

4. *Dermal branchiae*. The *dermal branchiae* or *gills* or *papulae* are very small, delicate, tubular or conical, finger-like or threadlike, thin walled processes protruding through minute pores in the integument between

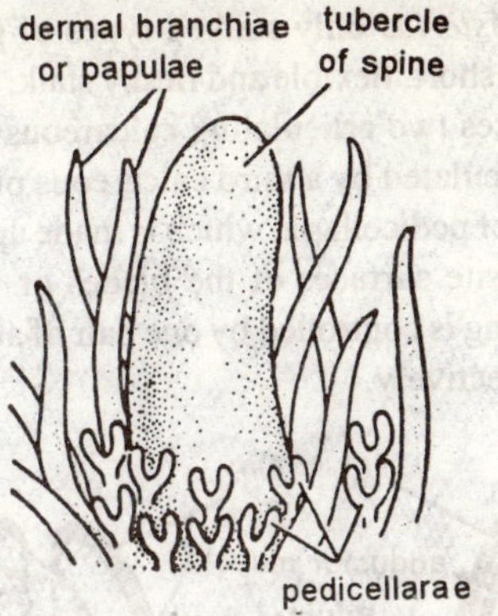


Fig. 24.3 *Asterias* A small area of body surface highly magnified.

the ossicles on the aboral surface. The papulae are hollow evaginations of the body wall and their lumen remain in continuation with the coelom. Internally they are lined with coelomic epithelium. They can be completely retracted into the body. They have respiratory as well as excretory functions.

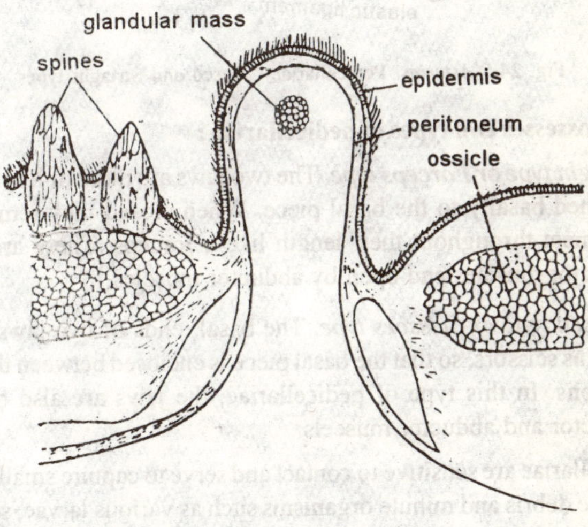


Fig. 24.4 Asterias Section through a Papula.

***Pedicellariae***. These structures are present scattered all over the body. These are found in the intervals between the spines of the aboral surface. But, on the oral surface, they are attached to the bases of spines and also in the spaces between them.

These are small whitish, jaw-like or pincer-like structures. They may be stalked or sessile. In *Asterias* only stalked types of pedicellariae are found. Each pedicellaria has a short, flexible and fleshy stalk, lacking internal calcareous support. It possesses two articulating calcareous blades or valves which are supported and articulated by a third calcareous plates is called *basilar* or *basal piece*. This type of pedicellaria, which is made up of three plates is called *forcipulate*. The opposite surfaces of the blades or valves are serrated and their opening and closing is controlled by one pair of abductor and two pairs of adductor muscles respectively.

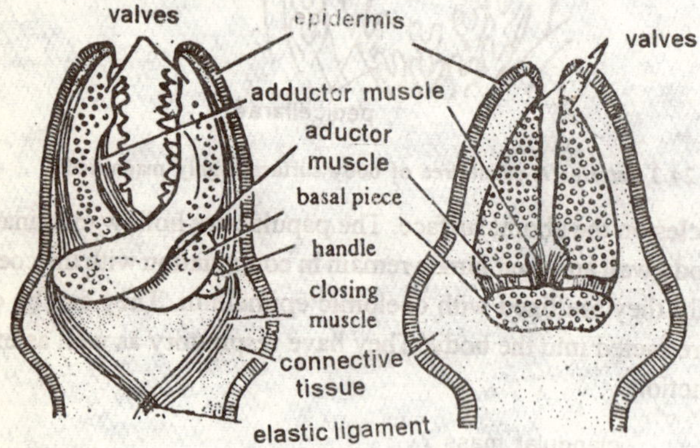


Fig. 24.5 *Asterias*. Pedicellariae Crossed and Striaght types.

**Asterias possesses two types of pedicellariae :**

(i) *Straight type or Forceps type.* The two jaws are more or less straight and attached basally to the basal piece. When closed they remain parallel and meet throughout their length like a forceps. These are closed by adductor muscles and open by abductor muscles.

(ii) *Crossed type or Scissors type.* The basal ends of two jaws cross each other as scissors, so that the basal piece is enclosed between their crossed portions. In this type of pedicellariae, the jaws are also operated by adductor and abductor muscels.

Pedicellariae are sensitive to contact and serve to capture small prey. They also remove debris and minute organisms such as various larvae, sponges and coelenterates which settle on body surface and disturb the respiration by covering the dermal branchiae and tube feet.

**Body Wall**

The body wall of *Asterias* is made up of epidermis, dermis, muscular coat and coelomic epithelium.

1. *Epidermis*. It is continued over all the external appendages. It is composed of ciliated columnar epithelium. The epithelium secretes outside itself a thin two-layered cuticle. Scattered in the epidermis are few gland cells and neurosensory cells. The gland cells are of two types (i) *goblet* or *mucous gland cells* which are filled with finely granular substances and provide the body surface with a protective mucous coating and (ii) *muriform cells*, which are filled with coarse spherules. Neurosensory cells are spindle shaped and are connected to nerve plexus or net situated below the epidermis.
2. *Dermis*. Below the epidermis is dermis which is separated from the former by a thin basement membrane. It is formed of fib ous connective tissue and can be divided into two regions (i) the outer and (ii) inner. In the

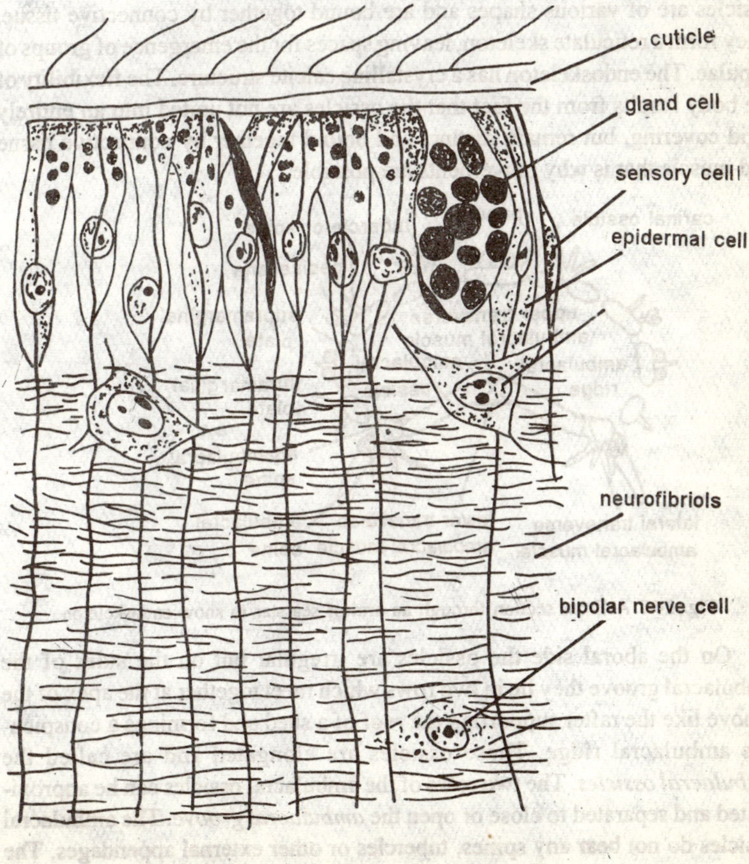


Fig. 24.6 Asterias. Section of body wall

outer region, there are self-secreted calcareous plates or ossicles which form the hard endoskeleton. It is to be specially noted that in other invertebrates the endoskeleton is ecotodermal in origin (if present) where as in *Asterias* it is mesodermal. The inner region of the dermis consists of smooth or involuntary muscle fibre. It is differentiated into an outer circular muscle layer and an inner longitudinal muscle layer. The longitudinal muscle is thicker and better developed on the aboral side than the oral side and brings about bending of the arms.

4. *Coelomic epithelium.* The inner most layer of the body wall is the parietal coelomic epithelium, consisting of cilliated epithelium.

**Endoskeleton**

The rigidity of body of *Asterias* is due to the presence of definite skeleton. As already stated, the endoskeleton consists of rod-like calcareous ossicles are of various shapes and are bound together by connective tissue. They form a reticulate skeleton, leaving spaces for the emergence of groups of papulae. The endoskeleton has a crystalline calcite structure. The flexibility of the body results from the fact that the ossicles are not united into an entirely rigid covering, but remain distinct and bound together by connective tissue and muscle that is why movements are possible.

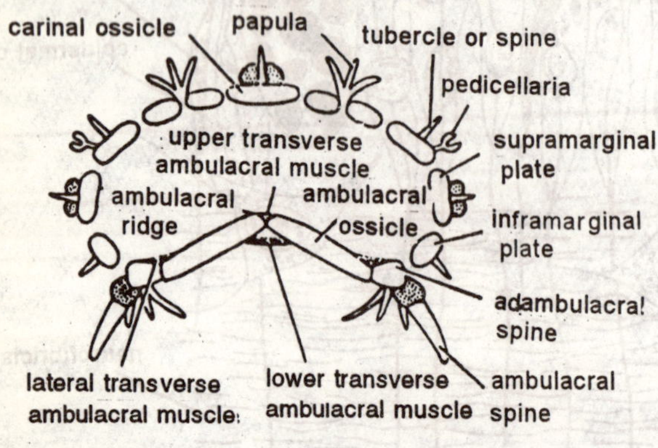


Fig. 24.7 A Cross section through an arm of sea star to show endoskeleton

On the aboral side the ossicles are irregular but on the sides of the ambulacral groove they lie in two rows which meet together at the apex of the groove like the rafter supporting the roof of a shed and formings a conspicuous ambulacral ridge. These ossicles are elongated and are called the *ambulacral ossicles*. The two rows of the ambulacral ossicles can be approximated and separated to close or open the *ambulacral groove*. The ambulacral ossicles do not bear any spines, tubercles or other external appendages. The two notches of the adjacent ossicles together form an oval aperture, the ***ambulacral pore***, for the passage of a tube-foot. The ambulacral pores are so arranged that they form two rows on each side of the ambulacral groove. At its outer end, each ambulacral groove articulates with one *ad-ambulacral ossicle* forming the edge of the groove and bearing two or three movable spines on small tubercles. Next to the adambulacral ossicle there are two rows of the ossicles called *supra* and *inframarginal ossicles*.

**Coelom**

The coelom is present in the form of a large cavity situated below the body wall and extends into the arms. It is lined by ciliated epithelium. There is

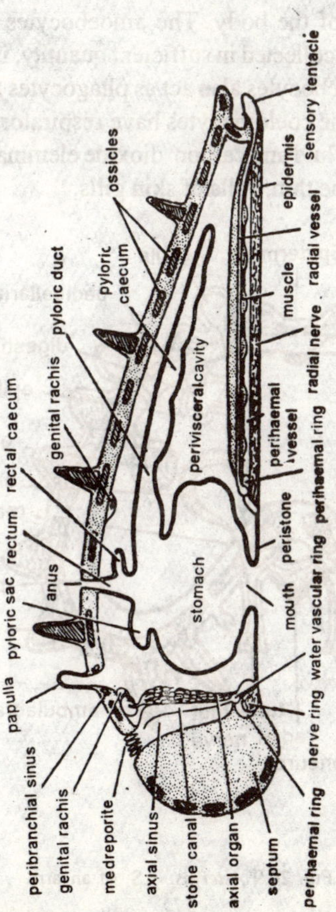


**Fig. 24.8** ***Asterias*. Diagramatic V.S. of disc and an arm.**

**a large *perivisceral coelom* extending into the central disc and the rays and** surrounding the visceral organs. Besides, there are also a number of other coelomic compartments such as the *water vascular system*, *axial sinus*, *perihaemal canals* and *sinuses*, and the *genital sinuses* etc. There are spaces in the dermis which are without definit linings and, therefore, not regarded as true cloelomic origin. The coelomic epithelium is not only applied to the inner surface of the body wall as parietal layer but also invest the visceral organs as visceral layer.

The coelom is filled with *coelomic fluid* containing *amoeboid corpuscles*. The composition of this fluid is like that sea water. The coelomic fluid, like haemolymph of arthropods, serves to bathe the tissue in the absence of true blood. The cilia of coelomic epithelium serve for transportation of materials between various parts of the body. The amoebocytes collect the excretory substances which, when collected in sufficient quantity, is expelled out through some papulae. The amoebocytes also act as phagocytes for removing harmful foreign substances. Some coelomocytes have respiratory pigment. Oxygen is taken into the coelomic fluid and carbon dioxide eleminated into the surrounding sea-water through the thin walls of skin gills.

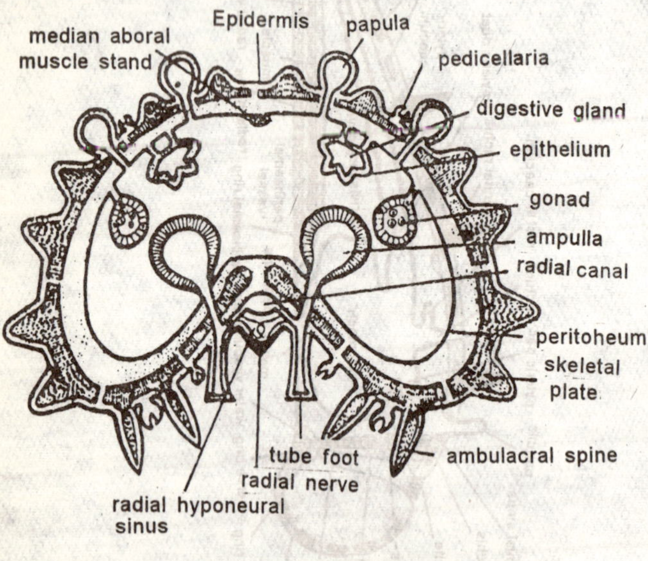


Fig. 24.9 *Asterias*. T.S. of an arm

**Digestive System**

**The digestive system consists of alimentary canal and digestive glands.**

### I. Alimentary canal.

The alimentary canal is complete extending from oral to aboral side. Therefore, it is very short and wide at places. It comprises mouth, oesophagus, stomach, intestine and anus.

1. *Mouth.* The mouth is the anterior most aperture of alimentary canal and it is situated in the centre of the peristomial membrane on the oral surface. It is provided with a sphincter muscle and radial fibres and is capable of great expansion and retraction.The mouth leads upward into the oesophagus.

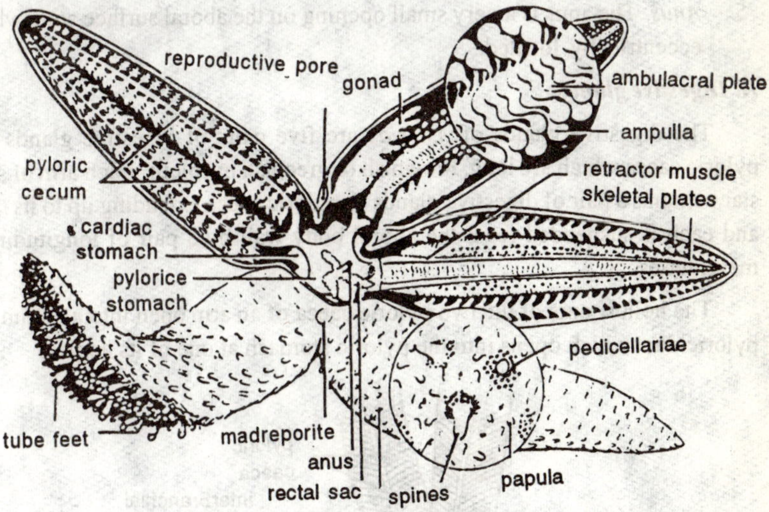


Fig. 24.10 *Asterias.* Showing anatomy.

2. *Oesophagus.* The mouth lead into oesophagus which is short and wide. It extends ventrally and open into the stomach.

3. *Stomach.* The stomach is a broad sac and fills the interior of the disc. It is typically divided by a horizontal constriction into a voluminous oral part, the *cardiac stomach* and a flattened aboral part the *pyloric stomach.*

   (a) *Cardiac stomach.* It is the larger oral part connected with the oesophagus. Its wall is thin and greatly folded. The cardiac stomach is held in position by five pairs of mesenteries or retractor muscles or the gastric filaments which are formed of connective tissue and muscles. A pair of gastric ligaments connect the cardiac stomach to the ambulacral ridge in each arm. The stomach is capable of everting out through mouth by the action of muscles extending from the body wall.

(b) *Pyloric stomach.* It is a small, flat pentagonal sac opening aborally into the intestine. The angles of the pentagon lie along the radii and each receives a duct, called the pyloric duct, from the corresponding pyloric caeca or digestive glands.

4. *Intestine.* It is a short, narrow five sided tube which is continued to the aboral surface where it opens into the anus. It gives out at interradial position two or three small branched and brownish appendages called the *intestinal* or *rectal caeca*. The intestinal caeca secrete a kind of brownish fluid which is supposed to be excretory in function.

5. *Anus.* The anus is a very small opening on the aboral surface somewhat eccentrically located.

***II. Digestive glands.***

The digestive glands of *Asterias* are five pairs of digestive glands or pyloric caeca which are long, brownish or greenish in colour. Each arm of sea star contains a pair of digestive glands or pyloric caeca extending up to its tip, and each is suspended from the aboral body wall by a pair of longitudinal mesenteries.

The hollow axis of the two pyloric caeca of an arm open into a common pyloric duct which opens into the pyloric stomach at one of its angles.

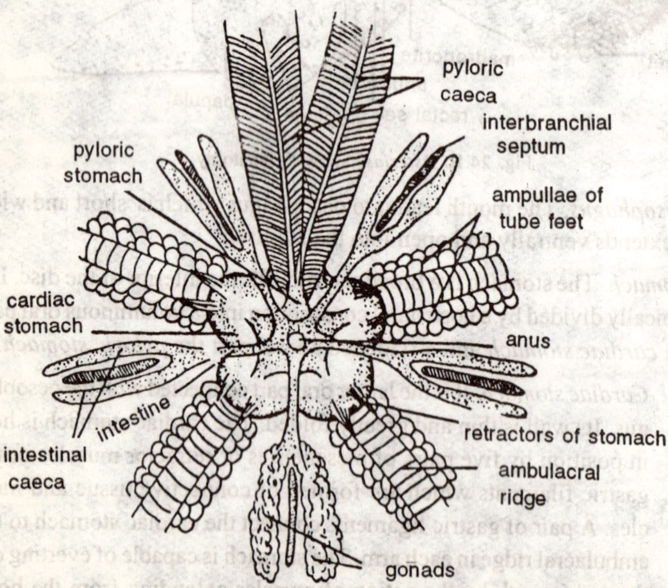


Fig. 24.11 *Asterias.* Showing digestive and reproductive system.

The pyloric caeca are lined by ciliated columnar epithelium having four types of cells :

(i) *Secretory or granular cells* which produce proteolytic and lipolytic enzymes.

(ii) *Mucous cells* which secrete mucus.

(iii) *Storage cells* which store reserve foods as lipids, glycogen and protein-polysaccharide complex.

(iv) *Current producing cells* which bear longer flagella that maintain a steady flow of enzymes towards the stomach along the aboral side and of digested food into the pyloric caeca along the oral side.

**Food**

The *Asterias* are carnivorous and feed upon all sorts of invertebrates, especially snails, bivalves, crustaceans, polychaetes, other echinoderms and even fishes. They also consume the bodies of dead animals encountered on the bottom.

***Feeding mechanism, Digestion and Absorption.*** In some sea-stars, the prey is swallowed whole and digested within the stomach, although the stomach wall must be in contact with the tissues being digested. Shell and other indigestible materials are then eleminated out of the mouth. In *Asterias* through the contraction of the body wall muscles, the coelomic fluid exerts pressure on the cardiac stomach, causing it to be everted through the mouth. The everted stomach, which is anchored by gastric ligaments, engulf the prey. The prey then may be brought into the stomach by retraction, or digestion may begin outside the body. The digested food in then passed into the body in ciliated

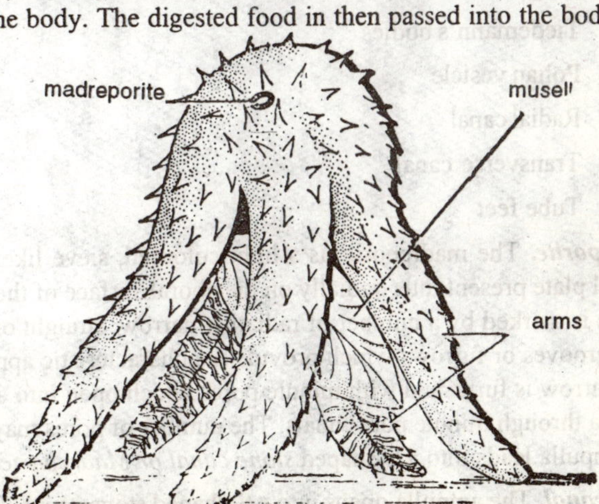


Fig. 24.12 *Asterias.* Feeding on a mussel.

gutters. When digestion is completed, the stomach muscles contract, retracting the stomach into the interior of the disc. The clams are devoured in an interesting manner. During feeding the sea star extends itself over a clam, holding the gape of the clam upward against the mouth and applying the arms against the sides of the clam valves. The sea star inserts the everted stomach through minute openings between the imperfectly sealed edges of the valves. The gape is produced quite rapidly and not by causing the clam to fatigue over a long period. The everted stomach pours out powerful proteolytic enzymes which completely dissolve the soft viscera of clam. The cilia of stomach set up strong currents that carry the digested food into the stomach and then pyloric caeca. By the contraction of muscles the everted stomach is withdrawn into the interior of the disc.

Thus the digestion is *extracellular*. Some *intracellular* digestion is believed to take place in pyloric caeca. The food is chiefly absorbed in pyloric caeca, which also store the food material. Distribution of food of various parts of the body takes place through the coelomic fluid. The undigested materials are eleminated mainly through the mouth. Little, if any, egestion takes place through the anus.

**Water Vascular System or Ambulacral System.**

The ambulacral system is derived from the coelom, from which it separates during development of the animal. It can be distinguished into following parts :

1. Madreporite
2. Stone canal
3. Ring canal
4. Tiedemann's bodies
5. Polian vesicle
6. Radial canal
7. Transverse canal
8. Tube feet

1. ***Madreporite.*** The madreporite is a light coloured, sieve like, circular skeletal plate present inter-radially on the aboral surface of the disc. Its surface is marked by a number of radiating narrow, straight or slightly wavy grooves or furrows which provide it a characteristic appearance. Each furrow is furnished with minute pores which open into a sac-like ampulla through minute pore canals. The number of pores may be 250. The ampulla leads into a S-shaped *stone canal or Madreporic canal.*
2. ***Stone canal.*** The ampulla opens into a S-shaped stone canal. The stone canal extends downwards and opens into a ring canal, around the mouth.

The walls of stone canal is supported by a series of calcareous rings. The lumen of stone canal is lined by very tall flagellated cells. In embryonic stages and in young *Asterias*, the stone canal remains a simple tube but in adult, lumen of stone canal possesses a prominent ridge with two spirally rolled lamellae which by branching become more complicated in structure. During its course, the stone canal with axial organ is ensheathed by a wide, thin-walled tubular coelomic sac, called *axial sinus*. The three together called *axial complex*.

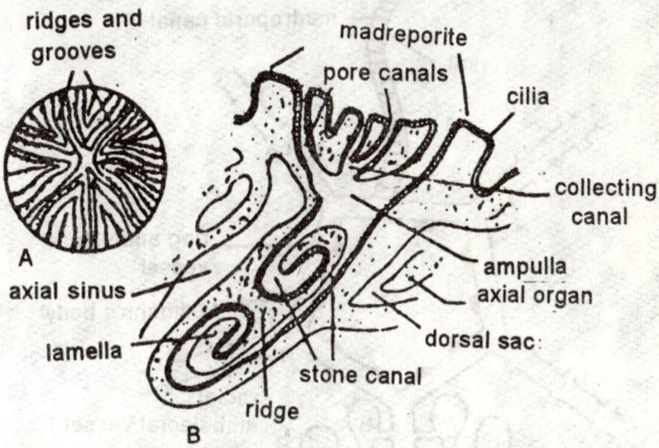


Fig. 24.13 *Asterias*. A= Madreporite seen from outside, B= V.S. through madrepodite

3. ***Ring canal.*** It is a wide, somewhat pentagonal canal forming a ring around the oesophagus. The angles of pentagon are situated in the radial position.

4. ***Tiedemann's bodies or Racemose glands.*** These are small rounded yellow, glandular sacs which open into a ring canal on its inner side, between each radius and interradius except where stone canal opens into the ring canal, there is one *Tiedemann's bodies*. Each consists of a peritoneum enclosing a stroma of connective tissue and muscle fibres having many radiating tubules. Recently, studies with the electron microscope pointed out that each cell of radiating tubules has flagellum in the midest of microvilli-like processes. The function of Tiedemann's bodies is not yet known. However, it is believed that they serve for filteration or as enzyme forming structure or lymphatic glands which produce phagocytic amoebocytes to be set free in the water vascular system.

5. ***Polian vesicle.*** In most sea-stars, but not in *Asterias*, the ring canal gives off in each interradius one, two or even four large, thin-walled, pear-

shaped sacs known as *Polian vesicles*. They are contractile structures and perform several functions. According to some workers they serve as central organs for regulating pressure in the ambulacral system. According to others they are some sort of lymphatic glands which give rise to amoebocytes of the ambulacral system.

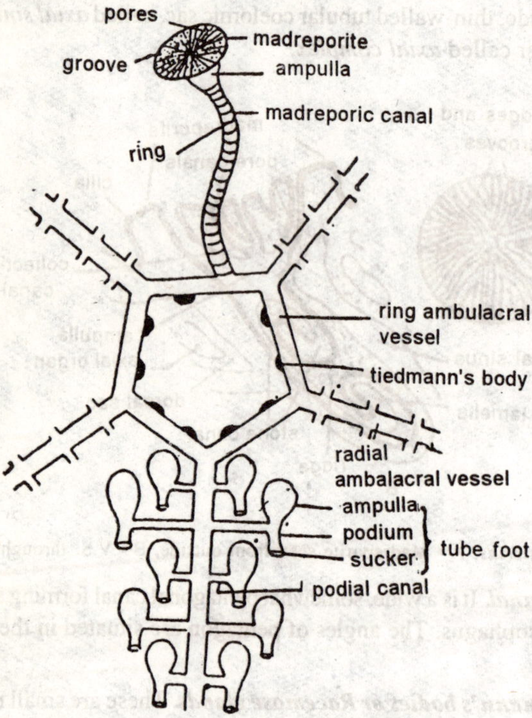


Fig. 24.14 *Asterias*. Ambulacral System.

6. ***Radial canals.*** From the outer surface of the ring canal are given out 5 radial canals, one to each arm upto its extremity through the ambulacral groove of the arm. Each runs passing above the radial nerve and terminating in the lumen of the terminal tentacle.

7. ***Lateral canal.*** In each arm, the radial canal gives out two series of short, narrow, transverse branches called *lateral or podial canals*. Each lateral canal is attached to the base of a tube foot and is provided with a valve to prevent backward flow of fluid in the radial canal.

8. ***Tube feet.*** There are two double rows of tube feet in each arm, one double row in ıelation to each series of alternately placed long and short lateral

canals. Each tube foot has the form of a closed thin-walled tube. It extends through a gap, called the *ambulacral pore*, which lies between two adjacent ambulacral ossicles. Each tube foot can be distinguished into three regions.

(i) *Ampulla*. It is a rounded sac-like structure situated above the ambulacral ossicles and projecting into the coelom.

(ii) *Podium*. It is a median tubular part which extends through the ambulacral groove.

(iii) *Sucker*. It is a cup-like structure situated at the lower end of podium.

The tube feet have strong longitudinal muscles.

The various parts of ambulacral system have a muscular wall lined internally by ciliated epithelium covered, according to their position, by epidermis or coelomic epithelium.

The whole system is filled with a watery fluid which is identical to the sea water. Water enters through the pores of madreporite and circulates through the system by the action of cilia.

***Functions.*** The ambulacral system chiefly helps in locomotion. It also helps in food capture and attachment of substratum.

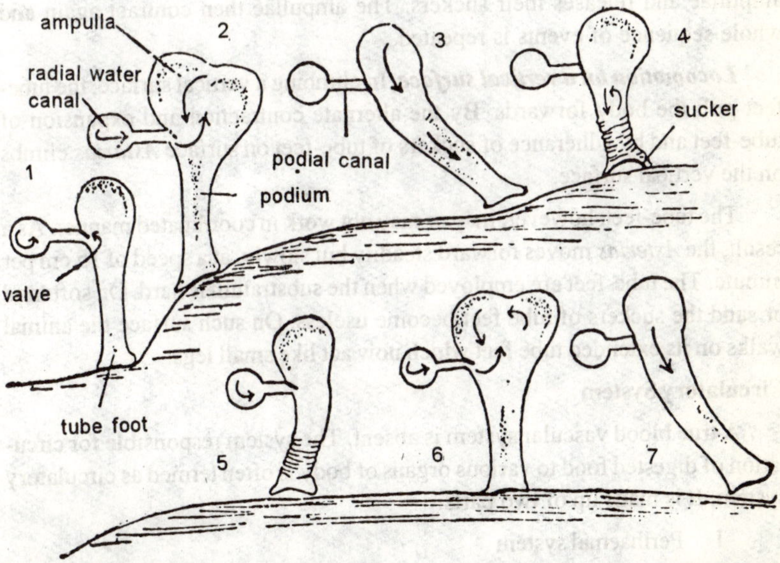


Fig. 24.15. Locomotion in a Star-fish. 1. Water enters the radial canal, 2. Water fills the ampulla, 3. podium swings forwards, 4. podium contracts and vaccum is created below the sucker, 5. sucker is released and podium is pulled back, 6. water fills the ampulla as before, and 7. podium swings forward to catch the substratum

**Locomotion**

In sea star, there is no head or anterior end. It can move in any direction and any arm can take the lead for locomotion. Locomotion is achieved with the help of water-vascular system which maintains a hydraulic pressure. The beating of cilia of the water vascular system causes the sea water to enter through the madreporite. After passing through the stone canal, ring canal, radial canals and lateral canals, it reaches the tube feet and their ampullae.

***Locomotion in a horizontal surface.*** When the sea star desires to move on horizontal surface, the tube feet push the body forwards. The ampullae of raised arms contract, the valves in the podial canals close and the water of ampullae is forced into the podia. The podia elongate in the general direction of movement due to hydrostatic pressure products by the influx of water into them.

The terminal suckers of the tube feet are brought in contact with the substratum and their central parts are withdrawn to form suction cups. The adhesion is further strengthened by mucous secreted by the tips of tube feet. The tube-feet now pivot forward on their attached suckers, assuming vertical position and thereby pushing the body forward. The tube-feet then shorten by contracting their longitudinal muscles and forcing some water back into their ampullae and releases their suckers. The ampullae then contract again and whole sequence of events is repeated.

***Locomotion on a vertical surface.*** In climbing a vertical surface, the tube-feet pull the body forwards. By the alternate contraction and expansion of tube-feet and by adherance of suckers of tube-feet on surface *Astarias* climbs on the vertical surface.

The tube-feet, however, may or may not work in coordinated manner. As a result, the *Asterias* moves forward steadily but slowly, at a speed of 15 cm per minute. The tube-feet are employed when the substratum is hard. On soft mud or sand the suckers of tube feet become useless. On such surface the animal walks on its extended tube feet which now act like small legs.

**Circulatory System**

A true blood vascular system is absent. The system responsible for circulation of digested food to various organs of body is often termed as circulatory system. It is made up of two parts.

I Perihaemal system

II Haemal System

*I.* ***The perihaemal system.*** It is also derived from the coelom and is lined with cuboidal ciliated epithelium. It comparises :

1. Axial sinus
2. Aboral ring sinus
3. Genital sinuses
4. Oral ring sinus
5. Radial perihaemal sinuses
6. Marginal sinuses
7. Peribranchial sinuses

1. *Axial sinus.* It is a wide, nearly vertical, thin-walled tube surrounding the stone canal and the axial gland. At its aboral end it communicates with the ampulla of stone canal and also with the exterior by some of the pore canals of the madreporite. At the same end, it opens into the aboral rings sinus. The three (stone canal, axial gland and axial sinus) together form an *axial complex.*

2. *Aboral ring sinus.* It is a pentagonal channel lying around the intestine, just beneath the central disc. It gives off 10 genital branches, two in each inter-radius. The aboral ring sinus and its genital branches contain genital rachis, a continuation of the axial gland.

3. *Genital sinuses.* Each gonad is surrounded by a small genital sinus which is connected with the aboral ring sinus by a short and slender branch.

4. *Oral ring sinus.* At its oral end, the axial sinus opens into the inner division of a circular channel, the oral ring sinus or peribuccal or hyponeural perihaemal ring sinus surrounding the mouth. It is a large tubular sinus and internally divided by an oblique septum (the haemal strand) into two-a smaller inner ring and a larger outer ring. From the outer part of the oral ring sinus arises five radial perihaemal sinuses.

5. *Radial perihaemal sinuses.* 5 such sinuses arise from oral ring sinus. Each one extends through each arm between the radial nerve and the radial ambulacral vessel. The radial sinuses are also divided by radial septa into two each. The radial perihaemal sinuses also give out fine channels into the tube feet.

6. *Marginal sinuses.* In each arm, two longitudinal marginal sinuses run longitudinally on each side just aboral to the marginal nerve cord. The fine lateral channels connect the marginal channels with the radial perihaemal sinuses.

7. *Perihaemal sinuses.* These are small spaces in the dermis of the body wall. These communicated with the radial hyponeural sinuses of the arms.

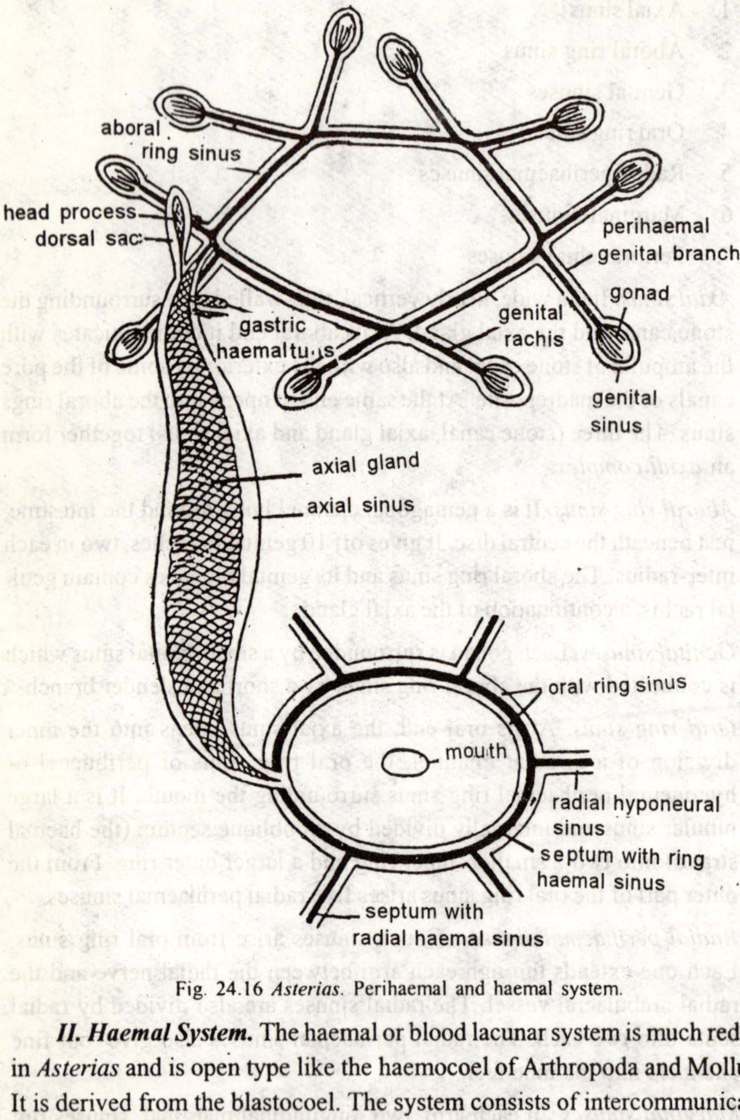


Fig. 24.16 *Asterias*. Perihaemal and haemal system.

***II. Haemal System.*** The haemal or blood lacunar system is much reduced in *Asterias* and is open type like the haemocoel of Arthropoda and Mollusca. It is derived from the blastocoel. The system consists of intercommunicating spaces which have no epithelial lining. These are enclosed in the coelomic spaces of perihaemal system. The channels are filled with coelomic fluid containing coelomic spaces of perihaemal system. The channels are filled with coelomic fluid containing coelomocytes. Histologically, the system is bounded by a layer of connective tissue, externally covered by coelomic epithelium. The haemal system comprises :

1. Axial haemal ring
2. Aboral haemal ring
3. Oral haemal ring
4. Radial haemal sinuses.

1. *Axial gland.* It is also known as *heart* or *brown gland* and forms the principle part of the haemal system. It has an external lining of peritoneum and its interior is filled with connective tissue containing many small inter-communicating spaces filled with a fluid with amoeboid coelomocytes with a brown pigment. The axial gland is connected with the oral haemal sinus at its oral end and with aboral haemal sinus at its aboral end. A small terminal head process arises from the aboral end of the axial gland. It communicates with a small dorsal sac in the madreporite. A pair of gastric tufts arising from the haemal sinuses in the wall of cardiac stomach open into the axial gland near its aboral end. The food digested in the stomach enters the haemal circulation through the gastric tufts.
2. *Aboral haemal ring.* The axial gland, at its aboral end connects with the ampulla of madreporic canal and joins a pentagonal ring canal situated beneath the aboral surface of the central disc. From this canal, five pairs of genital haemal strands are extended to gonads. These haemal strands are two in each inter-radius.
3. *Oral haemal ring.* It is the circular haemal sinus, located around the mouth just below the ring canal of the water vascular system.
4. *Radial haemal sinuses.* From the oral haemal ring arise five radial haemal sinuses or strands. One of these runs out through the septum of each hyponeural radial sinus. The radial haemal branches give out from their sides short branches to tube feet.

**Respiration**

In *Asterias*, respiration takes place by the *dermal branchiae* or *papulae* scattered over the aboral surface and by the tube feet projecting in the ambulacral groove. Both the respiratory structures are thin walled and are continuous with coelom. Oxygen dissolved in sea water diffuses into coelomic fluid through these structures. The coelomic fluid distributes oxygen to all internal tissue and brings carbon dioxide to the respiratory organs for outward diffusion.

**Excretory System**

There is no distinct excretory system in echinoderms. It is likely that there may be some organ in connection with the ambulacral system or with the axial sinus for excretion. The nitrogenous metabolic excretory waste usually contains ammonium compounds. The excretion is brought about with the agency of coelomocytes that pick up wastes and pass them out through the papulae. The rectal caecae serve in excretion to some extent. There is certain amount of diffusion of dissolved water through the gills and thin wall of tube-feet.

**Nervous System**

The nervous system of *Asterias* is of simple and primitive type. It is confined to the body wall. It consists of four parts :

1. Superficial or Ectoneural or Epidermal nervous system
2. Deeper hyponeural
3. Aboral or Coelomic nervous system
4. Visceral nervous system

**1. *Superficial or Ecotodermal nervous system.***

This nervous system is situated just beneath the epidermis. It is the main part of sensory nervous system. It includes circumoral nerve ring, radial nerve cords and subepidermal nerve plexus.

(i) *Circumoral nerve ring.* The nerve ring is pentagonal in shape and is circum-oral i.e., occurs around the mouth in the peristomial membrane. It supplies nerve fibres to the peristomial membrane and the oesophagus and at each radius gives off a radial nerve.

(ii) *Radial nerve cords.* The radial nerve cords arise from the angles of the nerve pentagon and extend outwards, one through each arm. Each radial nerve terminates as a sensory cushion on the aboral side of the terminal tentacle. Each nerve cord is 'V' shaped in cross-section and continuous on its outer side with the epidermis. It remains separated on its inner side from the hyponeural sinus only by a thin dermis and the coelomic epithelium.

(iii) *Subepidermal nerve-plexus.* It is in the form of an extensive network of nerve cells and nerve fibres situated below the epidermis all over the body surface including the papulae, pedicellariae and spines. In each nerve cord it is joined to radial nerve cord by five nerve fibres. In each nerve cord it forms a pair of marginal nerves out of which one traverses along each margin of the ambulacral groove and forms a nerve ring in the sucker of each tube feet.

The nerve pentagon and radial nerves contain many neurons, arranged in sensory, association and motor tracts. They receive and coordinate responses similar to that of central nervous system found in other animals.

**2. *Deeper nervous system.***

It is situated on the oral side just beneath the perihaemal sinuses and a little above the epidermal nervous system separated by a thin layer of connective tissue. It consists of a double nerve pentagon running round the mouth and sending a double radial nerves or Lange's nerve in each arm. Each nerve is continued as a plate of nervous tissue in the outer oral wall of the radial hyponeural sinus and gives out branches to the muscles of of the arm. The deep nervous system originates from the mesoderm. Its function is primarily motor.

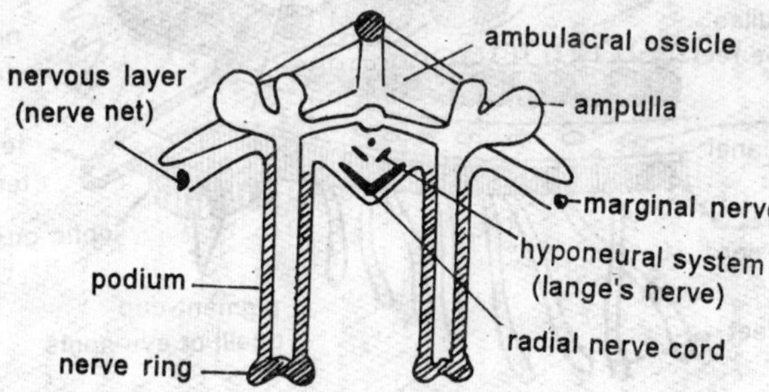


Fig. 24.17 *Asterias.* Nervous System in section.

**3. *Aboral or Coelomic nervous system.***

It is represented by a thin nerve plexus situated in the aboral body wall, just above the parietal layer of coelomic epithelium. It is somewhat thickened to form an anal nerve ring in the central disc and a nerve in each arm. The aboral nervous system is connected with the marginal nerves by several lateral nerves in each arm. This system is also mesodermal in origin and motor in nature.

**4. *Visceral nervous system.***

It is situated in the gut wall just outside the enteric epithelium. It has the form of a conspicuous plexus which has connections with the visceral receptors and also with the muscles layers of gut wall.

**Sense Organs**

The sense organs of *Asterias* are of three types :

1. Neurosensory cells
2. Eyes
3. Terminal tentacles.

1. *Neurosensory cells.* These are present here and there in epidermis. They are more or less spindle shaped containing the nucleus, a distal thread-like process reaching to the cuticle, and a proximal fibre entering the subepidermal nerve plexus. They may act as tectile receptor or

chaemoreceptors. They are especially numerous in the suckers of podia, at the bases of spines, pedicellariae and terminal tentacles.

2. *Eyes.* The eyes are small, bright-red spots, one under the base of the terminal of each arm. Each eye is also called *optic cushion* as it is really a

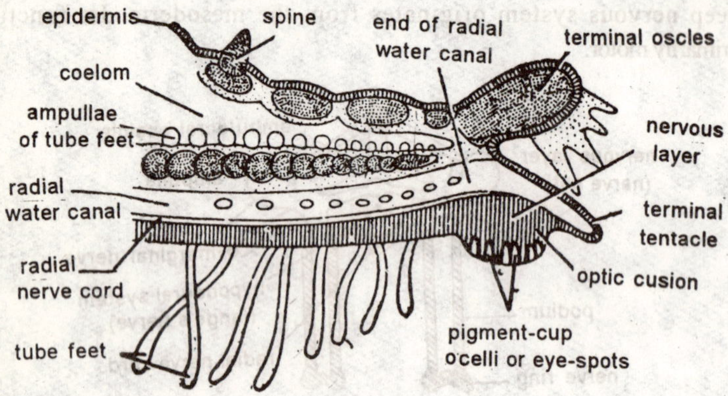


Fig. 24.18 *Asterias.* L.S. of an arm through terminal tentacle showing position of eye.

special thickening of radial nerve. Each eye comprises a number of cup-shaped concavities lined by alternating pigment cells and retinal cells. The pigment cells contain an orange pigment. The retinal cells are elongated cells with a distal bulbous enlargement projecting into the cavity of the cup and a proximal fibre passing into the underlying radial nerve. The number of ocelli in one eye ranges from 80-200 in different species.

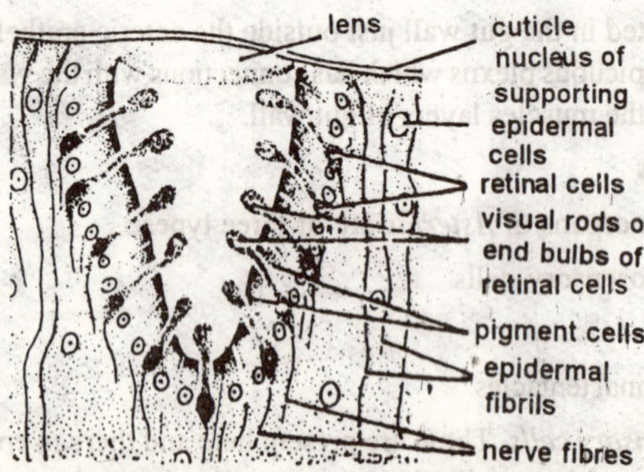


Fig. 24.19 *Asterias.* V.S. through an eye pit.

Each concavity is covered by a transparent cuticle and may contain a lens beneath the cuticle The lens is formed by epidermis. Eyes are sensitive to light and detect change in the intensity of light.

3. *Terminal tentacles.* The terminal tentacles have sensory cells which are tactile and also sensitive to food.

**Reproductive System.**

*Asterias* is unisexual i.e., dioecious. The male and female are separate but both the types of gonads are morphologically alike and situated in similar positions. They have only microscopic differences. The reproductive system is simple. Ovaries and testes consist of small rounded follicles like the bunches of grapes. They, however, differ in colour. The testes are pale-grey while ovaries from pink to orange. There are five pairs of gonads in either sex, a pair in each interradius. The gonads are attached at the bases of the arms and extend freely into the proximal parts of the arms on the sides of pyloric caeca. From the proximal end of each gonad arises a short dilated gonoduct which opens out laterally on the aboral surface by a minute gonopore. There are no special organs for copulation. The accessory glands and recepticles are also absent.

The sex-cells are believed to originate in the aboral part of the axial or from the wall of coelom near the proximal part of the axial gland and then migrate through the genital rachis to the gonads.

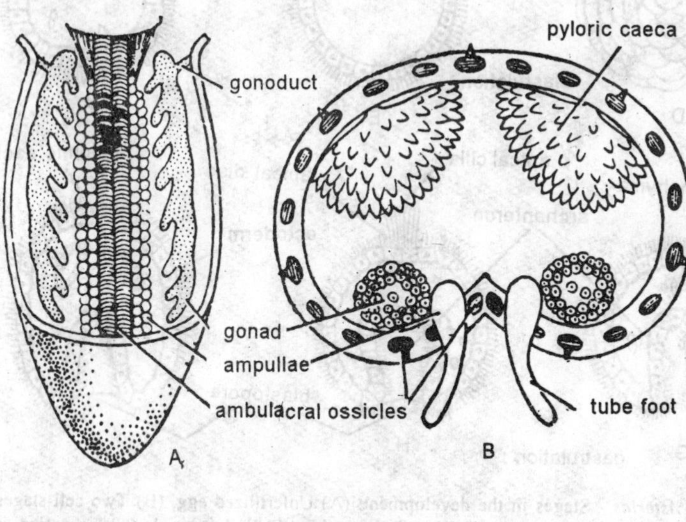


Fig. 24.20 *Asterias.* Gonads.

There is a single breeding season in sea-stars i.e., late spring. The ripe ova and sperms are simply shed into the sea water where fertilization takes place. Maturation and shedding of gametes are controlled by a secretion of neurosecretory cells in radial canal (*Chaet* and *McConnaughy*, 1959). According to *Binyon* (1964), the axial organ may serve as a chaemoreceptor for substances received through the madreporite, thus stimulating the release of gametes. A single female lays about 200 million eggs in a season.

***Fertilization.*** It is external and takes place in the sea water.

***Development.*** After fertilization the zygote divides into two equal blastomeres, the cleavage is *holoblastic*. Cleavage leads to the formation of blastula

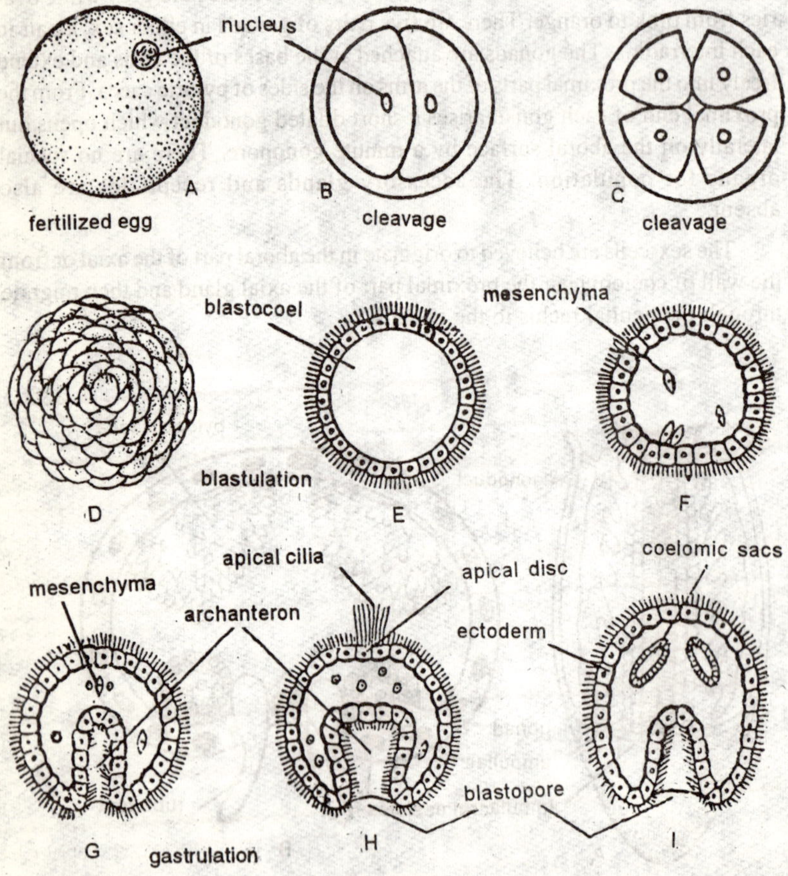


Fig.24.21 *Asterias* - Stages in the development. (A) Unfertilized egg. (B) Two cell stages (C) Four cell stages (D) Blastula (E) Blastula in section, (F) Early gastrula (G) Formation of mesenchyme, (H) Older gastrula, (I) Formation of coelomic sacs.

by second day. The blastula sooner or later develops a blastocoel and is called *coeloblastula*. After a free swimming existence of about a day the blastula undergoes embolic invagination and becomes a two layered, cup-like *gastrula*. Its outer layer is formed of ectoderm and inner layer of endoderm. The cavity of gastrula which is lined by endoderm is known as *archenteron*. It opens to the exterior by a wide aperature called *blastopore*. The archenteron, forms as a result of gastrulation, is narrow. When fully formed it gives rise to a coelomic sac, the *hydroenterocoel*, on each side. The two occupy the right and left of the archenteron and gradually elongate anteriorly in the plan of the long axis of the larva. From the left hydro-enterocoel arises a dorsal evagination which forms the water pore. Ventrally a stomodaeal inpushing is formed. This meets the archenteron and completes the digestive tract. The blastopore forms the anus. Now the alimentary canal is differentiated into oesophagus, stomach and intestine along with mouth and anus. The two hydroenterocoels continue to develop anteriorly where they fuse to form a U-shaped coelom. Finally the posterior parts of U are cut off from anterior axocoels forming right and left somatocoels. The uniform ciliation of the body is replaced by two ciliary wavy bands-*a-perioral band* surrounding the mouth and an *aboral band* lying inside the mouth. With these changes embryo develops into *dipleurula* larva capable of independent existence. It actively feeds on diatoms etc. The aboral band of cilia helps in collecting the food particles. The larva swims near the surface and rotates clock-wise with the help of perioral band of cilia.

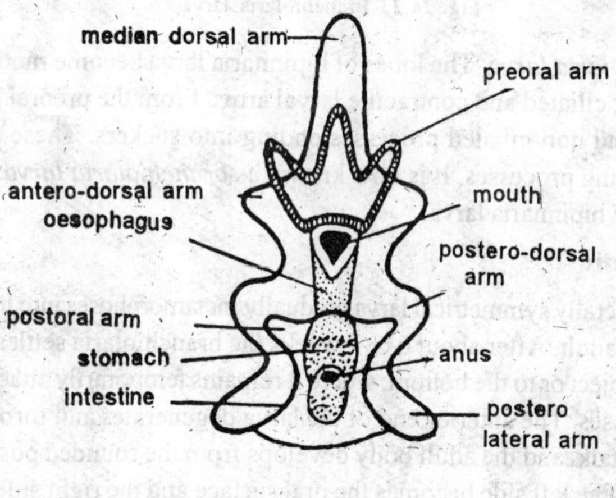


Fig. 24.22 A Bipinnaria larva

*Bipinnaria Larva*. The dipleurula larva forms on its front side a big preoral lobe, which is bordered by a preoral loop made up of cilia. Soon three lateral lobes are formed and these lobes are also bordered by a postoral loop of cilia. The larva is now called *bipinnaria.* It is also bilaterally symmetrical. It swims and feeds independently. After some weeks it is transformed into the next larva known as *branchiolaria larva.*

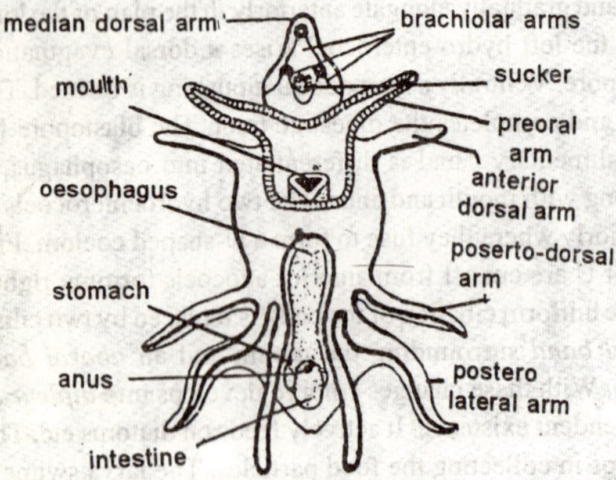


Fig. 24.23 Branchiolaria larva

*Branchiolaria larva.* The lobes of bipinnaria larva become modified into long, slender, ciliated and contractile larval arms. From the preoral lobe arise three short and non-ciliated processes ending into suckers. These processes are called fixing processes. It is now known as *branchiolaria larva*. It swims and feeds like bipinnaria larva.

**Metamorphosis**

The bilaterally symmetrical larva gradually metamorphoses into the radially symmetrical adult. After about 6 or 7 weeks the branchiolaria settles down on some solid object or to the bottom, where it remains temporarily attached by its fixing processes. The anterior end of the larva degenerates and forms only an attachment stalk, and the adult body develops from the rounded posterior end of the larva. The left side becomes the oral surface and the right side becomes aboral. The adult arms appear as extensions of the body. Internally the entire digestive tract degenerate. All these parts formed anew and in a position

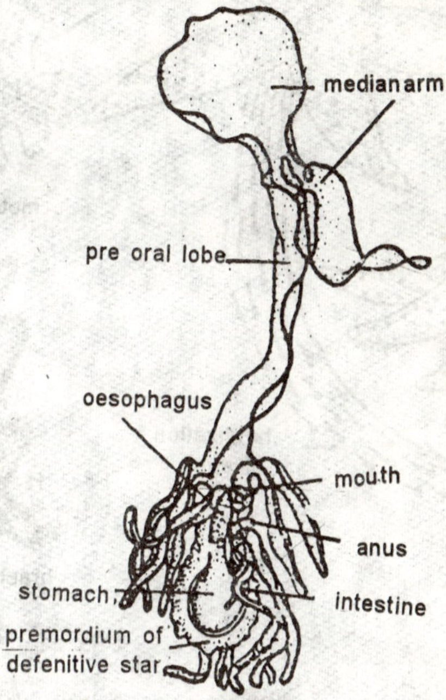


Fig. 24.24 Metamorphosis in larva

coinciling with the adult radial symmetry. The left axohydrocoel forms the water-vasculr system and from the hydrocoel develop five pairs of projections, two in each of the developing arms. These projections represent the cavity and coelomic lining of first pair of podia in each arms. At about this time, the skeletal system appears. The first adult ossicles to be formed are those around the aboral pole, which contains the anus. New ossicles are then added peripherally to the initial skeleton. The detached young sea star is not more than 1 mm in diameter.

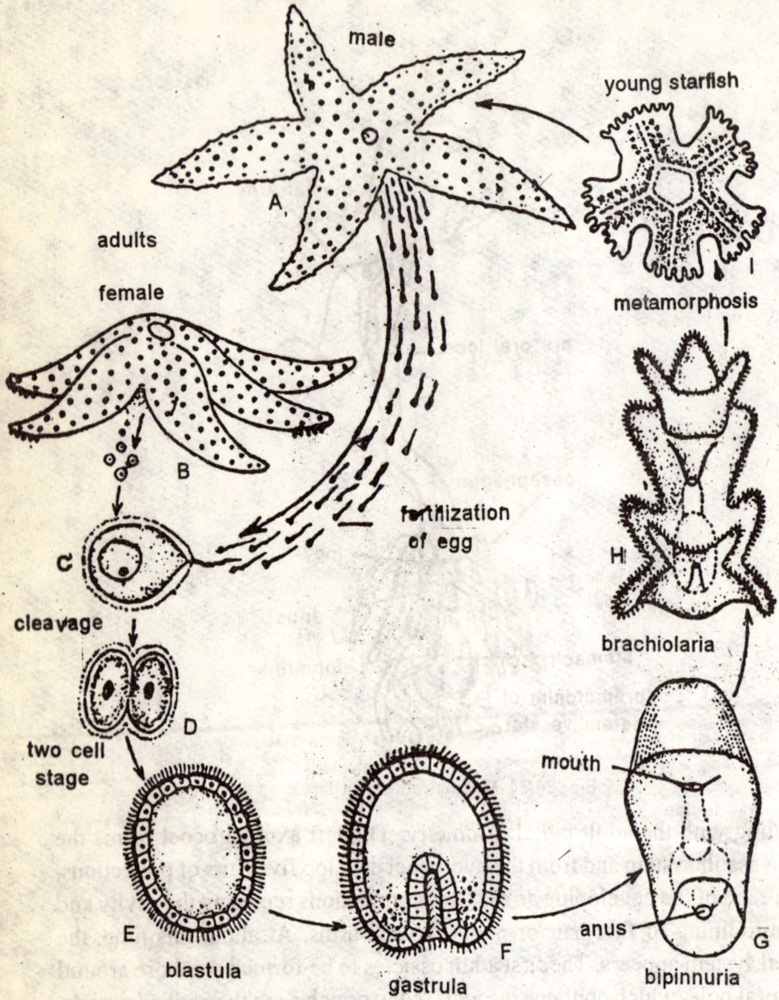


Fig. 24.25 *Asterias* Life History

◆◆◆